This book explores recent research underlining the remarkable connections between the algebraic and arithmetic world of Galois theory and the topological and geometric world of fundamental groups. Arising from an MSRI program held in the fall of 1999, it contains ten articles, all of which aim to present new results in a context of expository introductions to theories that are ramifications and extensions of classical Galois theory.

B. H. Matzat and M. van der Put introduce differential Galois theory and solve the differential inverse Galois problem over global fields in positive characteristic; D. Harbater gives a comparative exposition of formal and rigid patching starting from the familiar complex case. S. Mochizuki discusses aspects of Grothendieck's famous anabelian geometry, while the articles by R. Guralnick, A. Tamagawa, and F. Pop and M. Saïdi investigate the structure of the fundamental groups of curves over different kinds of characteristic p fields. M. Imbert and L. Schneps study the structure of the Hurwitz spaces and moduli spaces of curves, which are of great importance to Galois theory because of the Galois action on their fundamental groups. The first interesting such group is $SL_2(\mathbb{Z})$, a family of special subgroups of which is studied by F. Bogomolov and Y. Tschinkel. Finally, R. Hain and M. Matsumoto present their result proving part of a conjecture by Deligne on the structure of the Lie algebra associated to the Galois action on the fundamental group of the thrice-punctured projective plane.

Mathematical Sciences Research Institute
Publications

41

Galois Groups and Fundamental Groups

Mathematical Sciences Research Institute Publications

Volumes 1–4 and 6–27 are published by Springer-Verlag

Galois Groups and Fundamental Groups

Edited by

Leila Schneps
Institut de Mathématiques de Jussieu

The Mathematical Sciences Research Institute wishes to acknowledge support by the National Science Foundation. This material is based upon work supported by NSF Grant 9810361.

CAMBRIDGE UNIVERSITY PRESS
Cambridge, New York, Melbourne, Madrid, Cape Town,
Singapore, São Paulo, Delhi, Tokyo, Mexico City

Cambridge University Press
The Edinburgh Building, Cambridge CB2 8RU, UK

Published in the United States of America by Cambridge University Press, New York

www.cambridge.org
Information on this title: www.cambridge.org/9780521174572

© Mathematical Sciences Research Institute 2003

First published 2003
First paperback edition 2011

A catalogue record for this publication is available from the British Library

ISBN 978-0-521-80831-6 Hardback
ISBN 978-0-521-17457-2 Paperback

Galois Groups and Fundamental Groups
MSRI Publications
Volume 41, 2003

Contents

Introduction

This volume is the outcome of the MSRI special semester on *Galois Groups and Fundamental Groups*, held in the fall of 1999. Respecting the famous Greek requirements of unity of place, time, and action, the semester was an unforgettable, four-month-long occasion for all mathematicians interested in and responsible for the developments of the connections between Galois theory and the theory of fundamental groups of curves, varieties, schemes and stacks to interact, via a multitude of conferences, lectures and conversations.

Classical Galois theory has developed a number of extensions and ramifications into more specific theories, which combine it with other areas of mathematics or restate its main problems in different situations. Three of the most important of these extensions are *geometric Galois theory*, *differential Galois theory*, and *Lie Galois theory*, all of which have undergone very rapid development in recent years. Each of these theories can be developed in characteristic zero, over the field $\mathbb{C}$ of complex numbers, over number fields or p-adic fields, or in characteristic $p > 0$; various versions of the classical and inverse Galois problems can be posed in each situation. The purpose of this introduction is to give a brief overview of these three themes, which form the framework for all the articles contained in this book.

The main focus of study of **geometric Galois theory** is the theory of *curves* and the many objects associated to them: curves with marked points, their fields of moduli and their fundamental groups, covers of curves with their ramification information and their fields of moduli, and the finite quotients of the fundamental group which are the Galois groups of the covers, as well as the moduli spaces and Hurwitz spaces which parametrize all these objects.

To consider a curve X topologically is tantamount to considering it over the field of complex numbers $\mathbb{C}$. As an abstract group, the *topological fundamental group* of the curve depends only on the genus g and the number n of marked points chosen on the curve; it can be identified with the group of homotopy classes of loops on the curve based at a fixed (not marked) base point, and is presented by standard generators $a_1, \ldots, a_g, b_1, \ldots, b_g, c_1, \ldots, c_n$ subject to the unique relation

$$[a_1, b_1] \cdots [a_g, b_g] c_1 \cdots c_n = 1. \qquad (*)$$

Note that when $n \geq 1$, this group is actually free. The Galois covers of the curve correspond to the finite quotients of this group, which are exactly the finite groups generated by generators a_i, b_i and c_j satisfying $(*)$, so they are perfectly understood. The *algebraic fundamental group* of the curve is the Galois group of the compositum of all function fields of finite étale covers of the curve over the function field of the curve itself; in fact, it is exactly the profinite limit of the finite quotients of the topological fundamental group.

This simple situation leads or generalizes very naturally into new regions that contain all kinds of very difficult problems. We sketch some of them:

1. Fundamental groups in characteristic p.

When a curve X is defined over a field in characteristic p, all relations between its fundamental group and any topological notion of 'loops' must be forgotten. Over an algebraically closed field, one defines the algebraic fundamental group directly, exactly as above; it is the profinite limit of the Galois groups (monodromy groups) of finite étale covers of the curve. However, in this situation, for (g, n) different from $(0, 0)$ and $(1, 0)$, it is extremely difficult to determine the structure of the fundamental group, or even the weaker question of which finite groups can occur as its quotients. Indeed, this is one of the fundamental problems of geometric Galois theory in characteristic p. In the affine case $(n \geq 1)$, the complete solution to the weaker problem was conjectured by Abhyankar; this conjecture was proved over the affine line by M. Raynaud, and the proof extended to all curves by D. Harbater. However, the situation remains completely mysterious in the case of complete curves $(n = 0, g > 1)$.

Things are better if one considers only the quotients of order prime to p; then a result due to Grothendieck states that the groups of order prime to p which can occur are exactly the finite quotients of order prime to p of the topological fundamental group of type (g, n) defined in $(*)$, and that in fact the prime-to-p quotients of the fundamental groups over $\mathbb{C}$ and in characteristic p are isomorphic. In characteristic p, this group is a quotient of the *tame fundamental group*, which is the largest quotient of the fundamental group having inertia subgroups of order prime to p; this group (which is equal to the fundamental group when $n = 0$) is easier to work with than the full group for various purposes. But the structure of the tame fundamental group and the set of its finite quotients are absolutely unknown, except in the non-hyperbolic cases $(g, n) = (0, 0)$, $(0, 1)$, $(0, 2)$ and $(1, 0)$.

The articles by R. Guralnick, A. Tamagawa, and F. Pop and M. Saïdi all work in the situation of curves defined over an algebraically closed field of characteristic p. Guralnick works on the problem of determining which groups can occur as Galois groups (or their composition factors) of finite separable covers $f : Y \to X$, where Y is of fixed genus g, and seeks groups which can specifically be excluded. Tamagawa shows that given the tame fundamental group, it is possible to recover the type (g, n) of the curve (if $(g, n) \neq (0, 0)$ or $(0, 1)$).

Results in this and other papers by Tamagawa even tend to imply that in some cases, the tame fundamental group may determine the isomorphism class of the curve completely. This shows how different the characteristic p case is from the characteristic 0 case, in which as we saw, curves of many different types may have isomorphic fundamental groups (for instance, the fundamental groups of curves of type $(2, 2)$ and $(1, 4)$ are both free of rank 5). Finally, Pop and Saïdi address similar questions, proving, under certain hypotheses on the Jacobians, that at most a finite number of curves can have isomorphic fundamental groups.

2. Anabelian theory. We saw above that the isomorphism class of the topological or the algebraic fundamental group is very far from determining even the most basic information about a curve in characteristic 0, such as its type (g, n), whereas in characteristic p it determines much more if not all of the information about the specific curve. However, one can also consider the algebraic fundamental group equipped with its canonical outer Galois action, which should provide more information. Indeed, any variety (scheme, stack) defined over an algebraically closed field can actually be considered as defined over a subfield K, given by the coefficients of the equations of a defining model, say, and which is finitely generated over the prime field and not algebraically closed. Then there is an exact sequence

$$1 \to \pi_1(X \otimes \bar{K}) \to \pi_1(X) \to \mathrm{Gal}(\bar{K}/K) \to 1, \qquad (**)$$

where $\pi_1(X \otimes \bar{K})$ denotes the algebraic fundamental group. The *anabelian problem*, which was posed by Grothendieck in his famous letter to G. Faltings, asks which varieties are entirely determined by the group $\pi_1(X \otimes \bar{K})$ together with the action $\mathrm{Gal}(\bar{K}/K) \to \mathrm{Out}(\pi_1(X \otimes \bar{K}))$. Grothendieck called varieties which are thus determined *anabelian varieties*, and explicitly stated that hyperbolic curves should be anabelian. This is related to the hitherto unproven *section conjecture* for a hyperbolic curve X, which states that the sections

$$\mathrm{Gal}(\bar{K}/K) \to \pi_1(X)$$

of $(**)$ are in bijection with the rational points of X if X is complete, and this set together with the tangential base points if X is not complete.

S. Mochizuki proved that hyperbolic curves defined over sub-p-adic fields, that is, fields which are subfields of fields finitely generated over the p-adics, are indeed anabelian. In his article in this volume, he discusses various results related to this theorem, including a partial generalization to characteristic p and a discussion of the section conjecture over the field of real numbers.

3. Galois action on fundamental groups. In his *Esquisse d'un Programme*, completing the letter to Faltings, Grothendieck suggested that not only hyperbolic curves, but also the moduli spaces $\mathcal{M}_{g,n}$ of curves of type (g, n) should be examples of anabelian varieties, and that explicitly investigating the Galois action on their fundamental groups should provide information of an entirely new type about the elements of $\text{Gal}(\overline{\mathbb{Q}}/\mathbb{Q})$; this is known as *Grothendieck-Teichmüller theory*. The first non-trivial moduli space $\mathcal{M}_{g,n}$ is the case $(g,n) = (0,4)$; we have $\mathcal{M}_{0,4} = \mathbb{P}^1 - \{0,1,\infty\}$. Following the direction initiated by Grothendieck, the Galois action on the fundamental group of this space (the profinite free group $\widehat{F}_2$ on two generators) has been studied in a theory known as *dessins d'enfants*. The study has focused on coding the conjugacy classes of finite index subgroups of $\widehat{\pi}_1(\mathbb{P}^1 - \{0,1,\infty\})$, corresponding to the finite étale covers of $\mathbb{P}^1 - \{0,1,\infty\}$, as combinatorial objects (the dessins d'enfants), and using the combinatorics to look for invariants identifying the Galois orbits of these covers.

The only other moduli space of dimension 1 is $\mathcal{M}_{1,1}$, the moduli space of elliptic curves (genus one curves with one distinguished point). This space is the quotient of the Poincaré upper half-plane by the proper and discontinuous (but not free) action of $\text{SL}_2(\mathbb{Z})$. Finite-index subgroups of $\text{SL}_2(\mathbb{Z})$ correspond to covers of $\mathcal{M}_{1,1}$. As in the case of $\mathcal{M}_{0,4}$, many specific families of these subgroups have been studied in detail, most familiarly the modular subgroups $\Gamma(N)$. Using graphs in the spirit of the theory of dessins d'enfants, F. Bogomolov and Y. Tschinkel characterize another family of very special finite-index subgroups of $\text{SL}_2(\mathbb{Z})$, namely those corresponding to elliptic fibrations.

Before passing from these two curves to what can be said in the case of general moduli spaces $\mathcal{M}_{g,n}$, let us make a brief foray out of the geometric situation into the domain of **Lie Galois theory**, a subject that originates in the geometric situation but has been linearized by focusing on graded Lie algebras associated to the profinite fundamental groups rather than the groups themselves. A great deal of work has been done in this subject, mainly by Y. Ihara and his school, but we restrict ourselves here to discussing one conjecture which is a paradigm for the manner in which the problems in the domain arise in geometry, but raise their own interesting arithmetic questions.

Since as above, we have $\pi_1(\mathbb{P}^1 - \{0,1,\infty\}) \simeq \widehat{F}_2$, the exact sequence $(**)$ gives a canonical homomorphism

$$G_{\mathbb{Q}} \to \text{Out}(\widehat{F}_2). \qquad (***)$$

As an initial step, the passage from the geometric situation to the Lie situation involves replacing the profinite completions of fundamental groups by their pro-ℓ completions, that is, the completions with respect to all finite quotients which are ℓ-groups for a fixed prime ℓ. Denote the pro-ℓ completion of F_2 by $F_2^{(\ell)}$. This completion is a quotient of the profinite completion by a characteristic subgroup, so that $(***)$ yields a homomorphism $G_{\mathbb{Q}} \to \text{Out}(F_2^{(\ell)})$. Following Ihara, define a filtration on $G_{\mathbb{Q}}$ by setting

$$I^m G_{\mathbb{Q}} = \text{Ker}\{G_{\mathbb{Q}} \to \text{Out}(F_2^{(\ell)}/L^{m+1})\}$$

where L^m denotes the m-th term of the lower central series of $F_2^{(\ell)}$, and set

$$\text{Gr}^m G_{\mathbb{Q}} = I^m G_{\mathbb{Q}}/I^{m+1} G_{\mathbb{Q}}.$$

degree n with monodromy group S_n (or A_n) for $g = 0$, we will concentrate on Chevalley groups.

Thus, we let $\mathbf{E}_p(g)$ denote the set of genus g groups (in characteristic p) other than alternating groups. Similarly, let $\mathbf{E}_p^{ta}(g)$ denote the set of simple groups (other than alternating groups) which are composition factors of monodromy groups of tamely ramified covers $X \to Y$ with X of genus at most g.

By [40], this problem reduces to the case where f is indecomposable. It is also easy to see that the critical case is when Y has genus 0. If $p = 0$, there is a recent result answering a question posed in [40] (the final paper proving this result was done by Frohardt and Magaard [22]; other papers involved in the proof include [21], [32], [40], [58], [6], [49] and [51]) — since the proof really only uses the assumption that the cover is tame, the result can be stated as follows:

THEOREM 1.1. $\mathbf{E}_p^{ta}(g)$ is finite for each g.

Indeed, much more precise information is known and hopefully a complete determination of the monodromy groups of the tamely ramified indecomposable covers of genus zero (and in particular, indecomposable rational functions) will be available in the near future. In particular, there will be several infinite families and a finite list of other examples. There will be a similar result for any fixed genus g.

We mention two results which involve special cases of this analysis.

The first is a special case in [31]:

THEOREM 1.2. Let $f(x) \in \mathbb{Q}(x)$ be an indecomposable rational function. Suppose that f is bijective modulo p for infinitely many primes p. Aside from finitely many possibilities, the genus of the Galois closure of $\mathbb{Q}(x)/\mathbb{Q}(f)$ is at most 1.

A much more precise version of the theorem is in [31], where an essentially complete list of possibilities is given. After one solves the group theory problem, it is left to determine which possibilities actually arise. This involves a careful analysis of elliptic curves and results about torsion points and isogenies of elliptic curves over $\mathbb{Q}$.

The second result is a consequence of [32], [30] and [39].

THEOREM 1.3. Let $g \geq 4$ and $p = 0$. Let X be a generic curve of genus g. If $f : X \to \mathbb{P}^1$ is an indecomposable cover of degree n, then the monodromy group of f is either S_n with $n > (g+1)/2$ or A_n with $n > 2g$.

This was a problem originally studied by Zariski who proved that if $g > 6$ and $f : X \to \mathbb{P}^1$ with X generic of genus g, then the monodromy group of f is not solvable (this is a special case of the result above — using the observation of Zariski that any such cover is a composition of an indecomposable cover and covers from $\mathbb{P}^1$ to $\mathbb{P}^1$). A more precise statement of the theorem above is to say that the set of Riemann surfaces of genus $g \geq 4$ which have indecomposable covers of degree n to $\mathbb{P}^1$ with monodromy group other than A_n or S_n is contained in a proper closed subvariety of the moduli space of genus g curves. It is well known that S_n does occur as the monodromy group of the generic curve (for $n > (g+1)/2$). It has been recently shown [17] that A_n actually does occur for $n > 2g$, thus giving a fairly complete picture of the situation when $g > 3$.

If $g < 4$, there are more group theoretic possibilities. In unpublished work, Fried and Guralnick have considered some possibilities for $g = 2$. The recent work of Frey, Magaard and Völklein show that there are other examples when $g = 3$ (all the group theoretic possibilities for $g = 3$ are known by the results cited above).

Until now, it was not known that a single simple group in any positive characteristic could be shown not to be a genus 0 group. In this article, we show that there are infinitely many such groups. In particular, we show that:

THEOREM 1.4. *If p does not divide the order of $|S|$, then $S \in \mathbf{E}_p(g)$ implies that $S \in \mathbf{E}_p^{\mathrm{ta}}(g+2) \subseteq \mathbf{E}_0(g+2)$. In particular, for any odd prime p and any g, there are infinitely many simple groups not in $\mathbf{E}_p(g)$.*

We also show that there are infinitely many simple groups whose order is divisible by p which are not in contained in $\mathbf{E}_p(g)$ for a fixed p and g. Let $\mu_p(S)$ be the smallest g such that $S \in \mathbf{E}_p(g)$. Let $\mathrm{Chev}(r)$ denote the family of simple groups which are Chevalley groups in characteristic r. Let $\mathrm{Chev}_b(r)$ denote the groups in $\mathrm{Chev}(r)$ which have rank at most b. Indeed, we prove the following result.

THEOREM 1.5. *Let X be a fixed type of Chevalley group. Fix a nonnegative integer g. There are only finitely many pairs (p,q) with p a prime and q a prime power not divisible by p such that $X(q) \in \mathbf{E}_p(g)$. More precisely, $\mu_p(X(q)) \to \infty$ as $q \to \infty$ for $(p,q) = 1$ and $\mathbf{E}_p(g) \cap \left(\bigcup_{r \neq p} \mathrm{Chev}_b(r)\right)$ is finite for each g.*

The proof shows that typically $\mu_p(X(q))$ grows like a polynomial of degree close to b in q (as long as p does not divide q).

Abhyankar ([1], [2], [3], [4], [5]) has shown that many finite groups of Lie type (particularly the classical groups) are genus 0 groups in the natural characteristic and so the exclusion $p \neq r$ is necessary.

This led the author to make the following conjecture several years ago — the positive characteristic analog of the Guralnick–Thompson Let $\mathrm{Chev}(r)$ denote the set of finite simple groups which are finite groups of Lie type over a field of characteristic r.

CONJECTURE 1.6. $\mathbf{E}_p(g) \cap \left(\bigcup_{r \neq p} \mathrm{Chev}(r)\right)$ *is finite.*

Given the classification of finite simple groups, this conjecture says that there are only finitely many simple groups in $\mathbf{E}_p(g)$ other than Chevalley groups in characteristic p.

The previous theorem goes a long way towards proving the conjecture. Namely, the conjecture is true if we consider Chevalley groups of bounded dimension. The next step would be to prove the same result for fixed q and then finally to prove

The case $\ell = p$ is also interesting but in fact in that case V can be 0 (and in general $0 \le \dim V \le g(Z)$).

If $p = 0$, we could also use the module of holomorphic differentials on Z and remove the 2 in the formula.

We point out an interesting consequence. If H is a subgroup of G, let 1_H^G denote the permutation module for G over $\mathbb{C}$.

COROLLARY 3.2. *Let Z be a curve over k with G a finite group of automorphisms of Z. Suppose that H and K are subgroups of G such that 1_H^G is isomorphic to a submodule of 1_K^G. Then $g(Z/H) \le g(Z/K)$.*

PROOF. Let V denote the Tate module for some sufficiently large prime ℓ other than the characteristic of the curve. By Frobenius reciprocity, $\dim C_V(H) = \dim \mathrm{Hom}_G(1_H^G, V)$ and $\dim C_V(K) = \dim \mathrm{Hom}_G(1_K^G, V)$. Since 1_H^G is a direct summand of 1_K^G, $\dim \mathrm{Hom}_G(1_H^G, V) \le \dim \mathrm{Hom}_G(1_K^G, V)$, whence the result. $\square$

Here are some well known situations where the previous result applies.

(i) $G = S_n$ or A_n. Let H be the stabilizer of a subset of size j and K the stabilizer of a set of size j' with $1 \le j \le j' \le n/2$.

(ii) $PSL(n, q) \le G \le P\Gamma L(n, q)$. Let H be the stabilizer of a subspace of dimension j and K the stabilizer of the a subspace of dimension j' with $1 \le j \le j' \le n/2$.

(iii) G is a classical group and H is the stabilizer of a totally singular 1-space. Then we can take K to be the stabilizer of any totally singular space of less than maximal rank or usually the stabilizer of a nonsingular space as well. See [19] for a precise statement.

We now prove some easy representation theoretic facts that will be useful in estimating genera.

LEMMA 3.3. *Let G be a finite group with a normal subgroup E. Let H be a maximal subgroup of G which does not contain any normal subgroup of G contained in E. Assume that $E = X_1 \times \ldots \times X_t$ with the $X_i = X^{g_i}$ being the set of G-conjugates of $X := X_1$. Set $Y = X_2 \times \ldots \times X_t$. Let $N = N_G(X) = N_G(Y)$. If V is a finite dimensional $\mathbb{C}G$-module, then $\dim C_V(H) \ge \dim C_V(N_H(X)Y) - \dim C_V(N_G(X))$.*

PROOF. Since both sides of the inequality we are proving are additive over direct sums and since V is a completely reducible $\mathbb{C}G$-module, it suffices to prove the result for V irreducible. If V is trivial, there is nothing to prove. Suppose that E does not act faithfully on V. Let K denote the kernel of E on V. Since H is maximal and does not contain K, $G = HK$ and $N_G(X) = N_H(X)K$ and similarly for Y. In this case $0 = C_V(G) = C_V(HK) = C_V(H)$ and $C_V(N_G(X)) = C_V(N_H(X))$ whence we have equality.

So we may assume that E acts faithfully on V. Note that since $N_G(X) \ge E$, $C_V(N_G(X)) = 0$.

We may assume that $C_V(Y) = W$ is nonzero (or the result obviously holds). Let $Y_i = Y^{g_i}$. Note that $\sum_i C_V(Y_i)$ is a direct sum (for if $\sum v_i = 0$ with $0 \neq v_1$ and $v_i \in C_V(Y_i)$, then $v_1 \in C_V(Y) \cap \bigcap_{i>1} C_V(Y_i) = C_V(Y) \cap C_V(X) = C_V(E) = 0$).

Now $N_G(X)$ leaves W invariant (since $N_G(X)$ normalizes Y). As we have seen above the distinct images of W under G form a direct sum Also the stabilizer of W is $N_G(X)$ (for if $gW = W$ and g is not in $N_G(X)$, then $\langle Y, Y^g \rangle = E$ would imply that $W = C_V(E) = 0$). It follows that $V \cong W^G_{N_G(X)}$ and so $V \cong W^H_{N_H(X)}$ as H-modules (since as noted $G = HN_G(X)$). So by Frobenius reciprocity, $C_V(H) \cong C_W(N_H(X)) = C_V(N_H(X)Y)$. $\qquad\square$

The following variant of the previous result will also be useful.

LEMMA 3.4. *Let G be a finite group with a normal subgroup E. Let H be a maximal subgroup of G which does not contain any normal subgroup of G contained in E. Assume that $E = X_1 \times \ldots \times X_t$ with the $X_i = X^{g_i}$ being the set of G-conjugates of $X := X_1$. Let $\Delta = \{1, \ldots, t\}$. Let $\delta \subset \Delta$ and set $X_\delta = \prod_{i \in \delta} X_i$. Let $Y_\delta = X_{\delta'}$ where δ' is the complement of δ. Let $N_\delta = N_G(X_\delta)$. Let V be an irreducible $\mathbb{C}G$-module containing an E-submodule W of the form $W_1 \otimes \ldots \otimes W_t$ with W_i an irreducible X_i module with W_j trivial if and only if $j \in \delta'$. Then $\dim C_V(H) \geq \dim C_V(N_H(X_\delta)Y_\delta) - \dim C_V(N_G(X_\delta))$.*

PROOF. Note that $N_H(X_\delta)Y_\delta \leq N_G(X_\delta)$ and so each term on the righthand side of our desired inequality is non-negative.

First suppose that E does not act faithfully on V. Let K denote the kernel of E on V. Since K is normal in G, $G = KH$. Then $C_V(H) = C_V(HK) = C_V(G)$. If G acts trivially, then the lefthand side is 1 and the righthand side is 0.

Otherwise, the lefthand side is 0. Since $G = HK$, $N_G(X_\delta) = KN_H(X_\delta)$, whence the righthand side is also 0.

So we may assume that E acts faithfully on V. If $W = V$, Y_δ has no fixed points on V for δ any proper subset of Δ (for Y_δ contains some X_j and V restricted to X_j is a direct sum of copies of V_j). Thus, the righthand side of the equation is 0.

Let $U := U_\delta = C_V(Y_\delta)$. So $W \subseteq U$. By irreducibility, $V = \sum U_\gamma$ where γ is the orbit of δ. Note that this sum is in fact direct, since the terms are direct sums of irreducible E-modules which are not isomorphic (as they have different kernels). Moreover, the stabilizer in G of U_δ is precisely $N_G(X_\delta)$ (because of the permutation action on the X_i). Thus, V is isomorphic to the induced module, $U^G_{N_G(X_\delta)}$. Since $G = N_G(X_\delta)H$, this implies that $V_H \cong U^H_{N_H(X_\delta)}$ and so by Frobenius reciprocity, $\dim C_V(H) = \dim C_U(N_H(X_\delta))$. Since $U = C_V(Y_\delta)$, it follows tht $C_V(N_H(X_\delta))(Y_\delta) = C_U(N_H(X_\delta))$, whence the result. $\qquad\square$

We next deal with diagonal subgroups (see section 11 for terminology). The result is actually more general than we state — the condition that X is simple is not necessary.

LEMMA 3.5. *Let G be a finite group with a minimal normal subgroup $E = X_1 \times \ldots \times X_t$ with the X_i the set of G-conjugates of the nonabelian simple group $X = X_1$ and $t > 1$. Let H be a maximal subgroup of G not containing E such that $H \cap E$ is a diagonal subgroup of E. Let $\Delta = \{1, \ldots, t\}$. If $\delta \subset \Delta$, set $X_\delta = \prod_{i \in \delta} X_i$. Let $Y_\delta = X_{\delta'}$ where δ' is the complement of δ. Let $N_\delta = N_G(X_\delta)$. If V is a finite dimensional $\mathbb{C}G$-module, then $\dim C_V(H) \geq \dim C_V(N_H(X_{12})Y_{12}) - \dim C_V(N_G(X_{12}))$.*

PROOF. It suffices to assume that V is irreducible. Note that the righthand side is always non-negative (since $N_G(X_{12}) \geq N_H(X_{12})Y_{12}$).

If V is a trivial G-module, there is nothing to prove.

If E acts trivially on V, then $C_V(H) = C_V(HE) = 0$. On the other hand, since $G = HE$, we have $N_G(X_{12}) = EN_H(X_{12})$ and so $C_V(N_H(X_{12})Y_{12}) = C_V(N_G(X_{12}))$.

So assume that E acts nontrivially on V. In particular, this implies that $C_V(N_G(X_{12})) = 0$. If Y_{12} has no fixed points on V, then clearly the result holds (since the right hand side of the inequality is 0).

Suppose first that $C_V(Y_1)$ is nonzero. Then as in the previous result, V is the direct sum of the $C_V(Y_i) = [X_i, V]$. Since $H \cap E$ is a diagonal subgroup, this implies that $(H \cap E)Y_i \geq E$ and so $H \cap E$ has no fixed points on $[X_i, V]$ and so none on V. Since $H \cap E \leq N_H(X_{12})$, the right hand side is 0, whence the result.

Finally, assume that $W := C_V(Y_{12}) \neq 0$, but $C_V(Y_1) = 0$. This implies that every irreducible E-submodule of V is of the form $U_1 \otimes \ldots \otimes U_t$ with U_i an irreducible X_i-module with U_i nontrivial for precisely 2 terms. In particular, it follows that W is a sum of E-homogeneous components. Let $W_{ij} = C_V(Y_{ij})$. Since W_{ij} is also a sum of E-homogeneous components and there are no common irreducibles among the distinct W_{ij}, it follows that $V = \oplus W_{ij}$ and the nontrivial W_{ij} must be a single G-orbit. Clearly, W is invariant under $N_G(X_{12})$ and indeed, we see that this is the full stabilizer of W. Since $G = HE = HN_G(X_{12})$, H acts transitively on the W_{ij} as well. Thus, $V \cong W^H_{N_H(X_{12})}$ and so by Frobenius reciprocity, $C_V(H) = C_W(N_H(X_{12})) = C_V(N_H(X_{12})Y_{12})$. $\square$

The next lemma gives a bound in certain additional cases.

LEMMA 3.6. *Let A be a finite group and G a normal subgroup. Let $M = N_A(M)$ be a maximal subgroup of A such that $M \cap G$ is properly contained in the maximal subgroup J of G. Assume moreover that the intersection of any proper subset of M-conjugates of J properly contains $M \cap G$. Let V be an irreducible $\mathbb{C}A$-module. Then either G acts trivially on V or $\dim C_V(J) \leq \dim C_V(M)$.*

PROOF. Since M does not contain G, it follows that $A = GM$. Let $W = C_V(J)$. Choose a transversal $1 = x_1, \ldots x_t$ in M for A/G. We claim that $\sum x_i W$ is a direct sum. If not, then there exists a nonzero vector $v \in W$ with $v \in \sum_{I > 1} x_i W$. Thus, $v \in C_V(J) \cap C_V(\bigcap_{i > 1} J^{x_i})$. By assumption, $\bigcap_{i > 1} J^{x_i}$ is not contained in

J and so $G = \langle J, \bigcap_{i>1} J^{x_i} \rangle$. Hence $v \in C_V(G) = 0$. Thus, the map $w \mapsto \sum x_i w$ is an injection from W to $C_V(M)$. $\square$

Note that if $M \cap G$ is not maximal in G and J is a maximal self-normalizing subgroup of G containing $M \cap G$ and A/G has order 2, the hypotheses are always satisfied.

4. Upper Bounds for Genus

The first result is classical. Let k be an algebraically closed field of characteristic $p \geq 0$. All covers refer to curves over k.

LEMMA 4.1. *There exist covers* $f : \mathbb{P}^1 \to \mathbb{P}^1$ *of degree* n *with monodromy group* S_n.

PROOF. Let $f(x) = x^2 h(x)$, where h is a polynomial of degree $n-2$ with distinct nonzero roots. Choose h in addition so that f is separable and indecomposable (these are both open conditions on the coefficients of h). Then the monodromy group is a primitive group of degree n containing a transposition (consider the inertia group over 0). It is elementary to prove that a primitive permutation group of degree n containing a transposition is S_n. $\square$

LEMMA 4.2. *Let* S *be a nonabelian simple group and let* n *be the minimal index of a maximal subgroup of* S. *Then there exists a cover* $f : X \to \mathbb{P}^1$ *of degree* n *with* X *of genus* $n + 1$. *In particular,* $\mu(S) \leq n + 1$.

PROOF. By [63], there exists an unramified S-cover of a genus 2 curve Y. Thus, there exists a degree n cover X of Y with monodromy group S. By the Riemann–Hurwitz formula, X has genus $n + 1$. $\square$

With some effort, one should be able show that $\mu(S) < n/2$ at least for $p \neq 2$. One would need a slight generalization of a the generation result from [29] given below — something like given $1 \neq x \in S$, there exists $y \in S$ with $S = \langle x, y \rangle$ such that y and xy have order prime to p. If $p \neq 2$, this would give $\mu(S) \leq (n-1)/2$ and a slightly weaker bound for $p = 2$.

We give the proof of a slightly better result in characteristic zero only. One can do a bit better for most simple groups because they can be generated by elements of order 2 and 3, we can require in this case that (in characteristic 0) X has genus at most $n/6 + 1$ (and asymptotically for many families that is the best that can be done).

LEMMA 4.3. *Let* G *be an almost simple group with socle* S *acting faithfully on a set of cardinality* m. *If* x *is a nontrivial element of* G, *then there exists a Riemann surface* X *and* $f : X \to \mathbb{P}^1$ *of degree at most* m *with* X *of genus* $g \leq \mathrm{ind}(x)/2$ *and monodromy group* G_0 *with* $S \leq G_0 \leq G$. *In particular, if* n *is the minimal degree of a permutation representation of* S, *then there exists a Riemann surface* $f : X \to \mathbb{P}^1$ *of degree* m *with* X *of genus* $g \leq n/4$.

PROOF. By [29], there exists an element $y \in G$ so that $G_0 := \langle x, y \rangle \geq S$. By passing to a G_0-orbit if necessary we may assume that $G = G_0$. By Riemann's existence theorem, there exists a 3 branch point cover $f : X \to \mathbb{P}^1$ with inertia groups generated by x, y and $z := (xy)^{-1}$. Since $\text{ind}(y), \text{ind}(z) < m$, it follows that $g(X) \leq \text{ind}(x)/2$.

Apply this result to the case that $G = S$ and x is an involution to obtain the last statement. $\qquad\square$

LEMMA 4.4. [40] *If* $f : X \to Y$ *is a branched covering and* $f = f_1 \circ f_2$, *then any composition factor of the monodromy group of* f *is a composition factor of the monodromy group of* f_i *for* $i = 1$ *or* 2.

We will need the following result about minimal permutation representations of a groups with a given composition factor.

LEMMA 4.5. *Let* S *be a nonabelian simple group. Let* n *be the smallest cardinality of a faithful* G-set *for any group* G *with* S *as a composition factor of* G. *Then* $F^*(G) = S$ *and* n *is the index of the largest proper subgroup of* S.

PROOF. Let Ω be the given G-set of size n. Note first that G is transitive (for otherwise S is a composition factor in G/K or K where K is the subgroup acting trivially on some G-orbit and either group acts faithfully on a smaller G-set). We claim also that G is primitive on Ω. Otherwise, S is a composition factor of the stabilizer of a block or a composition factor of G acting on the blocks. In either case, we would have a smaller action with S as a composition factor.

Let H be a point stabilizer. If G has a normal abelian subgroup or 2 minimal normal subgroups, then H has a smaller faithful orbit and has S as a composition factor. So let L be a simple component of G and let $L_i, 1 \leq i \leq t$ be the G-conjugates of L. Thus, G has a unique minimal normal subgroup.

Suppose that S is not a component of G. Then S embeds in G/K where K is the subgroup normalizing each L_i. Thus, $t \geq n$, but (cf. section 11), $n \geq 5^t$, a contradiction.

So S is a component of G and G has a unique minimal normal subgroup. Let m be the index of the largest proper subgroup of S. By section 11, one of the following holds: $n \geq m^t$ or $n \geq |S| > m$. Thus, $t = 1$ (as claimed). $\qquad\square$

It is convenient to define $\text{md}(S)$ to be the smallest index of a proper subgroup of S. We remark that $\text{md}(S)$ is known for all S — cf [45].

LEMMA 4.6. *Let* $f : X \to Y$ *be a branched covering of degree* n *with* S *a nonabelian composition factor of the monodromy group* G *of* f. *Assume that* Y *has genus at least 1.*

(i) *If* Y *has genus at least 2, then* $g(X) - 1 \geq n \geq \text{md}(S)$;

(ii) *If* S *is not an alternating group, then* $g(X) - 1 > n/12 \geq \text{md}(S)/12$.

PROOF. By Lemma 4.4, we may assume that f is indecomposable. Let $h = g(Y)$. If $h > 1$, then the Riemann–Hurwitz formula yields $g - 1 \geq n(h - 1) \geq n$. By the previous lemma, $n \geq \mathrm{md}(S)$.

Suppose that $h = 1$. Since S is nonabelian, the cover must be ramified. Then the Riemann–Hurwitz formula yields that $g - 1 \geq (1/2)\rho(J)$ where J is a nontrivial inertia group. Clearly, $\rho(J) \geq \mathrm{ind}(J)$ and so by [49], $\mathrm{ind}(J) \geq n/6$ whence the second statement. $\qquad\square$

5. Regular Normal Subgroups

In this section, we show that the affine case can be reduced to the other cases. We first prove two general results.

LEMMA 5.1. *If G is a finite primitive permutation group with point stabilizer H and N is a regular normal subgroup, then $H \cap H^g$ contains no nontrivial normal subgroup of H for g any nontrivial element of $G \setminus N$.* ·

PROOF. Let K be a nontrivial normal subgroup of H in $H \cap H^g$. Since $G = HN$, we may assume that $g \in N$. Then $H \cap H^g = C_H(g)$ and so $N_G(K) \geq \langle H, g \rangle = G$ (by the maximality of H). This contradicts the fact that H contains no nontrivial normal subgroup of G. $\qquad\square$

LEMMA 5.2. *Let H be a group of automorphisms of a finite group N which is transitive on the set of all nontrivial elements of N. If H is not solvable, then H contains a normal cyclic subgroup C with $F^*(H/C)$ simple.*

PROOF. N is characteristically simple and must have all elements of the same order, whence N is an elementary abelian p-group for some prime p. The result follows easily from Aschbacher's theorem on subgroups of classical groups (see the appendix). $\qquad\square$

We now fix some notation. Fix a prime p. All curves will be smooth projective curves over an algebraically closed field of characteristic p. Let S be a finite nonabelian simple group. Fix a non-negative integer d. Let $\lambda(S, d)$ denote the smallest positive n such that there exists an indecomposable separable branched cover $f : X \to Y$ of degree n with monodromy group G such that S is a composition factor of G and X has genus at most d (if the characteristic is not clear, we write $\lambda_p(S, d)$).

Similarly, let $\lambda'(S, d)$ denote the smallest positive n' such that there exists an indecomposable separable branched cover of degree $f : X \to Y$ of degree n with monodromy group G such that S is a component of G and X has genus at most d.

Let $\lambda''(S, d)$ denote the smallest positive n such that there exists an indecomposable separable branched cover $f : X \to Y$ of degree n with monodromy group G such that $F(G) = 1$, X has genus at most d and S is a composition factor of G.

In particular, to say that any of these quantities is finite is to say that such covers exist.

THEOREM 5.3. *Let $f : X \to Y$ be an indecomposable separable nonconstant map of degree $n = \lambda(S, d)$ with X of genus $g \leq d$. Assume that S is a nonabelian composition factor of the monodromy group G of f. Assume that G contains a regular normal abelian subgroup N. Then Y has genus zero and one of the following holds:*

(i) *$g = 0$ and $\lambda'(S, 2) < n$;*
(ii) *$g = 0$ and $\lambda''(S, 1) < n$;*
(iii) *$g = d$, G acts transitively on the nontrivial elements of N via conjugation and $\lambda'(S, d + 1) < \lambda(S, d)$.*

PROOF. Let Ω denote the G-set of degree n corresponding to the cover. Let H be the stabilizer of a point ω and let Ω_i be the nontrivial H-orbits on Ω. Let H_i be the stabilizer of a point in Ω_i. Identifying H with G/N, we may identify Ω_i with the G-set G/NH_i.

Let Z denote the Galois closure of X/Y and consider the curves $X_i := Z/NH_i$ and let g_i denote the genus of X_i.

If $x \in G$, then write $x = yz$ with $y \in H$ and $z \in N$. Let $\mathrm{fix}(x, \Omega)$ denote the cardinality of the set of fixed points of x on Ω. We note that $\mathrm{fix}(x, \Omega) \leq 1 + \sum \mathrm{fix}(y, \Omega_i)$. For if x is conjugate to y, this is clear while if x is not conjugate to y, then $\mathrm{fix}(x, \Omega) = 0$.

Then $\mathrm{ind}(J, \Omega_i) = \mathrm{ind}(JN/N, \Omega_i)$. The previous paragraph shows that for any subgroup J of G, $\mathrm{ind}(J, \Omega) \leq 1 + \sum_i \mathrm{ind}(J, \Omega_i)$.

Let h denote the genus of Y. Now applying the Riemann–Hurwitz formula to the curves X and X_i, we obtain:

$$2(g - 1) = 2n(h - 1) + \sum_J \rho(J, \Omega) \geq 2n(h - 1) + \sum_{J,i} \rho(J, \Omega_i).$$

Here the sum is over the inertia groups (and higher ramification groups) J of the cover $X \to Y$.

Now

$$2(g_i - 1) = 2n_i(h - 1) + \sum_J \rho(J, \Omega_i),$$

and so since $n = 1 + \sum n_i$,

$$(g - 1) \geq (h - 1) + \sum(g_i - 1).$$

Note that the monodromy group of the cover of $X_i \to Y$ is $G/N \cong H$ (by Lemma 5.1). By minimality, it follows that $g_i > d \geq g$ for each i and so $h = 0$.

This implies that either $g = 0$ and each $g_i = 1$ or H has only one nontrivial orbit on Ω in which case $g = g_1 - 1 \geq d$ and so $g = d$.

Suppose that the second case occurs and $g \neq 0$. Now apply Lemma 5.2 to conclude that (3) holds.

Now suppose the first case holds. So $g = 0$. Now start over and choose an indecomposable cover $\phi : W \to \mathbb{P}^1$ of degree m with W of genus 1 and m minimal with S as a composition factor. Note that $m < n$. So this gives a primitive permutation group. If this group has a normal elementary abelian group, then we repeat the argument and obtain a cover as in the second case (with the genus at most 2) and so $\lambda'(S, 2) < n$. If not, then we conclude that $\lambda''(S, 1) < n$. □

LEMMA 5.4. *Let* $f : X \to Y$ *be an indecomposable separable nonconstant map of degree* $n = \lambda''(S, d)$ *with* X *of genus* $g \leq d$. *Assume that* S *is a nonabelian composition factor of the monodromy group* G *of* f. *Then* G *has a unique minimal normal subgroup or* $d = 0$ *and* $\lambda''(S, 1) < \lambda''(S, 0)$.

PROOF. Assume that G has more than one minimal normal subgroup. Let Ω be the G-set of size n associated with the cover. Let H be a point stabilizer. Let N be a minimal normal subgroup of G. Since there are 2 minimal normal subgroups of G, N is a regular normal nonabelian subgroup. So $G = HN$ is a semidirect product.

Define the curves X_i as in the previous proof. Arguing precisely as above, we have the same possibilities (and note that S is a composition factor of G/N and G/N has no normal abelian subgroups). In this case, H has more than one orbit on the nontrivial elements of N, eliminating that possibility. So it follows that $g = 0$ and $g_i = 1$ for each i. □

We note that for most S, the situations in the lemma cannot occur. For example, if $S = A_m, m \geq 5$, then there is a degree m cover from $\mathbb{P}^1$ to $\mathbb{P}^1$ and so $\lambda(S, d) = \lambda'(S, d) = \lambda''(S, d) = m$.

6. Minimal Genus for Composition Factors

Let S be a finite nonabelian simple group. As in the previous section, all curves considered are over an algebraically closed field of characteristic p. Let $\mu_p'(S) = \mu'(S)$ denote the minimal genus g of a curve X so that there exists a cover $f : X \to Y$ with f indecomposable and S is a component of the monodromy group of f.

Let $\mu''(S)$ denote the minimal genus g of a cover $f : X \to Y$ with f indecomposable such that the monodromy group has no normal abelian subgroup and S is a composition factor. Clearly, we have:

LEMMA 6.1. $\mu(S) \leq \mu''(S) \leq \mu'(S)$.

We rephrase Theorem 5.3 in this notation.

LEMMA 6.2. *Assume that* $\mu(S) < \mu''(S)$. *Then one of the following holds:*

(i) $\mu(S) = 0$ and $\mu'(S) \leq 2$;

(ii) $\mu(S) = 0$ and $\mu''(S) = 1$;

(iii) $\mu(S) > 0$, and $\mu'(S) = \mu(S) + 1$.

PROPOSITION 6.3. $\mu''(S) = \mu'(S)$ or $\mu'(S) \leq 2$.

PROOF. Let $f : X \to Y$ be a separable branched covering of degree $n = \lambda''(S, g)$ with X of genus $g = \mu''(S)$, S a composition factor of the monodromy group G of the cover. We may assume that G has no normal abelian subgroup and that $\mu'(S) > 2$.

Let Z denote the curve corresponding to the Galois closure. Let H be the subgroup with $X = Z/H$. If S is a component of G, $\mu''(S) = \mu'(S)$. So assume that this is not the case. It follows that S is a composition factor of $G/F^*(G)$.

Since G has no normal abelian subgroup, we can write $E = F^*(G)$ as a direct product of conjugates of a subgroup J. Let J' be the direct product of all the other distinct conjugates of J.

Let V denote a the complexification of a Tate module for Z. By Lemmas 3.3 and 3.4, $\dim C_V(H) \geq \dim C_V(N_H(J)J') - \dim C_V(N_G(J))$.

Note that S is a composition factor of the monodromy group of the cover $Z/N_G(J) \to Y$. It follows that $Z/N_G(J)$ has genus g' which is at least $\mu(S)$.

First suppose that $g' > 1$. Then the genus of $Z/(N_H(J)J')$ is at least $5(g' - 1) + 1$ (by the Riemann–Hurwitz formula and the fact that the degree of $Z/(N_H(J)J') \to Z/N_G(J)$ is at least 5). Hence $2g = \dim C_V(H) \geq 8(g' - 1)$.

Thus, $g = \mu''(S) \geq 4(g' - 1) \geq 4(\mu(S) - 1)$. Thus $\mu''(S) > \mu(S)$ and so the previous lemma applies. If $\mu(S) = 0$, then $\mu''(S) \leq 2$, contradicting the fact that $g \geq 4(g' - 1) \geq 4$. Otherwise, $\mu''(S) \leq \mu(S) + 1$ and so $\mu(S) + 1 \geq 4(\mu(S) - 1)$, whence $\mu(S) = 1$ and $g = \mu''(S) \leq 2$, contradicting the fact that $g \geq 4$.

Next consider the case that $g' \leq 1$. If $g' = 1$, since $Z/(N_H(X)Y) \to Z/N_G(X)$ is not an abelian, $Z/(N_H(J)J')$ has genus $g'' \geq 2$. Again, by Lemmas 3.3 and 3.4, $g \geq g'' - 1 \geq 1$.

It follows that $\mu(S) < \mu''(S)$ and $\lambda(S, \mu(S)) < \lambda''(S, \mu''(S))$ (because there is a smaller degree cover which yields a cover with composition factor S and genus no larger than g). Thus, the minimal degree cover achieving $\mu(S)$ must have an abelian normal subgroup. It follows by the proof of Theorem 5.3 that $\lambda''(S, \mu(S) + 1) < \lambda(S, \mu(S))$ or $\mu'(S) \leq 2$.

The first condition does not hold by the inequality at the start of the paragraph and the second does not hold by assumption. This completes the proof. $\square$

COROLLARY 6.4. Either $\mu(S) \leq \mu'(S) \leq 2$ or $\mu'(S) \leq \mu(S) + 1$.

PROOF. Assume $\mu'(S) > 2$. Then $\mu'(S) = \mu''(S)$. By Lemma 6.2 $\mu''(S) \leq \mu(S) + 1$. $\square$

The previous result allows us to concentrate on computing $\mu'(S)$ — i.e. a lower bound for $\mu'(S)$ is very close to the lower bound for $\mu(S)$. The next result essentially reduces this to the almost simple case.

THEOREM 6.5. *Let $f : X \to Y$ be an indecomposable degree n cover with monodromy group G with S a component of G. Assume that X has genus $g = \mu'(S)$. Moreover, assume that n is minimal with respect to these conditions. Then one of the following holds:*

(i) $F^*(G) = S$; *or*
(ii) $F^*(G) = S \times S$ *and* $H \cap F^*(G)$ *is a diagonal subgroup of* $F^*(G)$; *or*
(iii) $\mu'(S) \geq \mathrm{md}(S)/12 + 1$.

PROOF. Let $E = F^*(G)$. Let Z be the curve corresponding to the Galois closure of the cover. Let H be the subgroup of G of index n with $X = Z/H$.

Suppose that we can write $E = A_1 \times \ldots \times A_t$, where the A_i are all the conjugates of $A = A_1 \cong S^m$, $H \cap E$ is the direct product of the subgroups $H \cap A_i$ and $N_H(A)C(A)$ does not contain A. By Lemmas 3.3 and 3.4, $g \geq g(Z/(N_H(A)C(A)) - g(Z/N_G(A))$. If $h := g(X/N_G(A)) \geq 2$, then by the Riemann–Hurwitz formula, $g(Z/(N_H(A)C(A)) - 1 \geq \mathrm{md}(S)(h - 1)$ and so $g \geq (\mathrm{md}(S) - 1)(h - 1) \geq (\mathrm{md}(S) - 1)$.

If $h = 1$, then the same argument (together with Lemma 2.3 and the fact that we may assume that S is not an alternating group), implies that $g \geq \mathrm{md}(S)/12 + 1$.

So we may assume that $h = 0$ and so $g \geq g(Z/(N_H(A)C(A))$. Note that the monodromy group of the cover $Z/(N_H(A)C(A)) \to Z/N_G(A)$ is $N_G(A)/C_G(A)$ and so has S has a composition factor (since $C(A)N_H(A)$ does not contain A). Thus, by minimality, $A = E$. So we cannot decompose E in such a manner.

By the Aschbacher–O'Nan–Scott Theorem, this implies that either $F^*(G) = S$ or E is the unique minimal normal subgroup of G and $H \cap E$ is a full diagonal subgroup of E.

In the latter case, we apply Lemma 3.5. Arguing exactly as above, by minimality, we see that $F^*(G) = S \times S$. $\qquad\square$

Putting together the previous results, we obtain:

THEOREM 6.6. *Let S be a nonabelian simple group. If $\mu'(S) < (\mathrm{md}(S))/12 + 1$, then there exists an indecomposable cover $f : X \to \mathbb{P}^1$ with X of genus g and with S a composition factor of the monodromy group J of f such that $g \leq \mu(S) + 1$ or $g = 2$ and one of the following holds:*

(i) $F^*(J) = S$; *or*
(ii) $F^*(J) = S \times S$ *and* $H \cap F^*(J)$ *is a full diagonal subgroup of* $F^*(J)$.

If p is a prime that does not divide the order of $\mathrm{Aut}(S)$, the previous theorem essentially asserts that the minimal genus of any group involving S does not have order divisible by p, whence we can apply the results about tame covers and so we obtain the following result. Note this says nothing about characteristic 2.

THEOREM 6.7. *Let S be a nonabelian simple group. If p does not divide the order of* $\mathrm{Aut}(S)$, *then one of the following holds:*

(i) $\mu_p(S) \geq \mathrm{md}(S)/12 + 1$; or

(ii) $\mu_0(S) \leq 2$; or

(iii) $\mu_p(S) = \mu^{\mathrm{ta}}(S)$.

We can extend this result to the case that p does not divide the order of S (rather than $\mathrm{Aut}(S)$) by observing if that holds, then the Sylow p-subgroup P of $\mathrm{Aut}(S)$ is cyclic and $\mathrm{Aut}(S) = PG_0$ with G_0 being a p'-group. An easy analysis of this situation together with the results of this section yield Theorem 1.4.

Since for any odd prime there are infinitely many simple groups not divisible by p (for $p > 3$, consider $L_2(r)$ with r prime and p not dividing $r(r^2 - 1)$; for $p = 3$, consider the Suzuki groups), we have the following corollary.

COROLLARY 6.8. *If p is an odd prime and g is fixed, then there are infinitely many simple groups with $\mu_p(S) > g$.*

We will obtain much more precise results in the next few sections including results that hold for $p = 2$.

7. Composition Factors of Genus g Covers

We will use Theorem 6.6 to show that there are many groups which are not composition factors of genus g in characteristic p.

If G is a finite group, define $\gamma(G) := \min\{g(X) - 1 | G \leq \mathrm{Aut}(X)\}$. Of course, $\gamma(G)$ depends on the characteristic (although if p does not divide the order of G, then by Grothendieck, the description of G-covers is independent of characteristic).

We need the following result. There should be a more conceptual proof but we use the classification of finite simple groups. If S is a finite simple group, let $\mathrm{fpr}(S)$ be minimum of $\mathrm{fix}(x, \Omega)/|\Omega|$ as where Ω is a faithful G-set for some group G with $F^*(G) = S$ and $1 \neq x \in G$.

LEMMA 7.1. *Let S be a finite simple nonabelian group. If $x \in \mathrm{Aut}(S)$, the number of elements y in the coset xS with $y^2 = 1$ is at most $|S| \mathrm{fpr}(S)$.*

PROOF. If S is alternating, this is clear (since $\mathrm{fpr}(S)$ is very close to 1). The result follows by inspection for the sporadic groups.

So assume that S is a Chevalley group. The number of involutions in S (or xS) is at most $|S|/d$ where d is the smallest degree of a nontrivial complex representation of S. These degrees are known (see [66] and [52]). If S is a classical group, then $\mathrm{fpr}(S)$ is approximately $1/q$ whence we are not even close. If S is an exceptional group, all conjugacy classes of involutions are known and we can get an exact formula for the number of involutions and again the result holds easily.   □

THEOREM 7.2. *Let S be a finite nonabelian simple group. Then one of the following holds:*

(i) $\mu(S) \geq 1 + \mathrm{md}(S)/12$; *or*

(ii) $\mu(S) \geq -2\,\mathrm{md}(G)\,\mathrm{fpr}(G) + \mathrm{md}(G)(\gamma(G))/|G|)(1 - \mathrm{fpr}(G))$ *for some group G with $F^*(G) = S$.*

PROOF. We may assume that $\mu(S) < (\mathrm{md}(S) - 1)/6$. Then by Theorem 6.6, there exists an indecomposable cover $f : X \to \mathbb{P}^1$ with X either of genus 2 or genus at most $\mu(S) + 1$ such that the monodromy group G of the cover satisfies (1) or (2) of Theorem 6.6. Let n denote the degree of f.

Let Z denote the Galois closure of the cover and let H be such that $Z/H = X$. Let h denote the genus of Z.

Consider case (1) of 6.6 first.

Then $F^*(G) = S$ and $2 + (h - 1)/|G| = \sum_J \rho(J, G)/|G|$ and $2 + (g - 1)/n = \sum_J \rho(J, \Omega)/n$. Here the sum is over the inertia groups corresponding to branch points of the cover.

Now $\mathrm{ind}(J, \Omega)/n = 1 - |J|^{-1} - |J|^{-1} \sum_{g \in J\#} f(g, \Omega))/n) \geq (1 - |J|^{-1})(1 - \mathrm{fpr}(G))$.

It follows that $\rho(J, \Omega)/n \geq (\rho(J, G)/|G|)(1 - \mathrm{fpr}(G))$.

Thus, $(1 - \mathrm{fpr}(G))[2 + (h - 1)/|G|] \leq 2 + (g - 1)/n$, or

$$g - 1 \geq -2n\,\mathrm{fpr}(G) + n(\gamma(G))/|G|)(1 - \mathrm{fpr}(S)).$$

Since $n \geq \mathrm{md}(G)$ and $g \leq \mu(S) + 1$ or $\mu(S) = 0$ and $g = 2$, the result follows in this case.

Consider case (2). So $n = |S|$. In this case $f(g, \Omega)/n \leq \mathrm{fpr}(S)$ (by the previous lemma and [6]). By Lemma 5.4, we may assume that G does not normalize either component. Precisely as above, it follows that

$$(1 - \mathrm{fpr}(S))(h - 1)/|G| \leq -2\,\mathrm{fpr}(S) + (g - 1)/n.$$

Consider the curve Z/S where S is one of the components of G. Then $R := N_G(S)/S$ acts on this curve, whence it has genus at least $\gamma(R)$. Note that R is almost simple with socle S. Thus, $2(h - 1) \geq 2|S|\gamma(R)$ and so

$$(1 - \mathrm{fpr}(S))n\gamma(R)/|G| = (1 - \mathrm{fpr}(S))\gamma(R)/|R| \leq -2\,\mathrm{fpr}(S) + (g - 1)/n.$$

As above, this implies (ii) holds.   □

We restate this:

COROLLARY 7.3. *Let S be a nonabelian finite simple group with $n = \mathrm{md}(S)$. Then $(\mu(S) - 1) \geq n/12$ or $\mu(S)/n \geq -2\,\mathrm{fpr}(G) + (\gamma(G))/|G|)(1 - \mathrm{fpr}(G))$ for some group G with $F^*(G) = S$.*

In particular, if $\mathrm{fpr}(G)$ is small and $\gamma(G)$ is large, then $\mu(S)$ is large. We shall apply this to Chevalley groups of characteristic different from p. Here $\mathrm{fpr}(G)$ is roughly $1/q$ and $\gamma(G)$ is a constant times $|G| \log(q)$ (where the constant depends only on the type of G).

8. Estimates on Inertia Groups

Let k be an algebraically closed field of characteristic p. Let $f : X \to Y$ be a Galois cover with Galois group G. Let I be the inertia group of a point of X and let I_i be the higher ramification groups. We write $I = I_1 D$ with D cyclic. Let $C = C_D(I_1)$ and set $r = |D : C|$ and $s = |C|$. In this section, we obtain estimates for $\rho(I)/|G|$. The permutation representation is the regular representation for G since we have a Galois cover. Restricting this representation to I gives $|G : I|$ copies of the regular representation. Thus, $\rho(I)/|G|$ is independent of G and can be computed by considering any such cover with the same I (and higher ramification groups).

In particular, we want to reduce to the case that $I = G$. This can be done in several ways. We could just replace G by I and consider the cover $X \to X/I$ (and so the point with inertia group I is totally ramified). Alternatively, by Katz–Gabber [46], there exists a Galois cover $\psi : L \to \mathbb{P}^1$ ramified at precisely 2 points with inertia groups I (and the same higher ramification groups) and D. Let $g(L)$ denote the genus of L. Thus, $2(g(L) - 1)/|I| + 2 = \rho(I)/|I| + (|D| - 1)/|D|$ or $\rho(I)/|I| = 1 + 2(g(L) - 1)/|I| - 1/|D|$.

For the remainder of this section, we assume that $G = I$; i.e. there is a totally ramified point.

LEMMA 8.1. *If* $I_j \neq I_{j+1}$, *then* $j \equiv 0 \pmod{s}$.

PROOF. We may assume $j > 0$. Choose an element $x \in I_j$ with x not in I_{j+1} and pass to the abelian subgroup $C \times \langle x \rangle$. Now apply Hasse–Arf [61]. □

LEMMA 8.2. $\rho(I)/|I| \geq 1 + 1/r - (s + 1)/|I|$.

PROOF. By the previous result, $I_1 = \ldots = I_s$. Thus, $\rho(I)/|I| \geq 1 - 1/|I| + \sum_{i=1}^{s}(1/rs)(1 - 1/|I_i|)$, whence the result. □

Assume now that I_1 is abelian. We change our numbering scheme to keep track of distinct terms among the higher ramification subgroups and count multiplicities. Let $I_1 = J_1 > J_2 > \ldots > J_m = 1$ with the J_j being the *distinct* higher ramification groups. Let r_i denote the number of higher ramification groups equal to J_i. It follows from Hasse–Arf [61] that if $I_j \neq I_{j+1}$, then $|CI_1|$ divides $\sum_{i=1}^{j} |I_i|$ and in particular, r_{i+1} is a multiple of $|CI_1 : I_{i+1}|$.

Thus (recall our permutation representation is the regular representation)

$$\rho(I)/|I| = 1 - 1/|I| + |C|/|D| \sum_{i=1} \lambda_i(1 - 1/|J_i|),$$

where the λ_i are positive integers. In particular, we see that

$$\rho(I)/|I| > 1 + |C|/|D| - 1/|I| - |C|/|D||I_1|$$
$$\geq 1 + |C|/|D| - 1/|D||I_1| - |C|/|D||I_1|$$
$$\geq 1 + |C|/2|D|,$$

unless possibly $|I_1| = 2$ or $|I_1| = 3$ and $C = 1$.

If $|I_1| = 2$, then $C = D$. If $D = 1$, then $\rho(I)/|I|$ is a positive integer. If $D \neq 1$, then $|I| \geq 6$ and so $\rho(I)/|G| \geq 4/3$.

If $|I_1| = 3$ and $C = 1$, then either $D = 1 = I_2$ and $\rho(I)/|I| = 4/3$ or $\rho(I)/|I| \geq 1 + 1/|D| = 1 + |C|/|D|$. Summarizing we have the following:

LEMMA 8.3. *If I_1 is a nontrivial abelian group, then one of the following holds*:

(i) $\rho(I)/|I| \geq 1 + |C|/2|D|$;
(ii) $I = I_1$ *has order* 2, $I_2 = 1$ *and* $\rho(I)/|I| = 1$;
(iii) $|I_1| = 2$ *and* $\rho(I)/|I| \geq 4/3$; *or*
(iv) $I = I_1$ *has order* 3, $I_2 = 1$ *and* $\rho(I)/|I| = 4/3$.

Keep the assumption that I_1 is abelian. Set $\lambda = \sum \lambda_i$. Then

$$1 - 1/|I| + (|C|/|D|)(1 - 1/p)\lambda \leq \rho(I)/|I| < 1 - 1/|I| + |C|/|D|\lambda.$$

LEMMA 8.4. *Fix* $r = |D/C|$. *Assume that I_1 is abelian and* $p > 3$.

(i) *If* $\lambda \geq 2|D|/|C|$, *then* $\rho(I)/|I| \geq 12/5$;
(ii) *If p is sufficiently large, then* $\rho(I)/|I| > 1 + 1/3r$;
(iii) *Let* $d > 1$ *be a positive integer with* $\rho(I)/|I| > 1 + 1/d$. *If* $|I| \geq 8r^2$ *and p is sufficiently large, then* $\rho(I)/|I| > 1 + 1/d + 1/(9r^2)$; *and*
(iv) *If* $\rho(I)/|I| > 2$, $|I| \geq 8r^2$ *and p is sufficiently large, then* $\rho(I)/|I| > 2 + 1/(9r^2)$.

PROOF. The first statement follows immediately from the inequality above.

So we may assume that $\lambda \leq 2r$. Note that D/C is a cyclic group acting faithfully on I_1. Thus, $|I_1| > r$ and so $r - 1/|I| \geq 1/r - 1/r(r+1) \geq 1/2r$. Thus, for p sufficiently large, $\rho(I)/|I| > 1 + 1/3r$.

Suppose that $\rho(I)/|I| > 1 + 1/d$ for $d > 1$ an integer. If $d > 4r$, then $\rho(I)/|I| + 1 - 1/d > 2 + 1/(12r)$.

So assume that $d \leq 4r$ and $\rho(I)/|I| + 1 - 1/d > 2$. Thus, $1 - 1/|I| + \lambda/r > 1 + 1/d$, and $\lambda/r - 1/d \geq 1/rd \geq 1/(4r^2)$. Hence $\lambda/r - 1/d - 1/|I| \geq 1/(8r^2)$ and so for p sufficiently large, $\rho(I)/|I| > 1 + 1/d + 1/(9r^2)$.

The same argument yields the last statement. $\square$

If p is small or to get better bounds, one needs to analyze the above case more closely. However, we will not need this in this article.

We need to handle the remaining primes. We first show that $\rho(I)/|I|$ cannot be too close to $1 - 1/d$ if $|I_1|$ is small. If $|I|$ itself is small, this is clear since we have bounded the denominator and we can also bound d easily.

First note that an easy consequence of Hasse–Arf is the following:

LEMMA 8.5. *Each r_i is a multiple of* $|C|$. *Indeed, r_i is a multiple of* $|C_i|$, *where* $C_i = C_D(J_i/J_{i+1})$.

PROOF. By passing to the subgroup C_iJ_i, we may assume that $I = C_1I_1$ and prove the result for r_1. We can then pass to I/I_2 and assume that $I_2 = 1$, whence I is abelian and now Hasse–Arf applies. $\square$

LEMMA 8.6. *Fix $r = |D/C|$. Assume that I_1 contains an abelian subgroup I' of index at most m. Suppose that there are at least $t \geq 5rm$ distinct terms among the $I'_j = I_j \cap I'$. Then $\rho(I)/|I| > 5/2$.*

PROOF. Let $J_1, \ldots, J_t$ denote the smallest subgroups among the higher ramification groups with fixed intersection with I'. Let m_i denote the number of terms among the higher ramification groups which intersect I' in J_i. Then the Riemann–Hurwitz formula yields:

$$\rho(I)/|I| \geq 1 - 1/|I| + (1/|I|) \sum m_i(|J_i| - 1).$$

Now m_i is a multiple of $|C|$ and also by Hasse–Arf m_i is a multiple of $|I' : I'_j|$. Thus,

$$\rho(I)/|I| \geq 1 - 1/|I| + (1/rm) \sum (1 - 1/|J_i|) \geq 1 - 1/|I| + t/(2rm),$$

whence the result.   □

LEMMA 8.7. *Fix $r = |D/C|$. Assume that $|I_1| < N$.*

(i) *Let $d > 1$ be a positive integer with $\rho(I)/|I| > 1 + 1/d$. Then $\rho(I)/|I| > 1 + 1/d + 1/(32r^3N^2)$; and*

(ii) *If $\rho(I)/|I| > 2$, then $\rho(I)/|I| > 2 + 1/(2rN)$;*

(iii) *If $\rho(I)/|I| > 1$, then $\rho(I)/|I| > 1 + 1/(2rN)$.*

PROOF. By the previous lemma and the Riemann–Hurwitz formula, it follows that

$$\rho(I) = |I| - 1 + a_i \sum (|J_i| - 1),$$

with the a_i being positive integers that are multiples of s. Thus,

$$\rho(I)/|I| = 1 - 1/|I| + b/r|I_1|,$$

for some positive integer b.

We assume that $s > 1$. If $s = 1$, then $|I| = r|I_1|$ and $\rho(I)/|I| = 1 + b/r|I_1|$ and the argument we give below will also be valid.

If $\rho(I)/|I| > 2$, it follows that $b > r|I_1|$ and so $\rho(I)/|I| > 2 + 1/r|I_1| - 1/|I| \geq 2 + 1/(2r|I_1|)$. We get precisely the same estimate if $\rho(I)/|I| > 1$.

If $d \geq 4rN$ and $\rho(I)/|I| > 1 + 1/d > 1$, then $\rho(I)/|I| > 1 + 1/(2rN) \geq 1/d + 1/(4rN)$.

So assume that $d < 4rN$ and $\rho(I)/|I| > 1 + 1/d$. Thus, $b/r|I_1| > 1/d + 1/|I|$. It follows that $b/r|I_1| - 1/|I| \geq 1/d|I|$ and so $\rho(I)/|I| \geq 1 + 1/d + 1/d|I|$. If $s < 2d$, this implies that $\rho(I) > 1 + 1/d + 1/(32r^3N^2)$.

Also, $\rho(I) \geq 1 + 1/d + 1/rd|I_1| - 1/|I|$. If $s \geq 2d$, this implies $\rho(I) \geq 1 + 1/d + 1/2rd|I_1| > 1 + 1/(8r^2N^2)$.   □

9. Automorphism Groups of Curves

Let k be an algebraically closed field of characteristic $p \geq 0$. Let $f : Z \to Y$ be a G-Galois cover. Assume that Z has genus $h > 1$.

It is classical that $\mathrm{Aut}(Z)$ is finite. We sketch a slightly different proof of this than the standard ones. This proof came out of discussions with M. Zieve at MSRI. This topic will be more fully explored in further work by the author and Zieve.

First consider the case that Z is defined over a finite field. We first prove a weaker result that is valid for any genus curve.

THEOREM 9.1. *Let Z be a curve over k, the algebraic closure of a finite field. Then $\mathrm{Aut}(Z)$ is locally finite.*

PROOF. Let G be a finitely generated subgroup of $\mathrm{Aut}(Z)$. Then G is defined over some finite subfield k_0 of k. Let f be any nonconstant function on Z. We may enlarge k_0 and assume that f is defined over k_0 and all the poles and zeroes of f are k_0-rational points. Let H be the subgroup of G which fixes all k_0 rational points. Since there are only finitely many such points, H has finite index in G and so it suffices to prove that H is finite. Since $h \in H$ implies that h fixes the zeroes and poles of f, it follows that $f^h = a(h)f$ where $a : H \to k_0^*$ is a homomorphism. Thus, f^n is fixed by H for some n (for example, $n = |k_0^*|$). Thus, the fixed field F of H (acting on $k(Z)$) has transcendence degree 1 over k, whence $k(Z)/F$ is a finite extension and so H is finite.   □

THEOREM 9.2. *There exists a positive valued integral monotonic function $c(g)$ such that if Z has genus at least 2, then $|\mathrm{Aut}(Z)| \leq c(g)$. In particular, $\mathrm{Aut}(Z)$ is finite.*

PROOF. Let G be a subgroup of $\mathrm{Aut}(Z)$. It suffices to show the bound holds when G is finitely generated (for if every finitely generated subgroup has order less than $c(g)$, so does the whole group). So assume that G is finitely generated.

First we define $c(g)$ and show that if G is finite, then $|G| \leq c(g)$.

We can now apply the Riemann–Hurwitz formula and some relatively easy computations as in [64] or the previous section to obtain some bound (one easily gets cg^5; Stichtenoth obtains cg^4 with some extra effort) to see that $c(g)$ can be taken to be a polynomial in g.

Alternatively, let $W := W_\ell$ denote the set of ℓ-torsion points on the Jacobian of Z. As we have observed, if ℓ does not divide the order of G, then $2g(Z/G) = \dim W^G$ (this is really the character formula version of the Riemann–Hurwitz formula — see [61]). In particular, if H is the kernel of the action of G on T, we see that $g(Z/H) = g(Z)$. If Z has genus greater than 1, then the Riemann–Hurwitz formula shows that Z has no nontrivial separable maps to a curve of genus $g(Z)$, whence $H = 1$. Thus, G acts faithfully on W for any sufficiently large prime ℓ. In particular, $|G|$ divides the greatest common divisor of the orders of $GL(2g, \ell)$ for all sufficiently large ℓ. Note that this does not depend on how large we need to let ℓ be (this was observed by Minkowski in his proof of the bound on the orders of finite subgroups of $GL(n, \mathbb{Q})$ or more generally finite subgroups of $GL(n, \mathbb{C})$ with all traces rational). We can take $c(g)$ to be this greatest common divisor.

In conjunction with this previous theorem, this proves the result for k the algebraic closure of a finite field.

Consider the general case. Suppose that G has infinite order. We may write the function field $k(Z) = k(u, v)$. Choose a finitely generated subring R of k such that both Z and G are defined over R and that G acts on $S = R(u, v)$. We note that if M is a maximal ideal of R, then R/M is finite (by the Nullstullensatz). Moreover, taking a plane model for Z over R, we see that any reduction will have genus at most some fixed g' (in fact, one knows that the reduction will have genus at most g). Enlarging R if necessary (by inverting a finite number of elements), we can also assume that there are $c > c(g')$ distinct elements of G that remain distinct on $(R/M)(u, v)$ for any maximal ideal M of R and that the genus of $(R/M)(u, v)$ is at least g (all we need is at least 2).

Now we have G acting on $F(u, v)$ with F a finite field. Moreover, the image H of G in this action has order greater than $c(g')$ since the x_i are still distinct. This is a contradiction, whence G is finite. Then $|G| \leq c(g)$ and the result follows. $\square$

The standard Riemann–Roch argument shows that no nontrivial element of G fixes more than $2g + 2$ elements. So if we are over a finite field, we can enlarge the field to guarantee the existence of at least $2g + 3$ rational points and we see that G acts faithfully on these points and so is finite.

In this section, we consider subgroups of $\mathrm{Aut}(Z)$ that are isomorphic to Chevalley groups in a characteristic different from p and show that the genus of Z must be at least linear in the order of the group. The constant will depend only the type of Chevalley group and not the field. We will use the results of the previous section.

So fix a type of Chevalley group L. Let q be a prime power. Let $J(q)$ denote a group with $F^*(J(q)) = L(q)$ and $J(q)$ contained in the group of inner-diagonal automorphisms of $L(q)$ (for what we need we could just consider $L(q)$). Let W denote the Weyl group of L.

We need the following facts about $J(q)$.

LEMMA 9.3. *Let* $G = J(q)$ *and let* U *be a* p-subgroup of G with p not dividing q.

(i) *If* p *does not divide the order of the Weyl group of* L, *then the Sylow p-subgroup of G is abelian and if U has exponent p^d, then $|U| \leq p^{td}$ where t is the rank of the corresponding algebraic group.*

(ii) *If the exponent of U is at most p^d, then there exist constants e, t (depending only on L) such that $|U| \leq ep^{td}$.*

(iii) *There exists a constant δ (depending only on L) such that any p'-element of $N_G(U)/C_G(U)$ has order at most δ.*

PROOF. It suffices to prove this result for p-subgroups of the corresponding algebraic group $\mathbf{G}$.

If p does not divide the order of the Weyl group, then every p-subgroup of $\mathbf{G}$ is contained in maximal torus (see [24]). Thus, (i) and (ii) follow in this case. Similarly, it is known that W controls fusion in subgroups of tori and so $N(U)/C(U)$ embeds in W and (iii) holds as well for these primes.

Now consider a prime dividing the order of the Weyl group. There is no harm in embedding the simple algebraic $\mathbf{G}$ into $\mathrm{GL}(n, F)$ with F algebraically closed (over a finite field) and n depending only on L. Then every p-subgroup is conjugate to a subgroup of the monomial group M (since every absolutely irreducible representation of a p-group is induced from a 1-dimensional representation of a subgroup). So we may assume that $U \leq M$. If U has exponent p^d, then $U \cap T$ (with T the torus) has order at most p^{dn} and since $|U|$ divides $|U \cap T|n!$, (ii) follows.

Finally, we prove (iii). Again, it suffices to prove this for subgroups of $\mathrm{GL}(n, F) = \mathrm{GL}(V)$ and for primes dividing the order of the Weyl group. In particular, there are only finitely many primes (depending upon L) to consider. By a result of Thompson (see [23], 3.11), we may assume that $[U, U] \leq Z(U)$, $U/Z(U)$ is elementary abelian and the element is faithful on $U/Z(U)$. By elementary representation theory, we see that $U/Z(U)$ has rank at most n, whence the order of $U/Z(U)$ is bounded as a function of n. Thus (iii) holds. □

We will now prove:

THEOREM 9.4. *There exists a constant $c = c(L)$ such that if $J(q)$ is a subgroup of $\mathrm{Aut}(Z)$ for some curve Z of genus $g > 1$ defined over the algebraically closed field k of characteristic p with p not dividing q, then $g \geq c|J(q)|$.*

PROOF. Let $G = J(q)$. Consider Z/G. Let h be the genus of Z/G. If $h > 0$, the Riemann–Hurwitz formula together with the fact that $\mathrm{ind}(I) \geq |G|/2$ for any nontrivial inertia group I gives $g - 1 \geq |G|/4$. So we may assume that $h = 0$.

Let I be an inertia group with $I_1 \neq 1$. Write $I = DI_1$ with D a cyclic p'-group and let $C = C_D(I_1)$. By the previous results, we know that $|D/C| < r$ for some constant $r = r(L)$.

There are three possibilities for p.

First suppose that p is small (depending upon L) but $|I_1|$ is large (large enough so that it contains an abelian subgroup of exponent at least p^t with t satisfying the hypotheses of 8.6. Then $\rho(I)/|G| \geq 5/2$, whence $g - 1 \geq 5|G|/4$ and the result holds.

So assume either p is large (and in particular does not divide the order of the Weyl group, whence I_1 is abelian) or that $|I_1|$ is bounded.

If there is another wildly ramified point or at least 3 ramified branch points, then $2(g - 1)/|G| \geq -1 + \rho(I)/|I|$. It follows by Lemma 8.4 for p large and by Lemma 8.7 if $|I_1|$ is small, that $\rho(I)/|I| > 1 + \delta$ for some positive δ (depending only upon r which in turn is bounded in terms of L and the bound on $|I_1|$).

Suppose that there is only one ramified point. Then $2(g-1)/|G| = -2 + \rho(I)/|I|$. In particular, $\rho(I)/|I| > 2$ and Lemma 8.4 for p large and Lemma 8.7 if $|I_1|$ is small, that $\rho(I)/|I| > 2 + \delta$ for some positive δ (again depending only upon r).

So we may assume that there is precisely one more ramified point in the cover and it is tamely ramified with inertia group cyclic of order d. Thus, $2(g-1)/|G| = -1 - 1/d + \rho(I)/|I|$. In particular, $\rho(I)/|I| > 1 + 1/d$. Again, Lemma 8.4 for p large and Lemma 8.7 if $|I_1|$ is small imply that $\rho(I)/|I| > 1 + 1/d + \delta$ for some positive δ depending only upon r

This completes the proof.   □

COROLLARY 9.5. *Let L be a fixed type of Chevalley group. There exists a positive constant $c = c(L)$ such that $\mu_p(L(q)) \geq c\,\mathrm{md}(L(q))/\log q$ for q sufficiently large with q prime to p.*

PROOF. Let $f : X \to Y$ be a cover with monodromy group G involving $L(q)$ with X of genus $\mu_p(L(q))$. Let Z be the curve corresponding to the Galois closure. By Theorem 7.2, we may assume that $F^*(G) = L(q)$. Note that $|G| \leq 6|J(q)|\log(q)$, where $J(q)$ is the subgroup of G consisting of inner-diagonal automorphisms. By the previous result, this implies that $g(Z) \geq d(L)|G|/\log(q)$ for some constant $d(L)$. By [49] (excluding $L_2(q)$), $\mathrm{fpr}(G) \leq 4/3q$. There is an analogous result for $L_2(q)$. Now apply Corollary 7.3.   □

10. The Generalized Fitting Subgroup

Let G be a finite group. A subgroup H of G is *subnormal* in G if there is a chain of subgroups $H = G_0 < G_1 \ldots < G_m = G$ with G_i normal in G_{i+1}. A group G is called *quasisimple* if $G/Z(G)$ is simple and G is equal to its own commutator subgroup. A component of G is a quasisimple subnormal subgroup. It is not difficult to show that any two distinct components of G commute (see the next two lemmas). Let $F(G)$ denote the Fitting subgroup of G (the maximal normal nilpotent subgroup). Let $E(G)$ be the subgroup of G generated by the components of G. The generalized Fitting subgroup of G is defined to be $E(G)F(G)$ and is denoted by $F^*(G)$.

We first need an elementary result about commutators. This follows from the three subgroup lemma (see [23]).

LEMMA 10.1. *Suppose that H is a perfect subgroup of G, N is a subgroup of G and $[H, N]$ is centralized by H. Then H commutes with N.*

PROOF. Since H is perfect, $[H, N] = [[H, H], N]$. The three subgroup lemma asserts that $[[H, H], N] \leq [[H, N], H] = 1$ as claimed.   □

The next result is standard although the proof is not.

LEMMA 10.2. *Let H be a component of G.*

(i) *H commutes with every normal subgroup of G not containing H.*
(ii) *If H and K are distinct components, then H and K commute.*
(iii) *H commutes with F(G).*

PROOF. Let M be a normal subgroup of G minimal with respect to containing H. So M is generated by the conjugates of H. If $M = G$, then H is normal in G. If M is proper in G, then by induction, H commutes with all other components of M (any component of M is a component of G). So in either case, M is a central product of the components it contains and $H \triangleleft M$. Also, we see that $F(M) = Z(M)$.

Let N be a normal subgroup of G not containing H. Then $[M, N] \leq M \cap N \triangleleft M$. So either $[M, N] \leq Z(M)$ or $M \leq N$, whence either $H \leq N$ or H commutes with N by the previous result.

Since H is not contained in $F(G)$, it follows that H commutes with $F(G)$.

All that remains to show is that H commutes with any other distinct component K. Let N be a minimal normal subgroup containing K. So N is a central product of all its components, whence if $H \leq N$, the result holds. If not, then we have seen that $[H, N] = 1$ and so also $[H, K] = 1$. □

We next give a different characterization of $E(G)$.

LEMMA 10.3. *Let $D = Z(F(G))$. Let X/D be the product of all the minimal normal subgroups of $C_G(F(G))/D$. Then $E(G) = [X, X]$.*

PROOF. Note that X/D has no normal abelian subgroups, for if Y/D is abelian, then since D is central in Y, Y is nilpotent and so $Y \leq F(G) \cap X = Z(F(G)) = D$.

Thus, X/D is a direct product of nonabelian simple groups. In particular, $X = [X, X]D$ with $D \leq Z(X)$ and so $[X, X]$ is perfect and modulo its center is a direct product of nonabelian simple groups $S_1 \times \ldots \times S_t$. Let Q_i be the preimage of S_i in X. Then as above, we see that $[Q_i, Q_i]$ is perfect and simple modulo its center — i.e. a component. We also see that $[X, X]$ is the product of the $[Q_i, Q_i]$, whence $[X, X] \leq E(G)$.

By the previous lemma, every component centralizes $F(G)$ and modulo D is simple. Moreover, as we have already seen, its normal closure is a direct product of simple groups, whence X contains every component of G. Thus, $E(G) = [X, X]$. □

The important property of this subgroup is the following (cf [8]) result which follows immediately.

THEOREM 10.4. $C_G(F^*(G)) = Z(F(G)) = Z(F^*(G))$. *In particular, $F^*(G) \geq C_G(F^*(G))$.*

PROOF. Note that $D := Z(F^*(G)) = Z(F(G))$ (since every component commutes with $F(G)$). Let $C = C_G(F^*(G))$. Suppose that C properly contains D. Consider a minimal normal subgroup of $C_G(F(G)/D$ contained in C/D. By the previous result, this minimal normal subgroup is contained in $DE(G)$ and so is contained in the center of $DE(G)$, whence is contained in D, a contradiction. $\square$

In particular, this shows that there are only finitely many groups with a given generalized Fitting subgroup (for $G/Z(F(G))$ embeds in $\mathrm{Aut}(F^*(G))$ and so we have a bound on $|G|$).

11. Aschbacher–O'Nan–Scott Theorem

In this section, we give a proof a version of the structure theorem for primitive permutation groups which we have used extensively. See [9] for a more detailed version.

Recall that a group G is said to act primitively on a set Ω of cardinality greater than 1 if G preserves no nontrivial equivalence relations on Ω. In particular, this implies that G is transitive on Ω (consider the equivalence relation of being in the same G-orbit). With this added assumption, it is equivalent to the condition that a point stabilizer is maximal. We include a proof of this well known elementary fact.

LEMMA 11.1. *Let G act transitively on Ω. Then G is primitive if and only if the stabilizer of a point ω is a maximal subgroup of G.*

PROOF. Let H be the stabilizer of the point ω. Then Ω can be identified with G/H, the set of left cosets of H. If H is not maximal, consider the natural map $\pi : G/H \to G/M$ for M a maximal subgroup containing H (here $\pi(gH) = gM$). The fibers of π define a G-invariant equivalence relation on G/H.

Suppose that H is maximal. Let Γ be the equivalence class of ω in a nontrivial G-invariant equivalence relation. Since H fixes ω, H preserves Γ. The same is true for every point of Γ. Since G does not preserve Γ and H is maximal this implies that H preserves each point of Γ. Since G is transitive, this implies that $N_G(H)$ is transitive on Γ. Since H is maximal, this implies that H is normal in G and so is trivial. Thus, G has prime order, a contradiction. $\square$

THEOREM 11.2. *Let G be a finite group acting primitively on a set Ω of cardinality n. Let H be the stabilizer of a point. Let A be the product of the minimal normal subgroups of G. Then $A = F^*(G)$ and one of the following holds:*

(i) *A is an elementary abelian p-group, $G = AH$ (semidirect) and H acts irreducibly on A via conjugation and $n = p^a = |A|$.*

(ii) *$A = A_1 \times A_2$ with $A_1 \cong A_2$ a direct product of $t \geq 1$ isomorphic nonabelian simple groups, $H \cap A = \{(a, \phi(a) | a \in A_1\}$ for some isomorphism $\phi : A_1 \to A_2$. Moreover, A_1 and A_2 are the two minimal normal subgroups of G and $n = |A_1|$.*

(iii) *A is the unique minimal normal subgroup of G, $A = L_1 \times \ldots \times L_t$ is the direct product of t copies of isomorphic nonabelian simple groups and one of the following:*

(i) $1 \neq H \cap A = H \cap L_1 \times \ldots \times H \cap L_t$ and $n = m^t$ with $m = |L_1 : H \cap L_1|$ and the $H \cap L_i$ all H-conjugate. Moreover, $N_H(L_1)C_G(L_1)/C_G(L_1)$ is maximal in $N_G(L_1)/C_G(L_1)$.

(ii) There exists a partition $\{\Delta_1, \ldots \Delta_s\}$ of $\{1, \ldots, t\}$ into $s < t$ subsets of size t/s and $A \cap H = K_1 \times \ldots \times K_s$ where $K_i \cong L_1$ is a full diagonal of the direct product of the $A_i := L_j, j \in \Delta_i$. In this case, $n = |L_1|^{t-s}$.

(iii) $A \cap H = 1$, $t > 1$ and $n = |L_1|^t$.

PROOF. Note that H contains no nontrivial normal subgroups (since a normal subgroup fixing 1 point fixes all points). Let B be any normal subgroup of G. Then $G = BH$ (since H is maximal). Thus, B is transitive, Now $C_H(B)$ is normal in H and normalized by B, whence is normal in G and trivial. Now let B be a minimal normal subgroup. Suppose that B is abelian. Then $B \leq C_G(B)$, so $C_G(B) = B$, and so B is the unique minimal normal subgroup. Thus we are in case (1).

So we may assume that there are no minimal normal abelian subgroups. So $F^*(G) = E(G)$. Let $A_1 = L_1 \times \ldots \times L_t$ be a minimal normal subgroup with L_i conjugate nonabelian simple groups (and components of G). Suppose that there is another minimal normal subgroup A_2. Then A_1 and A_2 commute. Then $G = HA_i$ and $H \cap A_i$ centralizes A_j for $j \neq i$. As noted above, $C_H(A_i) = 1$, whence $H \cap A_i = 1$ for $i = 1, 2$. On the other hand, $A_i \leq G = HA_j$ and so the projections of $H \cap A_1 A_2$ into A_j are both onto and injective. Thus, $H \cap A_1 A_2 = \{(a, \phi(a) | a \in A_1\}$ for some isomorphism $\phi : A_1 \to A_2$ as required. Thus, we are in case (2).

So we may assume that A is the unique minimal normal subgroup of G and $A = L_1 \times \ldots \times L_t$ with the L_i conjugate nonabelian simple groups (and components). If $H \cap A = 1$, there is nothing more to say (except to show that $t > 1$ — this requires the classification of finite simple groups in the form of the Schreier conjecture that outer automorphism groups are solvable — see [9] for details).

Suppose that $H_1 = H \cap L_1 \neq 1$. Since $G = AH$ and A normalizes L_1, it follows that H permutes the L_i transitively, whence $H_j = H \cap L_j$ is conjugate to H_1 via H. The maximality of H implies that H is the normalizer of $H \cap A$ and all that remains to be shown in this case is that $N_H(L_1)C_G(L_1)$ is maximal in $N_G(L_1)$.

We note the following — H_1 is the maximal $N_H(L_1)$ invariant subgroup of L_1 (otherwise H normalizes the direct product of the conjugates of this $N_H(L_1)$ invariant overgroup of H_1 contradicting the maximality of H). Suppose that K is a maximal subgroup of $N_G(L_1)$ containing $N_H(L_1)C_G(L_1)$. Let $K_1 = K \cap L_1$. Since $AN_H(L_1) = N_G(L_1)$, it follows that $|N_G(L_1) : N_H(L_1)C_G(L_1)| = |L_1 : H_1|$ and similarly $|N_G(L_1) : K| = |L_1 : K_1|$. Clearly, K_1 contains H_1 and is normalized by $N_H(L_1)$ whence $K_1 = H_1$ and $K = N_H(L_1)C_G(L_1)$ as required.

In the remaining case, $H \cap A \neq 1 = H \cap L_i$. Let π_i denote the projection from onto L_i. Then H normalizes the direct product of these projections and by the maximality of H, if these are proper H contains them. Thus, each projection is onto. Let K_i be the kernel of π_i. If $K_i = 1$, then $H \cap A \cong L_i$ and is a full diagonal subgroup (i.e. of the form $\{(x, \phi_2(x), \ldots, \phi_t(x)) | x \in L_1\}$ where ϕ_i an isomorphism from L_1 to L_i).

If $K_1 \neq 1$, let Δ_1 be those i such that $\pi_i(K_1) = 1$. All other projections of K_1 are surjective (because K_1 is normal in $H \cap A$ and the projections are surjective). By induction, there is a partition $\Delta_1, \ldots, \Delta_s$ such that K_1 is a direct product of full diagonal subgroups of $A_i, i = 2, \ldots, s$. Since K_1 is normal in $A \cap H$ and is self normalizing in $A_2 \times \ldots \times A_s$, it follows that the projection τ of $A \cap H$ into $A_2 \times \ldots \times A_s$ is the same as that of K_1. Thus, $A \cap H = \ker(\tau) \times K_1$. Since $(A \cap H)/K_1 \cong L_1$, this implies that $\ker(\tau)$ is a full diagonal subgroup of A_1. The only remaining point is to show that the Δ_i all have the same cardinality. This is clear since H normalizes $A \cap H$ and acts transitively on the L_j. $\qquad\square$

Much more can be said particularly in case (3) of the theorem. See [9].

12. Aschbacher's Subgroup Theorem

In this section, we prove a version of Aschbacher's Theorem about subgroups of classical groups over finite fields. Roughly, the theorem is that if G is a classical group on a vector space over a finite field F, then any subgroup either is (modulo its intersection with the center) almost simple (i.e. has a unique minimal normal subgroup which is a nonabelian simple group) or preserves some natural geometric structure on the space. By a natural geometric structure on the space, we include such things as a tensor product decomposition, a subspace, a direct sum decomposition or a field extension structure. We will make this more precise below.

In Aschbacher's statement, there are 8 families of structures considered. In fact, there are different ways of organizing the possible structures that one wants to consider (or equivalently, the possible subgroups — the stabilizers of these structures). Aschbacher [7] proved a slightly more general theorem in that he considered subgroups of automorphism groups of classical groups. See also [50] for an approach using Lang's theorem.

This section is based on notes for a course given at USC in 1998.

This theorem has become a standard and important tool in the analysis of subgroups of classical groups. See [33] for one example of how it is used.

When one is using the theorem, it can be important to consider as fine a stratification of the possible geometric structures as possible. Indeed, one category that Aschbacher did not consider was the case of tensor decompositions over extension fields.

However, the proof of the theorem can be organized in different ways. In particular, one does not need to consider all the structures considered by Aschbacher.

The theorem is quite a bit simpler to prove in the case that all irreducible submodules for normal subgroups of G are absolutely irreducible (eg, if one works over the algebraic closure). If that fails, then either G preserves a field extension structure on the space or preserves a direct sum decomposition. Thus, one can give a proof as in the case of an algebraically closed field except for adding one additional class. One can then study this extra class separately. If there is no form involved, then essentially no extra work is required.

We now give a proof of the theorem. In the following subsections, we analyze and classify the groups preserving a field extension structure (in the case that there is no form preserved, there is essentially nothing more to add). In the last subsection, we give some elementary results about groups preserving forms and other representation theory facts which are used in the proof.

Recall that a group G is almost simple if and only if it has a unique minimal normal subgroup which is a nonabelian simple group S (this is equivalent to $S \leq G \leq \mathrm{Aut}(S)$).

Let F be a field of characteristic $p \geq 0$ which is either finite of algebraically closed (there are variations of this result over more general fields). Let G be a finite subgroup of $\mathrm{GL}(V)$ where V is a vector space of dimension d over F (the statement is valid for algebraic groups as well in the case F is algebraically closed with an identical proof where many of the details become quite a bit easier).

Suppose that q is a quadratic form, a unitary form or an alternating form on V. We assume that either $q = 0$ or that q is nondegenerate (i.e. except for the case of quadratic forms in characteristic 2, the radical of the form is 0). Let $X(V, q)$ denote the subgroup of $\mathrm{GL}(V)$ which preserves q up to scalar multiplication. The nondegeneracy condition implies that $X(V, q)$ acts irreducibly on V. So $X(V, q)$ is one of $\mathrm{GL}(V)$, $\mathrm{GO}(V, q)$, $\mathrm{GSp}(V)$ or $\mathrm{GU}(V)$ in the case $q = 0$, q is a quadratic form, alternating form or unitary form respectively. We let $I(V, q)$ denote the isometry group of q (i.e. the subgroup preserving the form). So $I(V, q)$ is one of $\mathrm{GL}(V)$, $O(V, q)$, $Sp(V)$ or $U(V)$. Note that except for the case of quadratic forms, there is only one class of nondegenerate forms. In the case of quadratic forms, there are 2 classes if F is finite and 1 if F is algebraically closed. Moreover, if $\dim V$ is odd, the two classes of quadratic forms give rise to the same isometry groups.

We say that a group H acts homogeneously on V (over F) if V is a direct sum of isomorphic simple FH-modules. The homogeneous component corresponding to a simple FH-module W is the sum of all simple submodules isomorphic to W.

THEOREM 12.1. *Let G be a subgroup of $X(V, q)$ with V a vector space of dimension n over a field F which is either finite or algebraically closed. Let p denote the characteristic of F. Then one of the following holds:*

(R1) *G stabilizes a totally singular subspace;*

(R2) *G stabilizes a nondegenerate subspace;*

(R3) *if the characteristic is* 2 *and* q *is a quadratic form,* G *stabilizes a* 1-*dimensional nonsingular subspace;*

(D1) G *leaves invariant a decomposition* $V = \oplus_{i=1}^{2} V_i$ *with each* V_i *totally singular;*

(D2) G *leaves invariant a decomposition* $V = \oplus_{i=1}^{t} V_i$ *with each* V_i *nondegenerate;*

(T1) G *leaves invariant a tensor decomposition on* V; *i.e.* G *embeds in*

$$X(W_1, q_1) \otimes X(W_2, q_2)$$

where $n = w_1 w_2$ *with* $\dim W_i = w_i$ *and* $q = q_1 \otimes q_2$ *or* $p = 2$, *each* q_i *is a nondegenerate alternating form and* $X(W_1, q_1) \otimes X(W_2, q_2) \leq O(V, q)$ *for the unique (up to scalars) quadratic form vanishing on all simple tensors;*

(T2) G *leaves invariant a tensor structure on* V; *i.e.* G *embeds in* $X(W, q') \wr S_r$ *where* $n = w^r$, $q = q' \otimes \cdots \otimes q'$ (*r times*) *and* $\dim W = w$ *or* $p = 2$, q' *is alternating and* G *preserves a quadratic form on* V ;

(E) $F^*(G) = Z(G)E$ *where* E *is extraspecial of order* s^{1+2a} *with* s *prime,* $n = s^a$ *and* E *acts absolutely irreducibly (and so* G *is contained in the normalizer of* $EZ(G)$); *if* s *is odd,* G *preserves no alternating or quadratic form and if* $s = 2$, G *will preserve a form.*

(EXT) G *preserves an extension field structure*; *or*

(S) $G/Z(G)$ *is almost simple.*

PROOF. Suppose that G acts reducibly. So let U be a proper invariant subspace of minimal dimension. Then $\mathrm{rad}(U)$ is also invariant under G—thus, either $U \subseteq \mathrm{Rad}(U)$ or U is nondegenerate (the radical is taken with respect to the corresponding form if it exists—in the case q is a quadratic form and F has characteristic 2, we compute the radical with respect to the corresponding alternating form). If U is nondegenerate, then (R2) holds.

If U is contained in $\mathrm{Rad}(U)$, then U is totally singular unless possibly $p = 2$. In the latter case, the set of vectors with $q(u) = 0$ forms a G-invariant hyperplane of U. By minimality, it follows that either U is totally singular or U is 1-dimensional. Thus (R1) or (R3) hold.

So we assume that G acts irreducibly. Let N be any normal noncentral subgroup of G.

Suppose that N does not act homogeneously on V. Let V_1 be a homogeneous component of V for N. If V_1 is nondegenerate (or there is no form), then so is every component and so (D1) holds. Otherwise, V_1 is totally singular (since $\oplus \mathrm{Rad}(V_i)$ is G-invariant). The irreducibility of G implies that there is a unique component V_i' so that V_i is not perpendicular to V_i'. Thus, G permutes the nondegenerate subspaces $V_i \oplus V_i'$. If this is a proper subspace, then (D1) holds. If not, then (D2) holds.

So we may assume that every normal subgroup of G acts homogeneously. Let W be an irreducible constituent of N. If W is not absolutely irreducible, then the center of $\operatorname{End}_N(V)$ is a proper extension field E/F whence G preserves an extension field structure on V (this uses the fact that the Brauer group of F is trivial). We will consider this situation in more detail below.

So we may assume that every normal noncentral subgroup acts homogeneously and each irreducible constituent is absolutely irreducible. In particular, this means that we are assuming that G is absolutely irreducible.

Let G_0 be the normal subgroup of G which actually preserves the form (rather than up to a scalar multiple). Since $G/G_0 Z(G)$ is cyclic, G_0 cannot consist of scalars (unless $n = 1$).

Let N be a normal noncentral subgroup and suppose that N does not act irreducibly. Moreover, if a form is involved, we assume that $N \le G_0$. Let W be an irreducible constituent of V for N. Let G_1 be the normalizer in $\operatorname{GL}(W)$ of N. Then a straightforward computation (using the fact that the centralizer of N on W consists of scalars) shows the normalizer of N in $\operatorname{GL}(V)$ is $G_1 \times \operatorname{GL}_{n/d}(U)$ acting on $W \otimes U$ with U of dimension n/d. In particular, G embeds in this group.

If there is no form involved we are done (i.e. G preserves a tensor decomposition—and we note that there is a unique conjugacy class of such subgroups in GL depending only on the dimensions d and n/d). We now consider how the form behaves with respect to this tensor decomposition.

Case 1. q is a quadratic form.

Then V is self dual as an FN-module and hence so W is self-dual for N. Since N acts absolutely irreducibly on W, there is a unique (up to scalar multiple) bilinear form B on W which is N-invariant (if $p \ne 2$, this form is symmetric; if $p = 2$, the form is alternating).

An easy dimension computation shows that all the N-invariant bilinear forms on V are of the form $(W, B) \otimes (U, B')$.

Suppose that $p \ne 2$. If B is alternating, then $q = B \otimes B'$ where B' is an alternating form on B'. As we noted above, $G \le G_1 \times \operatorname{GL}(U)$ acting on $V = W \otimes U$ where G_1 normalizes N on W. Thus, $G_1 \le \operatorname{GSp}(W)$. It follows that $G \le X(V, q) \cap G_1 \times \operatorname{GL}(U) = \operatorname{GSp}(W) \otimes \operatorname{GSp}(U) < \operatorname{GO}(q)$. Moreover, there is a unique conjugacy class of such subgroups.

If B is symmetric, then similarly, we see that $G \le \operatorname{GO}(B) \otimes \operatorname{GO}(B') < \operatorname{GO}(q)$ with B' symmetric as well. There may be several conjugacy classes depending upon the dimension of W and the class of B.

If $p = 2$, we need to proceed in a slightly different manner and the answer is actually easier. Let Δ be the associated alternating bilinear form associated to q. As above, we see that $\Delta = B \otimes B'$ where B' is a bilinear form on U and $G \le \operatorname{GSp}(W) \times \operatorname{GSp}(U)$. Let T denote the image of $\operatorname{GSp}(W) \otimes \operatorname{GSp}(U)$ in $\operatorname{GL}(V)$. In fact, T is contained in the orthogonal group and not just the symplectic group. Indeed, this last group preserves a unique (up to scalar multiplication) quadratic form. Since T is transitive on all nonzero vectors of the form $w \otimes u$, the quadratic form would have to be constant on such vectors. It is straightforward to compute that there exists a unique quadratic form which vanishes on all such vectors and has the corresponding associated alternating form $B \otimes B'$. The uniqueness shows that the form is T-invariant. Since G is absolutely irreducible, it preserves a unique quadratic form, whence q is the form described above.

Case 2. q is alternating.

As above, we see that N leaves an essentially unique form B on W. Thus, G_1 does as well. Arguing precisely, as above, we see that $G \le X(W, B) \otimes X(U, B') \le X(V, q)$. If $p \ne 2$, then we see that B is symmetric and B' alternating or vice versa. If $p = 2$, then we may take both B and B' alternating and we see that in fact G preserves a quadratic form (indeed, in Aschbacher's theorem, this is one of the geometric structures allowed — a subform).

Case 3. q is unitary.

In this case F is a finite field of cardinality m^2. Let F_0 be the subfield of F of cardinality m. Since N acts absolutely irreducibly on W and homogeneously on V, it follows that N preserves a unique (up to F_0 multiple) unitary form h on W.

Arguing as above, we see that this implies that $q = h \otimes h'$ and $G \le X(W, h) \otimes X(U, h') \le X(V, q)$.

So now we may assume that every noncentral normal subgroup acts absolutely irreducibly. Let N be a minimal such subgroup. Thus, $C_G(N) = Z(G)$. It follows that $N/(N \cap Z(G))$ is characteristically simple (i.e. has no nontrivial characteristic subgroups). Thus $M := N/(N \cap Z(G))$ is either an elementary abelian s-group for some prime s or it is a direct product $L_1 \times \ldots \times L_t$ where $L_i \cong L$ is a nonabelian simple group.

Suppose that M is an elementary abelian s-group. Since N', the derived group of N, is contained in the center of N, it follows that N' is either trivial or has order s. In the first case, N is abelian and noncentral. Since it acts absolutely irreducibly, it follows that $n = 1$. So we may assume that N' has order s. Since N acts absolutely irreducibly, $Z(N) \le Z(G)$.

It follows easily that if s is odd, then N is an extraspecial s-group of order s^{1+2a} and $n = s^a$. If s is even, then N is of symplectic type (i.e. either N is extraspecial or $Z(N)$ has order 4 and $N = Z(N)E$ with E extraspecial).

Thus, (E) holds.

So we may assume that M is a product of t isomorphic nonabelian simple groups. It follows that N is a central product of components $Q_1, \ldots, Q_t$. By minimality, each of the Q_i are conjugate in G. Also, we may assume that every minimal normal noncentral subgroup has this form. Since $C_G(N) = Z(G)$, it follows that N is unique. So if $t = 1$, we see that (S) holds. So assume that $t > 1$.

Since N acts absolutely irreducibly on V, it follows that $V = W_1 \otimes \ldots \otimes W_t$ and N embeds in $\hat{Q}_1 \times \ldots \times \hat{Q}_t \le \mathrm{GL}(W_1) \times \ldots \times \mathrm{GL}(W_t)$ where $\hat{Q}_i \cong \hat{Q}$ is a covering group of Q_i. Since the Q_i are conjugate, we may assume that W_i is an absolutely irreducible $\hat{Q}$-module and the W_i are isomorphic as $\hat{Q}$-modules. In particular, they have the same dimension. This is easily seen to be true over the algebraic closure. However, since the character of G is defined over F, the same is true for $\hat{Q}$ acting on W_i.

It follows that the normalizer of N in $\mathrm{GL}(V)$ is precisely $(R_1 \times \ldots \times R_t)\mathrm{Sym}_t$ where R_i is the normalizer of $\hat{Q}_i$ in $\mathrm{GL}(W_i)$ and Sym_t acts on $W_1 \otimes \ldots \otimes W_t$ by permuting the coordinates. In particular, G is contained in this product and so G preserves a tensor structure on V.

If G preserves a form on V, then so does N and since V restricted to Q_i is homogeneous (as $N = Q_i C_N(Q_i)$) and so $\hat{Q}_i$ preserves a form on W_i (and the type is the same for each i).

So if q is unitary, it follows that $G \le X(W, h) \wr S_t \le X(V, q)$ with h unitary. If V is self dual for N, then it follows that $G \le X(W, f) \le X(V, q)$. If $p \ne 2$, then for t even, necessarily q is symmetric. If t is odd, then q and f are either both symmetric or are both alternating. If $p = 2$, then we see that G does preserve a quadratic form (necessarily unique) and so we may always take f to be alternating.

This completes the proof. $\square$

Note that one can state the previous theorem in a different manner. Namely, we have produced natural families of subgroups so that any finite subgroup of $X(V, q)$ is either contained in one of those subgroups or is almost simple (modulo the center). One can of course add to this family some almost simple groups and so when analyzing the subgroup structure of $X(V, q)$ (or related groups), one can use this result. See [45] for an analysis of which of these subgroups are maximal.

The theorem above gives a very specific list of possibilities and one can analyze the conjugacy classes of such subgroups quite easily and produce some natural invariants. In particular, we note that two irreducible subgroups of $X(V, q)$ are conjugate in $X(V, q)$ if and only if the representations are equivalent (up to an outer automorphism) if and only if the characters are the same (up to an outer automorphism).

The one family that we did not analyze so carefully in the proof above is the case where G preserves an extension field structure on V. If F is algebraically closed, this cannot occur. If $q = 0$, then it is straightforward to see that the group preserving a field extension structure corresponding to a field extension is precisely $\mathrm{GL}_{n/d}(E).\mathrm{Gal}(E/F)$ where $d = [E : F]$—i.e. the subgroup of E-

semilinear transformations on V. Note that the only invariant for the conjugacy class is d (and d must divide n).

In the next section, we will analyze the case where $q \neq 0$ and G preserves an extension field structure on V.

12.1. Field Extension Structures

We now make more precise the family of overgroups occurring in the extension field case. We will break the proof up into the various cases depending upon the type of q.

So we assume that F is a finite field of order $m = p^a$ and that V is a vector space of dimension n over F. As usual, let q denote a form (zero, quadratic, alternating or unitary) on V and we assume that $G \leq X(V, q)$. We may also assume that G is irreducible on V and that G preserves a field extension structure on V. More precisely, there is an F-subalgebra $E \subset \operatorname{End}_F(V)$ so that G preserves E (and we have an homomorphism from G into $\operatorname{Gal}(E/F)$).

Note that if the isomorphism class of E is fixed, then E is uniquely determined up to conjugation in $\operatorname{GL}(V)$ (because E has a unique representation of fixed dimension). Since the norm map is surjective for finite fields, in fact this conjugation can always be realized in $SL(V)$. So in the case where $q = 0$, G preserves an E-structure on V if and only if G is a subgroup of $\operatorname{Aut}_E(V).\operatorname{Gal}(E/F)$.

So we assume that (V, q) is nondegenerate. We also make the blanket assumption throughout this section that every normal subgroup of G acts homogeneously on V (or G will satisfy (D1) or (D2)).

Let E/F be an extension of finite fields of degree $d > 1$. Suppose that B is a nondegenerate bilinear form on the vector space U over E. Let $V = U$ considered as a vector space over F. Then $q = \operatorname{tr} \circ B$ is a nondegenerate bilinear form on V and $X(U, B) \leq X(V, \operatorname{tr} \circ B)$.

We first discuss the case where q is either alternating or a quadratic form.

PROPOSITION 12.2. *Assume that q is a nondegenerate alternating or quadratic form on V where V is a vector space of dimension n over the finite field F. Let $G \leq X(V, q)$ be an irreducible subgroup. Assume that every normal subgroup of G acts homogeneously on V. Assume that G normalizes some proper field extension E/F where E is a subalgebra of $\operatorname{End}_F(V)$. Assume moreover that G preserves no additive decomposition of V. Then $G \leq X(U, q')$ where $U = V$ considered as a vector space over some nontrivial field extension E/F and q' is a form on U.*

PROOF. Let Z denote the subgroup of nonzero scalars in $GL(V)$. Let $H = G \cap I(X, q)$. Note that either $GZ = HZ$ or GZ/HZ has order 2. We claim that H acts irreducibly. If not, then $V = V_1 \oplus V_2$ with G permuting each of the H-invariant subspaces V_i. Since H is homogeneous on V, $V_1 \cong V_2$ as H-modules. Since V is self dual as an H-module, we may choose V_1 and V_2 nonsingular. Then G preserves the decomposition $V_1 \oplus V_2$, a contradiction.

Let $E = \operatorname{End}_H(V)$. So E/F is a proper field extension.

Let U denote V considered as an EH-module. Note that $V' := V \otimes_F E_0 \cong \oplus U^\sigma$ where the sum is taken over $\sigma \in \mathrm{Gal}(E/F)$. Since E is commutative, then centralizer of H in V' which is just $E \otimes_F E$ is also commutative. It follows that V' is multiplicity free, whence U^σ and U are nonisomorphic for all nontrivial σ. Since V' is self dual, it follows that $U^\tau \cong U^*$ for some τ. Thus, $U^{\tau^2} \cong U$ and so $\tau^2 = 1$.

Suppose first that $\tau = 1$, i.e. U is self dual. Then there exists a nondegenerate bilinear form B on U which is H-invariant. Moreover, B is unique up to scalar multiplication by E_0. Let $I = I(U, B)$. Note that this is independent of the choice of B. Since G acts naturally on the set of such forms, it normalizes I. Note that $I(U, B)$ preserves the form $\mathrm{tr}_{E/F} \circ B$. Note also that for $p \neq 2$, this form is the same type as q — i.e. B is alternating if and only if q is (because H and $I(U, B)$ have the same centralizer it follows that all forms stabilized by $I(U, B)$ are also stabilized by H and also are of the same type). Thus, $I(U, B) \leq I(X, q)$ and G is contained in the normalizer of $I(U, B)$ in $X(V, q)$.

If $p = 2$ and q is alternating, precisely the same argument suffices. All that remains to show is that if q is a quadratic form, then H preserves a quadratic form on U. Let B be an H-invariant alternating form on U. Then as above we may assume that $C = \mathrm{tr} \circ B$ where C is the alternating form on V associated to q.

Since the set of H-invariant forms on V has cardinality $|E|$ and the map $B \mapsto \mathrm{tr}_{E/F} \circ B$ is injective, we see that we may assume that $C := \mathrm{tr}_{E/F} \circ B$ is either q or if $p = 2$ and q is a quadratic form, C is the associated alternating form. Moreover, in the latter case, by considering $\mathrm{Sp}(B) \cap O(V, q)$, we see that $H \leq I(U, f)$ where f is a quadratic form on U whose associated alternating form is B. By a dimension argument, we may assume that $q = \mathrm{tr}_{E/F} \circ f$. Thus, we see that G is contained in the normalizer of $I(U, h)$ where h is a form of the same type as q (i.e. quadratic or alternating). Since $q = \mathrm{tr}_{E/F} \circ q$, it follows that $I(U, h) \leq I(X, q)$ and the result follows.

Next assume that $\tau \neq 1$. It follows that H preserves a unitary form h on U (note the assumption implies that $[E : F]$ is even). Let E_0 denote the fixed field of τ. Since H acts absolutely irreducibly on U, it is contained in precisely one unitary group on U, whence G normalizes this unitary group. $\qquad \square$

A minor variant on the previous argument shows that:

PROPOSITION 12.3. *Assume that h is a nondegenerate hermitian form on V where V is a vector space of dimension n over the finite field F. Let $G \leq X(V, h)$ be an irreducible subgroup. Assume that every normal subgroup of G acts homogeneously on V. Assume that G normalizes some proper field extension of F where contained in $\mathrm{End}_F(V)$. Assume moreover that G preserves no additive decomposition of V. Then $G \leq X(U, h')$ where $U = V$ considered as a vector space over some nontrivial field extension E/F and h' is a hermitian form on U.*

12.2. Some Elementary Representation Theory

LEMMA 12.4. *Let H be a normal subgroup of finite index in G and assume that V is a homogeneous FH-module and an irreducible FG-module. Let W be an irreducible constituent of V for H.*

(a) *If W is absolutely irreducible, then G embeds in $\mathrm{GL}(W) \otimes \mathrm{GL}(U)$ acting on $W \otimes U \cong V$.*

(b) *If G/H is abelian and V is absolutely irreducibly as a G-module, then V is an irreducible FH-module.*

(c) *If G/H is abelian of exponent m and all mth roots of 1 are in F, then V irreducible as an FG-module implies that V is irreducible as an FH-module.*

PROOF. (a) is well known and is straightforward by computing the normalizer of H in $\mathrm{GL}(V)$.

We now prove (b). By Frobenius reciprocity, V embeds in the induced module W_H^G. This has the same composition factors as $W \otimes F[G/H]$ (for example, we can compute the Brauer character). Since G/H is abelian, over the algebraic closure, we can find a chain of G-submodules of W_H^G so that all quotients are isomorphic to W as H-modules. Since V is absolutely irreducible, this implies that $\dim V \leq \dim W$, whence $V = W$ as required.

If G/H has exponent m and F contains all mth roots of 1, then the same argument shows that each FG-composition factor of W_H^G has dimension at most $\dim W$, whence $V = W$. Thus, (c) holds. $\square$

LEMMA 12.5. *Let G be a finite group.*

(i) *If V is an irreducible FG-module, then G preserves a nondegenerate symmetric or alternating form on V if and only if V is self dual. Moreover, any two forms are in the same C-orbit where C is the group of units in the centralizer of G in $\mathrm{End}(V)$.*

(ii) *If V is an irreducible FG-module and $p = 2$, then G preserves a nondegenerate alternating form if and only if V is self dual. If G preserves a nondegenerate quadratic form on V, then there is a single C-orbit of such quadratic forms.*

(iii) *If V is a homogeneous self dual module and W is an irreducible submodule of dimension m and $\dim \mathrm{End}_{FG}(W) = d$, then the dimension of the G-invariants on the space of bilinear forms is $d(n/m)^2$.*

PROOF. G leaves invariant a nonzero bilinear form on V if and only if there are nonzero fixed points on $V \otimes V$. Any nonzero invariant form must be nondegenerate (since $\mathrm{Rad}(V)$ would be invariant). Such a form gives an FG-isomorphism between V and V^*.

So we may assume that V is self dual. In that case $V \otimes V \cong V \otimes V^* \cong \mathrm{End}(V)$. The nonzero G invariants on $\mathrm{End}(V)$ are precisely C, whence there is a single C-orbit of invariants.

If $p \neq 2$, then $V \otimes V$ is the direct sum of alternating forms and symmetric forms so if G has a fixed point, there must be one that is either symmetric or alternating. Since C preserves both spaces (as $\mathrm{GL}(V)$ does), it follows that all invariant forms are either symmetric or alternating.

If $p = 2$, then the composition factors on $V \otimes V$ (as a $\mathrm{GL}(V)$-module) are $\wedge^2(V)$, V' and $\wedge^2(V)$ where V' is a twist of V (by the Frobenius automorphism). In any case, V' is an irreducible G-module, whence if G has fixed points, then G must have a fixed point on $\wedge^2(V)$. Thus, G always preserves an alternating form. Arguing as above, we see that the nontrivial G fixed points is a single C-orbit.

If G preserves a quadratic form q, we claim that the only quadratic forms which are G-invariant are Cq. Let B_q denote the associated alternating form (so $B_q(v, v') = q(v + v') + q(v) + q(v')$). We know that the set of G-invariant alternating forms is a single C-orbit. Suppose q' is G-invariant. Then replacing q' by an element in Cq', we may replace q' by something in its orbit so that $B_q = B_{q'}$. An elementary computation shows that the set of elements with $q(v) = q'(v)$ is a proper linear G-invariant subspace. Since G is irreducible, it must be 0, whence $q = q'$ and the result follows.

If V is homogeneous, then we compute the invariants as above. $\square$

LEMMA 12.6. *Let F be the finite field of order q^2. Let V be an n-dimensional vector space over F such that V restricted to G is a homogeneous module with irreducible constituent W. Then G fixes a unitary form on W if and only if it does so on V if and only if $\chi(g^q) = \chi(g^{-1})$ for all $g \in G$ where χ is the Brauer character associated to W. If V is irreducible, these conditions are also equivalent to $\chi(g^q) = \chi(g^{-1})$ for all $g \in G$ where χ is the character associated to W. Moreover, if G preserves a unitary form on W, and W is absolutely irreducible, then the space of unitary forms on V which are G-invariant is a vector space over F_q of dimension m^2 where $m = \dim V / \dim W$ is the multiplicity of W.*

PROOF. Let ρ denote the field automorphism $x \to x^q$ on F. By definition of the unitary group, we know that $\rho(g)$ is similar to g^{-1}, viewing G as a subgroup of the unitary group on V. Since V is homogeneous as a G-module, this same condition holds in $\mathrm{GL}(W)$. Thus, to complete the first part of the proof, we need only show that if the character of W satisfies the hypothesis, then W supports a G-invariant form.

The easiest proof is to use Lang's Theorem. Let ϕ denote the given representation of G into $\mathrm{GL}(V)$. We need to find $S \in \mathrm{GL}_n(\bar{F})$ so that $S\phi(g)S^{-1}$ is in the unitary group. This is equivalent to the equation

$$(S\phi(g)S^{-1})^{-T} = \rho((S\phi(g)S^{-1})).$$

By hypothesis, there exists $U \in \mathrm{GL}_n(F)$ with $U\phi = \phi'U$. By Lang's Theorem, $U = S\rho(S^{-T})$ for some $S \in \mathrm{GL}_n(\bar{F})$. This implies that S satisfies the equation above. $\square$

13. Abelian Supplements

Let p be a prime. Let $p(G)$ denote the normal subgroup of the finite group generated by all its Sylow p-subgroups. A finite group G is said to be a quasi p-group if $G = p(G)$ or equivalently if G has no nontrivial p'-quotients. This notion has become quite important in studying fundamental groups of varieties in characteristic p. See [43] and [60] for the solution of the Abhyankar conjecture about fundamental groups of affine curves.

Certain two dimensional varieties are considered in [44]. For these varieties, Abhyankar observed that for any finite image of the fundamental group we have

$$1 \to p(G) \to G \to A \to 1,$$

where A is abelian generated by 2 elements. Abhyankar also conjectured that these conditions were sufficient. Harbater and Van der Put [44] showed that in fact it must also be the case that $G = p(G)B$ for some abelian subgroup B of G. In the appendix of [44], we developed a theory about this situation and gave examples. Combining the examples with the results of [44] shows that the Abhyankar conjecture does not hold.

In this section, we give a simpler form of the example. We also give some examples of a similar phenomenon when A is cyclic of bounded order. Recall that if $H \leq G$, then B is called a supplement to H in G if $G = HB$ and a complement to B if in addition $H \cap B = 1$.

Of course, if $G/p(G)$ is cyclic, we can always write $G = p(G)B$ where B is a cyclic p'-group. If $p(G)$ is a p-group (i.e. there is a unique Sylow p-subgroup of G), then the short exact sequence above splits and so abelian supplements will always exist. The apparent hope was that quasi p-groups have cohomological properties similar to p-groups. However, that is not the case as the examples in the appendix of [44] show.

If $G/p(G)$ is abelian of rank larger than 2, it is quite easy to write down a plethora of examples where there is no abelian supplement to $p(G)$. It is a bit harder to find such examples with the quotient that is abelian of rank 2.

A generalization of the following result appears in the thesis of the author. See also [25].

LEMMA 13.1. *Let r be an odd prime. Let $R := R_d$ be the free group on $2d$ generators subject to $x^r = 1 = [[x,y],z]$ for all $x,y,z \in R$. Then there exist elements in $[R,R]$ which are a product of d commutators but no fewer. Moreover, $Z(R) = [R,R]$ is elementary abelian of order $p^{d(2d-1)}$ and $R/Z(R)$ is elementary abelian of order p^{2d}.*

PROOF. It is straightforward to verify that $|R| = r^{d(2d+1)}$ and that $Z(R) = [R,R]$ and $R/Z(R)$ are elementary abelian. If we choose generators $x_i, 1 \leq i \leq 2d$ for R and set $y_{ij} = [x_i, x_j]$ for $i < j$, then any element in $w \in [R,R]$ can be expressed uniquely as $\prod_{i<j} y_{ij}^{c_{ij}}$ where c_{ij} may be viewed as an element of $\mathbb{F}_r$.

This gives a bijection between $[R, R]$ and the set of $2d \times 2d$ skew symmetric matrices over $\mathbb{F}_r$ (by sending w to the unique skew symmetric matrix whose i, j entry to be c_{ij} for $i < j$). On the other hand, a commutator will correspond to a rank two skew symmetric matrix. So if w corresponds to a nonsingular skew symmetric matrix, w is a product of d commutators but no fewer. This also shows that $[R, R]$ has the order mentioned above. $\qquad \square$

We first give the example when $p = 2$ because it is so simple.

PROPOSITION 13.2. *Let $p = 2$ and r be any odd prime. Let H be the semidirect product of R_2 and a cyclic group generated by an involution τ, where τ inverts each generator x_i. Note that τ centralizes $Z := Z(R_2)$. Pick a subgroup $Y = \langle y \rangle$ of Z of order r which contains noncommutators. Let E be the extraspecial group of order r^3 and exponent r. Let G be the central product of H and E identifying $Z(E)$ and Y (precisely, $G = (H \times E)/\langle (y, w) \rangle$ where w generates $Z(E)$). Then*

(i) $p(G) = H$;

(ii) *$G/p(G)$ is elementary abelian of order r^2; and*

(iii) *there is no abelian supplement to H in G.*

PROOF. Since $R = [\tau, R]$, it follows that the normal closure of τ is H and so the first assertion holds. Clearly $G/H \cong E/W$ is elementary abelian of order r^2, whence the second statement holds.

Suppose that B is an abelian supplement to H in G. There is no harm in assuming that B is an r-group (pass to the Sylow r-subgroup of B) and is generated by two elements u, v (pass the subgroup generated by a pair of elements which generate modulo H). Then $u = ah_1$ and $v = bh_2$ where a, b generate E/W and $h_i \in R$ (clearly, we can take $h_i \in H$ but since u and v are r-elements, so are the h_i). Then $1 = [u, v] = [ah_1, bh_2] = [a, b][h_1, h_2]$. This implies that $y^j = [h_1, h_2]$ for some nontrivial j (since $[a, b]$ is a nontrivial power of w which we identify with that same power of y). However, y^j is not a commutator in R. This contradiction completes the proof. $\qquad \square$

We now show how to modify the construction for an arbitrary p. Let r be a prime congruent to 1 modulo p. Let $c, d \in F_r^*$ of order p with $cd = 1$. Then there is an automorphism τ of order p of R_2 which sends x_i to x_i^c if i is even and x_i^d if i is odd. Moreover, τ centralizes $y = y_{14}y_{23}$. Note that $y \in Z(R)$ and y is not a commutator in R. Let H be the semidirect product R and the group generated by τ. Let E be as above and define G to be the central product of H and E — identify y with a nontrivial central element of E. The proof of the previous result shows that:

PROPOSITION 13.3. (i) $p(G) = H$;

(ii) *$G/p(G)$ is elementary abelian of order r^2; and*

(iii) *there is no abelian supplement to H in G.*

More recently, Harbater has become interested in another fundamental group problem. In this case, G is a group with $G/p(G)$ cyclic of order dividing a fixed m (with m prime to p). The question is whether there is a cyclic supplement of order dividing m. If $G/p(G)$ cyclic of order exactly m, then any cyclic supplement of order dividing m would have to be a complement of order m. If m is infinite, then of course one can also find such a supplement (just choose any cyclic subgroup which generates $G/p(G)$). It is not hard to show that for any fixed m there are examples with no cyclic supplement of order dividing m.

The first example that comes to mind is $G = M_{10}$ and $m = 2$. Let p be 3 or 5. Then $p(G) = A_6$ and $G/p(G)$ has order 2. However, $p(G)$ contains all involutions of G and so there is no supplement of order 2.

For convenience, let us take m an odd prime (different from p). An obvious modification of the construction gives examples for any m prime to p. Let S be an extraspecial m-group of order m^{1+2d} such that S admits an automorphism τ of order p with $C_S(\tau) = Z(S)$. The existence of such an automorphism amounts to finding an element of order p in $Sp(V)$ that has no trivial eigenvalues. Let H be the semidirect product of S and τ. Let G be the central product of H and $J = \langle w \rangle$ with J cyclic of order m^2 (where we identify the center of S with the subgroup of J of order m). Any cyclic supplement of order m is generated by an element of the form wh for some $h \in S$. Since m is odd, it is straightforward to compute that $(wh)^m = w^m$ for all $h \in S$, whence any cyclic supplement has order a multiple of m^2. Clearly $p(G) = H$ and G/H is cyclic of order m.

References

[1] S. Abhyankar, Nice equations for nice groups. Israel J. Math. 88 (1994), 1–23.

[2] S. Abhyankar, Symplectic groups and permutation polynomials, Part I, preprint.

[3] S. Abhyankar, Symplectic groups and permutation polynomials, II, Finite Fields Appl. 8 (2002), 233–255.

[4] S. Abhyankar, Orthogonal groups and permutation polynomials, preprint.

[5] S. Abhyankar and N. Inglis, Galois groups of some vectorial polynomials, Trans. Amer. Math. Soc. 353 (2001), no. 7, 2941–2869.

[6] M. Aschbacher, On conjectures of Guralnick and Thompson, J. Algebra 135 (1990), 277–343.

[7] M. Aschbacher, On the maximal subgroups of the finite classical groups. Invent. Math. 76 (1984), 469–514.

[8] M. Aschbacher, Finite Group Theory, Cambridge University Press, Cambridge, 1986.

[9] M. Aschbacher and L. Scott, Maximal subgroups of finite groups, J. Algebra 92 (1985), 44–80.

[10] S. Cohen, Permutation Polynomials in Shum, Kar-Ping (ed.) et al. Algebras and combinatorics, Papers from the international congress, ICAC'97, Hong Kong, August 1997, Singapore, Springer, 133–146 (1999).

[11] S. Cohen, and R. Matthews, A class of exceptional polynomials, Trans. Amer. Math. Soc. 345 (1994), no. 2, 897–909.

[12] N. Elkies, Linearized algebra and finite groups of Lie type. I. Linear and symplectic groups. Applications of curves over finite fields (Seattle, WA, 1997), 77–107, Contemp. Math., 245, Amer. Math. Soc., Providence, RI, 1999.

[13] W. Feit, On symmetric balanced incomplete block designs with doubly transitive automorphism groups, J. Combinatorial Theory Ser. A 14 (1973), 221–247.

[14] M. D. Fried, Galois groups and complex multiplication, Trans. Amer. Math. Soc. 235 (1978), 141–162.

[15] M. D. Fried, On a Theorem of MacCluer, Acta Arith. XXV (1974), 122–127.

[16] M. Fried, R. Guralnick, and J. Saxl, Schur covers and Carlitz's conjecture, Israel J. Math. 82 (1993), 157–225.

[17] G. Frey, K. Magaard and H. Voelklein, The monodromy group of a function on a general curve, preprint.

[18] M. Fried and H. Völklein, Unramified abelian extensions of Galois covers, Proceedings of the Summer Research Institute on Theta Functions (Gunning and Ehrenpreis, editors), Proceedings of Symposia in Pure Mathematics 49 (1989), 675–693.

[19] D. Frohardt, R. Guralnick and K. Magaard, Incidence matrices, permutation characters, and the minimal genus of a permutation group. J. Combin. Theory Ser. A 98 (2002), 87–105.

[20] D. Frohardt and K. Magaard, Monodromy composition factors among exceptional groups of Lie type in Group Theory, Proceedings of the Biennial Ohio State-Denison Conference, 134–143(eds. Sehgal and Solomon), World Scientific, Singapore, 1993.

[21] D. Frohardt and K. Magaard, Grassmanian fixed point ratios, Geometriae Dedicata 82 (2000), 21–104.

[22] D. Frohardt and K. Magaard, Composition factors of monodromy groups, Ann. of Math. 154 (2001), 327–345.

[23] D. Gorenstein, Finite Groups, Harper and Row, New York, 1968.

[24] D. Gorenstein, R. Solomon and R. Lyons, The classification of finite simple groups. Number 3, Part 1. Chapter A, Almost simple K-groups, Amer. Math. Soc., Providence, RI, 1998.

[25] R. Guralnick, On a result of Schur, J. Algebra 59 (1979), 302–310.

[26] R. Guralnick, Subgroups of prime power index in a simple group. J. Algebra 81 (1983), 304–311.

[27] R. Guralnick, Monodromy groups of rational functions which are Frobenius groups, preprint (1998).

[28] R. Guralnick, The genus of a permutation group, in Groups, Combinatorics and Geometry, edited by M. Liebeck and J. Saxl, LMS Lecture Note Series 165, Cambridge University Press, London, 1992.

[29] R. Guralnick and W. Kantor, Probabilistic generation of finite simple groups. Special issue in honor of Helmut Wielandt. J. Algebra 234 (2000), 743–792.

[30] R. Guralnick and P. Müller, Exceptional polynomials of affine type. J. Algebra 194 (1997), 429–454.

[31] R. Guralnick, P. Müller, and J. Saxl, The rational function analogue of a question of Schur and exceptionality of permutation representations, Mem. Amer. Math. Soc., to appear.

[32] R. Guralnick and M. Neubauer, Monodromy groups of branched coverings: The generic case, in Recent developments in the inverse Galois problem (Seattle, WA 1993) 325–352, Comtemp. Math.,186, Amer. Math. Soc., Providence, RI, 1995.

[33] R. Guralnick, T. Pentilla, C. Praeger, and J. Saxl, Linear groups with orders having certain large prime divisors. Proc. London Math. Soc. (3) 78 (1999), 167–214.

[34] R. Guralnick, J. Rosenberg, and M. Zieve, A new class of exceptional polynomials in characteristic 2, preprint (2000).

[35] R. Guralnick and J. Saxl, Monodromy groups of polynomials. Groups of Lie type and their geometries (Como, 1993), 125–150, London Math. Soc. Lecture Note Ser., 207, Cambridge Univ. Press, Cambridge, 1995.

[36] R. Guralnick and J. Saxl, Exceptional polynomials over arbitary fields, to appear.

[37] R. Guralnick, J. Saxl and M. Zieve, in preparation.

[38] R. Guralnick and K. Stevenson, Prescribing ramification in Arithmetic fundamental groups and noncommutative algebra, Proceedings of Symposia in Pure Mathematics, 70 (2002) editors M. Fried and Y. Ihara, 1999 von Neumann Conference on Arithmetic Fundamental Groups and Noncommutative Algebra, August 16–27, 1999 MSRI.

[39] R. Guralnick and J. Shareshian, Genus of symmetric and alternating groups actions I., preprint (2002).

[40] R. Guralnick and J. Thompson, Finite Groups of Genus Zero, J. Algebra 131 (1990) 303–341.

[41] R. Guralnick and D. Wan, Bounds for fixed point free elements in a transitive group and applications to curves over finite fields, Israel J. Math. 101 (1997), 255–287.

[42] R. Guralnick and M. Zieve, Polynomials with monodromy $PSL(2, q)$, preprint (2000).

[43] D. Harbater, Abhyankar's conjecture on Galois groups over curves, Inventiones Math., 117 (1994), 1–25.

[44] D. Harbater and M. van der Put with an appendix by R. Guralnick, Valued fields and covers in characteristic p, in "Valuation Theory and its Applications", Fields Institute Communications, vol. 32, edited by F.-V. Kuhlmann, S. Kuhlmann and M. Marshall, 2002, 175–204.

[45] P. Kleidman and M. Liebeck, The subgroup structure of the finite classical groups. London Mathematical Society Lecture Note Series, 129. Cambridge University Press, Cambridge, 1990.

[46] N. Katz, Local-to-global extensions of representations of fundamental groups, Ann. Inst. Fourier (Grenoble) 36 (1986), 69–106.

[47] H. Lenstra, H. W., Jr. and M. Zieve, A family of exceptional polynomials in characteristic three, Finite fields and applications (Glasgow, 1995), 209–218, London Math. Soc. Lecture Note Ser., 233, Cambridge Univ. Press, Cambridge, 1996.

[48] M. Liebeck, C. Praeger and J. Saxl, On the O'Nan-Scott theorem for finite primitive permutation groups, J. Austral. Math. Soc. Ser. A 44 (1988), 389–396.

[49] M. Liebeck and J. Saxl, Minimal degrees of primitive permutation groups, with an application to monodromy groups of covers of Riemann surfaces, Proc. London Math. Soc. (3) 63 (1991), 266–314.

[50] M. W. Liebeck and G. Seitz, On the subgroup structure of classical groups, Invent. Math. 134 (1998), 427–453.

[51] M. Liebeck and A. Shalev, Simple groups, permutation groups and probability, J. Amer. Math. Soc. 12 (1999), 497–520.

[52] F. Lübeck, F., Smallest degrees of representations of exceptional groups of Lie type, Comm. Algebra 29 (2001), 2147–2169.

[53] K. Magaard, Monodromy and Sporadic Groups, Comm. Algebra 21 (1993), 4271–4297.

[54] G. Malle, Explicit realization of the Dickson groups $G_2(q)$ as Galois groups in their defining characteristic, Pacific J. Math., to appear.

[55] P. Müller, Primitive monodromy groups of polynomials, Recent developments in the inverse Galois problem (Seattle, WA, 1993), 385–401, Contemp. Math., 186, Amer. Math. Soc., Providence, RI, 1995.

[56] S. Nakajima, p-ranks and automorphism groups of algebraic curves, Trans. Amer. Math. Soc. 303 (1987), 595–607.

[57] S. Nakajima, On abelian automorphism groups of algebraic curves, J. London Math. Soc. (2) 36 (1987), 23–32.

[58] M. Neubauer, On monodromy groups of fixed genus, J. Algebra 153 (1992), 215–261.

[59] M. Neubauer, On primitive monodromy groups of genus zero and one. I. Comm. Algebra 21 (1993), 711–746.

[60] M. Raynaud, Revêtements de la droite affine en caractéristique $p > 0$ et conjecture d'Abhyankar, Invent. Math. 116 (1994), 425–462.

[61] J.-P. Serre, Local fields. Translated from the French by Marvin Jay Greenberg. Graduate Texts in Mathematics, 67. Springer-Verlag, New York-Berlin, 1979.

[62] Shih, T., A note on groups of genus zero, Comm. in Alg. 19 (1991), 2813–2826.

[63] K. Stevenson, Galois groups of unramified covers of projective curves in characteristic p, J. Algebra 182 (1996), 770–804.

[64] H. Stichtenoth, Über die Automorphismengruppe eines algebraischen Funktionenkörpers von Primzahlcharakteristik. I. Eine Abschätzung der Ordnung der Automorphismengruppe, Arch. Math. (Basel) 24 (1973), 527–544.

[65] H. Stichtenoth, Algebraic function fields and codes. Universitext. Springer-Verlag, Berlin, 1993.

[66] P. H. Tiep and A. Zalesskii, Minimal characters of the finite classical groups, Comm. Algebra 24 (1996), 2093–2167.

ROBERT GURALNICK
DEPARTMENT OF MATHEMATICS
UNIVERSITY OF SOUTHERN CALIFORNIA
LOS ANGELES, CA 90089-1113
UNITED STATES
guralnic@math.usc.edu

Galois Groups and Fundamental Groups
MSRI Publications
Volume 41, 2003

On the Tame Fundamental Groups of Curves over Algebraically Closed Fields of Characteristic > 0

AKIO TAMAGAWA

ABSTRACT. We prove that the isomorphism class of the tame fundamental group of a smooth, connected curve over an algebraically closed field k of characteristic $p > 0$ determines the genus g and the number n of punctures of the curve, unless $(g, n) = (0, 0), (0, 1)$. Moreover, assuming $g = 0$, $n > 1$, and that k is the algebraic closure of the prime field $\mathbb{F}_p$, we prove that the isomorphism class of the tame fundamental group even completely determines the isomorphism class of the curve as a scheme (though not necessarily as a k-scheme). As a key tool to prove these results, we generalize Raynaud's theory of theta divisors.

Introduction

Let k be an algebraically closed field of characteristic $p > 0$, and U a smooth, connected curve over k. (A curve is a separated scheme of dimension 1.) We denote by X the smooth compactification of U and put $S = X - U$. We define non-negative integers g and n to be the genus of X and the cardinality of the point set S, respectively.

In [T2], we proved that the isomorphism class of the (profinite) fundamental group $\pi_1(U)$ of U determines the pair (g, n), and that, when $g = 0$ and k is the algebraic closure $\overline{\mathbb{F}}_p$ of the prime field $\mathbb{F}_p$, the isomorphism class of $\pi_1(U)$ even completely determines the isomorphism class of the curve as a scheme.

The aim of the present paper is to generalize these results to the case that $\pi_1(U)$ is replaced by its quotient $\pi_1^t(U)$, the tame fundamental group of U (see [SGA1], Exp. XIII and [GM]), as the author announced in [T2], Note 0.3. Thus the main results of the present paper are the following.

THEOREM (0.1). (See (4.1).) *The isomorphism class of the profinite group* $\pi_1^t(U)$ *determines the pair* (g, n), *unless* $(g, n) = (0, 0), (0, 1)$.

THEOREM (0.2). (See (5.9).) *Assume $g = 0$, $n > 1$, and either $k = \overline{\mathbb{F}}_p$ or $n \leq 4$. Then the isomorphism class of the profinite group $\pi_1^t(U)$ completely determines the isomorphism class of the scheme U.*

More precisely, for two such curves U_i/k $(i = 1, 2)$, $\pi_1^t(U_1) \simeq \pi_1^t(U_2)$ if and only if $U_1 \simeq U_2$ as schemes.

Since it is rather easy to see that the quotient $\pi_1^t(U)$ of $\pi_1(U)$ can be recovered group-theoretically from $\pi_1(U)$ ([T2], Corollary 1.5), the results of the present paper are stronger than those of [T2].

REMARK (0.3). (i) When g and n are small (more precisely, when $2g + n \leq 4$), (0.1) has been settled by Bouw. See [B] for this and other related results.

(ii) In [T1], a result similar to (0.2) was proved for U affine, smooth, geometrically connected curve (of arbitrary genus) over a finite field $\mathbb{F}$. In this case, the (arithmetic) tame fundamental group $\pi_1^t(U)$ is an extension of the absolute Galois group $\mathrm{Gal}(\overline{\mathbb{F}}/\mathbb{F})$ by the geometric tame fundamental group $\pi_1^t(U \otimes_{\mathbb{F}} \overline{\mathbb{F}})$, and we exploited the (outer) Galois action on the geometric tame fundamental group. (0.2) above shows that (for $g = 0$) the geometric tame fundamental group, without the Galois action, is enough to recover the moduli of the curve.

REMARK (0.4). For a profinite group Π, let Π_A denote the set of isomorphism classes of all finite quotients of Π. It is known that the subset Π_A of the set of isomorphism classes of all finite groups completely determines the isomorphism class of the profinite group Π, if Π is finitely generated ([FJ], Proposition 15.4).

We shall write $\pi_A^t(U)$ instead of $\pi_1^t(U)_A$. Then, since $\pi_1^t(U)$ is finitely generated, the information carried by $\pi_A^t(U)$ is equivalent to the information carried by (the isomorphism class of) $\pi_1^t(U)$. Therefore, we can restate the above theorems in terms of $\pi_A^t(U)$. Moreover, as for (0.1), we can say how $\pi_A^t(U)$ determines the pair (g, n) explicitly, by looking carefully at the proofs in the present paper. For this, see [T3].

In [T2], the result corresponding to (0.1) followed from a quick argument combining the Hurwitz formula and the Deuring–Shafarevich formula, which involves wild ramification. However, in our case, we cannot resort to wild ramification, and we need another strategy.

In order to explain our strategy to prove (0.1), first we shall assume $n = 0$, or, equivalently, $U = X$. (Under this assumption, it is elementary to prove (0.1), though. See (4.3)(i).) Note that then we have $\pi_1^t(U) = \pi_1(X)$. In this case, all the ingredients of our strategy are given by Raynaud's theory of theta divisors ([R1]).

(i) The p-rank (or Hasse–Witt invariant) γ_X of X is defined to be the dimension of the $\mathbb{F}_p$-vector space $\mathrm{Hom}(\pi_1(X), \mathbb{F}_p)$. More generally, for each surjective homomorphism $\rho : \pi_1(X) \twoheadrightarrow G$, where G is a finite cyclic group of order N prime to p, $\mathrm{Ker}(\rho)$ may be identified with $\pi_1(Y)$, where $Y \to X$ is the finite étale G-covering corresponding to ρ. Then, G acts on $\mathrm{Hom}(\pi_1(Y), \mathbb{F}_p)$, and

$\text{Hom}(\pi_1(Y), \mathbb{F}_p) \otimes k$ admits a canonical decomposition as a direct sum, corresponding to the decomposition of the group algebra $k[G]$ as the direct product of N copies of k, each of which corresponds to a character $G \to k^\times$. Now, the dimension of each direct summand of $\text{Hom}(\pi_1(Y), \mathbb{F}_p) \otimes k$ is the so-called generalized Hasse–Witt invariant (see [Ka], [Na], and [B]).

(ii) On the other hand, it is well-known that the set of connected finite étale G-coverings of X is in one-to-one correspondence with the set of isomorphism classes of line bundles on X of order N, or, equivalently, the set of points of order N of the Jacobian variety J of X. More precisely, this one-to-one correspondence is given by fixing an isomorphism $G \xrightarrow{\sim} \mu_N(k)$. For each line bundle L of order N, let f be the order of p mod N in $(\mathbb{Z}/N\mathbb{Z})^\times$. Then, taking the composite of the p^f-th power map $L \to L^{\otimes p^f}$ and the isomorphism $L^{\otimes p^f} = L \otimes L^{\otimes(p^f-1)} \xrightarrow{\sim} L$, we get a map $L \to L$, which induces a p^f-linear map $\varphi_{[L]} : H^1(X, L) \to H^1(X, L)$. Now, the generalized Hasse–Witt invariant $\gamma_{[L]}$ with respect to L and $G \xrightarrow{\sim} \mu_N(k) \subset k^\times$ coincides with the dimension of the k-vector space $\bigcap_{r \geq 1} \text{Im}((\varphi_L)^r)$.

(iii) Raynaud ([R1]) defined a certain divisor Θ_B of J (more naturally, of the Frobenius twist J_1 of J) in a canonical way depending only on X, such that $[L] \in J$ belongs to Θ_B if and only if the p-linear map $H^1(X, L) \to H^1(X, L^{\otimes p})$ induced by the p-th power map $L \to L^{\otimes p}$ is an isomorphism. In particular, if $[L]$ is a torsion element of order N prime to p as above, $\varphi_{[L]}$ is an isomorphism (or, equivalently, $\gamma_{[L]} = \dim_k(H^1(X, L))$), if and only if $[L^{\otimes p^i}] \notin \Theta_B$ for all $i = 0, 1, \ldots, f - 1$.

(iv) More precisely, Raynaud defined a vector bundle B with rank $p - 1$, degree $(p - 1)(g - 1)$, and Euler–Poincaré characteristic 0 to be the cokernel of (the linearization of) the p-th power map $\mathcal{O}_X \to \mathcal{O}_X$. Then, he defined Θ_B as the theta divisor of B. (That is to say, $[L] \notin \Theta_B$ if and only if $H^0(X, B \otimes L) = H^1(X, B \otimes L) = 0$.) It is easy to see that Θ_B is a closed subscheme of codimension ≤ 1 of J, and the main difficulty consists in proving that Θ_B does not coincide with J. To prove this, Raynaud resorted to a ring-theoretic argument involving the Koszul complex over the (regular) local ring at the origin of J.

(v) By using intersection theory, Raynaud proved $\#(\Theta_B \cap J[N]) = O(N^{2g-2})$. From this, we obtain $\#\{[L] \in J[N] \mid \exists i, \text{ s.t. } [L^{\otimes i}] \in \Theta_B\} = O(N^{2g-1})$. So, as a conclusion, we can roughly say that, for 'most' prime-to-p-cyclic (finite étale) coverings of X, the generalized Hasse–Witt invariants are as large as possible. In other words, $\gamma_{[L]} = g - 1$ holds for 'most' L (unless $g = 0$). Since the generalized Hasse–Witt invariants are encoded in $\pi_1(X)$ by definition, this gives a group-theoretic characterization of the invariant $g - 1$.

For Raynaud's theory of theta divisors, see also [R2] (a generalization) and [Mad] (an exposition).

In this paper, we generalize these arguments to the (possibly) ramified case $n > 0$. Here, a cyclic (finite étale) covering of U of degree N prime to p corresponds to a pair of a line bundle L and an effective divisor D (whose support is contained in S) satisfying certain conditions. (In particular, $L^{\otimes N} \simeq \mathcal{O}_X(-D)$ is required.) Then, as in (ii) above, we can describe the corresponding generalized Hasse–Witt invariant $\gamma_{([L],D)}$ in terms of a p^f-linear map $\varphi_{([L],D)} : H^1(X,L) \to H^1(X,L)$. Note that, unlike in the case $n = 0$ (and $L \not\simeq \mathcal{O}_X$), the dimension of $H^1(X,L)$ depends on L, since $\deg(L)$ varies (among $0, -1, \ldots, -(n-1)$). Then, under certain assumptions on D, we can define a vector bundle B_D^f depending on f and D, which yields a closed subscheme of J (more naturally, of the f-th Frobenius twist J_f of J). Now, the main result (2.5) says that this closed subscheme is a divisor if $\deg(D) = p^f - 1$. As a corollary, we have that, if $N = p^f - 1$, for 'most' pairs $([L], D)$ with $\deg(L) = -1$, the generalized Hasse–Witt invariant $\gamma_{([L],D)}$ is as large as possible, i.e., coincides with $\dim_k(H^1(X,L)) = g$ (if $n > 1$). On the other hand, by a combinatorial argument, we prove that, if $n > 1$ and $N = p^f - 1$, for 'most' pairs $([L], D)$, there exists an $i = 0, 1, \ldots, f-1$ such that $\deg(L_{p^i}) = -1$. (Here, L_j is a certain modification of $L^{\otimes j}$, so that the cyclic (ramified) covering of X corresponding to $([L], D)$ is the spectrum of $\bigoplus_{j=0}^{N-1} L_j$.) Combining these, we can conclude that for 'most' prime-to-p-cyclic coverings of U (with degree in the form $p^f - 1$), the generalized Hasse–Witt invariants coincide with g. This gives a group-theoretic characterization of g. Since it is easy to see that the Euler–Poincaré characteristic $2 - 2g - n$ can be recovered group-theoretically from $\pi_1^t(U)$, this completes the proof of (0.1).

Just as in [T2], we can then prove that (for U hyperbolic) the set of inertia subgroups of $\pi_1^t(U)$ can be recovered group-theoretically from $\pi_1^t(U)$, by using (0.1) (see (5.2)). Now, what is missing to prove (0.2) along the lines of [T2] is only to recover the 'additive structures' of the inertia subgroups (see Section 5, (B)). In [T2], this was done by studying wild ramification again. In our case, another usage of our generalization of Raynaud's theory settles the problem. This completes the proof of (0.2).

We shall explain briefly the content of each section of the present paper, and show in which section each part of the above arguments is contained. The order of the sections does not necessarily follow the order of the above arguments, because it is natural to present the main theorems (which assure the existence of the theta divisor associated with B_D^f) as early as possible, and then to present the main results concerning $\pi_1(X)$ as corollaries of the main theorems.

In Section 1, we give a generalization of Raynaud's ring-theoretic argument in (iv) above. The main result is (1.12). In fact, Raynaud's original argument is sufficient for the proofs of the main results of the present paper. However, we include this generalization since it is done by just replacing the Koszul complex in Raynaud's proof with the so-called Eagon–Northcott complex, and since it is likely that this generalization will be applied to other related problems concerning coverings and fundamental groups.

In Section 2, after quickly reviewing Raynaud's theory concerning the theta divisor Θ_B, we define the vector bundle B_D^f, and prove (by using (1.12)) the main result (2.5) which assures the existence of the theta divisor associated with B_D^f under certain assumptions. (2.5), together with a slight generalization (2.6), plays a crucial role in the group-theoretic characterization of the genus. Moreover, with another variant (2.13), we investigate the case $n \leq 3$ in more detail (2.21). This plays a central role in the group-theoretic characterization of the additive structures of inertia groups.

In Section 3, we give a review of generalized Hasse–Witt invariants, a description (3.5) of prime-to-p-cyclic coverings of X that are unramified on U in terms of line bundles and divisors on X, and a reinterpretation of generalized Hasse–Witt invariants via this description. Then, after presenting some inputs from intersection theory (3.10), we prove the main numerical results (3.12) and (3.16) concerning the generalized Hasse–Witt invariants of prime-to-p-cyclic coverings of U, by using the results of Section 2. Note that so far the effective divisor D is fixed. Now, combining these results with a combinatorial result (3.18) (which enables D to vary), we finally establish the summarizing result (3.20) to the effect that most generalized Hasse–Witt invariants coincides with g', where $g' \overset{\text{def}}{=} g$ (resp. $g' \overset{\text{def}}{=} g - 1$) for $n > 1$ (resp. $n \leq 1$).

In Section 4, we apply the results of Section 3 and give a group-theoretic characterization of g' in an effective way (4.10) and in an ineffective but impressive way (4.11). The latter can be stated as follows. Here, for each profinite group Π and a natural number m, we denote by $\Pi(m)$ the kernel of $\Pi \twoheadrightarrow \Pi^{\text{ab}} \otimes \mathbb{Z}/m\mathbb{Z}$.

THEOREM (0.5). (See (4.8), (4.11), and (4.12).) *We have*

$$\lim_{f \to \infty} \gamma_{p^f - 1}^{\text{av}} = g',$$

unless $(g, n) = (0, 0), (0, 1)$, *where*

$$\gamma_N^{\text{av}} = \frac{\dim_{\mathbb{F}_p} \left(\pi_1^{\text{t}}(U)(N)^{\text{ab}} \otimes \mathbb{F}_p \right)}{\#(\pi_1^{\text{t}}(U)^{\text{ab}} \otimes \mathbb{Z}/N\mathbb{Z})}.$$

Moreover, after settling a few more minor technical problems (for example, the problem that the results above do not give a characterization of g but only give a characterization of g'), we obtain the group-theoretic characterization (4.1) of the pair (g, n).

Finally, in Section 5, we give group-theoretic characterizations of the inertia subgroups (5.2) and the 'additive structures' of inertia subgroups (5.3), and present anabelian-geometric results (5.8) and (5.9) for $g = 0$.

In the Appendix, we give a proof of a partial generalization (4.17) of the limit formula (4.11). Here, the theory of uniform distribution (especially, Stegbuchner's higher-dimensional version of LeVeque's inequality) plays a key role.

Acknowledgement. When the author presented the results of [T2] in Kyoto (May, 1996) and in Oberwolfach (June, 1997), Shinichi Mochizuki and Michel Raynaud, respectively, asked the author about the possibility of generalizing the results of [T2] to the case that the fundamental group is replaced by the tame fundamental group. Although the author had already taken an interest in such possibility, their questions stimulated and encouraged the author very much. The author would like to thank them very much. Also, when the author was trying to prove a more general limit formula (4.16) of the genus, Makoto Nagata suggested the possibility of applying the theory of uniform distribution, which is essential in the Appendix of the present paper. The author would like to thank him very much.

1. A Generalization of the Ring-Theoretic Part of Raynaud's Theory

In this section, we shall give a generalization of the ring-theoretic part of Raynaud's theory ([R1], 4.2). The statement of our main result is more general than [R1], Lemme 4.2.3, but the proof is rather similar to Raynaud's proof, if we replace the Koszul complex by the so-called Eagon–Northcott complex.

Now, let Y be a connected, noetherian scheme and $f : X \to Y$ a proper morphism whose fibers are of dimension ≤ 1. Let $\mathcal{F}$ be a coherent $\mathcal{O}_X$-module flat over Y. For each $y \in Y$ and $i = 0, 1$, define $h^i(y) = h^i(\mathcal{F}, y)$ to be the dimension of the $k(y)$-vector space $H^i(X_y, \mathcal{F} \otimes k(y))$, where X_y denotes the scheme-theoretic fiber $X \otimes k(y)$ of f at y and $\mathcal{F} \otimes k(y)$ denotes the $\mathcal{O}_{X_y}$- module obtained as the pull-back of $\mathcal{F}$ to X_y. By the local constancy of the Euler–Poincaré characteristic ([Mu], §5, Corollary on p.50),

$$\chi_{\mathcal{F}} \overset{\text{def}}{=} h^0(y) - h^1(y)$$

is independent of y.

DEFINITION. (i) For each $i \in \mathbb{Z}_{\geq 0}$, we denote by $Z_i = Z_i(\mathcal{F})$ the closed subscheme of Y defined by the i-th Fitting ideal $\text{Fitt}_i(R^1 f_*(\mathcal{F}))$ of $\mathcal{O}_Y$.

(ii) We put $W(\mathcal{F}) \overset{\text{def}}{=} Z_{(-\chi_{\mathcal{F}})^+}(\mathcal{F})$, where $x^+ \overset{\text{def}}{=} \max(x, 0)$.

REMARK (1.1). For the definition and the properties of Fitting ideals, we refer to [E], Chapter 20, where only Fitting ideals of modules over rings are treated. However, since the formation of Fitting ideals commutes with localization ([E], Corollary 20.5), we can define and treat Fitting ideals of coherent sheaves on schemes without any extra efforts. (See [SGA7I], Exp. VI, §5.)

LEMMA (1.2). *For each $y \in Y$, we have*

$$y \notin Z_i(\mathcal{F}) \iff h^1(y) \leq i.$$

In particular,

$$y \notin W(\mathcal{F}) \iff \min(h^0(y), h^1(y)) = 0.$$

PROOF. The first assertion follows from [E], Proposition 20.6 and [Mu], § 5, Corollary 3 on p. 53. The second assertion follows from the first, together with the identity

$$(1.3) \qquad h^1(y) - (-\chi_{\mathcal{F}})^+ = \min(h^0(y), h^1(y)). \qquad \square$$

In the special case that $h^0(y) = 1$, we have:

LEMMA (1.4). *Let y be a point of Y, and assume that $h^0(y) = 1$ and that $h^1(y) \geq 1$. Then, in a certain open neighborhood of y, $W(\mathcal{F})$ is the maximal closed subscheme W on which $R^1 f_*(\mathcal{F})|_W$ is locally free of rank $h^1(y)$.*

PROOF. By the assumption, we have $(-\chi_{\mathcal{F}})^+ = h^1(y) - 1$, hence $W(\mathcal{F}) = Z_{h^1(y)-1}(\mathcal{F})$. Put $U = Y - Z_{h^1(y)}(\mathcal{F})$, which is an open neighborhood of y by (1.2). Then, by [E], Proposition 20.8, $W(\mathcal{F}) \cap U$ is the maximal closed subscheme W_U of U on which $R^1 f_*(\mathcal{F})|_{W_U}$ is locally free of rank $h^1(y)$, as desired. $\square$

Next, we shall describe $W(\mathcal{F})$ by using the theory of perfect complexes.

DEFINITION. For a homomorphism $\phi : A \to B$ in an abelian category, we denote by $(A \xrightarrow{\phi} B)$ or simply by $(A \to B)$ the complex

$$\cdots \to 0 \to 0 \to A \xrightarrow{\phi} B \to 0 \to 0 \to \cdots,$$

where A (resp. B) is placed in degree 0 (resp. 1).

LEMMA (1.5). *Let y be a point of Y. Then, in some open neighborhood of y, the object $Rf_*(\mathcal{F})$ (in the derived category of $\mathcal{O}_Y$-modules) can be represented by a complex in the form $(\mathcal{O}_Y^{h^0(y)} \xrightarrow{\phi} \mathcal{O}_Y^{h^1(y)})$ with $\phi \otimes k(y) = 0$.*

Moreover, for each homomorphism $\mathcal{F} \to \mathcal{F}'$ between coherent $\mathcal{O}_X$-modules $\mathcal{F}$ and $\mathcal{F}'$ flat over Y, the corresponding morphism $Rf_*(\mathcal{F}) \to Rf_*(\mathcal{F}')$ can be represented by a homomorphism of complexes in the form

$$(\mathcal{O}_Y^{h^0(\mathcal{F},y)} \xrightarrow{\phi} \mathcal{O}_Y^{h^1(\mathcal{F},y)}) \to (\mathcal{O}_Y^{h^0(\mathcal{F}',y)} \xrightarrow{\phi'} \mathcal{O}_Y^{h^1(\mathcal{F}',y)}),$$

that is:

$$
\begin{array}{ccc}
\vdots & & \vdots \\
\downarrow & & \downarrow \\
\mathcal{O}_Y^{h^0(\mathcal{F},y)} & \to & \mathcal{O}_Y^{h^0(\mathcal{F}',y)} \\
\phi \downarrow & & \phi' \downarrow \\
\mathcal{O}_Y^{h^1(\mathcal{F},y)} & \to & \mathcal{O}_Y^{h^1(\mathcal{F}',y)} \\
\downarrow & & \downarrow \\
\vdots & & \vdots
\end{array}
$$

PROOF. Zariski locally on Y, $Rf_*(\mathcal{F})$ can be represented by a perfect complex in the form $((\mathcal{O}_Y)^{n_0} \to (\mathcal{O}_Y)^{n_1})$. (See [Mu], § 5, the second Theorem on p.46.

We can take a complex in this form since $h^i(y) = 0$ for $y \in Y$ and $i > 1$. See also [SGA6], Exp. I–III.) In particular, we have the exact sequence

$$0 \to R^0 f_*(\mathcal{F}) \to (\mathcal{O}_Y)^{n_0} \to (\mathcal{O}_Y)^{n_1} \to R^1 f_*(\mathcal{F}) \to 0. \tag{1.6}$$

On the other hand, consider a minimal free resolution of the $\mathcal{O}_{Y,y}$-module $R^1 f_*(\mathcal{F})_y$:

$$\cdots \to (\mathcal{O}_{Y,y})^{m_0} \to (\mathcal{O}_{Y,y})^{m_1} \to R^1 f_*(\mathcal{F})_y \to 0.$$

By [E], Theorem 20.2, we can see that the exact sequence

$$(\mathcal{O}_{Y,y})^{n_0} \to (\mathcal{O}_{Y,y})^{n_1} \to R^1 f_*(\mathcal{F})_y \to 0$$

obtained by localizing (1.6) is isomorphic to the direct sum of (part of) the minimal resolution

$$(\mathcal{O}_{Y,y})^{m_0} \to (\mathcal{O}_{Y,y})^{m_1} \to R^1 f_*(\mathcal{F})_y \to 0 \tag{1.7}$$

and the complex $((\mathcal{O}_{Y,y})^{a+b} \overset{\text{proj.}}{\to} (\mathcal{O}_{Y,y})^a)$ for some $a, b \geq 0$. Now, taking the direct sum of (1.7) and the complex $((\mathcal{O}_{Y,y})^b \to 0)$, we obtain a new complex $((\mathcal{O}_{Y,y})^{m_0'} \to (\mathcal{O}_{Y,y})^{m_1})$, where $m_0' = m_0 + b$, which is homotopically equivalent to $((\mathcal{O}_{Y,y})^{n_0} \to (\mathcal{O}_{Y,y})^{n_1})$, by definition. Since we are dealing with only finite number of modules and homomorphisms, we can extend this homotopy equivalence to one between $((\mathcal{O}_Y)^{n_0} \to (\mathcal{O}_Y)^{n_1})$ and $((\mathcal{O}_Y)^{m_0'} \to (\mathcal{O}_Y)^{m_1})$, if we replace Y by a suitable open neighborhood of y.

By the definition of the minimality, the homomorphism $(\mathcal{O}_{Y,y})^{m_1} \to R^1 f_*(\mathcal{F})_y$ becomes an isomorphism after being tensored with $\boldsymbol{k}(y)$. Then, by [Mu], § 5, Corollary 3 on p.53, we obtain $m_1 = h^1(y)$. Thus $Rf_*(\mathcal{F})$ is represented by the complex $((\mathcal{O}_Y)^{m_0'} \to (\mathcal{O}_Y)^{h^1(y)})$. Finally, by tensoring this complex with $\boldsymbol{k}(y)$ again, we obtain $m_0' = h^0(y)$ and $\phi \otimes \boldsymbol{k}(y) = 0$, as desired.

The second assertion follows from the first assertion and a standard fact in the theory of derived categories (see, e.g., [E], Exercise A3.54). $\square$

COROLLARY (1.8). *In some neighborhood of* y, $W(\mathcal{F})$ *is defined by the ideal generated by the maximal minors of an* $h^0(y) \times h^1(y)$ *matrix representing* ϕ *in* (1.5).

PROOF. This follows from the definition of Fitting ideal, together with identity (1.3). $\square$

THEOREM (MACAULAY [Mac]). (See [E], Exercise 10.9.) *Let* R *be a noetherian ring, and let* F *and* G *be free* R-*modules of finite rank. Let* ϕ *be an* R-*homomorphism* $F \to G$, *and choose a matrix* A *with coefficients in* R *that represents* ϕ. *Let* I *be the ideal of* R *generated by the maximal minors of* A. *Then, for each minimal prime ideal* $\mathfrak{p}$ *containing* I, *we have* $\mathrm{ht}(\mathfrak{p}) \leq |\mathrm{rk}(F) - \mathrm{rk}(G)| + 1$. $\square$

DEFINITION. In Macaulay's theorem, if, moreover, the equality $\text{ht}(\mathfrak{p}) = |\text{rk}(F) - \text{rk}(G)| + 1$ holds for every minimal prime ideal $\mathfrak{p}$ containing I, we say that ϕ is determinantal. (See, e.g., [E], 18.5.) Equivalently, ϕ is determinantal if and only if either $I = R$ or $\text{ht}(I) = |\text{rk}(F) - \text{rk}(G)| + 1$.

COROLLARY (1.9). *The codimension of each irreducible component of $W(\mathcal{F})$ does not exceed $|\chi_{\mathcal{F}}| + 1$.*

PROOF. This follows from (1.8) and Macaulay's theorem above. □

DEFINITION. We say $\mathcal{F}$ is determinantal, if the codimension of every irreducible component of $W(\mathcal{F})$ coincides with $|\chi_{\mathcal{F}}| + 1$. (Equivalently, $\mathcal{F}$ is determinantal if and only if either $W(\mathcal{F}) = \varnothing$ or $\text{codim}(W(\mathcal{F})) = |\chi_{\mathcal{F}}| + 1$.)

Before presenting the main result of this section, we shall establish the following key lemma, which is purely in commutative ring theory. In the special case that R is regular, $\text{rk}(F) = \text{rk}(F') = \dim(R)$ and $\text{rk}(G) = \text{rk}(G') = 1$, this can be seen in [R1], Lemme 4.2.3.

LEMMA (1.10). *Let R be a Cohen–Macaulay local ring. Let*

$$\begin{array}{ccc} F & \xrightarrow{f} & F' \\ \phi \downarrow & & \phi' \downarrow \\ G & \xrightarrow{g} & G' \end{array} \qquad (1.11)$$

be a commutative diagram of R-modules, where F, G, F', G' are free R-modules of finite rank. Assume that:

(a) $\text{rk}(F) - \text{rk}(G) \geq 0$ *and ϕ is determinantal;*
(b) *g is surjective;*
(c) *either $\text{rk}(F) - \text{rk}(G) = 0$ or ϕ' is not surjective; and*
(d) $\text{rk}(F) - \text{rk}(G) \geq \text{rk}(F') - \text{rk}(G')$.

Then:

(i) $\text{rk}(F) - \text{rk}(G) = \text{rk}(F') - \text{rk}(G')$.
(ii) *ϕ' is determinantal.*
(iii) *The fiber product $F_1 \overset{\text{def}}{=} G \times_{G'} F'$ is a free R-module of rank $\text{rk}(F)$, and the determinant of the natural homomorphism $F \to F_1$ is not zero.*

PROOF. By replacing R with its completion, we may assume that R is complete. In particular, we may assume that R admits a canonical module ω (see [BH], Corollary 3.3.8).

From now on, we write χ and χ' instead of $\text{rk}(F) - \text{rk}(G)$ and $\text{rk}(F') - \text{rk}(G')$, respectively. Moreover, we denote by I and I' the ideals of R generated by the maximal minors of ϕ and ϕ', respectively.

First, we treat the easier case that ϕ' is surjective. Then we must have $\chi' \geq 0$. On the other hand, by (c) and (d), we have $\chi' \leq \chi = 0$. Thus $\chi' = \chi = 0$, which implies (i). Since ϕ' is surjective with $\chi' = 0$, ϕ' must be an isomorphism. In particular, we have $I' = R$, hence (ii) holds. Next, the natural map $F_1 \to G$ is an isomorphism, so we have $\mathrm{rk}(F_1) = \mathrm{rk}(G) = \mathrm{rk}(F)$. Moreover, the natural map $F \to F_1$ can be identified with ϕ. Now, since ϕ is determinantal, the determinant of ϕ is non-zero. This complete the proof in the case that ϕ' is surjective.

Next, assume that ϕ' is not surjective. Put $M \overset{\mathrm{def}}{=} \mathrm{Coker}(\phi)$ and $M' \overset{\mathrm{def}}{=} \mathrm{Coker}(\phi')$. Since $\chi \geq 0$ by the first half of (a), we have $I = \mathrm{Fitt}_0(M)$ by definition, and I annihilates M by [E], Proposition 20.7a. By (b), the natural map $M \to M'$ is also surjective, hence I annihilates M'. In particular, $I \neq R$ as $M' \neq 0$, so, by (a), we have $\mathrm{ht}(I) = \chi + 1 \geq 1$. Now, since M' is annihilated by I with $\mathrm{ht}(I) \geq 1$, we obtain $\chi' \geq 0$. (To see this, for example, tensor the (right) exact sequence $F' \to G' \to M' \to 0$ with the residue field at any minimal prime ideal of R.) Thus we have $I' = \mathrm{Fitt}_0(M')$.

Note that g is surjective by (b) and G' is free. Accordingly, g is split surjective, or, equivalently, if we put $K \overset{\mathrm{def}}{=} \mathrm{Ker}(g)$, g is a composite of an isomorphism $G \overset{\sim}{\to} K \times G'$ that restricts to the identity on K and the projection $K \times G' \to G'$. From this, we can easily see that the fiber product $F_1 = G \times_{G'} F'$ is free and fits naturally into the following commutative diagram whose columns are all exact:

$$
\begin{array}{ccccc}
F & \overset{f_1}{\to} & F_1 & \twoheadrightarrow & F' \\
\phi \downarrow & & \phi_1 \downarrow & & \phi' \downarrow \\
G & = & G & \twoheadrightarrow & G' \\
\downarrow & & \downarrow & & \downarrow \\
M & \twoheadrightarrow & M' & = & M' \\
\downarrow & & \downarrow & & \downarrow \\
0 & & 0 & & 0.
\end{array}
$$

Moreover, we have $\mathrm{rk}(F_1) - \mathrm{rk}(G) = \chi' \geq 0$. Thus, calculating $\mathrm{Fitt}_0(M')$ by using ϕ_1, we see $\mathrm{Fitt}_0(M) \subset \mathrm{Fitt}_0(M')$, or, equivalently, $I \subset I'$. Moreover, since I' annihilates $M' \neq 0$ by [E], Proposition 20.7a, we have $I' \neq R$. Thus, we have

$$\chi + 1 = \mathrm{ht}(I) \leq \mathrm{ht}(I') \leq \chi' + 1,$$

where the last inequality follows from Macaulay's theorem. Combining this with (d), we obtain $\mathrm{ht}(I') = \chi' + 1$ and $\chi' = \chi$. The former implies (ii), and the latter implies both (i) and the first half of (iii). Note that ϕ_1 is also determinantal.

To see the second half of (iii), we shall compare the Eagon–Northcott complexes associated with ϕ and ϕ_1. (For Eagon–Northcott complexes, see [E], A2.6.) So, consider the following commutative diagram whose first (resp. second) row is the Eagon–Northcott complex canonically associated with ϕ (resp. ϕ_1):

$$
\begin{array}{ccccccccccc}
0 & \to & D_\chi \otimes \bigwedge^r F & \to \cdots \to & D_0 \otimes \bigwedge^s F = \bigwedge^s F & \overset{\bigwedge^s \phi}{\to} & \bigwedge^s & \to & \bigwedge^s \otimes R/I & \to & 0 \\
& & \downarrow & & \downarrow & & \| & & \downarrow \\
0 & \to & D_\chi \otimes \bigwedge^r F_1 & \to \cdots \to & D_0 \otimes \bigwedge^s F_1 = \bigwedge^s F_1 & \overset{\bigwedge^s \phi_1}{\to} & \bigwedge^s & \to & \bigwedge^s \otimes R/I' & \to & 0,
\end{array}
$$

where $r = \mathrm{rk}(F) = \mathrm{rk}(F_1)$, $s = \mathrm{rk}(G)$, $D_i = (S_iG)^*$, and $\bigwedge^s = \bigwedge^s G \ (\simeq R)$. Since ϕ and ϕ_1 are determinantal, the two rows are exact and both $\bigwedge^s \otimes R/I$ and $\bigwedge^s \otimes R/I'$ are Cohen–Macaulay R-modules, by [E], Corollary A2.13.

Now, suppose that the determinant map $\bigwedge^r F \to \bigwedge^r F_1$ is zero. Then, the first vertical arrow $D_\chi \otimes \bigwedge^r F \to D_\chi \otimes \bigwedge^r F_1$ is also zero. So, calculating $\mathrm{Ext}_R^{\chi+1}(-, \omega)$ by using the Eagon–Northcott complexes, we see that the map $\mathrm{Ext}_R^{\chi+1}(\bigwedge^s \otimes R/I', \omega) \to \mathrm{Ext}_R^{\chi+1}(\bigwedge^s \otimes R/I, \omega)$ associated with the natural surjection $\bigwedge^s \otimes R/I \to \bigwedge^s \otimes R/I'$ must be also zero. However, since $\mathrm{ht}(I') = \mathrm{ht}(I) = \chi + 1$, the duality theory (see [BH], Theorem 3.3.10, (a)$\Rightarrow$(c)) tells us that this implies that the original surjection $\bigwedge^s \otimes R/I \to \bigwedge^s \otimes R/I'$ is zero. This is absurd, since $\bigwedge^s \simeq R$ and $I' \neq R$. This completes the proof. $\qquad\square$

The following is the main result of this section.

THEOREM (1.12). *Let Y be a Cohen–Macaulay, noetherian, integral scheme. Let $f : X \to Y$ be a proper morphism whose fibers are of dimension ≤ 1. Let $\mathcal{F}_i$ ($i = 1, 2, 3$) be coherent $\mathcal{O}_X$-modules flat over Y, and $0 \to \mathcal{F}_1 \to \mathcal{F}_2 \to \mathcal{F}_3 \to 0$ an exact sequence of $\mathcal{O}_X$-modules. Assume that:*

(a) *$\mathcal{F}_2$ is determinantal (in the sense of the Definition following (1.9));*
(b) *one of the following three conditions holds: $\chi_{\mathcal{F}_2} < 0, W(\mathcal{F}_1) \neq \varnothing$; $\chi_{\mathcal{F}_2} = 0$;*
  $\chi_{\mathcal{F}_2} > 0, W(\mathcal{F}_3) \neq \varnothing$; and
(c) *$\chi_{\mathcal{F}_1} \cdot \chi_{\mathcal{F}_3} \geq 0$.*

Then:

(i) *$\chi_{\mathcal{F}_1} \cdot \chi_{\mathcal{F}_3} = 0$.*
(ii) *$\mathcal{F}_1$ and $\mathcal{F}_3$ are determinantal.*

PROOF. First, we shall treat the case that $\chi_{\mathcal{F}_2} \geq 0$. By (1.5), in some neighborhood of each $y \in Y$, the objects $Rf_*(\mathcal{F}_2)$, $Rf_*(\mathcal{F}_3)$, and the morphism $Rf_*(\mathcal{F}_2) \to Rf_*(\mathcal{F}_3)$ can be represented by complexes $(\mathcal{O}_Y^{h^0(\mathcal{F}_2, y)} \to \mathcal{O}_Y^{h^1(\mathcal{F}_2, y)})$, $(\mathcal{O}_Y^{h^0(\mathcal{F}_3, y)} \to \mathcal{O}_Y^{h^1(\mathcal{F}_3, y)})$, and a commutative diagram

$$
\begin{array}{ccc}
\mathcal{O}_Y^{h^0(\mathcal{F}_2, y)} & \to & \mathcal{O}_Y^{h^0(\mathcal{F}_3, y)} \\
\downarrow & & \downarrow \\
\mathcal{O}_Y^{h^1(\mathcal{F}_2, y)} & \to & \mathcal{O}_Y^{h^1(\mathcal{F}_3, y)}
\end{array}
\tag{1.13}
$$

respectively. Put $R = \mathcal{O}_{Y,y}$ and $k = k(y)$. Localizing (1.13) at y, we obtain a commutative diagram (1.11) of free R-modules of finite rank, where $\mathrm{rk}(F) = h^0(\mathcal{F}_2, y)$, $\mathrm{rk}(G) = h^1(\mathcal{F}_2, y)$, $\mathrm{rk}(F') = h^0(\mathcal{F}_3, y)$, and $\mathrm{rk}(G') = h^1(\mathcal{F}_3, y)$. In particular, we have $\mathrm{rk}(F) - \mathrm{rk}(G) = \chi_{\mathcal{F}_2} \geq 0$ and $\mathrm{rk}(F') - \mathrm{rk}(G') = \chi_{\mathcal{F}_3}$. We shall check conditions (a)–(d) of (1.10) by using our assumptions (a)–(c). (For conditions (c) and (d) of (1.10), we need an extra assumption on y. See below.)

Condition (a) of (1.10) follows directly from our assumption that $\chi_{\mathcal{F}_2} \geq 0$ and our assumption (a). Next, since the fiber X_y is of dimension ≤ 1, we see that the map $g \otimes k : G \otimes k = H^1(X_y, \mathcal{F}_2 \otimes k(y)) \to H^1(X_y, \mathcal{F}_3 \otimes k(y)) = G' \otimes k$

is surjective, hence g is surjective by Nakayama's lemma. Thus condition (b) of (1.10) holds.

If $\chi_{\mathcal{F}_2} = 0$, condition (c) of (1.10) clearly holds. Moreover, in this case, our assumption (c) says $-(\chi_{\mathcal{F}_3})^2 \geq 0$, or, equivalently, $\chi_{\mathcal{F}_3} = 0$. Thus condition (d) of (1.10) holds.

If $\chi_{\mathcal{F}_2} > 0$, we shall put an extra assumption that $y \in W(\mathcal{F}_3)$. Then, by (1.2), we have $h^1(\mathcal{F}_3, y) > 0$. Since $\phi' \otimes k = 0$, this implies that $\phi' \otimes k$ is not surjective, hence ϕ' is not surjective. Thus condition (c) of (1.10) holds. Moreover, suppose that condition (d) of (1.10) does not hold. Then, we have $\chi_{\mathcal{F}_3} > \chi_{\mathcal{F}_2} \geq 0$, and $\chi_{\mathcal{F}_1} = \chi_{\mathcal{F}_2} - \chi_{\mathcal{F}_3} < 0$. Thus $\chi_{\mathcal{F}_1} \cdot \chi_{\mathcal{F}_3} < 0$, which contradicts our assumption (c).

Now, we may apply (1.10). Conclusion (i) of (1.10) implies $\chi_{\mathcal{F}_1} = 0$, hence (i) of (1.12) (and $\chi_{\mathcal{F}_3} \geq 0$). Conclusion (ii) of (1.10) implies that each irreducible component of $W(\mathcal{F}_3)$ passing through y has codimension $|\chi_{\mathcal{F}_3}| + 1 = \chi_{\mathcal{F}_3} + 1$. So, (considering all points $y \in W(\mathcal{F}_3)$) we obtain that $\mathcal{F}_3$ is determinantal. Moreover, conclusion (iii) of (1.10) implies that the map $F \to G \times_{G'} F'$ is injective, since R is an integral domain by the assumption that Y is integral. Accordingly, the map

$$R^0 f_*(\mathcal{F}_2)_y = \mathrm{Ker}(\phi) \to \mathrm{Ker}(\phi') = R^0 f_*(\mathcal{F}_3)_y$$

is injective, or, equivalently, $R^0 f_*(\mathcal{F}_1)_y = 0$. Thus, in particular, we obtain $h^0(\mathcal{F}_1, \eta) = 0$, where η is the generic point of the integral scheme Y. (Here, we have used our assumption (b) for the first time to choose a point y.) Thus, $\eta \notin W(\mathcal{F}_1)$. Since Y is integral and $|\chi_{\mathcal{F}_1}| + 1 = 1$, this implies that $\mathcal{F}_1$ is determinantal.

Next, we shall treat the case that $\chi_{\mathcal{F}_2} < 0$. In this case, by (1.5), we can take (Zariski locally) a commutative diagram

$$
\begin{array}{ccc}
\mathcal{O}_Y^{h^0(\mathcal{F}_1, y)} & \to & \mathcal{O}_Y^{h^0(\mathcal{F}_2, y)} \\
\downarrow & & \downarrow \\
\mathcal{O}_Y^{h^1(\mathcal{F}_1, y)} & \to & \mathcal{O}_Y^{h^1(\mathcal{F}_2, y)}
\end{array}
\tag{1.14}
$$

representing the morphism $Rf_*(\mathcal{F}_1) \to Rf_*(\mathcal{F}_2)$. Then, localizing (1.14) at $y \in Y$ and taking the dual ($= \mathrm{Hom}_R(-, R)$, where $R = \mathcal{O}_{Y,y}$) of the diagram, we obtain a commutative diagram of free R-modules such as

$$
\begin{array}{ccc}
G' & \leftarrow & G \\
\phi' \uparrow & & \phi \uparrow \\
F' & \leftarrow & F
\end{array},
$$

where $\mathrm{rk}(G') = h^0(\mathcal{F}_1, y)$, $\mathrm{rk}(F') = h^1(\mathcal{F}_1, y)$, $\mathrm{rk}(G) = h^0(\mathcal{F}_2, y)$, and $\mathrm{rk}(F) = h^1(\mathcal{F}_2, y)$. If we regard this diagram as (1.11), the proof in the case $\chi_{\mathcal{F}_2} < 0$ can be done just in parallel with that of $\chi_{\mathcal{F}_2} > 0$. This completes the proof. □

2. Generalizations of Raynaud's Theorem

Let k be an algebraically closed field and X a proper, smooth, connected curve of genus g over k.

For a vector bundle E on X (regarded as a locally free $\mathcal{O}_X$-module of finite rank), let $\mathrm{rk}(E)$, $\deg(E)$ and $h^i(E)$ $(i = 0, 1)$ denote the rank of E, the degree of E (which is defined to be the degree of the line bundle $\det(E) \stackrel{\mathrm{def}}{=} \overset{\mathrm{rk}(E)}{\wedge} E$), and the dimension (as a k-vector space) of the i-th cohomology group $H^i(X, E)$. The Riemann–Roch theorem implies the following formula for the Euler–Poincaré characteristic $\chi(E) \stackrel{\mathrm{def}}{=} h^0(E) - h^1(E)$ of E:

$$\chi(E) = \deg(E) - (g - 1)\,\mathrm{rk}(E). \tag{2.1}$$

In [R1], Raynaud investigated the following property of a vector bundle E on X.

DEFINITION (CONDITION $(\star)$). We say that E satisfies $(\star)$ if there exists a line bundle L of degree 0 on X such that $\min(h^0(E \otimes L), h^1(E \otimes L)) = 0$.

First, we shall see the relation between condition $(\star)$ and the contents of Section 1. So, Let J be the Jacobian variety of X, and let $\mathcal{L}$ be a universal line bundle on $X \times J$. Let pr_X and pr_J denote the projections $X \times J \to X$ and $X \times J \to J$, respectively. Regarding $\mathrm{pr}_J : X \times J \to J$ and $(\mathrm{pr}_X)^*(E) \otimes \mathcal{L}$ as $f : X \to Y$ and $\mathcal{F}$ in Section 1, respectively, we can apply definitions and results to our situation.

DEFINITION. We denote by Θ_E the closed subscheme $W((\mathrm{pr}_X)^*(E) \otimes \mathcal{L})$ of J.

We have the following first properties of Θ_E.

PROPOSITION (2.2). *Let the notations be as above.*

(i) *The definition of Θ_E is independent of the choice of $\mathcal{L}$.*

(ii) *Let L be a line bundle of degree 0 on X, and let $[L]$ denote the point of J corresponding to L. Then, $[L] \notin \Theta_E$ if and only if*

$$\min(h^0(E \otimes L), h^1(E \otimes L)) = 0.$$

(iii) *We have the following implications:*

$$\Theta_E = \varnothing \ \textit{or}\ \mathrm{codim}(\Theta_E) = |\chi(E)| + 1 \iff (\mathrm{pr}_X)^*(E) \otimes \mathcal{L} \textit{ is determinantal}$$
$$\Downarrow$$
$$\Theta_E \neq J \iff E \textit{ satisfies } (\star).$$

Moreover, if $\chi(E) = 0$, the above four conditions are all equivalent.

PROOF. (i) It is known that the difference of two choices of $\mathcal{L}$ comes from a line bundle on J. So, Zariski locally on J, the difference is resolved. Since the definition of Fitting ideal is of local nature (see (1.1)), this shows the desired well-definedness of Θ_E.

(ii) The fiber $(\mathrm{pr}_J)^{-1}([L])$ is naturally identified with X, and the restriction of $(\mathrm{pr}_X)^*(E) \otimes \mathcal{L}$ to this fiber is nothing but $E \otimes L$. Now, (ii) is just the second half of (1.2).

(iii) The first $\iff$ is just the definition, if we note that $\chi_{(\mathrm{pr}_X)^*(E) \otimes \mathcal{L}}$ defined in Section 1 coincides with $\chi(E)$. The second $\Rightarrow$ is trivial, and the third $\iff$ follows from (ii). Finally, if $\chi(E) = 0$, Macaulay's theorem (see Section 1) says that either $\Theta_E = \varnothing$ or $\mathrm{codim}(\Theta_E) \leq 1$. So, in this case, the converse of the second $\Rightarrow$ also holds. $\square$

From now on, we assume that k is of characteristic $p > 0$. For an $\mathbb{F}_p$-scheme S, we shall denote by F_S the absolute Frobenius endomorphism $S \to S$. We define X_1 to be the pull-back of X by $F_{\mathrm{Spec}(k)}$, and denote by $F_{X/k}$ the relative Frobenius morphism $X \to X_1$ over k.

We put $B = ((F_{X/k})_*(\mathcal{O}_X))/\mathcal{O}_{X_1}$, which is a vector bundle on X_1 with $\mathrm{rk}(B) = p - 1$ and $\chi(B) = 0$. In [R1], Raynaud proved, among other things, the following:

THEOREM. ([R1], Théorème 4.1.1.) *The vector bundle B on X_1 satisfies $(\star)$.* $\square$

As an application of this theorem, Raynaud proved, roughly speaking, that the p-ranks of the Jacobian varieties of 'most' (prime-to-p-)cyclic étale coverings of X are as large as can be expected. In order to generalize such a result to ramified coverings, we need to modify the vector bundle B, so that it involves a divisor whose support is in the ramification locus. So, the aim of this section is to generalize Raynaud's theorem along these lines, and, in the next section, the application to cyclic ramified coverings will be given.

Let $q = p^f$ be a power of p ($f \geq 1$). We define X_f to be the pull-back of X by $(F_{\mathrm{Spec}(k)})^f$, and define $F_{X/k}^f : X \to X_f$ to be the composite of the f relative Frobenius morphisms: $F_{X/k}^f \overset{\mathrm{def}}{=} F_{X_{f-1}/k} \circ \cdots \circ F_{X_1/k} \circ F_{X/k}$.

Let $D = \sum_{P \in X} n_P P$ be an effective divisor on X (i.e., $n_P \geq 0$ for all P). We shall write $\mathrm{ord}_P(D)$ instead of n_P, which is a non-negative integer. Then, by definition, $\deg(D) = \sum_{P \in X} \mathrm{ord}_P(D)$.

DEFINITION. We put

$$B_D^f = ((F_{X/k}^f)_*(\mathcal{O}_X(D)))/\mathcal{O}_{X_f}.$$

LEMMA (2.3). (i) B_D^f *is a vector bundle on X_f if and only if the torsion-freeness condition*

$$\mathrm{ord}_P(D) < q \text{ for each } P \in X \tag{TF}$$

holds.

(ii) *Assume that* (TF) *holds. Then we have*

$$\mathrm{rk}(B_D^f) = q - 1, \ \deg(B_D^f) = \deg(D) + (g-1)(q-1), \ \chi(B_D^f) = \deg(D).$$

More generally, for a line bundle L on X_f, we have

$$\mathrm{rk}(B_D^f \otimes L) = q - 1, \ \deg(B_D^f \otimes L) = \deg(D) + (g - 1 + \deg(L))(q - 1),$$

and

$$\chi(B_D^f \otimes L) = \deg(D) + \deg(L)(q - 1). \tag{2.4}$$

PROOF. (i) Since B_D^f is a coherent sheaf on the (smooth) curve X_f, it is a vector bundle if and only if the stalk $(B_D^f)_{P_f}$ is a torsion-free $\mathcal{O}_{X_f, P_f}$-module for each $P_f \in X_f$. By definition, we have $(B_D^f)_{P_f} = (\mathfrak{m}_{X,P})^{-\operatorname{ord}_P(D)}/\mathcal{O}_{X_f, P_f}$, where P is the unique point of X above P_f and $\mathfrak{m}_{X,P}$ denotes the maximal ideal of the local ring $\mathcal{O}_{X,P}$. So, $(B_D^f)_{P_f}$ is torsion-free if and only if $(\mathfrak{m}_{X,P})^{-\operatorname{ord}_P(D)} \cap k(X_f) = \mathcal{O}_{X_f, P_f}$, which turns out to be equivalent to $\operatorname{ord}_P(D) < q$. This completes the proof.

(ii) We have

$$\mathrm{rk}(B_D^f) = \mathrm{rk}((F_{X/k}^f)_*(\mathcal{O}_X(D))) - \mathrm{rk}(\mathcal{O}_{X_f}) = q - 1$$

and

$$\chi(B_D^f) = \chi((F_{X/k}^f)_*(\mathcal{O}_X(D))) - \chi(\mathcal{O}_{X_f}) = (\deg(D) + 1 - g) - (1 - g) = \deg(D).$$

From these, $\deg(B_D^f)$ can be calculated by using (2.1).

Moreover, for a line bundle L on X_f, we have

$$\mathrm{rk}(B_D^f \otimes L) = \mathrm{rk}(B_D^f) = q - 1$$

and

$$\deg(B_D^f \otimes L) = \deg(B_D^f) + \mathrm{rk}(B_D^f)\deg(L) = \deg(D) + (g - 1 + \deg(L))(q - 1).$$

From these, $\chi(B_D^f \otimes L)$ can be calculated by using (2.1). $\qquad\square$

Now, the following is one of the main results of this section.

THEOREM (2.5). *Assume $\deg(D) = q - 1$, and let L_{-1} be a line bundle of degree -1 on X_f. Then $B_D^f \otimes L_{-1}$ is a vector bundle on X_f with $\chi = 0$, and satisfies $(\star)$.*

Before proving (2.5), we shall give a slight generalization (which will be used later), assuming (2.5):

COROLLARY (2.6). *Let s be a non-negative integer. We assume that $\deg(D) = s(q - 1)$ and that*

$$\#\{P \in X \mid \operatorname{ord}_P(D) = q - 1\} \geq s - 1.$$

Let L_{-s} be a line bundle of degree $-s$ on X_f. Then $B_D^f \otimes L_{-s}$ is a vector bundle on X_f with $\chi = 0$, and satisfies $(\star)$.

PROOF. Let D_0 be an effective divisor on X, and Q a point of X which is not contained in the support of D_0. We put $D_1 = D_0 + (q-1)Q$, and consider the following commutative diagram with two rows exact:

$$0 \to \mathcal{O}_{X_f}(-Q_f) \to (F^f_{X/k})_*(\mathcal{O}_X(D_1)) \otimes \mathcal{O}_{X_f}(-Q_f) \to B^f_{D_1} \otimes \mathcal{O}_{X_f}(-Q_f) \to 0$$

$$\|$$

$$\cap \qquad\qquad (F^f_{X/k})_*(\mathcal{O}_X(D_0 - Q)) \qquad\qquad \vdots$$

$$\cap \qquad\qquad\qquad\qquad \vee$$

$$0 \to \quad \mathcal{O}_{X_f} \quad \to \quad (F^f_{X/k})_*(\mathcal{O}_X(D_0)) \quad \to \quad B^f_{D_0} \quad \to 0,$$

where Q_f denotes $F^f_{X/k}(Q)$ ($\in X_f$). From this, we can see

$$B^f_{D_1} \otimes \mathcal{O}_{X_f}(-Q_f) \xrightarrow{\sim} B^f_{D_0}.$$

Using this isomorphism repeatedly, our assumption

$$\#\{P \in X \mid \operatorname{ord}_P(D) = q - 1\} \geq s - 1$$

enables us to reduce the problem to the case $s \leq 1$. The case $s = 1$ is just (2.5). The case $s = 0$ can be reduced to the case $s = 1$ again by using this isomorphism. (Choose any $Q \in X$.) $\qquad\qquad\qquad\qquad\qquad\qquad\qquad\qquad\qquad\qquad\square$

Note that (2.6) includes Raynaud's original theorem as the case that $f = 1$ and $s = 0$.

PROOF OF (2.5). Since $\deg(D) = q - 1 < q$, (TF) clearly holds, hence B^f_D is a vector bundle by (2.3)(i). Moreover, we have $\chi(B^f_D \otimes L_{-1}) = \deg(D) + \deg(L_{-1})(q - 1) = 0$ by (2.4).

We would like to prove that $B^f_D \otimes L_{-1}$ satisfies $(\star)$ by using (1.12). To do this, let J_f be the Jacobian variety of X_f, and let $\mathcal{L}_f$ be a universal line bundle on $X_f \times J_f$. Let pr_{X_f} and pr_{J_f} denote the projections $X_f \times J_f \to X_f$ and $X_f \times J_f \to J_f$, respectively.

The main difficulty is that, unlike Raynaud's original case, we cannot apply (1.12) directly to the exact sequence on $X_f \times J_f$ obtained by taking $(\mathrm{pr}_{X_f})^*(-) \otimes \mathcal{L}_f$ of the exact sequence

$$0 \to L_{-1} \to (F^f_{X/k})_*(\mathcal{O}_X(D)) \otimes L_{-1} \to B^f_D \otimes L_{-1} \to 0$$

on X_f, because $W((\mathrm{pr}_{X_f})^*(L_{-1}) \otimes \mathcal{L}_f) = \varnothing$ and condition (b) of (1.12) is not satisfied (unless $g = 0$). (Note that $h^0(L_{-1} \otimes L) = 0$ for all line bundle L of degree 0 on X_f.) This leads us to the following procedure.

Since the validity of (2.5) is independent of the choice of the line bundle L_{-1} of degree -1, we may and shall assume $L_{-1} = \mathcal{O}_{X_f}(-Q_f)$, where we fix any

$Q \in X$ and put $Q_f = F^f_{X/k}(Q)$. By definition, we have the exact sequence

$$0 \to \mathcal{O}_{X_f}(-Q_f) \to (F^f_{X/k})_*(\mathcal{O}_X(D)) \otimes \mathcal{O}_{X_f}(-Q_f) \to B^f_D \otimes \mathcal{O}_{X_f}(-Q_f) \to 0.$$
$$\|$$
$$(F^f_{X/k})_*(\mathcal{O}_X(D - qQ))$$

Let $E^f_{D,Q}$ be the sum of $\mathcal{O}_{X_f}$ and $(F^f_{X/k})_*(\mathcal{O}_X(D - qQ))$ in $(F^f_{X/k})_*(\mathcal{O}_X(D))$. This coincides with the amalgamated sum with respect to $\mathcal{O}_{X_f}(-Q_f)$, since $\mathcal{O}_{X_f} \cap (F^f_{X/k})_*(\mathcal{O}_X(D - qQ)) = \mathcal{O}_{X_f}(-Q_f)$. Thus the vector bundle $E^f_{D,Q}$ fits into the following commutative diagram with two rows exact:

$$\begin{array}{ccccccccc}
0 & \to & \mathcal{O}_{X_f}(-Q_f) & \to & (F^f_{X/k})_*(\mathcal{O}_X(D - qQ)) & \to & B^f_D \otimes \mathcal{O}_{X_f}(-Q_f) & \to & 0 \\
& & \cap & & \cap & & \| & & \\
0 & \to & \mathcal{O}_{X_f} & \to & E^f_{D,Q} & \to & B^f_D \otimes \mathcal{O}_{X_f}(-Q_f) & \to & 0.
\end{array} \qquad (2.7)$$

If $\Theta_{B^f_D \otimes \mathcal{O}_{X_f}(-Q_f)} = \varnothing$, the assertion of (2.5) clearly holds. So, from now on, we assume $\Theta_{B^f_D \otimes \mathcal{O}_{X_f}(-Q_f)} \neq \varnothing$. Then, in order to prove that $\Theta_{B^f_D \otimes \mathcal{O}_{X_f}(-Q_f)} \neq J_f$, we shall apply (1.12) to the exact sequence on $X_f \times J_f$ obtained by taking $(\mathrm{pr}_{X_f})^*(-) \otimes \mathcal{L}_f$ of the second row of (2.7).

First, we shall check condition (b) of (1.12). Note that

$$\chi_{(\mathrm{pr}_{X_f})^*(E^f_{D,Q}) \otimes \mathcal{L}_f} = \chi(E^f_{D,Q}) = \chi(\mathcal{O}_{X_f}) + \chi(B^f_D \otimes \mathcal{O}_{X_f}(-Q_f)) = 1 - g.$$

If $g = 0$, condition (b) is equivalent to $\Theta_{B^f_D \otimes \mathcal{O}_{X_f}(-Q_f)} \neq \varnothing$, which we have just assumed. If $g = 1$, condition (b) automatically holds. If $g > 1$, condition (b) is equivalent to $\Theta_{\mathcal{O}_{X_f}} \neq \varnothing$, which follows from $\Theta_{\mathcal{O}_{X_f}} \ni 0$. (Note that $h^0(\mathcal{O}_{X_f}) = 1$ and $h^1(\mathcal{O}_{X_f}) = g$.)

Moreover, since

$$\chi_{(\mathrm{pr}_{X_f})^*(B^f_D \otimes \mathcal{O}_{X_f}(-Q_f)) \otimes \mathcal{L}_f} = \chi(B^f_D \otimes \mathcal{O}_{X_f}(-Q_f)) = 0,$$

condition (c) of (1.12) holds.

Finally, we shall check that condition (a) holds, or, equivalently, that either $\Theta_{E^f_{D,Q}} = \varnothing$ or $\mathrm{codim}(\Theta_{E^f_{D,Q}}) = |1 - g| + 1$. Namely, we have to prove that $\Theta_{E^f_{D,Q}}$ is finite over k (resp. empty) if $g > 0$ (resp. $g = 0$). This is the most difficult part of our proof. (In Raynaud's original case, this part was immediate. See (2.11)(ii).)

LEMMA (2.8). *We have*

$$h^0_{\max} \stackrel{\text{def}}{=} \max\{h^0(E^f_{D,Q} \otimes L) \mid [L] \in J_f\} = 1$$

and

$$h^1_{\max} \stackrel{\text{def}}{=} \max\{h^1(E^f_{D,Q} \otimes L) \mid [L] \in J_f\} = g.$$

PROOF. Since $\chi(E^f_{D,Q} \otimes L) = 1 - g$, it is sufficient to prove the first equality.

By (2.7), we get an exact sequence

$$0 \to (F^f_{X/k})_*(\mathcal{O}_X(D - qQ)) \to E^f_{D,Q} \to k(Q_f) \to 0.$$

So, we have

$$h^0(E^f_{D,Q} \otimes L) \leq h^0((F^f_{X/k})_*(\mathcal{O}_X(D - qQ)) \otimes L) + 1$$
$$= h^0(\mathcal{O}_X(D - qQ) \otimes (F^f_{X/k})^*(L)) + 1 = 1,$$

the last equality following from the fact that $\deg(\mathcal{O}_X(D - qQ) \otimes (F^f_{X/k})^*(L)) = -1 < 0$. Therefore, we have $h^0_{\max} \leq 1$. On the other hand, since $E^f_{D,Q}$ contains $\mathcal{O}_{X_f}$, we have

$$1 \leq h^0(E^f_{D,Q}) \leq h^0_{\max}.$$

This completes the proof.   □

We shall return to the proof of (2.5). Put $W \overset{\text{def}}{=} \Theta_{E^f_{D,Q}}$ ($\subset J_f$) for simplicity. If $g = 0$, we have $W = \varnothing$ by (1.2) and (2.8), as desired. So, we shall assume $g > 0$ and prove that W is finite over k. Note that, again by (1.2) and (2.8), we have

$$W = \{[L] \in J_f \mid h^0(E^f_{D,Q} \otimes L) = 1\}$$
$$= \{[L] \in J_f \mid h^1(E^f_{D,Q} \otimes L) = g\},$$

set-theoretically.

We define the divisor D' on X to be the 'prime-to-Q part' of D, namely,

$$D' \overset{\text{def}}{=} \sum_{P \in X, P \neq Q} \mathrm{ord}_P(D)P,$$

and let d' denote the degree of D'. Then, by using the definition of $E^f_{D,Q}$, we see that the following exact sequence exists:

$$0 \to (F^f_{X/k})_*(\mathcal{O}_X(D - qQ)) \to E^f_{D,Q} \oplus (F^f_{X/k})_*(\mathcal{O}_X(D' - Q))$$
$$\to (F^f_{X/k})_*(\mathcal{O}_X(D')) \to 0.$$

For each $[L] \in W$, we tensor this sequence with L and take the global sections. Then we will obtain

$$H^0(X_f, E^f_{D,Q} \otimes L) \quad \oplus \quad H^0(X_f, (F^f_{X/k})_*(\mathcal{O}_X(D' - Q)) \otimes L)$$
$$\wr\wr \qquad\qquad\qquad\qquad \|$$
$$k \qquad\qquad H^0(X, \mathcal{O}_X(D' - Q) \otimes (F^f_{X/k})^*(L))$$

$$\hookrightarrow \quad H^0(X_f, (F^f_{X/k})_*(\mathcal{O}_X(D')) \otimes L)$$
$$\|$$
$$H^0(X, \mathcal{O}_X(D') \otimes (F^f_{X/k})^*(L)).$$

That is to say, each $[L] \in W$ defines a point of

$$| \mathcal{O}_X(D') \otimes (F_{X/k}^f)^*(L) | - | \mathcal{O}_X(D' - Q) \otimes (F_{X/k}^f)^*(L) | .$$

Here, for a line bundle M on X, $| M |$ denotes the (schematized) projective space $(H^0(X, M) - \{0\})/k^\times$.

In other words, consider the following diagram:

$$
W \subset J_f \xrightarrow{V^f} J \xrightarrow[\sim]{+(D'-Q)} J^{(d'-1)} \xrightarrow[\sim]{+Q} J^{(d')}
$$
$$
\begin{array}{ccc}
& \uparrow & \uparrow \\
X^{(d'-1)} \xrightarrow[\hookrightarrow]{+Q} & X^{(d')} & \qquad\qquad (2.9)\\
& \cup & \\
& X^{(d')} - X^{(d'-1)} &
\end{array}
$$

where V^f denotes the composite of the f Verschiebungs, i.e., $V^f : [L] \mapsto [(F_{X/k}^f)^*(L)]$, $J^{(r)}$ denotes the degree r part of the Picard variety of X (hence, in particular, $J^{(0)} = J$), and $X^{(r)}$ denotes the r-th symmetric power of X. (We put $X^{(0)} = \mathrm{Spec}(k)$ and $X^{(-1)} = \varnothing$.) In this setting, the above observation tells us that there exists a natural set-theoretic map $W \to X^{(d')} - X^{(d'-1)}$ over $J^{(d')}$.

Now, assume that this map $W \to X^{(d')} - X^{(d'-1)}$ can be regarded as a morphism (as $J^{(d')}$-schemes). Then, since $W \to J^{(d')}$ is a finite morphism, so is $W \to X^{(d')} - X^{(d'-1)}$. Now, since $X^{(d')} - X^{(d'-1)} = (X - \{Q\})^{(d')}$ is affine, so is W. On the other hand, W is proper over k as a closed subscheme of J_f. Thus W must be finite over k.

So, it suffices to prove that the above set-theoretic map $W \to X^{(d')} - X^{(d'-1)}$ is a morphism. To do this, we name the morphisms involved as follows, for the sake of simplicity:

$$
\begin{array}{ccc}
X & \xleftarrow{\xi} & X \times W \\
& & \qquad \searrow^{\eta} \\
\downarrow \pi \quad \square & \downarrow \pi_W \quad \circlearrowleft & \qquad W, \qquad\qquad (2.10) \\
& & \nearrow_{\eta'} \\
X_f & \xleftarrow{\xi'} & X_f \times W
\end{array}
$$

where $\pi = F_{X/k}^f$, π_W is the base change of π, and ξ, ξ', η and η' are projections.

We shall apply the functor (from the category of $\mathcal{O}_{X_f}$-modules to that of $\mathcal{O}_{X \times W}$-modules)

$$\eta^* \eta'_*(\xi'^*(-) \otimes (\mathcal{L}_f)_W)$$

to the natural map $E_{D,Q}^f \hookrightarrow \pi_*(\mathcal{O}_X(D'))$. By the flat base change theorem and the projection formula, we have

$$\xi'^*(\pi_*(\mathcal{O}_X(D'))) \otimes (\mathcal{L}_f)_W = (\pi_W)_*(\xi^*(\mathcal{O}_X(D'))) \otimes (\mathcal{L}_f)_W$$
$$= (\pi_W)_*(\xi^*(\mathcal{O}_X(D')) \otimes (\pi_W)^*(\mathcal{L}_f)_W).$$

Thus we obtain

$$\eta^*\eta_*'(\xi'^*(E_{D,Q}^f) \otimes (\mathcal{L}_f)_W) \to \eta^*\eta_*(\xi^*(\mathcal{O}_X(D')) \otimes (\pi_W)^*(\mathcal{L}_f)_W),$$

and taking the composite with the natural map $\eta^*\eta_*(-) \to (-)$, we obtain

$$\eta^*\eta_*'(\xi'^*(E_{D,Q}^f) \otimes (\mathcal{L}_f)_W) \to \xi^*(\mathcal{O}_X(D')) \otimes (\pi_W)^*(\mathcal{L}_f)_W.$$

By (1.2) and (2.8), we have $Z_g(\xi'^*(E_{D,Q}^f) \otimes \mathcal{L}_f) = \varnothing$. So, by (1.4) and its proof, $R^1\eta_*'(\xi'^*(E_{D,Q}^f) \otimes (\mathcal{L}_f)_W)$ is a locally free $\mathcal{O}_W$-module of rank g. (Note that we have $R^1\eta_*'(\xi'^*(E_{D,Q}^f) \otimes (\mathcal{L}_f)_W) = R^1\eta_*'(\xi'^*(E_{D,Q}^f) \otimes (\mathcal{L}_f))|_W$.) By this and (1.5), we see that $R\eta_*'(\xi'^*(E_{D,Q}^f)\otimes(\mathcal{L}_f)_W)$ can be represented Zariski locally by the complex $(\mathcal{O}_W \xrightarrow{0} \mathcal{O}_W^g)$. In particular, $\eta_*'(\xi'^*(E_{D,Q}^f) \otimes (\mathcal{L}_f)_W)$ is a locally free $\mathcal{O}_W$-module of rank 1, and, for each $z \in W$, $\eta_*'(\xi'^*(E_{D,Q}^f) \otimes (\mathcal{L}_f)_W) \otimes k(z) \xrightarrow{\sim} H^0((X_f)_{k(z)}, E_{D,Q}^f \otimes L_z)$, where L_z is the line bundle on $(X_f)_{k(z)}$ corresponding to $z \in W \subset J_f$. (We denote the base change from k to $k(z)$ by means of a subscript $k(z)$.) Hence the pull-back of the $\mathcal{O}_{X \times W}$-module $\eta^*\eta_*'(\xi'^*(E_{D,Q}^f) \otimes (\mathcal{L}_f)_W)$ to $X_{k(z)}$ can be identified with

$$H^0((X_f)_{k(z)}, E_{D,Q}^f \otimes L_z) \otimes_{k(z)} \mathcal{O}_{X_{k(z)}}.$$

On the other hand, the pull-back of $\xi^*(\mathcal{O}_X(D')) \otimes (\pi_W)^*(\mathcal{L}_f)_W$ to $X_{k(z)}$ is $\mathcal{O}_X(D') \otimes (\pi_{k(z)})^*(L_z)$, and we can check that the resulting map

$$H^0((X_f)_{k(z)}, E_{D,Q}^f \otimes L_z) \otimes_{k(z)} \mathcal{O}_{X_{k(z)}} \to \mathcal{O}_X(D') \otimes (\pi_{k(z)})^*(L_z)$$

is the composite of

$$\left\{ \begin{array}{c} H^0((X_f)_{k(z)}, E_{D,Q}^f\otimes L_z) \to H^0((X_f)_{k(z)}, \pi_*(\mathcal{O}_X(D'))\otimes L_z) \\ \| \\ H^0(X_{k(z)}, \mathcal{O}_X(D') \otimes (\pi_{k(z)})^*(L_z)) \end{array} \right\} \otimes_{k(z)} \mathcal{O}_{X_{k(z)}}$$

and the natural map

$$H^0(X_{k(z)}, \mathcal{O}_X(D') \otimes (\pi_{k(z)})^*(L_z)) \otimes_{k(z)} \mathcal{O}_{X_{k(z)}} \to \mathcal{O}_X(D') \otimes (\pi_{k(z)})^*(L_z).$$

Here, as in the previous argument, the map

$$H^0((X_f)_{k(z)}, E_{D,Q}^f \otimes L_z) \to H^0(X_{k(z)}, \mathcal{O}_X(D') \otimes (\pi_{k(z)})^*(L_z))$$

is injective. Since $H^0((X_f)_{k(z)}, E_{D,Q}^f \otimes L_z)$ is a one-dimensional $k(z)$-vector space, we conclude that

$$(\eta^*\eta_*'(\xi'^*(E_{D,Q}^f) \otimes (\mathcal{L}_f)_W)) \otimes k(z) \to (\xi^*(\mathcal{O}_X(D')) \otimes (\pi_W)^*(\mathcal{L}_f)_W) \otimes k(z)$$

is injective. Now, by [EGA4], Proposition (11.3.7), the map

$$\eta^*\eta_*'(\xi'^*(E_{D,Q}^f) \otimes (\mathcal{L}_f)_W) \to \xi^*(\mathcal{O}_X(D')) \otimes (\pi_W)^*(\mathcal{L}_f)_W$$

is injective and its cokernel is flat over W. Hence it equips $\xi^*(\mathcal{O}_X(D')) \otimes (\pi_W)^*(\mathcal{L}_f)_W$ with a structure of relative effective Cartier divisor on $X \times W/W$.

Thus we are given the morphism $W \to X^{(d')}$ over $J_f^{(d')}$, whose underlying map coincides with the set-theoretic map $W \to X^{(d')} - X^{(d'-1)}$ in the previous argument. (See [Mi], §3, especially Proposition 3.13 there.) This completes the proof of the finiteness of W.

Now, we can apply (1.12) and conclude that $(\mathrm{pr}_{X_f})^*(B_D^f \otimes \mathcal{O}_{X_f}(-Q_f)) \otimes \mathcal{L}_f$ is determinantal, or, equivalently, that $\Theta_{B_D^f \otimes \mathcal{O}_{X_f}(-Q_f)} \ (= W((\mathrm{pr}_{X_f})^*(B_D^f \otimes \mathcal{O}_{X_f}(-Q_f)) \otimes \mathcal{L}_f) \ ) \neq J_f$. This finally completes the proof of (2.5). $\qquad\square$

REMARK (2.11). (i) When $D = (q-1)Q$, $E_{D,Q}^f$ coincides with $(F_{X/k}^f)_*(\mathcal{O}_X)$. In general, $E_{D,Q}^f$ is not isomorphic to the direct image of a line bundle on X. In fact, suppose that $E_{D,Q}^f$ is isomorphic to $(F_{X/k}^f)_*(M)$ for some line bundle M on X. Then we have

$$\chi(M) = \chi((F_{X/k}^f)_*(M)) = \chi(\mathcal{O}_{X_f}) + \chi(B_D^f \otimes \mathcal{O}_{X_f}(-Q_f)) = 1 - g,$$

hence $\deg(M) = 0$. On the other hand, since $\mathcal{O}_{X_f} \subset (F_{X/k}^f)_*(M)$, M admits a non-trivial global section. Thus $M \simeq \mathcal{O}_X$. By the definition of $E_{D,Q}^f$, we can see that

$$\det((F_{X/k}^f)_*(M)) \simeq \det((F_{X/k}^f)_*(\mathcal{O}_X(D - qQ))) \otimes \mathcal{O}_{X_f}(Q_f),$$

and since $M \simeq \mathcal{O}_X$, we have

$$\det((F_{X/k}^f)_*(\mathcal{O}_X(D - qQ))) \otimes \det((F_{X/k}^f)_*(\mathcal{O}_X))^{-1} \otimes \mathcal{O}_{X_f}(Q_f) \simeq \mathcal{O}_{X_f}.$$

Here the left-hand side is known to be isomorphic to

$$\mathcal{O}_{X_f}(D_f - qQ_f) \otimes \mathcal{O}_{X_f}(Q_f) \simeq \mathcal{O}_{X_f}(D_f - (q-1)Q_f),$$

where $D_f \overset{\text{def}}{=} \sum_{P \in X} \mathrm{ord}_P(D)P_f$. Thus it follows that the divisor $D_f - (q-1)Q_f$ on X_f should be principal, which does not hold in general.

(ii) When $D = (q-1)Q$, the subscheme W of J_f is nothing but $\mathrm{Ker}(V^f)$, hence its degree over k is p^f. In general, the author does not know much about the finite k-scheme W. For example, he does not know its degree over k.

In later sections, we use a slight generalization (2.13) of (2.6). First, we shall prove the following:

LEMMA (2.12). *Let s be a non-negative integer.*

(i) *Let f be a natural number and D an effective divisor of degree $s(p^f - 1)$ on X satisfying* (TF) *with respect to $q = p^f$. Let f_1 be a natural number. Then the vector bundle $B_D^f \otimes L_{-s}$ on X_f satisfies condition* ($\star$) *of page 59 for some (or, equivalently, all) line bundle L_{-s} of degree $-s$ on X_f, if and only if the vector bundle $B_{D_{f_1}}^f \otimes L_{1,-s}$ on X_{f_1+f} satisfies condition* ($\star$) *for some (or, equivalently, all) line bundle $L_{1,-s}$ of degree $-s$ on X_{f_1+f}.*

(ii) *For each $i = 0, 1$, we let f_i be a natural number, D_i an effective divisor of degree $s(p^{f_i} - 1)$ satisfying (TF) with respect to $q = p^{f_i}$. Then $D \overset{\text{def}}{=} p^{f_1} D_0 + D_1$ becomes an effective divisor of degree $s(p^f - 1)$ satisfying (TF) with respect to $q = p^f$, where $f \overset{\text{def}}{=} f_0 + f_1$. Moreover, $B_D^f \otimes L_{-s}$ satisfies condition $(\star)$ for some (or all) line bundle L_{-s} of degree $-s$ on X_f, if and only if, for each $i = 0, 1$, $B_{D_i}^{f_i} \otimes L_{i,-s}$ satisfies condition $(\star)$ for some (or all) line bundle $L_{i,-s}$ of degree $-s$ on X_{f_i}.*

PROOF. (i) Under the natural (p^{f_1}-linear) isomorphism $X_{f_1} \overset{\sim}{\to} X$ of schemes, $B_{D_{f_1}}^f$ corresponds to B_D^f and the line bundles of degree $-s$ correspond to the line bundles of degree $-s$.

(ii) First the numerical conditions can be checked as follows:

$$\deg(D) = p^{f_1} \deg(D_0) + \deg(D_1) = p^{f_1} s(p^{f_0} - 1) + s(p^{f_1} - 1) = s(p^f - 1),$$

$$\operatorname{ord}_P(D) = p^{f_1} \operatorname{ord}_P(D_0) + \operatorname{ord}_P(D_1) \le p^{f_1}(p^{f_0} - 1) + (p^{f_1} - 1) = p^f - 1.$$

For simplicity, we shall denote by π, π_1 and π_{0,f_1} the relative Frobenius morphisms $F_{X/k}^f : X \to X_f$, $F_{X/k}^{f_1} : X \to X_{f_1}$ and $F_{X_{f_1}/k}^{f_0} : X_{f_1} \to X_f$, respectively, so that $\pi = \pi_{0,f_1} \circ \pi_1$. We have natural homomorphisms

$$\mathcal{O}_{X_f} \to (\pi_{0,f_1})_*(\mathcal{O}_{X_{f_1}}((D_0)_{f_1})) \to \pi_*(\mathcal{O}_X(D))$$

of $\mathcal{O}_{X_f}$-modules, and obtain the following exact sequence:

$$0 \to B_{(D_0)_{f_1}}^{f_0} \to B_D^f \to (\pi_{0,f_1})_*(B_{D_1}^{f_1} \otimes \mathcal{O}_{X_{f_1}}((D_0)_{f_1})) \to 0.$$

From this, the first assertion follows. Moreover, tensoring this exact sequence with L_{-s}, we obtain

$$0 \to B_{(D_0)_{f_1}}^{f_0} \otimes L_{-s} \to B_D^f \otimes L_{-s}$$

$$\to (\pi_{0,f_1})_*(B_{D_1}^{f_1} \otimes \mathcal{O}_{X_{f_1}}((D_0)_{f_1}) \otimes (\pi_{0,f_1})^*(L_{-s})) \to 0.$$

Now, since

$$\deg(\mathcal{O}_{X_{f_1}}((D_0)_{f_1}) \otimes (\pi_{0,f_1})^*(L_{-s})) = s(p^{f_0} - 1) + p^{f_0}(-s) = -s,$$

the second assertion follows from the associated long exact sequence (and (i)). $\square$

COROLLARY (2.13). *Let s be a non-negative integer, D an effective divisor of degree $s(p^f - 1)$ on X. Assume the following condition:*

CONDITION (2.14). *There exist natural numbers f_i and effective divisors D_i of degree $s(p^{f_i} - 1)$ $(i = 0, 1, \ldots, k)$, such that $f = \sum_{i=0}^{k} f_i$, $D = \sum_{i=0}^{k} p^{f_{>i}} D_i$, where $f_{>i} \overset{\text{def}}{=} \sum_{j=i+1}^{k} f_j$ $(f_{>k} = 0)$, and that, for each $i = 0, 1, \ldots k$,*

$$\#\{P \in X \mid \operatorname{ord}_P(D_i) = p^{f_i} - 1\} \ge s - 1.$$

Then, for a line bundle L_{-s} of degree $-s$ on X_f, $B_D^f \otimes L_{-s}$ is a vector bundle on X_f with $\chi = 0$, and satisfies $(\star)$.

PROOF. Use (2.6) for each D_i and apply (2.12) repeatedly. □

What can we expect for a more general effective divisor D? For the time being, we are interested in vector bundles with $\chi = 0$. So, considering (2.4), we shall assume that $\deg(D) = s(q-1)$ for some natural number s, and consider $B_D^f \otimes L_{-s}$ for a line bundle L_{-s} of degree $-s$ on X_f. Moreover, we have to assume the torsion-freeness condition (TF): $\mathrm{ord}_P(D) < q$ for each $P \in X$, which does not hold automatically this time. Under these assumptions, can we expect that $B_D^f \otimes L_{-s}$ satisfies $(\star)$?

In general, the answer is no. In fact, as Raynaud remarked in [R1], §0, condition $(\star)$ for a vector bundle E with $\chi(E) = 0$ implies that E is semi-stable, in the sense that $\deg(F)/\mathrm{rk}(F) \leq \deg(E)/\mathrm{rk}(E)$ for all vector subbundles F of E. So, if $B_D^f \otimes L_{-s}$ satisfies $(\star)$, then $B_D^f \otimes L_{-s}$ is semi-stable, hence so is B_D^f.

DEFINITION. Let D be an effective divisor on X.

(i) For each natural number n, we put

$$[D/n] \overset{\mathrm{def}}{=} \sum_{P \in X} [\mathrm{ord}_P(D)/n]P,$$

which is an effective divisor on X.

(ii) For each natural number i, we put

$$D_i \overset{\mathrm{def}}{=} \sum_{P \in X} \mathrm{ord}_P(D)P_i,$$

where P_i denotes $F_{X/k}^i(P) \in X_i$. This is an effective divisor on X_i.

(iii) For $n = 0, 1, \ldots, p^f - 1$, let $n = \sum_{j=0}^{f-1} n_j p^j$ be the p-adic expansion with $n_j = 0, \ldots, p-1$. Identifying $\{0, 1, \ldots, f-1\}$ with $\mathbb{Z}/f\mathbb{Z}$ naturally, we put $n^{(i)} \overset{\mathrm{def}}{=} \sum_{j=0}^{f-1} n_{i+j} p^j$. Now, assume that D satisfies (TF) with respect to $q = p^f$.

Then, we put

$$D^{(i)} \overset{\mathrm{def}}{=} \sum_{P \in X} \mathrm{ord}_P(D)^{(i)} P,$$

which is an effective divisor on X.

LEMMA (2.15). Assume that $\deg(D) = s(q - 1)$ for some natural number s and that D satisfies (TF) with respect to $q = p^f$. Then, if B_D^f is semi-stable, we have

$$\deg(D^{(i)}) \geq \deg(D) \text{ for each } i = 0, 1, \ldots, f - 1. \tag{NSS}$$

('NSS' means 'necessary condition for semi-stability'.)

PROOF. The vector bundle B_D^f on X_f admits the vector subbundles

$$B_{[D/p^i]_i}^{f-i} \overset{\mathrm{def}}{=} ((F_{X_i/k}^{f-i})_*(\mathcal{O}_{X_i}([D/p^i]_i)))/\mathcal{O}_{X_f}$$

for $i = 0, 1, \ldots, f - 1$. (Note that $(X_i)_{f-i} = X_f$.) We have

$$
\frac{\deg(B^{f-i}_{[D/p^i]_i})}{\mathrm{rk}(B^{f-i}_{[D/p^i]_i})} = \frac{\deg([D/p^i]_i) + (g-1)(p^{f-i} - 1)}{p^{f-i} - 1}
$$

$$
= \frac{\deg([D/p^i])}{p^{f-i} - 1} + g - 1,
$$

so we must have

$$
\frac{\deg([D/p^i])}{p^{f-i} - 1} \le \frac{\deg(D)}{p^f - 1} \text{ for each } i = 0, 1, \ldots, f - 1, \tag{2.16}
$$

since B^f_D is assumed to be semi-stable. Now, it is elementary to check that (2.16) is equivalent to (NSS). $\qquad\qquad\square$

REMARK (2.17). We have $\deg(D^{(i)}) \equiv p^{f-i} \deg(D) \equiv 0 \pmod{p^f - 1}$. So, if $\deg(D) = p^f - 1$, (NSS) automatically holds. (Of course, by (2.5) and [R1], § 0, we know that B^f_D is then semi-stable.)

Now, we are tempted to ask the following:

QUESTION (2.18). *Let s be a natural number. Let D be an effective divisor of degree $s(q - 1)$ on X satisfying* (TF) *and* (NSS), *and let L_{-s} be a line bundle of degree $-s$ on X_f. Then, $B^f_D \otimes L_{-s}$ is a vector bundle on X_f with $\chi = 0$. Does it satisfy* $(\star)$?

However, in general this fails, as the following example shows.

EXAMPLE (2.19). We assume $p \ne 2$ and let $X = \mathbf{P}^1$. We put $f = 1$ and let $D = \dfrac{p-1}{2}\{(0) + (1) + (\lambda) + (\infty)\}$, where $\lambda \in k - \{0, 1\}$, so that $s = 2$. Then $B^1_D \otimes L_{-2}$ satisfies $(\star)$(if and) only if the elliptic curve $y^2 = x(x - 1)(x - \lambda)$ is ordinary. (We omit the proof, which uses some contents of the next section.)

Considering Bouw's work ([B]), we might hope that the following is affirmative.

QUESTION (2.20). *Is* (2.18) *true for U generic (in the moduli space)?*

Finally, the following proposition shows to what extent our results can be applied, in the case where $\#(\mathrm{Supp}(D))$ is small. This analysis is a key to recover 'additive structures' of inertia subgroups of tame fundamental groups in Section 5 (B). Here, for each natural number N, we denote by I_N the set $\{0, 1, \ldots, N-1\}$.

PROPOSITION (2.21). *Let s be a non-negative integer and D an effective divisor of degree $s(p^f - 1)$ satisfying* (TF) *with respect to $q = p^f$. We assume that D can be written as $D = n_1 P_1 + n_2 P_2 + n_3 P_3$, where P_1, P_2, P_3 are three distinct points of X. Then:*

(i) *$n_h \in I_{p^f}$ holds for each $h = 1, 2, 3$, and $0 \le s \le 3$ holds.*
(ii) *If $s \ne 2$, D satisfies* (2.14).

(iii) *Assume $s = 2$. Then, (NSS) is equivalent to*

$$n_{1,j} + n_{2,j} + n_{3,j} = 2(p - 1) \text{ for each } j = 0, 1, \ldots, f - 1,$$

where $n_h = \sum_{j=0}^{f-1} n_{h,j} p^j$ is the p-adic expansion with $n_{h,j} \in I_p$ $(h = 1, 2, 3)$.

(iv) *Assume $s = 2$ and (NSS).*

(iv–a) *If $n_1 \in p^{I_f} \stackrel{\text{def}}{=} \{p^b \mid b \in I_f\}$, then either $n_2 = p^f - 1$ or $n_3 = p^f - 1$ holds, and D satisfies (2.14).*

(iv–b) *If $n_1 \in I_{p-1} p^{I_f} \stackrel{\text{def}}{=} \{ap^b \mid a \in I_{p-1}, b \in I_f\}$, then D satisfies (2.14) if and only if either $n_2 = p^f - 1$ or $n_3 = p^f - 1$.*

(iv–c) *For each*

$$n_1 \notin p^{I_f} \cup I_{p-1} p^{I_f} = \begin{cases} I_{p-1} p^{I_f}, & \text{if } p \neq 2, \\ I_p p^{I_f}, & \text{if } p = 2, \end{cases}$$

there exist $n_2, n_3 \in I_{p^f - 1}$ such that $D = n_1 P_1 + n_2 P_2 + n_3 P_3$ satisfies (2.14).

PROOF. (i) The first assertion just says that D is effective and satisfies (TF). The second assertion follows from the first, since $s(q-1) = \deg(D) = n_1 + n_2 + n_3$.

(ii) If $s \leq 1$, (2.14) requires nothing. If $s = 3$, then we must have $n_1 = n_2 = n_3 = q - 1$, which implies (2.14). (Take $k = 0$, $f = f_0$, and $D = D_0$.)

(iii) (NSS) is equivalent to saying that

$$n_1^{(j)} + n_2^{(j)} + n_3^{(j)} \geq n_1 + n_2 + n_3 = 2(p^f - 1)$$

holds for $j = 0, 1, \ldots, f-1$. Here, by definition, the left-hand side is congruent to the right-hand side modulo $p^f - 1$, hence it is a multiple of $p^f - 1$. On the other hand, it is less than or equal to $3(p^f - 1)$. Moreover, if it is equal to $3(p^f - 1)$, each of $n_1^{(j)}, n_2^{(j)}, n_3^{(j)}$ must be $p^f - 1$, which implies that each of n_1, n_2, n_3 is $p^f - 1$. This contradicts the assumption $n_1 + n_2 + n_3 = 2(p^f - 1)$. Thus (NSS) turns out to be equivalent to saying that

$$n_1^{(j)} + n_2^{(j)} + n_3^{(j)} (= n_1 + n_2 + n_3) = 2(p^f - 1)$$

holds for $j = 0, 1, \ldots, f - 1$. Now, put $\nu \stackrel{\text{def}}{=} n_1 + n_2 + n_3 = 2(p^f - 1)$ and $\nu_j \stackrel{\text{def}}{=} n_{1,j} + n_{2,j} + n_{3,j}$. Since we have

$$n_h^{(j+1)} = \frac{n_h^{(j)} - n_{h,j}}{p} + n_{h,j} p^{f-1} = \frac{1}{p} n_h^{(j)} + \frac{p^f - 1}{p} n_{h,j},$$

we see that

$$\nu = \frac{1}{p} \nu + \frac{p^f - 1}{p} \nu_j, \text{ i.e., } \nu_j = 2(p - 1)$$

is a necessary condition for (NSS). It is clear that this condition is also sufficient for (NSS).

(iv–a) If $n_1 = p^b$ ($b \in I_f$), we must have

$$n_{2,j} + n_{3,j} = \begin{cases} 2(p-1), & \text{if } j \neq b, \\ 2(p-1) - 1, & \text{if } j = b, \end{cases}$$

by (iii), or, equivalently,

$$(n_{2,j}, n_{3,j}) = \begin{cases} (p-1, p-1), & \text{if } j \neq b, \\ (p-1, p-2) \text{ or } (p-2, p-1), & \text{if } j = b. \end{cases}$$

From this (iv–a) follows.

(iv–b) If $n_1 = ap^b$ ($a \in I_{p-1}, b \in I_f$), we have $n_{1,j} = 0$ (resp. a) for $j \neq b$ (resp. $j = b$). Accordingly, by (iii), we must have $n_{2,j} = n_{3,j} = p - 1$ for $j \neq b$ and $n_{2,b} + n_{3,b} = 2(p-1) - a$. First, if either n_2 or n_3 coincides with $p^f - 1$, then it is clear that D satisfies (2.14) for $k = 0$. Conversely, suppose that there exist k, f_i and D_i as in (2.14). Since $\deg(D_i) = 2(p^{f_i} - 1)$ and $\operatorname{ord}_P(D_i) = p^{f_i} - 1$ for some $P = P_1, P_2, P_3$, we have $\operatorname{ord}_P(D_i) \leq p^{f_i} - 1$ for all $P = P_1, P_2, P_3$. From this, (considering the p-adic expansion of $n_h = \operatorname{ord}_{P_h}(D)$) we conclude that $\operatorname{ord}_{P_h}(D_i)$ should coincide with $\sum_{j=0}^{f_i - 1} n_{h, f_{>i} + j} p^j$ for each $h = 1, 2, 3$. Thus, for some $h = 1, 2, 3$ (depending on i), we must have $n_{h, f_{>i} + j} = p - 1$ for $j = 0, \ldots, f_i - 1$. Now, taking the unique i such that $f_{>i} \leq b < f_{>i-1}$, we see that $n_{h,b} = p - 1$ holds for some h. Since $n_{1,b} = a < p - 1$, we must have either $n_{2,b} = p - 1$ or $n_{3,b} = p - 1$, which implies $n_2 = p^f - 1$ or $n_3 = p^f - 1$, respectively. This completes the proof of (iv–b).

(iv–c) Assume $n_1 \notin p^{I_f} \cup I_{p-1} p^{I_f}$. In particular, $n_1 \neq 0$, hence there exists $b = 0, 1, \ldots, f - 1$ with $n_{1,b} > 0$. Now, we put

$$(n_{2,j}, n_{3,j}) \stackrel{\text{def}}{=} \begin{cases} (p-1, p-1-n_{1,j}), & \text{if } j \neq b, \\ (p-1-n_{1,b}, p-1), & \text{if } j = b \text{ and } n_{1,b} < p-1, \\ (p-2, 1), & \text{if } j = b \text{ and } n_{1,b} = p-1. \end{cases}$$

(Note that (NSS) holds by (iii).) Since $p - 1 \in \{n_{1,j}, n_{2,j}, n_{3,j}\}$ for each $j = 0, 1, \ldots, f - 1$, we see that D satisfies (2.14). Finally, since $n_{2,b} < p - 1$ by definition, we have $n_2 < p^f - 1$. On the other hand, suppose $n_3 = p^f - 1$. Then we must have $p - 1 - n_{1,j} = p - 1$ for $j \neq b$, and $1 = p - 1$ if $n_{1,b} = p - 1$. Namely, we have $n_1 = n_{1,b} p^b$ and either $p = 2$ or $n_{1,b} < p - 1$. This contradicts the assumption $n_1 \notin p^{I_f} \cup I_{p-1} p^{I_f}$. This completes the proof of (iv–c). $\qquad\square$

3. The p-Ranks of p'-Cyclic Ramified Coverings

As in Section 2, let k be an algebraically closed field of characteristic $p > 0$ and X a proper, smooth, connected curve of genus g over k. Let S be a finite (possibly empty) set of closed points of X and denote by n the cardinality of S. We put $U = X - S$. In this section, we investigate the p-ranks of the Jacobian varieties of p'-cyclic coverings of X, étale over U and possibly ramified over S.

Cyclic coverings and generalized Hasse–Witt invariants. Let N be a natural number prime to p. We consider the elements of the étale cohomology group $H^1_{\text{ét}}(U, \boldsymbol{\mu}_N)$, where $\boldsymbol{\mu}_N = \boldsymbol{\mu}_N(k)$ is the group of N-th roots of unity. In terms of fundamental groups,

$$H^1_{\text{ét}}(U, \boldsymbol{\mu}_N) = \text{Hom}(\pi_1(U), \boldsymbol{\mu}_N) = \text{Hom}(\pi_1^{\text{t}}(U), \boldsymbol{\mu}_N),$$

and, in terms of torsors, $H^1_{\text{ét}}(U, \boldsymbol{\mu}_N)$ can be identified with the set of isomorphism classes of (étale) $\boldsymbol{\mu}_N$-torsors of U. We shall consider the p-ranks for such $\boldsymbol{\mu}_N$-torsors, or $\boldsymbol{\mu}_N$-coverings.

Let V be a $\boldsymbol{\mu}_N$-torsor of U and $[V]$ the corresponding element of $H^1_{\text{ét}}(U, \boldsymbol{\mu}_N)$. Let Y be the normalization of X in V, to which the $\boldsymbol{\mu}_N$-action on V extends uniquely. We define the p-rank (or the Hasse–Witt invariant) $\gamma_{[V]}$ to be the dimension of the $\mathbb{F}_p$-vector space $H^1_{\text{ét}}(Y, \mathbb{F}_p)$.

To obtain finer invariants, we consider the following canonical decomposition of the group algebra $k[\boldsymbol{\mu}_N]$:

$$
\begin{array}{ccc}
k[\boldsymbol{\mu}_N] & \overset{\sim}{\to} & \prod_{i \in \mathbb{Z}/N\mathbb{Z}} k, \\
\cup & & \\
\boldsymbol{\mu}_N & & \uplus \\
\uplus & & \\
\zeta & \mapsto & (\zeta^i)_{i \in \mathbb{Z}/N\mathbb{Z}}.
\end{array}
\tag{3.1}
$$

Corresponding to this decomposition, each $k[\boldsymbol{\mu}_N]$-module M admits a canonical decomposition $M = \bigoplus_{i \in \mathbb{Z}/N\mathbb{Z}} M_i$, where $\zeta \in \boldsymbol{\mu}_N$ acts on M_i as the ζ^i-multiplication. We shall denote by $\gamma_i(M)$ the dimension of the k-vector space M_i. Moreover, for an $\mathbb{F}_p[\boldsymbol{\mu}_N]$-module M, we shall write $\gamma_i(M)$ instead of $\gamma_i(M \otimes_{\mathbb{F}_p} k)$. In the latter case, $\gamma_{p^a i}(M) = \gamma_i(M)$ holds for each integer a. (Observe that the p-th power map of k maps $(M \otimes k)_i$ isomorphically onto $(M \otimes k)_{pi}$.)

Now, since $\boldsymbol{\mu}_N$ naturally acts on the $\mathbb{F}_p$-vector space $H^1_{\text{ét}}(Y, \mathbb{F}_p)$, we can define as follows:

DEFINITION. $\gamma_{[V],i} \overset{\text{def}}{=} \gamma_i(H^1_{\text{ét}}(Y, \mathbb{F}_p))$.

These invariants essentially coincide with the so-called generalized Hasse–Witt invariants (see [Ka], [Na], and [B]). Of course, we have

$$\gamma_{[V]} = \sum_{i \in \mathbb{Z}/N\mathbb{Z}} \gamma_{[V],i}.$$

We shall present another description of these invariants, which is also well-known. Let ψ denote the structure morphism $Y \to X$. Corresponding to the decomposition (3.1), we obtain a decomposition of the sheaf $\psi_*(\mathcal{O}_Y)$ on X:

$$\psi_*(\mathcal{O}_Y) = \bigoplus_{i \in \mathbb{Z}/N\mathbb{Z}} L_i. \tag{3.2}$$

Let f be the order of $p \bmod N$ in the multiplicative group $(\mathbb{Z}/N\mathbb{Z})^{\times}$. The p-th power map of $\mathcal{O}_Y$ sends L_i into L_{pi}, hence the p^f-th power map of $\mathcal{O}_Y$ sends L_i into itself, which induces a p^f-linear map

$$\varphi_{[V],i} : H^1(X, L_i) \to H^1(X, L_i)$$

on the Zariski cohomology group $H^1(X, L_i)$. We denote by $\gamma'_{[V],i}$ the dimension of the k-vector space $\bigcap_{r \geq 1} \operatorname{Im}((\varphi_{[V],i})^r)$. Then, Artin–Schreier theory, together with the well-known properties of p^f-linear maps, implies $\gamma'_{[V],i} = \gamma_{[V],i}$

Cyclic coverings and line bundles. Next, in order to apply the results of Section 2, we shall give a description of $\boldsymbol{\mu}_N$-torsors of U in terms of line bundles and divisors on X, which is essentially widely known (possibly in slightly different forms).

We denote by $\operatorname{Pic}(X)$ the Picard group of X and by $\mathbb{Z}[S]$ the group of divisors whose supports are contained in S, which can be identified with the free $\mathbb{Z}$-module with basis S. We denote by $\mathbb{Z}/N\mathbb{Z}[S]$ the free $\mathbb{Z}/N\mathbb{Z}$-module with basis S, hence $\mathbb{Z}/N\mathbb{Z}[S] = \mathbb{Z}[S]/N\mathbb{Z}[S]$. Let $(\mathbb{Z}/N\mathbb{Z})^{\sim}$ denote the set $\{0, 1, \dots, N-1\}$, and $(\mathbb{Z}/N\mathbb{Z})^{\sim}[S]$ the subset of $\mathbb{Z}[S]$ consisting of the elements whose 'coefficients' are contained in $(\mathbb{Z}/N\mathbb{Z})^{\sim}$.

Consider the following (short) complex of abelian groups:

$$\mathbb{Z}[S] \overset{\alpha_N}{\to} \operatorname{Pic}(X) \oplus \mathbb{Z}[S] \overset{\beta_N}{\to} \operatorname{Pic}(X), \tag{3.3}$$

where $\alpha_N(D) = ([\mathcal{O}_X(-D)], ND)$ and $\beta_N(([L], D)) = [L^{\otimes N} \otimes \mathcal{O}_X(D)]$.

DEFINITION. We define the abelian group $P_N = P_N(X, S)$ to be the homology group $\operatorname{Ker}(\beta_N)/\operatorname{Im}(\alpha_N)$ of the complex (3.3).

We can easily see that the following exact sequence exists:

$$0 \to \operatorname{Pic}(X)[N] \overset{a_N}{\to} P_N \overset{b_N}{\to} \mathbb{Z}/N\mathbb{Z}[S] \overset{c_N}{\to} \mathbb{Z}/N\mathbb{Z}, \tag{3.4}$$

where $[N]$ means the N-torsion subgroup, and

$$a_N([L]) = ([L], 0) \bmod \operatorname{Im}(\alpha_N),$$

$$b_N(([L], D) \bmod \operatorname{Im}(\alpha_N)) = D \bmod N,$$

$$c_N(D \bmod N) = \deg(D) \bmod N.$$

From this, P_N turns out to be isomorphic to $(\mathbb{Z}/N\mathbb{Z})^{\oplus 2g+n-1+b^{(2)}}$, where

$$b^{(2)} \overset{\text{def}}{=} \begin{cases} 1, & \text{if } n = 0, \\ 0, & \text{if } n > 0, \end{cases}$$

is the second Betti number of U.

We shall define two maps

$$i_N : P_N \to H^1_{\text{ét}}(U, \boldsymbol{\mu}_N), \quad j_N : H^1_{\text{ét}}(U, \boldsymbol{\mu}_N) \to P_N.$$

To do this, we need some more notations. First, we denote by $\mathbb{Z}/N\mathbb{Z}[S]^0$ the kernel of c_N in (3.4) and by $(\mathbb{Z}/N\mathbb{Z})^{\sim}[S]^0$ the subset of $(\mathbb{Z}/N\mathbb{Z})^{\sim}[S]$ corresponding to $\mathbb{Z}/N\mathbb{Z}[S]^0$ under the natural bijection $(\mathbb{Z}/N\mathbb{Z})^{\sim}[S]\tilde{\to}\mathbb{Z}/N\mathbb{Z}[S]$. We define $\tilde{P}_N$ to be the inverse image of $(\mathbb{Z}/N\mathbb{Z})^{\sim}[S]^0$ under the projection $\mathrm{Ker}(\beta_N) \to \mathbb{Z}[S]$ (see (3.3)). Then we can easily see that the modulo-$\mathrm{Im}(\alpha_N)$ map $\tilde{P}_N \to P_N$ is a bijection. We denote by $\tilde{b}_N$ the projection $\tilde{P}_N \to (\mathbb{Z}/N\mathbb{Z})^{\sim}[S]^0$:

Now, first, take $([L], D)$ in $\tilde{P}_N$. We have $L^{\otimes N}\otimes\mathcal{O}_X(D) \simeq \mathcal{O}_X$, or, equivalently, $L^{\otimes N} \simeq \mathcal{O}_X(-D)$. These isomorphisms are unique up to multiplication by an element of $k^{\times}$. We fix such an isomorphism $L^{\otimes N}\tilde{\to}\mathcal{O}_X(-D)$, which induces an isomorphism $(L|_U)^{\otimes N}\tilde{\to}\mathcal{O}_X(-D)|_U = \mathcal{O}_U$. Then, by using this isomorphism, we can equip the locally free $\mathcal{O}_U$-module $\bigoplus_{i\in(\mathbb{Z}/N\mathbb{Z})^{\sim}}(L|_U)^{\otimes i}$ with a structure of étale $\mathcal{O}_U$-algebra, as usual. This $\mathcal{O}_U$-algebra admits a $\boldsymbol{\mu}_N$-action: $\zeta \in \boldsymbol{\mu}_N$ acts on $(L|_U)^{\otimes i}$ as the ζ^i-multiplication. The finite U-scheme corresponding to this $\mathcal{O}_U$-algebra, together with this $\boldsymbol{\mu}_N$-action, defines an étale $\boldsymbol{\mu}_N$-torsor of U. It is easy to check (by using the surjectivity of the N-th power map $k^{\times} \to k^{\times}$) that the isomorphism class of the $\boldsymbol{\mu}_N$-torsor we have just constructed is independent of the choice of the isomorphism $L^{\otimes N}\tilde{\to}\mathcal{O}_X(-D)$. This gives the definition of a map $\tilde{i}_N : \tilde{P}_N \to H^1_{\text{ét}}(U, \boldsymbol{\mu}_N)$. Composing this with the canonical bijection $P_N\tilde{\leftarrow}\tilde{P}_N$, we obtain $i_N : P_N \to H^1_{\text{ét}}(U, \boldsymbol{\mu}_N)$.

Next, take a $\boldsymbol{\mu}_N$-torsor V/U, and let $\psi : Y \to X$ be the normalization of $V \to U$, as above. Then, as we have seen, the locally free $\mathcal{O}_X$-module $\psi_*(\mathcal{O}_Y)$ can be canonically decomposed as a direct sum $\bigoplus_{i\in\mathbb{Z}/N\mathbb{Z}}L_i$. Using the fact that V is a $\boldsymbol{\mu}_N$-torsor of U, we can see that each L_i is a line bundle on X and that $L_0 = \mathcal{O}_X$. Since $\boldsymbol{\mu}_N$ acts on $\psi_*(\mathcal{O}_Y)$ as an $\mathcal{O}_X$-algebra, the multiplication of L_i and $L_{i'}$ is contained in $L_{i+i'}$. In particular, we are given an $\mathcal{O}_X$-linear map $L_1^{\otimes N} \to L_0 = \mathcal{O}_X$. Since V is a $\boldsymbol{\mu}_N$-torsor of U, the restriction $(L_1|_U)^{\otimes N} \to \mathcal{O}_U$ is an isomorphism. Therefore the map $L_1^{\otimes N} \to \mathcal{O}_X$ is injective, and it factors as $L_1^{\otimes N}\tilde{\to}\mathcal{O}_X(-D) \subset \mathcal{O}_X$ for some (uniquely determined) effective divisor $D \in \mathbb{Z}[S]$. We claim that the effective divisor D belongs to $(\mathbb{Z}/N\mathbb{Z})^{\sim}[S]$. This comes from the fact that Y is the normalization of X. In fact, the N-th power of a (local) section of $L_1([D/N])$ belongs to $\mathcal{O}_X(N[D/N] - D) \subset \mathcal{O}_X$, hence should belong to $\psi_*(\mathcal{O}_Y)$. (See Section 2 for the definition of the divisor $[D/N]$.) Considering the $\boldsymbol{\mu}_N$-action, we have $L_1([D/N]) \subset L_1$, which implies $[D/N] = 0$, or, equivalently, $D \in (\mathbb{Z}/N\mathbb{Z})^{\sim}[S]$. Now, since $([L_1], D)$ falls in the kernel of β_N by definition, $([L_1], D)$ is an element of $\tilde{P}_N$. This gives the definition of a map $\tilde{j}_N : H^1_{\text{ét}}(U, \boldsymbol{\mu}_N) \to \tilde{P}_N$. Composing this with the canonical bijection $\tilde{P}_N\tilde{\to}P_N$, we obtain $j_N : H^1_{\text{ét}}(U, \boldsymbol{\mu}_N) \to P_N$.

PROPOSITION (3.5). *The canonical maps* i_N, j_N *are group isomorphisms, which are inverse to each other.*

PROOF. By the definition of multiplication of two torsors, we see that i_N is a group homomorphism. (See, for example, [Mi1], III, Remark 4.8(b).)

In the above construction concerning j_N, the canonical map $L_1^{\otimes i} \to L_i$ ($i \in$ $(\mathbb{Z}/N\mathbb{Z})^{\sim}$) becomes an isomorphism after restricting to U: $(L_1|_U)^{\otimes i} \xrightarrow{\sim} L_i|_U$, since V is a $\boldsymbol{\mu}_N$-torsor of U. Using this fact, we can check that $i_N \circ j_N = \mathrm{id}$.

Finally, for $([L], D) \in \tilde{P}_N$, let V be the corresponding $\boldsymbol{\mu}_N$-torsor of U, and $\psi : Y \to X$ its normalization. Then, since the N-th power of each (local) section of L belongs to $\mathcal{O}_X(-D) \subset \mathcal{O}_X$, $\psi_*(\mathcal{O}_Y)$ should contain L, by the definition of normalization. Moreover, observing the $\boldsymbol{\mu}_N$-action, we can conclude that L should be contained in $L_1 \subset \psi_*(\mathcal{O}_Y)$. Now, the N-th power map induces a commutative diagram

$$
\begin{array}{ccc}
L^{\otimes N} & \xrightarrow{\sim} & \mathcal{O}_X(-D) \\
\cap & & \cap \\
L_1^{\otimes N} & \to & \mathcal{O}_X.
\end{array}
$$

Since $D \in (\mathbb{Z}/N\mathbb{Z})^{\sim}[S]$, this implies that $L = L_1$, hence that $\tilde{j}_N \circ \tilde{i}_N = \mathrm{id}$, or, equivalently $j_N \circ i_N = \mathrm{id}$. From this j_N is also a group isomorphism. (To prove $j_N \circ i_N = \mathrm{id}$, we may also resort to the fact $\#(P_N) = \#(H^1_{\text{ét}}(U, \boldsymbol{\mu}_N))$, which equals $N^{2g+n-1+b^{(2)}}$. $\qquad\square$

Generalized Hasse–Witt invariants via line bundles. Now, we can describe the "generalized Hasse–Witt invariants" in terms of P_N, as follows. For an element $([L], D)$ of $\tilde{P}_N$, fix an isomorphism $L^{\otimes N} \xrightarrow{\sim} \mathcal{O}_X(-D)$ (unique up to $k^{\times}$-multiplication). Taking the composite of the p^f-th power map $L \to L^{\otimes p^f}$ and

$$
L^{\otimes p^f} = L \otimes L^{\otimes (p^f - 1)} \xrightarrow{\sim} L \otimes \mathcal{O}_X\left(-\frac{p^f - 1}{N} D\right) \hookrightarrow L,
$$

we get a map $L \to L$, which induces a p^f-linear map

$$
\varphi_{([L], D)} : H^1(X, L) \to H^1(X, L).
$$

We denote by $\gamma_{([L], D)}$ the dimension of the k-vector space $\bigcap_{r \geq 1} \mathrm{Im}((\varphi_{([L], D)})^r)$. Then, by the various definitions, we see

$$
\gamma_{([L], D)} = \gamma_{[V], 1}, \tag{3.6}
$$

where $[V] = \tilde{i}_N(([L], D))$.

REMARK (3.7). By the Riemann–Roch theorem, we have

$$
\begin{aligned}
\dim_k(H^1(X, L)) &= g - 1 - \deg(L) + \dim_k(H^0(X, L)) \\
&= g - 1 + \frac{1}{N} \deg(D) + \dim_k(H^0(X, L)) \\
&\leq g - 1 + \left[\frac{n(N-1)}{N}\right] + \dim_k(H^0(X, L)) \\
&= g + n - 1 + \left[-\frac{n}{N}\right] + \dim_k(H^0(X, L)).
\end{aligned}
$$

From this, we obtain the following rough estimate:

$$\gamma_{([L],D)} \le \begin{cases} g, & \text{if } ([L],D) = ([\mathcal{O}_X],0), \\ g+n-2+b^{(2)}, & \text{otherwise.} \end{cases}$$

More generally, we can describe $\gamma_{[V],i}$ in terms of line bundles and divisors. First, we shall determine the $\mathbb{Z}$-action on $\tilde{P}_N$ induced by the natural $\mathbb{Z}$-action on the abelian group P_N. In P_N, i times $([L],D)$ is $([L^{\otimes i}],iD)$ (mod $\mathrm{Im}(\alpha_N)$) for each $i \in \mathbb{Z}$. Since the element of $(\mathbb{Z}/N\mathbb{Z})^{\sim}[S]$ that is equivalent to iD modulo N is $iD - N[iD/N]$, we can see that the i-action on $\tilde{P}_N$ is given by

$$([L],D) \mapsto (L^{\otimes i}([iD/N]), iD - N[iD/N]).$$

We shall denote $L^{\otimes i}([iD/N])$ and $iD - N[iD/N]$ by $L(i)$ and $D(i)$, respectively.

Now, let V be the μ_N-torsor of U corresponding to $([L],D)$. Then we have the following generalization of (3.6):

CLAIM (3.8). $\gamma_{([L(i)],D(i))} = \gamma_{[V],i}$.

In fact, consider the decomposition (3.2). By the definition of i_N, Y is the normalization of X in the finite X-scheme corresponding to the $\mathcal{O}_X$-algebra $\bigoplus_{i \in (\mathbb{Z}/N\mathbb{Z})^{\sim}} L^{\otimes i}$, hence we have the canonical injection $L^{\otimes i} \hookrightarrow L_i$, for each $i \in (\mathbb{Z}/N\mathbb{Z})^{\sim}$. We have more: $L(i) = L^{\otimes i}([iD/N]) \hookrightarrow L_i$, since the N-th power of a (local) section of $L^{\otimes i}([iD/N])$ is contained in $\mathcal{O}_X(N[iD/N] - iD) = \mathcal{O}_X(-D(i)) \subset \mathcal{O}_X$. (Note that Y is normal.) In fact, we have $L(i) = L_i$. Otherwise, L_i would be strictly bigger than $L(i)$, hence we could find a (local) section of L_i whose N-th power would not belong to $\mathcal{O}_X$. This is absurd. Thus, we can identify $\varphi_{([L(i)],D(i))}$ with $\varphi_{[V],i}$, which implies our claim.

Digression: Torsion points on divisors of abelian varieties. As in [R1], we need some inputs from intersection theory to deduce our main (numerical) results concerning p-ranks of cyclic coverings from the results of Section 2.

LEMMA (3.9). *Let A be an abelian variety of dimension $d > 0$ over an algebraically closed field k, and D an effective divisor on A. For each natural number N not divisible by $\mathrm{char}(k)$, we put*

$$c^{\mathrm{fin}}(D,N) = \min\{(D . C) \,|C\colon \text{irreducible, reduced curve in } A,$$

$$\text{such that } C \ni 0 \text{ and } D \cap N_A^{-1}(C) \text{ is finite.}\}$$

and

$$c^{\mathrm{irr}}(D,N) = \min\{(D . C) \,|C\colon \text{irreducible, reduced curve in } A,$$

$$\text{such that } C \ni 0 \text{ and } N_A^{-1}(C) \text{ is irreducible.}\}$$

Then:

(i) *For each irreducible, smooth curve C in A such that $\pi_1(C)$ surjects onto $\pi_1(A)$, we have $c^{\mathrm{irr}}(D,N) \le (D . C)$.*

(ii) *For each very ample divisor H on A, we have $c^{\mathrm{fin}}(D, N) \leq (D \cdot H^{d-1})$, where H^r denotes the r-th self-intersection product $\underbrace{H \cdot \ldots \cdot H}_{r\ \text{times}}$.*

(iii) *We have $\#(\mathrm{Supp}(D) \cap A[N]) \leq c^{\mathrm{fin}}(D, N)N^{2d-2}$.*

(iv) *If $c^{\mathrm{fin}}(D, N) < N^2$, then we have $c^{\mathrm{fin}}(D, N) \leq c^{\mathrm{irr}}(D, N)$.*

PROOF. (See [R1], Lemme 4.3.5 and the proof of [R1], Théorème 4.3.1.)

(i) The condition $\pi_1(C) \twoheadrightarrow \pi_1(A)$ and the number $(D \cdot C)$ do not change if C is translated by an element of A. So, we may assume that C passes through 0. Now, since $\pi_1(C) \twoheadrightarrow \pi_1(A)$, $N_A^{-1}(C)$ must be irreducible. Thus the inequality holds.

(ii) Since H is very ample, a general member C_1 of H^{d-1} that passes through 0 has finite intersection with $N_A(D)$, hence $D \cap N_A^{-1}(C_1)$ is also finite. Let C be an irreducible component of C_1 that passes through 0. Then, regarding C as a reduced scheme, we obtain

$$c^{\mathrm{fin}}(D, N) \leq (D \cdot C) \leq (D \cdot C_1) = (D \cdot H^{d-1}).$$

(iii) Take an irreducible, reduced curve C in A with $(D \cdot C) = c^{\mathrm{fin}}(D, N)$ such that $C \ni 0$ and that $N_A^{-1}(C) \cap D$ is finite. Then we have

$$\#(\mathrm{Supp}(D) \cap A[N]) \leq (D \cdot N_A^{-1}(C)) = N^{2d-2}(D \cdot C) = c^{\mathrm{fin}}(D, N)N^{2d-2}.$$

Here, The inequality follows from the fact that $C \ni 0$ and that $N_A^{-1}(C) \cap D$ is finite, and the first equality follows from intersection theory as in the proof of [R1], Lemme 4.3.5.

(iv) Take an irreducible, reduced curve C in A with $(D \cdot C) = c^{\mathrm{irr}}(D, N)$ such that $C \ni 0$ and that $N_A^{-1}(C)$ is irreducible. By (iii) and the assumption $c^{\mathrm{fin}}(D, N) < N^2$, we have $\#(\mathrm{Supp}(D) \cap A[N]) < N^{2d}$, hence $A[N] \not\subset D$, and, a fortiori, $N_A^{-1}(C) \not\subset D$. Since $N_A^{-1}(C)$ is irreducible by assumption, this implies that $D \cap N_A^{-1}(C)$ is finite. Thus we have $c^{\mathrm{fin}}(D, N) \leq (D \cdot C) = c^{\mathrm{irr}}(D, N)$. $\square$

COROLLARY (3.10). *Let X be a proper, smooth, connected curve of genus g over k and J the Jacobian variety of X. Let E be a vector bundle on X with $\chi(E) = 0$, and assume that E satisfies $(\star)$ of page 59. Then, Θ_E is a divisor on J, and:*

(i) *If $g > 0$, we have $c^{\mathrm{irr}}(\Theta_E, N) \leq g\,\mathrm{rk}(E)$.*

(ii) *If $g > 0$, we have $c^{\mathrm{fin}}(\Theta_E, N) \leq 3^{g-1}g!\,\mathrm{rk}(E)$.*

(iii) *We have $\#(\mathrm{Supp}(\Theta_E) \cap J[N]) \leq 3^{g-1}g!\,\mathrm{rk}(E)N^{2g-2}$.*

(iv) *If $3^{g-1}g!\,\mathrm{rk}(E) < N^2$, we have $\#(\mathrm{Supp}(\Theta_E) \cap J[N]) \leq g\,\mathrm{rk}(E)N^{2g-2}$.*

PROOF. First, we note that, by [R1], Proposition 1.8.1 (2), Θ_E is algebraically equivalent to $\mathrm{rk}(E)\Theta$, where Θ is the classical theta divisor (the image of $X^{(g-1)}$ in J).

(i) This is obtained by applying (3.9)(i) to $C = X$, which is embedded into $A = J$ by means of an Albanese morphism. In fact, then $\pi_1(X) \twoheadrightarrow \pi_1(J)$ is well-known, and we have

$$(\Theta_E \cdot X) = \mathrm{rk}(E)(\Theta \cdot X) = \mathrm{rk}(E)g.$$

(ii) We first note that 3Θ is very ample by [Mu], Section 17, Theorem, since Θ is ample. Now, (ii) is obtained by applying (3.9)(ii) to $H = 3\Theta$. In fact, then we have

$$(\Theta_E \cdot (3\Theta)^{g-1}) = \mathrm{rk}(E)3^{g-1}(\Theta^g) = \mathrm{rk}(E)3^{g-1}g!.$$

(iii) We may assume $g > 0$, since $\Theta_E = \varnothing$ for $g = 0$. Then, (iii) follows from (ii) and (3.9)(iii).

(iv) We may assume $g > 0$, as in (iii). Then, (iv) follows from (i), (ii) and (3.9)(iii)(iv). $\qquad\square$

REMARK (3.11). In [R1], Lemme 4.3.5, it was necessary to assume that $D \cap l_A^{-1}(C)$ is finite. (Counterexample: C: a one-dimensional abelian subvariety of A, D: the inverse image in A of a divisor of A/C that contains the whole $(A/C)[l]$.) Accordingly, in Théorème 4.3.1, loc. cit., the condition $l + 1 \geq (p-1)g$ had to be modified. (For example, $l + 1 \geq (p-1)3^{g-1}g!$ is sufficient.) Similarly, in [T1], Lemma (1.9), the condition $l^m > \frac{l^{2g}-l^{2g-1}}{l^{2g}-1}(p-1)g$ had to be modified as $l^m > \frac{l^{2g}-l^{2g-1}}{l^{2g}-1}(p-1)3^{g-1}g!$, and, in its proof, we should have assumed that $D \cap (l_A^m)^{-1}(C)$ is finite.

Main numerical consequences. Until the end of this section, with the exception of (3.17), we restrict ourselves to the case that $N = q - 1$, where q is a (positive) power of p. (Note that, in this case, we have $p^f = q$.) Then, we get some numerical consequences of the results of Section 2, as follows.

We note that, for each element D of $(\mathbb{Z}/N\mathbb{Z})^\sim[S]^0$, $\deg(D) = s(D)N$ for some integer $s(D)$ with $0 \leq s(D) \leq n - 1 + b^{(2)}$, and that the cardinality of $\tilde{b}_N^{-1}(D)$ is N^{2g}.

THEOREM (3.12). *Put*

$$C(g) \overset{\mathrm{def}}{=} \begin{cases} 0, & \text{if } g = 0, \\ 3^{g-1}g!, & \text{if } g > 0. \end{cases} \tag{3.13}$$

Then, for each $D \in (\mathbb{Z}/N\mathbb{Z})^\sim[S]^0$ with $s(D) \leq 1$, the following statements hold.

(i) *We have*

$$\#\{[L] \in \mathrm{Pic}(X) \mid ([L], D) \in \tilde{P}_N \text{ and } \varphi_{([L],D)} \text{ is bijective}\} \geq N^{2g} - C(g)N^{2g-1}.$$

(ii) *We have*

$$\#\{[L] \in \mathrm{Pic}(X) \mid ([L], D) \in \tilde{P}_N \text{ and } \gamma_{([L],D)} \geq g - 1 + s(D)\}$$
$$\geq N^{2g} - C(g)N^{2g-1}$$

and

$$\#\{[L] \in \mathrm{Pic}(X) \mid ([L], D) \in \tilde{P}_N \text{ and } \gamma_{([L],D)} = g - 1 + s(D)\}$$
$$\geq \begin{cases} N^{2g} - C(g)N^{2g-1} - 1, & \text{if } s(D) = 0, \\ N^{2g} - C(g)N^{2g-1}, & \text{if } s(D) = 1. \end{cases}$$

PROOF. For simplicity, we shall write s instead of $s(D)$. Since the degree of $L \in \tilde{b}_N^{-1}(D)$ is $-s$, we see that

$$\gamma_{([L],D)} \leq \dim_k(H^1(X, L)) = \begin{cases} g, & \text{if } s = 0 \text{ and } L \simeq \mathcal{O}_X, \\ g - 1 + s, & \text{otherwise,} \end{cases} \tag{3.14}$$

as in (3.7). In particular, the statements clearly hold for $g = 0$. From now on, we shall assume $g > 0$.

(i) First, recall the following commutative diagram (see Section 2):

$$
\begin{array}{ccc}
X & = & X \\
F_{X/k}^f \downarrow & \circlearrowleft & \downarrow F_X^f \\
X_f & \overset{\sim}{\to} & X \\
\downarrow & \underset{(F_{\mathrm{Spec}(k)})^f}{\square} & \downarrow \\
\mathrm{Spec}(k) & \overset{\sim}{\to} & \mathrm{Spec}(k).
\end{array}
$$

We shall denote by ι the q-linear isomorphism $X_f \overset{\sim}{\to} X$ in this diagram. Note that the pullback by F_X^f of a line bundle L on X is canonically isomorphic to $L^{\otimes q}$. In fact, we can easily check that the $\mathcal{O}_X$-linear map $(F_X^f)^*(L) \to L^{\otimes q}$ induced by the q-th power map $L \to L^{\otimes q}$ is an isomorphism.

Take any $[L'_{-s}] \in \tilde{b}_N^{-1}(D)$. Then we have

$$(L'_{-s})^{\otimes N} \simeq \mathcal{O}_X(-D) \text{ (hence } \deg(L'_{-s}) = -s)$$

and

$$\tilde{b}_N^{-1}(D) = \{[L' \otimes L'_{-s}] \mid [L'] \in \mathrm{Pic}(X)[N]\}. \tag{3.15}$$

We put $L_{-s} \overset{\mathrm{def}}{=} \iota^*(L'_{-s})$. Then we have $\deg(L_{-s}) = -s$ and

$$(F_{X/k}^f)^*(L_{-s}) = (L'_{-s})^{\otimes q} \simeq \mathcal{O}_X(-D) \otimes L'_{-s}.$$

Now, by (2.6), the vector bundle $E \overset{\mathrm{def}}{=} B_D^f \otimes L_{-s}$ on X_f satisfies condition $(\star)$ of page 59. So, applying (3.10), we get

$$\#\{[L] \in \mathrm{Pic}(X_f)[N] \mid h^0(E \otimes L) = h^1(E \otimes L) = 0\} \geq N^{2g} - C(g)N^{2g-1}.$$

By the definition of B_D^f, the condition

$$h^0(E \otimes L) = h^1(E \otimes L) = 0$$

implies

$$H^1(X_f, L_{-s} \otimes L) \xrightarrow{\sim} H^1(X_f, (F^f_{X/k})_*(\mathcal{O}_X(D)) \otimes L_{-s} \otimes L)$$
$$\|$$
$$H^1(X, \mathcal{O}_X(D) \otimes (F^f_{X/k})^*(L_{-s} \otimes L)).$$

In terms of $L' = \iota_*(L)$ (hence $L = \iota^*(L')$), this is equivalent to:

$$H^1(X, L'_{-s} \otimes L') \xrightarrow{\sim} H^1(X, \mathcal{O}_X(D) \otimes (L'_{-s} \otimes L')^{\otimes q}).$$

Considering (3.15), these imply the inequality in (i).

(ii) Immediate from (i) and (3.14). (The term -1 in the case $s = 0$ comes from the trivial line bundle.) $\qquad\square$

We have the following slight generalization of (3.12). See Section 2 for the definition of the divisor $D^{(i)}$.

COROLLARY (3.16). *Let $N = p^f - 1$ and $D \in (\mathbb{Z}/N\mathbb{Z})^\sim[S]^0$. Assume that there exists $i \in \{0, 1, \ldots, f-1\}$ such that $s(D^{(i)}) = 1$. Then we have*

$$\#\{[L] \in \mathrm{Pic}(X) \mid ([L], D) \in \tilde{P}_N \text{ and } \gamma_{([L],D)} = g\} \geq N^{2g} - C(g)N^{2g-1}.$$

PROOF. Notations being as in (3.8), we can deduce

$$\gamma_{([L(p^i)], D(p^i))} = \gamma_{[V], p^i} = \gamma_{[V], 1} = \gamma_{([L], D)},$$

where the first and the third equalities follows from (3.8) and the second from the remark just before the definition of $\gamma_{[V], i}$.

Since $N = p^f - 1$, we have the coincidence $D^{(i)} = D(p^i)$. Thus, we obtain (3.16) by applying (3.12) to $D^{(i)} = D(p^i)$. (Note that the p^i-action on $\tilde{P}_N$ is bijective.) $\qquad\square$

REMARK (3.17). In this remark, we do not assume $N = p^f - 1$.

(i) For general N, the same argument as in the proof of (3.12) shows that (3.12) holds if we replace $N^{2g} - C(g)N^{2g-1}$ by $N^{2g} - C(g)(p^f - 1)N^{2g-2}$. (Apply (2.6) to $\frac{p^f - 1}{N}D$.) Similarly, (3.16) holds if we replace $D^{(i)}$ by $D(p^i)$, and $N^{2g} - C(g)N^{2g-1}$ by $N^{2g} - C(g)(p^f - 1)N^{2g-2}$.

However, the resulting inequalities say nothing, unless $N^2 > C(g)(p^f - 1)$. For $g > 0$, the last condition forces N to be a rather big divisor of $p^f - 1$ ($N > \sqrt{p^f - 1}$).

(ii) Following [R1], we can improve the inequalities for $s(D) = 0$ in (3.12) (for N general). This can be achieved by considering the p-linear maps $L_1 \to L_p, L_p \to L_{p^2}, \ldots, L_{p^{f-1}} \to L_{p^f} = L_1$ step by step, instead of considering the whole p^f-linear map $L_1 \to L_1$ at a time. Then, the right-hand sides of both the inequality of (3.12)(i) and the first inequality of (3.12)(ii) become $N^{2g} - C(g)(p-1)fN^{2g-2}$.

This time, the results say something nontrivial, if $N^2 > C(g)(p-1)f$. The last condition is satisfied for $N > C(g)(p-1)$. In particular, they say something nontrivial for almost all N.

(iii) What is the counterpart of (ii) above in the case $s(D) = 1$ (or $s(D(p^i)) = 1$)? To state it, put

$$\{a_1, \ldots, a_m\} = \{a = 0, 1, \ldots, f - 1 \mid s(D(p^a)) = 1\},$$

where we assume $(0 \leq)a_1 < a_2 < \cdots < a_m(\leq f - 1)$. We define the natural numbers f_i by

$$f_i \overset{\text{def}}{=} \begin{cases} a_{i+1} - a_i & \text{if } 1 \leq i < m, \\ a_1 + f - a_m & \text{if } i = m, \end{cases}$$

so that $\sum_{i=1}^{m} f_i = f$. Now, by considering the p^{f_i}-linear maps $L_{p^{a_i}} \to L_{p^{a_{i+1}}}$ step by step as in (ii), we obtain the improvement (for N general)

$$\geq N^{2g} - C(g)\left(\sum_{i=1}^{m}(p^{f_i} - 1)\right)N^{2g-2}$$

in the statements of (3.12)(i)(ii) and (3.16). (For (3.12), we assume $a_1 = 0$.)

The improved inequalities say something nontrivial, if N^2 is greater than $C(g)\left(\sum_{i=1}^{m}(p^{f_i} - 1)\right)$. However, the right-hand side of the last condition depends not only on p, g and f but also on the coefficients of $D \in \mathbb{Z}[S]$.

Assuming $N = p^f - 1$ again, we shall give a rough estimate of the number of D to which (3.12) or (3.16) can be applied. (See Appendix for a related result for general N.) Note that

$$\#((\mathbb{Z}/N\mathbb{Z})^{\sim}[S]^0) = N^{n-1+b^{(2)}} = \begin{cases} 1 & \text{if } n \leq 1, \\ N^{n-1} & \text{if } n > 1. \end{cases}$$

PROPOSITION (3.18). (i) If $n \leq 1$, the value s for the unique element of $(\mathbb{Z}/N\mathbb{Z})^{\sim}[S]^0$ is 0.

(ii) If $n > 1$, there exist $M > 0$ and $0 \leq \alpha < 1$ depending only on p and n, such that

$$\#\{D \in (\mathbb{Z}/N\mathbb{Z})^{\sim}[S]^0 \mid s(D^{(i)}) = 1 \text{ for some } i = 0, 1, \ldots f - 1\}$$
$$\geq N^{n-1}(1 - \alpha^f) - 1$$

for all $f \geq M$.

More precisely, let k be any positive integer $\geq \log_p(n-1)$ and ε any positive real number < 1. Then we can take

$$M = \frac{k}{\varepsilon}, \quad \alpha = \left(1 - \frac{1}{p^{k(n-1)}}\binom{p^k}{n-1}\right)^{(1-\varepsilon)/k}.$$

PROOF. (i) Clear. $((\mathbb{Z}/N\mathbb{Z})^{\sim}[S]^0$ consists of the trivial divisor.)

(ii) We choose any $Q \in S$ and put $S' = S - \{Q\}$, whose cardinality is $n' \overset{\text{def}}{=} n - 1$. The projection $(\mathbb{Z}/N\mathbb{Z})^{\sim}[S]^0 \to (\mathbb{Z}/N\mathbb{Z})^{\sim}[S'], D \mapsto D'$ is bijective. We have

$$s(D) = \frac{1}{N}\deg(D) = \left[\frac{1}{N}\deg(D')\right],$$

where $\lceil x \rceil$ denotes the smallest integer $\geq x$. Therefore, we have

$$s(D) \leq 1 \iff \deg(D') \leq N.$$

When $n' = 1$, $\deg(D') \leq N$ for all D'. Then the statements clearly hold ($\alpha = 0$). (Note that the term -1 comes from the trivial divisor.) So, from now on, we shall assume $n' > 1$.

Let k be a positive integer $\geq \log_p(n')$ and ε any positive real number < 1. We assume $f \geq M \overset{\text{def}}{=} \frac{k}{\varepsilon}$. Let D' be any element of $(\mathbb{Z}/N\mathbb{Z})^{\sim}[S']$, and, for each $P \in S'$, consider the p-adic expansion

$$\operatorname{ord}_P(D') = \sum_{j=0}^{f-1} n_{P,j} p^j,$$

with $n_{P,j} \in \{0, 1, \ldots, p-1\}$.

If we have

$$\sum_{P \in S'} \sum_{j=0}^{k-1} n_{P,f-k+j} p^j \leq p^k - n',$$

then we obtain

$$\begin{aligned}
\deg(D') &= \sum_{P \in S'} \left(\sum_{j=0}^{f-k-1} n_{P,j} p^j \right) + \sum_{j=f-k}^{f-1} n_{P,j} p^j \\
&\leq n'(p^{f-k} - 1) + p^{f-k}(p^k - n') \\
&= p^f - n' \leq p^f - 1 = N.
\end{aligned}$$

In the same way, if we have

$$\sum_{P \in S'} \sum_{j=0}^{k-1} n_{P,f-hk+j} p^j \leq p^k - n'$$

for some $h = 1, 2, \ldots, \left[\frac{f}{k} \right]$, then we obtain

$$\deg((D')^{(-(h-1)k)}) \leq N.$$

In other words, if we suppose that $\deg((D')^{(i)}) > N$ for all $i = 0, 1, \ldots, f-1$, we must have

$$\sum_{P \in S'} \sum_{j=0}^{k-1} n_{P,f-hk+j} p^j > p^k - n' \tag{3.19}$$

for all $h = 1, \ldots, \left[\frac{f}{k} \right]$. Now, since

$$\#\{E \in (\mathbb{Z}/p^k\mathbb{Z})^{\sim}[S'] \mid \deg(E) \leq p^k - n'\} = \binom{(p^k - n') + n'}{n'} = \binom{p^k}{n'}$$

("repeated combination"), we obtain

$$\#\{D \in (\mathbb{Z}/N\mathbb{Z})^{\sim}[S]^0 \mid s(D^{(i)}) > 1 \text{ for all } i = 0, 1, \ldots, f-1\}$$

$$\leq \left(p^{kn'} - \binom{p^k}{n'}\right)^{\left[\frac{f}{k}\right]} p^{(f-k[\frac{f}{k}])n'} - (p^{fn'} - (p^f - 1)^{n'}).$$

Here, the term $(p^{fn'} - (p^f - 1)^{n'})$ is the cardinality of the set

$$(\mathbb{Z}/p^f\mathbb{Z})^{\sim}[S'] - (\mathbb{Z}/N\mathbb{Z})^{\sim}[S'].$$

(Note that each element of $(\mathbb{Z}/p^f\mathbb{Z})^{\sim}[S'] - (\mathbb{Z}/N\mathbb{Z})^{\sim}[S']$ automatically satisfies (3.19), since we have assumed $n' > 1$.) By applying the identity

$$\frac{B}{B'}A' - \{A' - (B' - B)\} = \frac{(B' - A')(B' - B)}{B'}$$

to $B = (p^f - 1)^{n'}$, $B' = p^{fn'}$ and $A' = \left(p^{kn'} - \binom{p^k}{n'}\right)^{\left[\frac{f}{k}\right]} p^{(f-k[\frac{f}{k}])n'}$, we obtain

$$\#\{D \in (\mathbb{Z}/N\mathbb{Z})^{\sim}[S]^0 \mid s(D^{(i)}) > 1 \text{ for all } i = 0, 1, \ldots f-1\}$$

$$\leq \frac{(p^f - 1)^{n'}}{p^{fn'}} \left(p^{kn'} - \binom{p^k}{n'}\right)^{\left[\frac{f}{k}\right]} p^{(f-k[\frac{f}{k}])n'}$$

$$= N^{n'} \left(1 - \frac{1}{p^{kn'}}\binom{p^k}{n'}\right)^{\left[\frac{f}{k}\right]}$$

$$\leq N^{n'} \left(1 - \frac{1}{p^{kn'}}\binom{p^k}{n'}\right)^{(1-\varepsilon)f/k},$$

where the last inequality follows from our assumption $f \geq k/\varepsilon$:

$$\left[\frac{f}{k}\right] \geq \frac{f}{k} - 1 \geq \frac{(1-\varepsilon)f}{k}.$$

Now, the statements of (ii) follow immediately.   □

Finally, we shall summarize (3.12), (3.16) and (3.18) in terms of $\boldsymbol{\mu}_N$-torsors, via (3.5) and (3.6). Recall that we are assuming $N = p^f - 1$.

THEOREM (3.20). *Let* $C(g)$ *be as in* (3.13).

(i) *If* $n \leq 1$, *we have*

$$\#\{[V] \in H^1_{\text{ét}}(U, \boldsymbol{\mu}_N) \mid \gamma_{[V],1} = g - 1\} \geq N^{2g} - C(g)N^{2g-1} - 1$$

(ii) *If* $n > 1$, *we have*

$$\#\{[V] \in H^1_{\text{ét}}(U, \boldsymbol{\mu}_N) \mid \gamma_{[V],1} = g\} \geq (N^{2g} - C(g)N^{2g-1})\{N^{n-1}(1 - \alpha^f) - 1\}$$

for $f \geq M$, *where* M *and* α *are as in* (3.18)(ii).   □

Roughly speaking, (3.20) says that the generalized Hasse–Witt invariants for 'most' $(p^f - 1)$-cyclic coverings are g (resp. $g - 1$) if $n > 1$ (resp. $n \leq 1$).

4. A Group-Theoretic Characterization of Genera

In this section, we shall prove that the genus of a curve over an algebraically closed field of characteristic > 0 can be recovered group-theoretically from the tame fundamental group of the curve. More precisely, we shall prove the following:

THEOREM (4.1). *For each $i = 1, 2$, let p_i be a prime number, k_i an algebraically closed field of characteristic p_i, X_i a proper, smooth, connected curve of genus g_i over k_i, S_i a finite (possibly empty) set of closed points of X_i with cardinality n_i, and $U_i = X_i - S_i$. If $\pi_1^{\mathrm{t}}(U_1) \simeq \pi_1^{\mathrm{t}}(U_2)$ (as topological groups), then we have:*

(i) $p_1 = p_2$, *unless* $g_i = 0$, $n_i \leq 1$ *for* $i = 1, 2$;
(ii) $g_1 = g_2$; *and*
(iii) $n_1 = n_2$, *unless* $\{(g_1, n_1), (g_2, n_2)\} = \{(0, 0), (0, 1)\}$.

REMARK (4.2). To specify the notion of being recovered group-theoretically, we need to introduce two curves (see [T2], § 1, Definition). However, the following proof involves only one curve, and what we shall do is to extract its various invariants from its tame fundamental group by purely group-theoretic procedure.

Now, let p, k, X, g, S, n, U be as in Section 2 and Section 3. Recall that the i-th Betti number $b^{(i)}$ of U, defined as the $\mathbb{Z}_l$-rank of the l-adic étale cohomology group $H_{\text{ét}}^i(U, \mathbb{Z}_l)$ (l: a prime number $\neq p$), is given in terms of (g, n) as:

$$b^{(0)} = 1, \ b^{(1)} = 2g + n - 1 + b^{(2)}, \ b^{(2)} = \begin{cases} 1 & \text{if } n = 0, \\ 0 & \text{if } n > 0. \end{cases}$$

First, we shall settle some minor things.

LEMMA (4.3). (i) *The invariant $b^{(1)}$ can be recovered group-theoretically from $\pi_1^{\mathrm{t}}(U)$.*
(ii) *We have*

$$\left. \begin{array}{l} b^{(1)} = 0 \iff (g, n) = (0, 0), (0, 1) \\ b^{(1)} = 1 \iff (g, n) = (0, 2) \end{array} \right\} \Rightarrow g = 0.$$

(iii) *Except for the case $b^{(1)} = 0$, the invariant p can be recovered from $\pi_1^{\mathrm{t}}(U)$ group-theoretically.*

PROOF. (ii) is trivial. As is well-known, $\pi_1^{\mathrm{t}}(U)^{\mathrm{ab}}$ is isomorphic to

$$\prod_{l \neq p} \mathbb{Z}_l^{b^{(1)}} \times \mathbb{Z}_p^{\gamma},$$

where γ is the p-rank of (the Jacobian variety of) X (see [T1], Corollary (1.2)). From this, (i) follows.
  Since

$$(0 \leq) \gamma \leq g \leq 2g \leq 2g + n - 1 + b^{(2)} = b^{(1)},$$

$\gamma = b^{(1)}$ holds (if and) only if $g = n - 1 + b^{(2)} = 0$ holds, or, equivalently, $b^{(1)} = 0$. In other words, except for the case $b^{(1)} = 0$, $\gamma < b^{(1)}$ holds, so we can extract the invariant p from the above description of $\pi_1^t(U)^{ab}$ (see [T2], Proposition (1.2)). Thus, (iii) follows.   □

By this lemma, we obtain (4.1)(i). Moreover, by (i) and (ii) of this lemma, we may assume that $b^{(1)} > 1$ when we prove (4.1)(ii). In particular, we may use the invariant p freely.

The essence of (4.1)(ii) is in (3.20), which says, roughly speaking, that the generalized Hasse–Witt invariants for 'most' $(p^f - 1)$-cyclic coverings are g (resp. $g - 1$) if $n > 1$ (resp. $n \leq 1$).

However, a few problems remain. The first problem is that, strictly speaking, $\boldsymbol{\mu}_N = \boldsymbol{\mu}_N(k)$ is not a group-theoretic object. Namely, even if we are given an isomorphism $\pi_1^t(U_1) \simeq \pi_1^t(U_2)$, we are not given any natural isomorphism $\boldsymbol{\mu}_N(k_1) \simeq \boldsymbol{\mu}_N(k_2)$, a priori, hence we do not have any natural isomorphism

$$H^1_{\text{ét}}(U_1, \boldsymbol{\mu}_N) \simeq H^1_{\text{ét}}(U_2, \boldsymbol{\mu}_N).$$

Moreover, in order to define $\gamma_{[V],1}$ for each $[V] \in H^1_{\text{ét}}(X, \boldsymbol{\mu}_N)$, we have used not only the group $\boldsymbol{\mu}_N$ but also the natural embedding $\boldsymbol{\mu}_N \hookrightarrow k$ and the field structure of k, which are also not group-theoretic objects.

In fact, by means of (3.8), any fixed isomorphism $\boldsymbol{\mu}_N(k_1) \simeq \boldsymbol{\mu}_N(k_2)$ will turn out to work for our purpose. However, to avoid confusion and to make things clear, in this section, we will use the set of open normal subgroups H of $\pi_1^t(U)$ such that $\pi_1^t(U)/H$ is a cyclic group of order dividing N, instead of using $H^1_{\text{ét}}(U, \boldsymbol{\mu}_N)$, and rewrite (3.20) in purely group-theoretic terms.

The second problem is that, if $n \leq 1$, the generalized Hasse–Witt invariant for a general $(p^f - 1)$-covering is $g - 1$, and, if $n > 1$, it is g, and that we do not know, a priori, in which case we are.

We overcome this second problem by considering not only the base curve U but also suitable (tame) coverings of U.

Now, we shall start with the first problem.

Assume that a cyclic group G of order prime to p and an $\mathbb{F}_p[G]$-module M are given. As in Section 3, as soon as we are given a character $\chi : G \to k^\times$ for some field k of characteristic p, we can define $\gamma_\chi(M) \overset{\text{def}}{=} \dim_k((M \otimes k)(\chi))$, where

$$(M \otimes k)(\chi) \overset{\text{def}}{=} \{x \in M \otimes k \mid \sigma \cdot x = \chi(\sigma)x \text{ for all } \sigma \in G\}.$$

However, the case is that only G and M are given. In this situation, only certain sums of $\gamma_\chi(M)$ can be well-defined, as follows.

DEFINITION. We define the primitive part of M by

$$M^{\text{prim}} \overset{\text{def}}{=} M/(\sum_{\sigma \neq 1} M^{\langle \sigma \rangle}),$$

where $M^{\langle \sigma \rangle} \overset{\text{def}}{=} \{x \in M \mid \sigma \cdot x = x\}$ for each $\sigma \in G$. We put

$$\gamma^{\text{prim}}(M) \overset{\text{def}}{=} \dim_{\mathbb{F}_p}(M^{\text{prim}}).$$

REMARK (4.4). (i) Let k be a field of characteristic p containing all $\#(G)$-th roots of unity. Then we can check:

$$\gamma^{\text{prim}}(M) = \sum_{\chi: G \hookrightarrow k^\times} \gamma_\chi(M).$$

(ii) Assume that M is finite-dimensional as an $\mathbb{F}_p$-vector space. We can naturally regard the dual vector space $M^* \overset{\text{def}}{=} \text{Hom}_{\mathbb{F}_p}(M, \mathbb{F}_p)$ as a G-module by $(\sigma \cdot \phi)(x) = \phi(\sigma^{-1} \cdot x)$, where $\sigma \in G$, $\phi \in M^*$, and $x \in M$. Then we can check:

$$\gamma^{\text{prim}}(M^*) = \gamma^{\text{prim}}(M).$$

(One way to check this: $\gamma_\chi(M^*) = \gamma_{\chi^{-1}}(M)$.)

We return to the tame fundamental group $\pi_1^t(U)$. Let H be an open normal subgroup of $\pi_1^t(U)$ such that $\pi_1^t(U)/H$ is cyclic of order prime to p. The conjugation induces an action of $\pi_1^t(U)/H$ on the $\mathbb{F}_p$-vector space H^{ab}/p, whose dimension we denote by γ_H.

DEFINITION. $\gamma_H^{\text{prim}} \overset{\text{def}}{=} \gamma^{\text{prim}}(H^{\text{ab}}/p)$.

For each natural number N prime to p, we define $\mathcal{H}_N$ to be the set of open normal subgroups H of $\pi_1^t(U)$ such that $\pi_1^t(U)/H$ is cyclic of order dividing N.

DEFINITION. $\gamma_N^{\text{av}} \overset{\text{def}}{=} \dfrac{1}{N^{b(1)}} \sum_{H \in \mathcal{H}_N} \gamma_H^{\text{prim}}$.

In this definition, "av" means "average". In fact, we have a reinterpretation of γ_N^{av}:

LEMMA (4.5). *We have*

$$\gamma_N^{\text{av}} = \underset{[V] \in H_{\text{ét}}^1(U, \boldsymbol{\mu}_N)}{\text{Average}} \gamma_{[V],1}$$

$$\overset{\text{def}}{=} \frac{1}{\#(H_{\text{ét}}^1(U, \boldsymbol{\mu}_N))} \sum_{[V] \in H_{\text{ét}}^1(U, \boldsymbol{\mu}_N)} \gamma_{[V],1}.$$

PROOF. By (4.4)(ii),

$$\gamma_H^{\text{prim}} = \gamma^{\text{prim}}(H^{\text{ab}}/p) = \gamma^{\text{prim}}((H^{\text{ab}}/p)^*)$$
$$= \gamma^{\text{prim}}(\text{Hom}(\pi_1^t(U_H), \mathbb{F}_p)) = \gamma^{\text{prim}}(H_{\text{ét}}^1(X_H, \mathbb{F}_p)),$$

where U_H is the tame covering of U corresponding to $H \subset \pi_1^t(U)$ and X_H is the normalization of X in U_H, and then by (4.4)(i),

$$\gamma_H^{\text{prim}} = \sum_{\chi: \pi_1^t(U)/H \hookrightarrow k^\times} \gamma_\chi(H_{\text{ét}}^1(X_H, \mathbb{F}_p)).$$

On the other hand, we have the following bijection:

$$\{(H,\chi) \mid H \in \mathcal{H}_N,\ \chi : \pi_1^{\mathrm{t}}(U)/H \hookrightarrow k^\times\} \overset{\sim}{\to} \mathrm{Hom}(\pi_1^{\mathrm{t}}(U), \boldsymbol{\mu}_N), \qquad (4.6)$$

where (H, χ) goes to $(\pi_1^{\mathrm{t}}(U) \twoheadrightarrow \pi_1^{\mathrm{t}}(U)/H \overset{\chi}{\hookrightarrow} \boldsymbol{\mu}_N)$. Now, let $[V]$ be the element of $H^1_{\text{ét}}(U, \boldsymbol{\mu}_N) = \mathrm{Hom}(\pi_1^{\mathrm{t}}(U), \boldsymbol{\mu}_N)$ corresponding to (H, χ). Then we claim

$$\gamma_{[V],1} = \gamma_\chi(H^1_{\text{ét}}(X_H, \mathbb{F}_p)). \qquad (4.7)$$

In fact, let Y be the normalization of X in V. Then, since U_H is the $\mathrm{Im}(\chi)$-torsor of U corresponding to $(\pi_1^{\mathrm{t}}(U) \twoheadrightarrow \pi_1^{\mathrm{t}}(U)/H \overset{\chi}{\overset{\sim}{\to}} \mathrm{Im}(\chi))$, we see that Y coincides with $\boldsymbol{\mu}_N \times_{\mathrm{Im}(\chi)} X_H$, the quotient of $\boldsymbol{\mu}_N \times X_H$ by the $\mathrm{Im}(\chi)$-action $(\zeta, x)^{\zeta_0} = (\zeta_0^{-1}\zeta, x^{\zeta_0})$. From this, we can deduce that $H^1_{\text{ét}}(Y, \mathbb{F}_p)$ is the induced $\mathbb{F}_p[\boldsymbol{\mu}_N]$-module of the $\mathbb{F}_p[\mathrm{Im}(\chi)]$-module $H^1_{\text{ét}}(X_H, \mathbb{F}_p)$. Now our claim (4.7) follows immediately.

Now, bijection (4.6), identity (4.7) and the fact $H^1_{\text{ét}}(U, \boldsymbol{\mu}_N) \simeq (\mathbb{Z}/N\mathbb{Z})^{b^{(1)}}$ complete the proof. $\qquad \square$

REMARK (4.8). Here is another simple reinterpretation of γ_N^{av}. For a profinite group Π and a natural number m, we shall denote by $\Pi(m)$ the kernel of $\Pi \twoheadrightarrow \Pi^{\mathrm{ab}}/(\Pi^{\mathrm{ab}})^m$, or, equivalently, $\Pi(m)$ is the topological closure of the subgroup $[\Pi, \Pi]\Pi^m$ of Π. Moreover, we shall denote by $U(m)$ the tame covering of U corresponding to the subgroup $\pi_1^{\mathrm{t}}(U)(m)$ of $\pi_1^{\mathrm{t}}(U)$, so that $\pi_1^{\mathrm{t}}(U(m)) = \pi_1^{\mathrm{t}}(U)(m)$, and $X(m)$ the normalization of X in $U(m)$. (Note that this last notation is somewhat confusing: $X(m)$ does not coincide with the (étale) covering of X corresponding to $\pi_1(X)(m)$ in general.) Then we have

$$\gamma_N^{\mathrm{av}} = \frac{\dim_{\mathbb{F}_p}(\pi_1^{\mathrm{t}}(U)(N)/(\pi_1^{\mathrm{t}}(U)(N))(p))}{(\pi_1^{\mathrm{t}}(U) : \pi_1^{\mathrm{t}}(U)(N))}.$$

In fact, the denominator of the right-hand side is $N^{b^{(1)}}$, while the numerator is $\dim_{\mathbb{F}_p}(H^1_{\text{ét}}(X(N), \mathbb{F}_p))$. Since $\pi_1^{\mathrm{t}}(U)/\pi_1^{\mathrm{t}}(U)(N) = \pi_1^{\mathrm{t}}(U)^{\mathrm{ab}}/N$ is abelian of order prime to p, we see that the following canonical decomposition exists:

$$H^1_{\text{ét}}(X(N), \mathbb{F}_p) = \bigoplus_{H \in \mathcal{H}_N} (H^1_{\text{ét}}(X(N), \mathbb{F}_p)^{H/\pi_1^{\mathrm{t}}(U)(N)})^{(\pi_1^{\mathrm{t}}(U)/H)\text{-prim}},$$

where $(\pi_1^{\mathrm{t}}(U)/H)$-prim means the primitive part as a $(\pi_1^{\mathrm{t}}(U)/H)$-module. Since $H^1_{\text{ét}}(X(N), \mathbb{F}_p)^{H/\pi_1^{\mathrm{t}}(U)(N)} = H^1_{\text{ét}}(X_H, \mathbb{F}_p)$ (see the proof of (4.5) for the definition of X_H), we obtain the desired equality

$$\dim_{\mathbb{F}_p}(H^1_{\text{ét}}(X(N), \mathbb{F}_p)) = \sum_{H \in \mathcal{H}_N} \gamma_H^{\mathrm{prim}}$$

(see the beginning of the proof of (4.5)).

The following is a variant of (3.20). Recall that we are assuming $b^{(1)} > 1$.

THEOREM (4.9). Assume $N = p^f - 1$. Let $C(g)$ be as in (3.13).

(i) *If $n \leq 1$, we have*

$$g - 1 - \frac{C(b^{(1)}/2)(b^{(1)}/2 - 1)}{N} \leq \gamma_N^{\mathrm{av}} \leq g - 1 + \frac{1}{N^{b^{(1)}}}.$$

(ii) *If $n > 1$, let k be any positive integer $\geq \log_p(b^{(1)})$, ε any positive real number < 1, and put*

$$M = \frac{k}{\varepsilon}, \quad \alpha = \left(1 - \frac{1}{p^{kb^{(1)}}}\left(\frac{p^k}{b^{(1)}}\right)\right)^{(1-\varepsilon)/k}.$$

Then, we have

$$g - \left\{ \left(C\left(\left[\frac{b^{(1)}}{2} \right] \right) \right) \left[\frac{b^{(1)}}{2} \right] + 1 \right) \frac{1}{N} + \left[\frac{b^{(1)}}{2} \right] \alpha^f \right\} \leq \gamma_N^{\mathrm{av}} \leq g + (b^{(1)} - 1)\alpha^f$$

for all $f \geq M$.

PROOF. (i) We first note that $b^{(1)} = 2g$ holds in this case. By (3.7)(and (3.6)), we have

$$\gamma_{[V],1} \leq \begin{cases} g, & \text{if } [V] = 0, \\ g - 1, & \text{otherwise,} \end{cases}$$

for each $[V] \in H^1_{\text{ét}}(U, \boldsymbol{\mu}_N)$. From this,

$$\sum_{[V] \in H^1_{\text{ét}}(U, \boldsymbol{\mu}_N)} \gamma_{[V],1} \leq (N^{b^{(1)}} - 1)(g - 1) + g = N^{b^{(1)}}(g - 1) + 1,$$

hence

$$\gamma_N^{\mathrm{av}} \leq g - 1 + \frac{1}{N^{b^{(1)}}}.$$

On the other hand, by (3.12)(ii) and (3.18)(i), we have

$$\sum_{[V] \in H^1_{\text{ét}}(U, \boldsymbol{\mu}_N)} \gamma_{[V],1} \geq (g - 1)(N^{b^{(1)}} - C(g)N^{b^{(1)} - 1}),$$

hence

$$\gamma_N^{\mathrm{av}} \geq g - 1 - \frac{C(g)(g - 1)}{N}.$$

Since $g = b^{(1)}/2$, this completes the proof.

(ii) Dividing the sum $\sum_{[V] \in H^1_{\text{ét}}(U, \boldsymbol{\mu}_N)} \gamma_{[V],1}$ into the three parts: (0) $D = 0$; (1) $s(D^{(i)}) = 1$ for some $i = 0, 1, \ldots, f - 1$; (2) otherwise, we obtain

$$\sum_{[V] \in H^1_{\text{ét}}(U, \boldsymbol{\mu}_N)} \gamma_{[V],1}$$

$$\leq ((g - 1)N^{2g} + 1) + gN^{2g}(N^{n-1}(1 - \alpha^f) - 1) + (g + n - 2)N^{2g}N^{n-1}\alpha^f$$

$$= gN^{b^{(1)}} - (N^{2g} - 1) + (n - 2)N^{b^{(1)}}\alpha^f$$

$$\leq gN^{b^{(1)}} + (b^{(1)} - 1)N^{b^{(1)}}\alpha^f$$

for $f \geq M$, by (3.7) and (3.18)(ii). (Here, note that

$$k \geq \log_p(b^{(1)}) \geq \log_p(n-1)$$

and

$$\alpha = \left(1 - \frac{1}{p^{kb^{(1)}}}\binom{p^k}{b^{(1)}}\right)^{(1-\varepsilon)/k} \leq \left(1 - \frac{1}{p^{k(n-1)}}\binom{p^k}{n-1}\right)^{(1-\varepsilon)/k},$$

since $n - 1 \leq b^{(1)}$.) Therefore

$$\gamma_N^{\mathrm{av}} \leq g + (b^{(1)} - 1)\alpha^f.$$

Similarly, considering the cases (0) and (1), we have

$$\sum_{[V] \in H^1_{\text{ét}}(U, \boldsymbol{\mu}_N)} \gamma_{[V],1}$$

$$\geq (g-1)(N^{2g} - C(g)N^{2g-1}) + g(N^{2g} - C(g)N^{2g-1})(N^{n-1}(1-\alpha^f) - 1)$$

$$= gN^{b^{(1)}} - C(g)gN^{b^{(1)}-1} - gN^{b^{(1)}}\alpha^f + C(g)gN^{b^{(1)}-1}\alpha^f - N^{2g} + C(g)N^{2g-1}$$

$$\geq gN^{b^{(1)}} - C(g)gN^{b^{(1)}-1} - gN^{b^{(1)}}\alpha^f - N^{2g}$$

$$\geq gN^{b^{(1)}} - (C(g)g + 1)N^{b^{(1)}-1} - gN^{b^{(1)}}\alpha^f$$

$$\geq gN^{b^{(1)}} - \left(C\left(\left[\frac{b^{(1)}}{2}\right]\right)\left[\frac{b^{(1)}}{2}\right] + 1\right)N^{b^{(1)}-1} - \left[\frac{b^{(1)}}{2}\right]N^{b^{(1)}}\alpha^f$$

for $f \geq M$, by (3.16), etc. (For the last inequality, note that $g \leq [b^{(1)}/2]$ and that $C(g)$ is monotone increasing.) Therefore, we have

$$\gamma_N^{\mathrm{av}} \geq g - \left\{\left(C\left(\left[\frac{b^{(1)}}{2}\right]\right)\left[\frac{b^{(1)}}{2}\right] + 1\right)\frac{1}{N} + \left[\frac{b^{(1)}}{2}\right]\alpha^f\right\}$$

for all $f \geq M$.   □

DEFINITION. We define

$$g' \overset{\text{def}}{=} \begin{cases} g - 1, & \text{if } n \leq 1, \\ g, & \text{if } n > 1. \end{cases}$$

The following (together with (4.3)) gives a group-theoretic characterization of the invariant g'.

COROLLARY (4.10). (We are assuming $b^{(1)} > 1$.) Let M and α be as in (4.9), and $C(g)$ as in (3.13). Then

$$g' - \left\{\left(C\left(\left[\frac{b^{(1)}}{2}\right]\right)\left[\frac{b^{(1)}}{2}\right] + 1\right)\frac{1}{N} + \left[\frac{b^{(1)}}{2}\right]\alpha^f\right\} \leq \gamma_N^{\mathrm{av}}$$

$$\leq g' + \max\left(\frac{1}{N^{b^{(1)}}}, (b^{(1)} - 1)\alpha^f\right)$$

for all $f \geq M$.

In particular, if f is sufficiently large (e.g., if

$$f \geq \max\left(M, \frac{\log(3b^{(1)})}{\log(\alpha^{-1})}, \log_p\left(C\left(\left[\frac{b^{(1)}}{2}\right]\right)b^{(1)} + 5\right)\right)$$

holds), then g' can be characterized as the unique integer in the interval

$$\left[\gamma_N^{\text{av}} - \max\left(\frac{1}{N^{b^{(1)}}}, (b^{(1)} - 1)\alpha^f\right),\right.$$

$$\left.\gamma_N^{\text{av}} + \left\{\left(C\left(\left[\frac{b^{(1)}}{2}\right]\right)\left[\frac{b^{(1)}}{2}\right] + 1\right)\frac{1}{N} + \left[\frac{b^{(1)}}{2}\right]\alpha^f\right\}\right].$$

PROOF. By (4.9), g' falls in the interval for $f \geq M$.

Assume $f \geq \max\left(M, \dfrac{\log(3b^{(1)})}{\log(\alpha^{-1})}, \log_p\left(C\left(\left[\dfrac{b^{(1)}}{2}\right]\right)b^{(1)} + 5\right)\right)$. Then the length λ of the interval satisfies:

$$\lambda = \max\left(\frac{1}{N^{b^{(1)}}} + \left\{\left(C\left(\left[\frac{b^{(1)}}{2}\right]\right)\left[\frac{b^{(1)}}{2}\right] + 1\right)\frac{1}{N} + \left[\frac{b^{(1)}}{2}\right]\alpha^f\right\},\right.$$

$$\left.(b^{(1)} - 1)\alpha^f + \left\{\left(C\left(\left[\frac{b^{(1)}}{2}\right]\right)\left[\frac{b^{(1)}}{2}\right] + 1\right)\frac{1}{N} + \left[\frac{b^{(1)}}{2}\right]\alpha^f\right\}\right)$$

$$< \frac{1}{N^{b^{(1)}}} + (b^{(1)} - 1)\alpha^f + \left\{\left(C\left(\left[\frac{b^{(1)}}{2}\right]\right)\left[\frac{b^{(1)}}{2}\right] + 1\right)\frac{1}{N} + \left[\frac{b^{(1)}}{2}\right]\alpha^f\right\}$$

$$< \left(C\left(\left[\frac{b^{(1)}}{2}\right]\right)\left[\frac{b^{(1)}}{2}\right] + 2\right)\frac{1}{N} + \frac{3b^{(1)}}{2}\alpha^f$$

$$\leq \tfrac{1}{2} + \tfrac{1}{2} = 1.$$

This implies the desired uniqueness. $\qquad\square$

At the cost of sacrificing effectivity, we also obtain the following more impressive characterization of g'.

COROLLARY (4.11). $\displaystyle\lim_{f\to\infty} \gamma_{p^f - 1}^{\text{av}} = g'$. $\qquad\square$

REMARK (4.12). It is easy to see that (4.11) is also valid for $b^{(1)} = 1$. In order to include the case that $b^{(1)} = 0$, the formula should be modified as

$$\lim_{f\to\infty} \gamma_{p^f - 1}^{\text{av}} = (g')^+,$$

where $x^+ \overset{\text{def}}{=} \max(x, 0)$.

Now, we shall treat the second problem: the group-theoretic characterization above is not for g but for g'.

First, we introduce the following temporary invariant, which can be recovered group-theoretically from $\pi_1^{\text{t}}(U)$ (assuming $b^{(1)} > 0$):

$$n^{\text{temp}} \overset{\text{def}}{=} b^{(1)} - 2g' + 1.$$

Next, let m be a natural number prime to p, and $U(m)$ the tame covering of U as in (4.8). We define $n(m)$ (resp. $n^{\text{temp}}(m)$) to be the invariant n (resp. n^{temp}) for the curve $U(m)$. Note that $n^{\text{temp}}(m)$ can also be recovered group-theoretically from $\pi_1^{\text{t}}(U)$.

LEMMA (4.13). *Assume $b^{(1)} > 0$.*

(i) *We have*

$$n^{\text{temp}} = \begin{cases} 3, & \text{if } n \leq 1, \\ n, & \text{if } n > 1. \end{cases}$$

(ii) *If $m > 1$, we have*

$$n^{\text{temp}}(m) = \begin{cases} 3, & \text{if } n = 0, \\ m^{b^{(1)}}, & \text{if } n = 1, \\ m^{b^{(1)}-1}n, & \text{if } n > 1. \end{cases}$$

PROOF. (i) Immediate from the definitions.

(ii) By (i), we have

$$n^{\text{temp}}(m) = \begin{cases} 3, & \text{if } n(m) \leq 1, \\ n(m), & \text{if } n(m) > 1. \end{cases}$$

On the other hand, we have

$$n(m) = \begin{cases} m^{b^{(1)}}n, & \text{if } n \leq 1, \\ m^{b^{(1)}-1}n, & \text{if } n > 1. \end{cases}$$

These give the desired equality.   □

Finally, we can prove the following:

THEOREM (4.14). *Assume $b^{(1)} > 0$. Fix any natural number $m \neq 1, 3$ prime to p. Then*

$$n = \begin{cases} n^{\text{temp}}, & \text{if } n^{\text{temp}} \neq 3, \\ 0, & \text{if } n^{\text{temp}} = 3 \text{ and } n^{\text{temp}}(m) = 3, \\ 1, & \text{if } n^{\text{temp}} = 3 \text{ and } n^{\text{temp}}(m) = m^{b^{(1)}}, \\ 3, & \text{if } n^{\text{temp}} = 3 \text{ and } n^{\text{temp}}(m) = 3m^{b^{(1)}-1}. \end{cases}$$

In particular, the invariant n can be recovered group-theoretically from $\pi_1^{\text{t}}(U)$ (except for the case $b^{(1)} = 0$).

PROOF. The first statement follows from (4.13). Since n^{temp}, $n^{\text{temp}}(m)$ and $b^{(1)}$ can be recovered group-theoretically from $\pi_1^{\text{t}}(U)$, the second statement follows. (More precisely, we have to consider two cases separately. If $b^{(1)} = 1$, then we have $n = n^{\text{temp}} = 2$. Otherwise, i.e. if $b^{(1)} > 1$, then the numbers 3, $m^{b^{(1)}}$ and $3m^{b^{(1)}-1}$ are distinct from one another.)   □

END OF PROOF OF (4.1). As we have already seen, (4.3) implies (4.1)(i) and that we may assume $b^{(1)} > 0$ when we prove (4.1)(ii)(iii). Now, (4.14) implies (4.1)(iii). Since $b^{(2)}$ is determined by n, (4.14) and (4.3), together with the equality

$$b^{(1)} = 2g + n - 1 + b^{(2)},$$

implies (4.1)(ii). This completes the proof of (4.1).   □

REMARK (4.15). We might hope for the more general limit formula

$$\lim_{\substack{N \to \infty \\ p \nmid N}} \gamma_N^{\mathrm{av}} = g'. \tag{4.16}$$

For the present, we can only prove 'one half' of (4.16):

$$\limsup_{\substack{N \to \infty \\ p \nmid N}} \gamma_N^{\mathrm{av}} \le g', \tag{4.17}$$

by using a higher-dimensional version of LeVeque's inequality, due to Stegbuchner ([S]), in the theory of uniform distribution modulo 1. See Appendix for this. It may be interesting to ask if (3.17)(iii) gives an approach to the other half of (4.16).

REMARK (4.18). As in [T2], Remark (1.11), not only $\pi_1^{\mathrm{t}}(U)$ but also a suitable quotient is enough to determine g (and n). For example, $\pi_1^{\mathrm{t}}(U)/D(D(D(\pi_1^{\mathrm{t}}(U))))$ is enough, where, for a profinite group G, $D(G)$ denotes the (topological) commutator subgroup of G, or, equivalently, the kernel of $G \twoheadrightarrow G^{\mathrm{ab}}$.

Finally, as a direct consequence of (4.1), we have:

COROLLARY (4.19). *The quotient $\pi_1(X)$ of $\pi_1^{\mathrm{t}}(U)$ can be recovered group-theoretically from $\pi_1^{\mathrm{t}}(U)$.*

PROOF. As in [T2], Corollary 1.10, this follows from (4.1)(ii) and the Hurwitz formula.   □

5. Applications

(A) A group-theoretic characterization of inertia groups. Since we have established (4.1), we can prove, as in [T2], that the set of inertia subgroups of $\pi_1^{\mathrm{t}}(U)$ can be recovered group-theoretically from $\pi_1^{\mathrm{t}}(U)$ for U hyperbolic. (We say that the curve U is hyperbolic, if the Euler–Poincaré characteristic $b^{(0)} - b^{(1)} + b^{(2)} = 2 - 2g - n$ of U is negative.)

Let K be the function field $k(U) = k(X)$, and define $\tilde{K}^{\mathrm{t}}$ to be the maximal Galois extension of K in a fixed separable closure K^{sep}, unramified over U and at most tamely ramified over S. We may and shall identify $\pi_1^{\mathrm{t}}(U)$ with $\mathrm{Gal}(\tilde{K}^{\mathrm{t}}/K)$. We define $\tilde{X}^{\mathrm{t}}$ to be the normalization of X in $\tilde{K}^{\mathrm{t}}$ and $\tilde{S}^{\mathrm{t}}$ to be the inverse image of S in $\tilde{X}^{\mathrm{t}}$. For each $\tilde{P} \in \tilde{S}^{\mathrm{t}}$, we denote by $I_{\tilde{P}}$ the inertia subgroup of $\pi_1^{\mathrm{t}}(U)$

associated to $\tilde{P}$, i.e. the stabilizer of $\tilde{P}$. We have $I_{\tilde{P}} \neq \{1\}$ if and only if $(n > 0$ and) $(g, n) \neq (0, 1)$ (see [T1], Lemma (2.2)).

LEMMA (5.1). *Assume that U is hyperbolic.*

(i) *Let $\tilde{P}$ and $\tilde{Q}$ be two points of $\tilde{S}^t$ distinct from each other. Then the intersection of $I_{\tilde{P}}$ and $I_{\tilde{Q}}$ is trivial in $\pi_1^t(U)$. In particular, for any $\sigma \in \pi_1^t(U) - I_{\tilde{P}}$, the intersection of $I_{\tilde{P}}$ and $\sigma I_{\tilde{P}} \sigma^{-1}$ is trivial.*

(ii) *The map $\tilde{S}^t \to \mathrm{Sub}(\pi_1^t(U))$, $\tilde{P} \mapsto I_{\tilde{P}}$ is injective, where, for a profinite group G, $\mathrm{Sub}(G)$ denotes the set of closed subgroups of G. Moreover, for each $\tilde{P} \in \tilde{S}^t$, the normalizer of $I_{\tilde{P}}$ in $\pi_1^t(U)$ is $I_{\tilde{P}}$ itself.*

PROOF. (i) [T2], Lemma (2.1). (ii) [T2], Corollary (2.2).   □

Let $\mathcal{I}^t$ be the set of inertia subgroups in $\pi_1^t(U)$, namely the image of the map $\tilde{S}^t \to \mathrm{Sub}(\pi_1^t(U))$, $\tilde{P} \mapsto I_{\tilde{P}}$.

THEOREM (5.2). *If U is hyperbolic, then the set $\mathcal{I}^t$ can be recovered group-theoretically from $\pi_1^t(U)$. More precisely, let the notations and the assumptions be as in (4.1), and assume further that $2 - 2g_i - n_i < 0$ for some $i = 1, 2$. Then, if an isomorphism $\pi_1^t(U_1) \simeq \pi_1^t(U_2)$ (as topological groups) is given, the induced bijection $\mathrm{Sub}(\pi_1^t(U_1)) \simeq \mathrm{Sub}(\pi_1^t(U_2))$ induces a bijection $\mathcal{I}_1^t \simeq \mathcal{I}_2^t$, where $\mathcal{I}_i^t$ denotes the set $\mathcal{I}^t$ for the curve U_i, for each $i = 1, 2$.*

PROOF. See [T2], Proposition (2.4) and Remark (2.6). (Use (4.1)(iii).)   □

(B) A group-theoretic characterization of 'additive structures' of inertia groups. Let $\tilde{P} \in \tilde{S}^t$. As is well-known, $I_{\tilde{P}}$ can be canonically identified with the Tate module

$$\widehat{\mathbb{Z}}'(1) \overset{\mathrm{def}}{=} \varprojlim_{p \nmid m} \boldsymbol{\mu}_m(k)$$

of the multiplicative group $k^\times$, where $\boldsymbol{\mu}_m(k)$ is the group of m-th roots of unity in k. So, $I_{\tilde{P}} \otimes_{\mathbb{Z}} (\mathbb{Q}/\mathbb{Z})'$, where $(\mathbb{Q}/\mathbb{Z})'$ denotes the prime-to-p part of $\mathbb{Q}/\mathbb{Z}$, can be canonically identified with

$$(\mathbb{Q}/\mathbb{Z})'(1) \overset{\mathrm{def}}{=} \bigcup \boldsymbol{\mu}_m(k) = F^\times,$$

where F denotes the algebraic closure of the prime field $\mathbb{F}_p$ in k. Thus, $F_{\tilde{P}} \overset{\mathrm{def}}{=} (I_{\tilde{P}} \otimes_{\mathbb{Z}} (\mathbb{Q}/\mathbb{Z})') \coprod \{*\}$ (where $\{*\}$ means a one-point set) can be identified with F, hence carries a structure of field, whose multiplicative group is $I_{\tilde{P}} \otimes_{\mathbb{Z}} (\mathbb{Q}/\mathbb{Z})'$ and whose zero element is $*$.

Now, we have the following proposition. Unlike in (A) above, the proof here is quite different from [T2], Proposition 2.8, even after we have established (4.1) and (5.2). (See [T2], Remark 2.10(ii).)

PROPOSITION (5.3). *Assume that U is hyperbolic. Then the field structure of $F_{\tilde{P}} = I_{\tilde{P}} \otimes_{\mathbb{Z}} (\mathbb{Q}/\mathbb{Z})' \coprod \{*\}$ can be recovered group-theoretically from $\pi_1^t(U)$.*

PROOF. We may assume $n > 0$.

First, we shall reduce the problem to the case $n \geq 3$. If $g = 0$, this follows automatically from the hyperbolicity condition. For $g > 0$, take any natural number m prime to p such that $m^{2g}n \geq 3$. Then, replacing $\pi_1^t(U)$ by the kernel of $\pi_1^t(U) \to \pi_1(X)^{ab}/m$ (whose index in $\pi_1^t(U)$ is m^{2g}), we have $n \geq 3$. (Note that $I_{\tilde{P}}$ is contained in the kernel.) Next, by (5.1)(i), the set $\mathcal{I}^t$ divided by the conjugacy action of $\pi_1^t(U)$ consists of n orbits. Choosing any 3 orbits among these n orbits such that one of them is the conjugacy class of the given $I_{\tilde{P}}$, and dividing $\pi_1^t(U)$ by the subgroup (topologically) generated by all members I of $\mathcal{I}^t$ whose conjugacy class is among the other $n - 3$ orbits, we may reduce the problem to the case $n = 3$. (Observe that these reduction steps are purely group-theoretic, by (4.1), (4.19) and (5.2).)

From now on, we assume $n = 3$ and we shall use (2.21). For each natural number f, let $\mathbb{F}_{p^f, \tilde{P}}$ denote the unique subfield of $F_{\tilde{P}}$ with cardinality p^f. Since $\mathbb{F}_{p^f, \tilde{P}}^{\times} = I_{\tilde{P}}/(p^f - 1)$, the subfield $\mathbb{F}_{p^f, \tilde{P}}$ can be recovered group-theoretically as a (multiplicative) submonoid. Fix any field $\mathbb{F}_{p^f}$ with cardinality p^f (unrelatedly to $\mathbb{F}_{p^f, \tilde{P}}$). Then the set $\mathrm{Hom}(\mathbb{F}_{p^f, \tilde{P}}^{\times}, \mathbb{F}_{p^f}^{\times}) = \mathrm{Hom}_{(\mathrm{groups})}(\mathbb{F}_{p^f, \tilde{P}}^{\times}, \mathbb{F}_{p^f}^{\times})$ is group-theoretic; recovering the field structure of $\mathbb{F}_{p^f, \tilde{P}}$ is equivalent to recovering $\mathrm{Hom}(\mathbb{F}_{p^f, \tilde{P}}, \mathbb{F}_{p^f}) = \mathrm{Hom}_{(\mathrm{fields})}(\mathbb{F}_{p^f, \tilde{P}}, \mathbb{F}_{p^f})$ as a subset of $\mathrm{Hom}(\mathbb{F}_{p^f, \tilde{P}}^{\times}, \mathbb{F}_{p^f}^{\times})$. Moreover, it is sufficient to recover this subset for f in a cofinal subset of $\mathbb{Z}_{>0}$ with respect to division.

To do this, we shall consider the two maps

$$\mathrm{Res}_f : \mathrm{Hom}(\pi_1^t(U)^{ab}/(p^f - 1), \mathbb{F}_{p^f}^{\times}) \to \mathrm{Hom}(\mathbb{F}_{p^f, \tilde{P}}^{\times}, \mathbb{F}_{p^f}^{\times})$$

and

$$\Gamma_f : \mathrm{Hom}(\pi_1^t(U)^{ab}/(p^f - 1), \mathbb{F}_{p^f}^{\times}) \to \mathbb{Z}_{\geq 0}.$$

The first map Res_f is the restriction with respect to the canonical inclusion $\mathbb{F}_{p^f, \tilde{P}}^{\times} = I_{\tilde{P}}/(p^f - 1) \hookrightarrow \pi_1^t(U)^{ab}/(p^f - 1)$. The second map Γ_f is defined to send $\chi \in \mathrm{Hom}(\pi_1^t(U)^{ab}/(p^f - 1), \mathbb{F}_{p^f}^{\times})$ to $\gamma_\chi(H_{\text{ét}}^1(X_H, \mathbb{F}_p))$, where $H \overset{\text{def}}{=} \mathrm{Ker}(\chi)$. (For the definitions of γ_χ and X_H, see Section 4, especially (4.5) and its preceding paragraphs. Strictly speaking, in Section 4, we use the notation γ_χ only for a character χ of a cyclic group. However, the same definition goes well for characters of general finite groups, or we can replace $\pi_1^t(U)^{ab}/(p^f - 1)$ by $\mathrm{Im}(\chi)$. See the proof of (4.5).)

Now we can state the following claim, which completes the proof of (5.3).

CLAIM (5.4). Let m_0 be the product of all prime numbers $\leq p-2$. (For $p = 2, 3$, $m_0 = 1$.) Let f_0 be the order of p mod m_0 in the multiplicative group $(\mathbb{Z}/m_0\mathbb{Z})^{\times}$. For each $f > \log_p(C(g) + 1)$ divisible by f_0, we have

$$\mathrm{Hom}(\mathbb{F}_{p^f, \tilde{P}}, \mathbb{F}_{p^f}) = \mathrm{Surj}(\mathbb{F}_{p^f, \tilde{P}}^{\times}, \mathbb{F}_{p^f}^{\times}) - \mathrm{Res}_f(\Gamma_f^{-1}(\{g + 1\}))$$
$$(\subset \mathrm{Hom}(\mathbb{F}_{p^f, \tilde{P}}^{\times}, \mathbb{F}_{p^f}^{\times})).$$

To prove this claim, we fix any embedding $\mathbb{F}_{p^f} \to k$ as fields. Then we have

$$\mathrm{Hom}(\pi_1^{\mathrm{t}}(U)^{\mathrm{ab}}/(p^f - 1), \mathbb{F}_{p^f}{}^\times) = H_{\text{ét}}^1(U, \boldsymbol{\mu}_N),$$

where $N \stackrel{\text{def}}{=} p^f - 1$. By (3.5), this can be identified with P_N (or $\tilde{P}_N$), in the notation of Section 3. On the other hand, $F_{\tilde{P}}$ can be canonically identified with the algebraic closure of $\mathbb{F}_p$ in k. This identification, together with the fixed embedding $\mathbb{F}_{p^f} \to k$, specifies one identification $\mathbb{F}_{p^f, \tilde{P}} = \mathbb{F}_{p^f}$. By using this, we obtain

$$\mathrm{Hom}(\mathbb{F}_{p^f, \tilde{P}}{}^\times, \mathbb{F}_{p^f}{}^\times) = \mathbb{Z}/N\mathbb{Z}.$$

By using various definitions, we see that the map $P_N \to \mathbb{Z}/N\mathbb{Z}$ coming from Res_f is nothing but the composite of $b_N : P_N \to \mathbb{Z}/N\mathbb{Z}[S]$ (see (3.4)) and $\mathbb{Z}/N\mathbb{Z}[S] \to \mathbb{Z}/N\mathbb{Z}$, $D \bmod N \mapsto \mathrm{ord}_P(D)$. Thus we can reformulate (5.4) as follows: For each $n \in (\mathbb{Z}/N\mathbb{Z})^\sim$,

$n \in \{p^b \mid b = 0, 1, \ldots, f - 1\} \iff$

$(n, N) = 1$ and $\not\exists([L], D) \in \tilde{P}_N$ s.t. $\mathrm{ord}_P(D) = n$ and $\gamma_{([L],D)} = g + 1$.   (5.5)

First, by (3.7), we see that $\gamma_{([L],D)} \leq g+1$ always holds and that $\gamma_{([L],D)} = g+1$ holds if and only only if $\gamma_{([L],D)} = \dim_k(H^1(X, L))$ (or, equivalently, $\varphi_{([L],D)}$ is bijective) and $\deg(D) = 2N$. Moreover, if $\varphi_{([L],D)}$ is bijective, then B_D^f should satisfy condition $(\star)$ of page 59, and, in particular, it should be a semi-stable vector bundle.

By this observation, the '$\Rightarrow$' part of (5.5) follows from (2.21)(iv–a). More specifically, let us denote by $P_1 = P, P_2, P_3$ the three points of S. Then (2.21)(iv–a) implies that either $\mathrm{ord}_{P_2}(D) = N$ or $\mathrm{ord}_{P_3}(D) = N$ holds, which is impossible as $D \in (\mathbb{Z}/N\mathbb{Z})^\sim[S]$.

To prove the '$\Leftarrow$' part of (5.5), let n be a natural number $\in (\mathbb{Z}/N\mathbb{Z})^\sim$ such that $(n, N) = 1$, and suppose that $n \notin \{p^b \mid b = 0, 1, \ldots, f - 1\}$. Then we have to prove that there exists $([L], D) \in \tilde{P}_N$ such that $\mathrm{ord}_P(D) = n$ and $\gamma_{([L],D)} = g + 1$. Since f is assumed to be divisible by f_0, N is divisible by all prime numbers $\leq p - 2$. Therefore, by the assumption $(n, N) = 1$, we must have $n \notin I_{p-1}p^{I_f} = \{ap^b \mid a = 0, 1, \ldots, p - 2, \ b = 0, 1, \ldots, f - 1\}$. Now, by (2.21)(iv–c), there exists $D \in (\mathbb{Z}/N\mathbb{Z})^\sim[S]$ with degree $2N$ which satisfies (2.14) and to which (2.13) can be applied. Then, as in the proof of (3.12), we obtain

$$\#\{[L] \in \mathrm{Pic}(X) \mid ([L], D) \in \tilde{P}_N \text{ and } \gamma_{([L],D)} = g+1\} \geq N^{2g} - C(g)N^{2g-1} > 0,$$

where the last inequality comes from the assumption $f > \log_p(C(g) + 1)$. This completes the proof. $\square$

REMARK (5.6). A similar technique as in the proof of (5.4) gives an alternative proof of (4.1). More precisely, fix an algebraic closure $\overline{\mathbb{F}}_p$ of the prime field $\mathbb{F}_p$, and put

$$\gamma^{\max} \stackrel{\text{def}}{=} \max\{\gamma_\chi(H_{\text{ét}}^1(X_{\mathrm{Ker}(\chi)}, \mathbb{F}_p)) \mid \chi \in \mathrm{Hom}(\pi_1^{\mathrm{t}}(U), \overline{\mathbb{F}}_p^\times), \ \chi \neq 1\}.$$

(For the sake of convenience, we shall define $\max \varnothing = -1$.) Then:

CLAIM (5.7). We have

$$\gamma^{\max} = g + n - 2 + b^{(2)}.$$

If we assume (5.7),

$$g = b^{(1)} - \gamma^{\max} - 1, \quad n' \stackrel{\text{def}}{=} n + b^{(2)} = 2\gamma^{\max} - b^{(1)} + 3$$

can be recovered group-theoretically. Moreover, to recover n (assuming $b^{(1)} > 0$), we have only to note that

$$n = \begin{cases} n', & \text{if } n'(m) > 1, \\ 0, & \text{if } n'(m) \leq 1, \end{cases}$$

where m is an arbitrary natural number > 1 prime to p and $n'(m)$ is the invariant n' for the curve $U(m)$. (See the paragraph preceding (4.13).)

For the proof of (5.7), first, the inequality $\gamma^{\max} \leq g + n - 2 + b^{(2)}$ follows from (3.7) (together with (3.6) and (4.7)). For the opposite inequality, let f be a natural number $> \max(n - 1, \log_p(C(g) + 1))$ and put $N = p^f - 1$. Write the set S of cardinality n as $\{P_{-1}, P_0, P_1, \ldots, P_{n-2}\}$ and define $D \in (\mathbb{Z}/N\mathbb{Z})^{\sim}[S]$ by

$$D \stackrel{\text{def}}{=} \sum_{i=-1}^{n-2} n_i P_i, \quad n_i = \begin{cases} \sum_{j=0}^{n-2} p^j, & \text{if } i = -1, \\ N - p^i, & \text{if } i = 0, 1, \ldots, n - 2. \end{cases}$$

($D = 0$ for $n \leq 1$.) Then, we see that D satisfies (2.14) (with $s = n - 1 + b^{(2)}$). Now, as in the proof of (3.12), we deduce that

$$\#\{[L] \in \mathrm{Pic}(X) \mid ([L], D) \in \tilde{P}_N \text{ and } \gamma_{([L],D)} = g + n - 2 + b^{(2)}\}$$
$$\geq N^{2g} - C(g)N^{2g-1} > 0,$$

where the last inequality comes from the assumption $f > \log_p(C(g) + 1)$. This completes the proof.

(C) The genus 0 case. By means of the above results, we can prove that the isomorphism class of the scheme U can be recovered group-theoretically from the tame fundamental group $\pi_1^{\mathrm{t}}(U)$ in the case where $g = 0$ and $k = \overline{\mathbb{F}}_p$. More precisely, we have:

THEOREM (5.8). *Let k be an algebraically closed field of characteristic > 0 and F the algebraic closure of $\mathbb{F}_p$ in k. Let U be a smooth, connected curve over k. For each given smooth, connected curve U_0 over F whose smooth compactification is of genus 0 and whose number of punctures is greater than 1, we can detect whether U is isomorphic to $U_0 \otimes_F k$ as a scheme or not, group-theoretically from $\pi_1^{\mathrm{t}}(U)$.*

COROLLARY (5.9). *For each $i = 1, 2$, let k_i be an algebraically closed field of characteristic > 0 and U_i a smooth, connected curve over k_i. Let (g_i, n_i) denote (g, n) for U_i. Assume $k_1 \simeq k_2$. For some $i = 1, 2$, assume that $g_i = 0$, $n_i > 1$,* and either (a) U_i is defined over F_i, the algebraic closure of $\mathbb{F}_p$ in k_i or (b) $n_i \leq 4$. Then $\pi_1^t(U_1)$ and $\pi_1^t(U_2)$ are isomorphic as topological groups if and only if U_1 and U_2 are isomorphic as schemes.

PROOF OF (5.8) AND (5.9). With (4.1), (5.2), (5.3), etc., the same proofs as those of [T2], Theorem 3.5 and Corollary 3.6 work for $\pi_1^t(U)$. $\square$

Appendix: Proof of (4.17)

First, we recall some notations in the text. Let k be an algebraically closed field of characteristic $p > 0$, and let U be a smooth, connected curve over k. We denote by X the smooth compactification of U and put $S = X - U$. We define non-negative integers g and n to be the genus of X and the cardinality of the point set S, respectively. We put

$$g' \overset{\text{def}}{=} \begin{cases} g - 1, & \text{if } n \leq 1, \\ g, & \text{if } n > 1. \end{cases}$$

Moreover, see Section 4 for the definition of the i-th Betti number $b^{(i)}$ of U.

In Section 4, we introduced the invariant γ_N^{av} of U for each natural number N prime to p, as a certain average of generalized Hasse–Witt invariants of N-cyclic étale coverings of U. Now, the following is the main result of this Appendix.

THEOREM (A.1).   Assume $b^{(1)} > 0$ (or, equivalently, $(g, n) \neq (0, 0), (0, 1)$). Then (4.17) holds, that is, we have

$$\limsup_{\substack{N \to \infty \\ p \nmid N}} \gamma_N^{\mathrm{av}} \leq g'.$$

We devote the rest of this Appendix to proving (A.1).

Let N be a natural number prime to p. Recall that for each divisor D in $(\mathbb{Z}/N\mathbb{Z})^{\sim}[S]^0$, $s(D) \overset{\text{def}}{=} \deg(D)/N$ is an integer with $0 \leq s(D) \leq n - 1 + b^{(2)}$. Moreover, for each integer a, we denote by $D(a)$ the element of $(\mathbb{Z}/N\mathbb{Z})^{\sim}[S]^0$ that is equivalent to aD modulo N. Let f be the order of $p \bmod N$ in the multiplicative group $(\mathbb{Z}/N\mathbb{Z})^{\times}$. We put

$$M_N \overset{\text{def}}{=} \#\{D \in (\mathbb{Z}/N\mathbb{Z})^{\sim}[S]^0 \mid s(D(p^j)) \leq 1 \text{ for some } j = 0, 1, \dots f - 1\}$$

and $E_N \overset{\text{def}}{=} \#((\mathbb{Z}/N\mathbb{Z})^{\sim}[S]^0) - M_N = N^{n-1+b^{(2)}} - M_N$. ($M$ and E mean "main term" and "error term".)

Now, we have the following:

LEMMA (A.2).   (i) If $n \leq 1$, we have

$$\gamma_N^{\mathrm{av}} \leq g - 1 + \frac{1}{N^{b^{(1)}}}.$$

(ii) If $n > 1$, we have

$$\gamma_N^{\mathrm{av}} \leq g + (b^{(1)} - 1)\frac{E_N}{N^{n-1}}.$$

PROOF. (i) Just the same as the first half of the proof of (4.9)(i).

(ii) Let $[V]$ be an element of $H^1_{\text{ét}}(U, \boldsymbol{\mu}_N)$, and $([L], D)$ the element of $\tilde{P}_N$ corresponding to $[V]$. (See Section 3, especially (3.5) and paragraphs preceding it.) Then, as in the proof of (3.16), the remark just before the definition of $\gamma_{[V],i}$ (at the beginning of Section 3) and (3.8) imply

$$\gamma_{[V],1} = \gamma_{[V],p^j} = \gamma_{([L(p^j)],D(p^j))}$$

for each i. So, if $s(D(p^j)) \leq 1$ for some $j = 0, 1, \ldots, f-1$, we have $\gamma_{[V],1} \leq g$, as in (3.7). From this, we obtain

$$\sum_{[V] \in H^1_{\text{ét}}(U,\boldsymbol{\mu}_N)} \gamma_{[V],1} \leq g N^{2g} M_N + (g + n - 2) N^{2g} E_N = g N^{b^{(1)}} + (n-2) N^{2g} E_N$$

by (3.4) and (3.7). Thus we have

$$\gamma_N^{\text{av}} \leq g + (n-2) \frac{E_N}{N^{n-1}} \leq g + (b^{(1)} - 1) \frac{E_N}{N^{n-1}},$$

as desired. □

(A.2)(i) settles the proof of (A.1) for $n \leq 1$, while (A.2)(ii) reduces the proof of (A.1) for $n > 1$ to

$$E_N = o(N^{n-1}), \quad \text{that is,} \quad \lim_{\substack{N \to \infty \\ p \nmid N}} \frac{E_N}{N^{n-1}} = 0. \tag{A.3}$$

Note that (A.3) depends only on the finite set S, so it no longer involves the geometry of U.

To prove (A.3), we need some knowledge of the theory of uniform distribution modulo 1, which we shall recall here. (For more details, see [KN].)

DEFINITION. (i) $I \overset{\text{def}}{=} [0, 1) = \{x \in \mathbb{R} \mid 0 \leq x < 1\}$. For each $x \in \mathbb{R}$, we denote by $\{x\}$ the fractional part $x - [x] \in I$ of x.

(ii) Let s be a positive integer. Let $\mathbf{a} = (a_1, \ldots, a_s)$ and $\mathbf{b} = (b_1, \ldots, b_s)$ be elements of $\mathbb{R}^s$. We say that $\mathbf{a} < \mathbf{b}$ (resp. $\mathbf{a} \leq \mathbf{b}$) if $a_i < b_i$ (resp. $a_i \leq b_i$) for each $i = 1, \ldots, s$. We put

$$[\mathbf{a}, \mathbf{b}) \overset{\text{def}}{=} \{\mathbf{x} \in \mathbb{R}^s \mid \mathbf{a} \leq \mathbf{x} < \mathbf{b}\}.$$

If $\mathbf{a} \leq \mathbf{b}$, the Lebesgue measure $\lambda([\mathbf{a}, \mathbf{b}))$ of $[\mathbf{a}, \mathbf{b})$ is given by $(b_1 - a_1) \ldots (b_s - a_s)$. Note that $I^s = [\mathbf{0}, \mathbf{1})$, where $\mathbf{0} = (0, \ldots, 0), \mathbf{1} = (1, \ldots, 1)$.

For each $\mathbf{x} = (x_1, \ldots, x_s) \in \mathbb{R}^s$, we put

$$\{\mathbf{x}\} \overset{\text{def}}{=} (\{x_1\}, \ldots, \{x_s\}),$$

$$\|\mathbf{x}\| \overset{\text{def}}{=} \max_{i=1,\ldots,s} |x_i|,$$

$$r(\mathbf{x}) \overset{\text{def}}{=} \prod_{\substack{i=1,\ldots,s \\ x_i \neq 0}} |x_i| \quad (r(\mathbf{0}) = 1).$$

For each $\mathbf{x} = (x_1, \ldots, x_s), \mathbf{y} = (y_1, \ldots, y_s) \in \mathbb{R}^s$, we put

$$\langle \mathbf{x}, \mathbf{y} \rangle \overset{\text{def}}{=} x_1 y_1 + \ldots + x_s y_s.$$

(iii) Let $\mathbf{x}_1, \ldots, \mathbf{x}_M$ be a sequence of length M of elements of $\mathbb{R}^s$. For each subset E of I^s, put

$$A(E; M; \mathbf{x}_1, \ldots, \mathbf{x}_M) \overset{\text{def}}{=} \#\{j = 1, \ldots, M \mid \{\mathbf{x}_j\} \in E\}.$$

Moreover, we define the discrepancy $\mathcal{D}_M$ of the sequence $\mathbf{x}_1, \ldots, \mathbf{x}_M$ by

$$\mathcal{D}_M = \mathcal{D}_M(\mathbf{x}_1, \ldots, \mathbf{x}_M) \overset{\text{def}}{=} \sup_J \left| \frac{A(J; M; \mathbf{x}_1, \ldots, \mathbf{x}_M)}{M} - \lambda(J) \right|,$$

where J runs over the subsets of I^s in the form $[\mathbf{a}, \mathbf{b})$ with $\mathbf{a}, \mathbf{b} \in \mathbb{R}^s, 0 \le \mathbf{a} < \mathbf{b} \le 1$. (Observe that $0 \le \mathcal{D}_M \le 1$.)

Now, we can state the following higher-dimensional version of LeVeque's inequality, due to Stegbuchner.

THEOREM (STEGBUCHNER [S]). *Let* $\mathbf{x}_1, \ldots, \mathbf{x}_M$ *be a sequence of length* M *of elements of* $\mathbb{R}^s$. *Then*

$$\mathcal{D}_M(\mathbf{x}_1, \ldots, \mathbf{x}_M) \le \left(C_s \sum_{\mathbf{h} \in \mathbb{Z}^s - \{\mathbf{0}\}} \frac{1}{r(\mathbf{h})^2} \left| \frac{1}{M} \sum_{j=1}^M e^{2\pi i \langle \mathbf{h}, \mathbf{x}_j \rangle} \right|^2 \right)^{1/(s+2)},$$

where C_s *is a positive constant depending only on* s. *More precisely, we may take* $C_s = s^{2(s+2)} 2^s 9^{s^2 + 3s + 1}$. □

For various improvements of the constant C_s, see, e.g., [GT], Theorem 3 and [DT], Theorem 1.28.

As in the proof of (3.18)(ii), we choose any $Q \in S$ and put $S' = S - \{Q\}$, whose cardinality is $n' \overset{\text{def}}{=} n - 1(> 0)$. The projection $(\mathbb{Z}/N\mathbb{Z})^\sim[S]^0 \to (\mathbb{Z}/N\mathbb{Z})^\sim[S'], D \mapsto D'$ is bijective. For each $a \in \mathbb{Z}$, we have $D(a)' = D'(a)$. Moreover, we see that

$$s(D) = \frac{1}{N} \deg(D) = \left\lceil \frac{1}{N} \deg(D') \right\rceil,$$

where $\lceil x \rceil$ denotes the smallest integer not less than x. Therefore, we have

$$s(D) \le 1 \iff \deg(D') \le N.$$

Taking S' as a basis, we may identify $(\mathbb{Z}/N\mathbb{Z})^\sim[S'] = ((\mathbb{Z}/N\mathbb{Z})^\sim)^{n'} \subset \mathbb{Z}^{n'}$.

Then, for each $D \in (\mathbb{Z}/N\mathbb{Z})^{\sim}[S]^0$, we can apply Stegbuchner's theorem to $s = n'$, $M = f$, and $\mathbf{x}_j = D(p^{j-1})'/N$ and obtain

$$\mathcal{D}_f(D) \overset{\text{def}}{=} \mathcal{D}_f\left(\frac{D(p^0)'}{N}, \ldots, \frac{D(p^{f-1})'}{N}\right)$$

$$\leq \left(C_{n'} \sum_{\mathbf{h} \in \mathbb{Z}^{n'}-\{\mathbf{0}\}} \frac{1}{r(\mathbf{h})^2} \left|\frac{1}{f}\sum_{j=0}^{f-1} e^{2\pi i \langle \mathbf{h}, \frac{D(p^j)'}{N}\rangle}\right|^2\right)^{1/(n'+2)}$$

$$\leq \left(C_{n'} \sum_{\mathbf{h} \in \mathbb{Z}^{n'}-\{\mathbf{0}\}} \frac{1}{r(\mathbf{h})^2} \left|\frac{1}{f}\sum_{j=0}^{f-1} e^{2\pi i \frac{p^j \langle \mathbf{h}, D'\rangle}{N}}\right|^2\right)^{1/(n'+2)}.$$

So

$$\sum_{D \in (\mathbb{Z}/N\mathbb{Z})^{\sim}[S]^0} \mathcal{D}_f(D)^{n'+2} \leq C_{n'} \sum_{D' \in ((\mathbb{Z}/N\mathbb{Z})^{\sim})^{n'}} \sum_{\mathbf{h} \in \mathbb{Z}^{n'}-\{\mathbf{0}\}} \frac{1}{r(\mathbf{h})^2} \left|\frac{1}{f}\sum_{j=0}^{f-1} e^{2\pi i \frac{p^j \langle \mathbf{h}, D'\rangle}{N}}\right|^2$$

$$= C_{n'} \sum_{\mathbf{h} \in \mathbb{Z}^{n'}-\{\mathbf{0}\}} \frac{1}{r(\mathbf{h})^2} \sum_{D' \in ((\mathbb{Z}/N\mathbb{Z})^{\sim})^{n'}} \left|\frac{1}{f}\sum_{j=0}^{f-1} e^{2\pi i \frac{p^j \langle \mathbf{h}, D'\rangle}{N}}\right|^2.$$

Here, we have

$$\left|\frac{1}{f}\sum_{j=0}^{f-1} e^{2\pi i \frac{p^j \langle \mathbf{h}, D'\rangle}{N}}\right|^2 = \left(\frac{1}{f}\sum_{j=0}^{f-1} e^{2\pi i \frac{p^j \langle \mathbf{h}, D'\rangle}{N}}\right)\overline{\left(\frac{1}{f}\sum_{j=0}^{f-1} e^{2\pi i \frac{p^j \langle \mathbf{h}, D'\rangle}{N}}\right)}$$

$$= \frac{1}{f^2}\sum_{j,j'=0}^{f-1} e^{2\pi i \frac{(p^j - p^{j'})\langle \mathbf{h}, D'\rangle}{N}}.$$

Now, since

$$\chi_{\mathbf{h},j,j'} : D' \bmod N \mapsto e^{2\pi i \frac{(p^j - p^{j'})\langle \mathbf{h}, D'\rangle}{N}}$$

is a character of the abelian group $(\mathbb{Z}/N\mathbb{Z})^{n'}$, we have

$$\sum_{D' \in ((\mathbb{Z}/N\mathbb{Z})^{\sim})^{n'}} \chi_{\mathbf{h},j,j'}(D') = \begin{cases} 0, & \text{if } \chi_{\mathbf{h},j,j'} \neq 1, \\ N^{n'}, & \text{if } \chi_{\mathbf{h},j,j'} = 1. \end{cases}$$

Moreover, we have

$$\chi_{\mathbf{h},j,j'} = 1 \iff (p^j - p^{j'})\mathbf{h} \equiv \mathbf{0} \pmod{N} \iff f_{\mathbf{h}} \mid (j - j').$$

Here, $f_{\mathbf{h}}$ is the order of p mod $N_{\mathbf{h}}$ in the multiplicative group $(\mathbb{Z}/N_{\mathbf{h}}\mathbb{Z})^{\times}$, where $N_{\mathbf{h}}$ is the order of $\mathbf{h}$ mod $N \in (\mathbb{Z}/N\mathbb{Z})^{n'}$. Thus, in summary, we get

$$\sum_{D \in (\mathbb{Z}/N\mathbb{Z})^{\sim}[S]^0} \mathcal{D}_f(D)^{n'+2}$$

$$\leq C_{n'} \sum_{\mathbf{h} \in \mathbb{Z}^{n'}-\{\mathbf{0}\}} \frac{1}{r(\mathbf{h})^2} \frac{1}{f^2} \sum_{j,j'=0}^{f-1} \sum_{D' \in ((\mathbb{Z}/N\mathbb{Z})^\sim)^{n'}} \chi_{\mathbf{h},j,j'}(D')$$

$$= C_{n'} \sum_{\mathbf{h} \in \mathbb{Z}^{n'}-\{\mathbf{0}\}} \frac{1}{r(\mathbf{h})^2} \frac{1}{f^2} \#(\{(j,j') \mid j,j' = 0,\ldots,f-1, f_{\mathbf{h}} \mid (j-j')\}) N^{n'}$$

$$= C_{n'} \sum_{\mathbf{h} \in \mathbb{Z}^{n'}-\{\mathbf{0}\}} \frac{1}{r(\mathbf{h})^2} \frac{1}{f^2} \frac{f^2}{f_{\mathbf{h}}} N^{n'}$$

$$= C_{n'} N^{n'} \sum_{\mathbf{h} \in \mathbb{Z}^{n'}-\{\mathbf{0}\}} \frac{1}{r(\mathbf{h})^2 f_{\mathbf{h}}}.$$

Taking a positive integer K (which we fix later), we divide the last infinite sum into the sum of the infinite sum with $\|\mathbf{h}\| > K$ and the finite sum with $\|\mathbf{h}\| \leq K$. For the former, we have

$$\sum_{\|\mathbf{h}\|>K} \frac{1}{r(\mathbf{h})^2 f_{\mathbf{h}}} \leq \sum_{\|\mathbf{h}\|>K} \frac{1}{r(\mathbf{h})^2}$$

$$\leq \sum_{i=1}^{n'} \sum_{\mathbf{h} \text{ s.t. } |h_i|>K} \frac{1}{r(\mathbf{h})^2}$$

$$= \sum_{i=1}^{n'} \left(\sum_{|h_i|>K} \frac{1}{|h_i|^2} \right) \prod_{j \neq i} \left(\sum_{h_j \in \mathbb{Z}} \frac{1}{\max(|h_j|,1)^2} \right)$$

$$\leq \sum_{i=1}^{n'} \left(2 \int_K^\infty \frac{dx}{x^2} \right) (1 + 2\zeta(2))^{n'-1}$$

$$= 2n'(1 + 2\zeta(2))^{n'-1} \frac{1}{K}.$$

For the latter, we need an estimate of $f_{\mathbf{h}}$. Since $N \mid N_{\mathbf{h}}\mathbf{h}$, we have $N_{\mathbf{h}}\|\mathbf{h}\| \geq N$, unless $\mathbf{h} = \mathbf{0}$. So, if $\|\mathbf{h}\| \leq K$ and $\mathbf{h} \neq \mathbf{0}$, we have $N_{\mathbf{h}} \geq N/K$. On the other hand, since $N_{\mathbf{h}} \mid p^{f_{\mathbf{h}}} - 1 < p^{f_{\mathbf{h}}}$, we have $f_{\mathbf{h}} \geq \log(N_{\mathbf{h}})/\log(p)$. These two inequalities imply

$$f_{\mathbf{h}} \geq \frac{\log(N/K)}{\log(p)}.$$

Therefore, we have (assuming $N/K > 1$)

$$\sum_{0<\|\mathbf{h}\|\leq K} \frac{1}{r(\mathbf{h})^2 f_{\mathbf{h}}} \leq \frac{\log(p)}{\log(N/K)} \sum_{0<\|\mathbf{h}\|\leq K} \frac{1}{r(\mathbf{h})^2}$$

$$\leq \frac{\log(p)}{\log(N/K)} \{(1 + 2\zeta(2))^{n'} - 1\}.$$

Now, fix any real number δ with $0 < \delta < 1$ and put $K = [N^\delta]$. Then we have

$$\frac{1}{K} \leq \frac{1}{N^\delta - 1} \text{ and } \frac{1}{\log(N/K)} \leq \frac{1}{\log(N/N^\delta)} = \frac{1}{(1-\delta)\log(N)}.$$

Thus we conclude that

$$\sum_{\mathbf{h} \in \mathbb{Z}^{n'} - \{0\}} \frac{1}{r(\mathbf{h})^2 f_{\mathbf{h}}} \leq 2n'(1 + 2\zeta(2))^{n'-1} \frac{1}{N^\delta - 1} + \frac{\log(p)\{(1 + 2\zeta(2))^{n'} - 1\}}{(1-\delta)\log(N)}.$$

From this, we finally obtain

$$\sum_{D \in (\mathbb{Z}/N\mathbb{Z})^{\sim}[S]^0} \mathcal{D}_f(D)^{n'+2} \leq \left(c_1(n') \frac{1}{N^\delta - 1} + c_2(n',p,\delta) \frac{1}{\log(N)} \right) N^{n'},$$

where $c_1(n')$ (resp. $c_2(n',p,\delta)$) is a positive constant depending only on n' (resp. n', p, and δ).

On the other hand, let D be an element of $(\mathbb{Z}/N\mathbb{Z})^{\sim}[S]^0$ such that $s(D(p^j)) > 1$ for all $j = 0, 1, \ldots f-1$. Then, in particular, $\frac{D(p^j)'}{N} \notin [0, 1/n')^{n'}$ for all such j. So, by the definition of discrepancy, we have

$$\mathcal{D}_f(D) \geq \lambda([0, 1/n')^{n'}) = 1/(n')^{n'}.$$

From this, we obtain

$$\sum_{D \in (\mathbb{Z}/N\mathbb{Z})^{\sim}[S]^0} \mathcal{D}_f(D)^{n'+2} \geq c_3(n') E_N,$$

where $c_3(n') = 1/(n')^{n'(n'+2)}$ is a positive constant depending only on n'.

Now, we can conclude that

$$0 \leq \frac{E_N}{N^{n'}} \leq d_1(n') \frac{1}{N^\delta - 1} + d_2(n',p,\delta) \frac{1}{\log(N)},$$

where $d_1(n') = c_1(n')/c_3(n')$ and $d_2(n',p,\delta) = c_2(n',p,\delta)/c_3(n')$ are positive constants independent of N. Since $\delta > 0$, this implies (A.3).

This completes the proof of (A.1).

References

[B] I. Bouw, "Tame covers of curves: p-ranks and fundamental groups", thesis, Univ. Utrecht, 1998.

[BH] W. Bruns and J. Herzog, *Cohen–Macaulay rings*, Cambridge Studies in Advanced Mathematics **39**, Cambridge University Press, Cambridge, 1993.

[DT] M. Drmota and R. F. Tichy, *Sequences, discrepancies and applications*, Lecture Notes in Mathematics **1651**, Springer, Berlin, 1997.

[E] D. Eisenbud, *Commutative algebra with a view toward algebraic geometry*, Graduate Texts in Mathematics **150**, Springer, New York, 1994.

[EGA4] A. Grothendieck, "Éléments de Géométrie Algébrique IV: Étude locale des schémas et des morphismes de schémas", *Publications Mathématiques de l'IHES* **20, 24, 28, 32**, 1964–1967.

[FJ] M. D. Fried and M. Jarden, *Field arithmetic*, Ergebnisse der Mathematik und ihrer Grenzgebiete (3. Folge) **11**, Springer, Berlin and New York, 1986.

[GT] P. J. Grabner and R. F. Tichy, "Remark on an inequality of Erdös–Turán–Koksma", *Anz. Österreich. Akad. Wiss. Math.-Natur. Kl.* **127** (1990), 15–22.

[GM] A. Grothendieck and J. P. Murre, *The tame fundamental group of a formal neighbourhood of a divisor with normal crossings on a scheme*, Lecture Notes in Mathematics **208**, Springer, Berlin and New York, 1971.

[Ka] H. Katsurada, "On generalized Hasse–Witt invariants and unramified Galois extensions of an algebraic function field", *J. Math. Soc. Japan* **31** (1979), 101–125.

[KN] L. Kuipers and H. Niederreiter, *Uniform distribution of sequences*, Wiley, New York-London-Sydney, 1974.

[Na] S. Nakajima, "On generalized Hasse–Witt invariants of an algebraic curve", pp. 69–88 in *Galois groups and their representations* (Nagoya, 1981), Advanced Studies in Pure Mathematics **2**, edited by Y. Ihara, North-Holland, Amsterdam, and Kinokuniya, Tokyo, 1983.

[Mac] F. S. Macaulay, *The algebraic theory of modular systems*, Cambridge University Press, Cambridge, 1916.

[Mad] D. A. Madore, "Theta divisors and the Frobenius morphism", pp. 279–289 in *Courbes semi-stables et groupe fondamental en géométrie algébrique* (Luminy, 1998) Progr. Math. **187**, edited by J.-B. Bost, F. Loeser and M. Raynaud, Birkhäuser, Basel, 2000.

[Mi1] J. S. Milne, *Étale cohomology*, Princeton Mathematical Series **33**, Princeton Univ. Press, Princeton, New Jersey, 1980.

[Mi2] ———, "Jacobian varieties", pp. 167–212 in *Arithmetic geometry* (Storrs, 1984), edited by G. Cornell and J. H. Silverman, Springer, New York-Berlin, 1986.

[Mu] D. Mumford, *Abelian varieties*, Oxford University Press, London, 1970.

[R1] M. Raynaud, "Sections des fibrés vectoriels sur une courbe", *Bull. Soc. math. France* **110** (1982), 103–125.

[R2] ———, "Revêtements des courbes en caractéristique $p > 0$ et ordinarité", *Compositio Math.* **123** (2000), 73–88.

[S] H. Stegbuchner, "Eine mehrdimensionale Version der Ungleichung von LeVeque", *Monatsh. Math.* **87** (1979), 167–169.

[SGA1] A. Grothendieck and Mme. M. Raynaud, *Revêtements étales et groupe fondamental, Séminaire de Géometrie Algébrique du Bois Marie* 1960–61 (SGA 1), Lecture Notes in Mathematics **224**, Springer, Berlin and New York, 1971.

[SGA6] P. Berthelot, A. Grothendieck and L. Illusie, *Théorie des intersections et théorème de Riemann–Roch, Séminaire de Géométrie Algébrique du Bois Marie* 1966–1967 (SGA 6), Lecture Notes in Mathematics **225**, Springer, Berlin and New York, 1971.

[SGA7I] A. Grothendieck, M. Raynaud and D. S. Rim, *Groupes de monodromie en géométrie algébrique, Séminaire de Géométrie Algébrique du Bois Marie 1967–1969* (SGA 7I), Lecture Notes in Mathematics **288**, Springer, Berlin and New York, 1972.

[T1] A. Tamagawa, "The Grothendieck conjecture for affine curves", *Compositio Math.* **109** (1997), 135–194.

[T2] _____, "On the fundamental groups of curves over algebraically closed fields of characteristic > 0", *Internat. Math. Res. Notices* (1999), no. 16, 853–873.

[T3] _____, "Fundamental groups and geometry of curves in positive characteristic", pp. 297–333 in *Arithmetic fundamental groups and noncommutative algebra* (Berkeley, 1999), Proceedings of Symposia in Pure Mathematics **70**, edited by M. D. Fried and Y. Ihara, American Mathematical Society, Providence, 2002.

[Y] Y. Yoshino, *Cohen–Macaulay modules over Cohen–Macaulay rings*, London Mathematical Society Lecture Note Series **146**, Cambridge University Press, Cambridge, 1990.

AKIO TAMAGAWA
RESEARCH INSTITUTE FOR MATHEMATICAL SCIENCES
KYOTO UNIVERSITY
KYOTO 606-8502
JAPAN
  tamagawa@kurims.kyoto-u.ac.jp

Galois Groups and Fundamental Groups
MSRI Publications
Volume 41, 2003

On the Specialization Homomorphism of Fundamental Groups of Curves in Positive Characteristic

FLORIAN POP AND MOHAMED SAïDI

Introduction

Recall that for proper smooth and connected curves of genus $g \geq 2$ over an algebraically closed field of characteristic 0 the structure of the étale fundamental group π_g is well known and depends only on the genus g. Namely it is the profinite completion of the topological fundamental group of a compact orientable topological surface of genus g. In contrast to this, the structure of the étale fundamental group of proper smooth and connected curves of genus $g \geq 2$ in positive characteristic is unknown, and it depends on the isomorphy type of the curve in discussion. The aim of this paper is to give new evidence for anabelian phenomena for proper curves over algebraically closed fields of characteristic $p > 0$.

Before going into the details of the results we are going to prove, we set some notation and recall well known facts. Let k be an algebraically closed field of characteristic $p > 0$. Let X be a projective smooth and connected curve of genus $g \geq 2$ over k, and let J be the Jacobian of X. We denote by $\pi_1(X)$, $\pi_1^p(X)$, and $\pi_1^{p'}(X)$ the étale fundamental group of X, its pro-p quotient, and its prime to p quotient. Then:

(1) The structure of $\pi_1^p(X)$ is given by Shafarevich's Theorem; see [Sh]. It is isomorphic to the pro-p free group on $r := r_X$ generators, where r_X is the p-rank of J.

(2) The structure of $\pi_1^{p'}(X)$ is well known by Grothendieck's Specialization Theorem; see [SGA-1]. It is the prime to p completion of the topological fundamental group of a compact orientable topological surface of genus g.

(3) In contrast to this, the structure of the whole fundamental group $\pi_1(X)$ is a big mystery! Its structure is not known in any single case. However, by Grothendieck's Specialization Theorem we know that $\pi_1(X)$ is the quotient of

107

the profinite completion Π_g of the topological fundamental group of a compact orientable topological surface of genus g. In particular $\pi_1(X)$ is topologically finitely generated. Since such groups are completely determined by the set of their finite quotients, another interpretation of (1) is the following:

- If two curves as above have the same p-rank, then there is a bijection between the set of their Galois étale covers with Galois group a p-group.

- In the same way, the interpretation of (2) is that for two curves of the same genus there is a bijection between the set of their Galois étale covers having a Galois group of order prime to p.

In order to approach the complexity of π_1 of proper curves in positive characteristic we introduce the following: Let $M_g \to \operatorname{Spec} \mathbb{F}_p$ be the coarse moduli space of proper and smooth curves of genus g in characteristic p. It is well known that M_g is a quasi-projective and geometrically irreducible variety. Let k be an algebraically closed field of characteristic p; thus $M_g(k)$ is the set of isomorphism classes of curves of genus g over k. For $\bar{x} \in M_g(k)$ let $C_{\bar{x}} \to \operatorname{Spec} k$ be a curve classified by $\bar{x}$, and let $x \in M_g$ such that $\bar{x} : \operatorname{Spec} k \to M_g$ factors through x. We set

$$\pi_1(x) := \pi_1(C_{\bar{x}}), \quad \pi_1^p(x) := \pi_1^p(C_{\bar{x}}), \quad \pi_1^{p'}(x) = \pi_1^{p'}(C_{\bar{x}}).$$

We remark that the structure of $\pi_1(x)$ as a profinite group depends only on x and not on the concrete geometric point $\bar{x} \in M_g(k)$ used to define it. Indeed, let κ be the algebraic closure of the residue field $\kappa(x)$ at x in k. Then, if C_x is the curve classified by $\operatorname{Spec} \kappa \to M_g$, then $C_{\bar{x}}$ is the base change $C_{\bar{x}} \simeq C_x \times_\kappa k$ of C_x to k. Hence $\pi_1(C_{\bar{x}}) \simeq \pi_1(C_x)$ by the geometric invariance of the fundamental group for proper varieties; see [SGA-1]. Second, the isomorphy type of C_x as an $\mathbb{F}_p$-scheme does depend only on x, and not the concrete choice of the algebraic closure κ of $\kappa(x)$.

We further remark that by the comments above, if $J_{\bar{x}}$ is the Jacobian of $C_{\bar{x}}$, then the p-rank of $J_{\bar{x}}$ as well as $J_{\bar{x}}$ being a simple abelian variety depends only on x and not on the geometric point $\bar{x}$. Indeed, in the notations above, if J_x is the Jacobian of C_x, then $J_{\bar{x}} \simeq J_x \times_\kappa k$; and for different choices of the algebraic closure of $\kappa(x)$, the corresponding curves are isomorphic as $\mathbb{F}_p$-schemes. Hence their Jacobians too are isomorphic as $\mathbb{F}_p$-schemes.

Coming back to the fundamental group we thus have maps

$$\pi_1 : M_g \to (\text{Prof.groups}), \quad x \to \pi_1(x),$$

and the induced maps

$$\pi_1^p : M_g \to (\text{Prof.groups}), \quad x \to \pi_1^p(x)$$

and

$$\pi_1^{p'} : M_g \to (\text{Prof.groups}), \quad x \to \pi_1^{p'}(x),$$

where (Prof.groups) are the objects of the category of profinite groups. The last two maps are not very interesting: first, the isomorphy type of the images of π_1^p depends only on the p-rank; and second, the isomorphy type is constant on the image of $\pi_1^{p'}$.

To finish our preparation we remark that for points $x, y \in M_g$ such that x is a specialization of y, by Grothendieck's specialization theorem there exists a surjective continuous homomorphism $\mathrm{Sp} : \pi_1(y) \to \pi_1(x)$. In particular, if η is the generic point of M_g, then C_η is the generic curve of genus g; and every point x of M_g is a specialization of η. Thus, for every $x \in M_g$, there is a surjective homomorphism $\mathrm{Sp}_x : \pi_1(\eta) \to \pi_1(x)$ which is determined up to Galois-conjugacy by the choice of the local ring of x in the algebraic closure of $\kappa(\eta)$. For every $x \in M_g$ we fix such a map once for all; in particular, if x is a specialization of y, there exists a specialization homomorphism $\mathrm{Sp}_{y,x} : \pi_1(y) \to \pi_1(x)$ such that $\mathrm{Sp}_{y,x} \circ \mathrm{Sp}_y = \mathrm{Sp}_x$. (In order to obtain $\mathrm{Sp}_{y,x}$ one has to choose the local ring of x to be contained in the local ring of y.)

Finally, let $S^{\mathrm{a.s.}} \subset M_g$ be the set of *closed points* corresponding to curves C_x having an absolutely simple Jacobian J_x. Further, let $S^{\mathrm{a.s.}}_{\geq g-1} \subset S^{\mathrm{a.s.}}$ be the subset of points $x \in S^{\mathrm{a.s.}}$ such that the p-rank of C_x equals g or $g-1$. Concerning the set $S^{\mathrm{a.s.}}$, Chai and Oort proved the following (see [Se-1] for facts concerning Dirichlet density):

THEOREM ([CH-OO]). *The subset $S^{a.s.}$ is non empty and has a positive Dirichlet density. In particular, $S^{a.s.}$ is Zariski dense.*

We now come to the main results of the present article. We remark that for genus $g = 2$, even stronger results were proven by Raynaud. This is *Raynaud's theory of the theta divisor* of the sheaf of locally exact differentials for curves in positive characteristic; see [Ra-1] the main tool that we use in our approach.

THEOREM A. *For all points $s \in S^{a.s.}$, the specialization homomorphism $\mathrm{Sp}_s : \pi_1(\eta) \to \pi_1(s)$ is not an isomorphism.*

  More precisely, every cyclic étale cover of X_η of order prime to p is ordinary, whereas there exist such covers of C_s that are not ordinary.

THEOREM B. *If a point $y \in M_g$ specializes to some point $s \in S^{a.s.}_{\geq g-1}$ with $s \neq y$, then the specialization homomorphism $\mathrm{Sp}_{y,s} : \pi_1(y) \to \pi_1(s)$ is not an isomorphism.*

  In particular, for a given point $s \in S^{a.s.}_{\geq g-1}$ there exist only finitely many points $s' \in S^{a.s.}_{\geq g-1}$ such that $\pi_1(s') \simeq \pi_1(s)$.

As an application we have the following corollary answering a question raised by David Harbater:

COROLLARY. *There is no nonempty open subset $U \subset M_g$ such that the isomorphy type of the geometric fundamental group $\pi_1(x)$ is constant on U.*

We conclude with a question:

QUESTION. *Is it true that the specialization homomorphism*

$$\mathrm{Sp} : \pi_1(y) \to \pi_1(x)$$

to points $y \neq x$ with x closed is never an isomorphism?

If this is the case the same proof as that of Corollary 4.4 below would imply the following finiteness result: Given a closed point $x \in M_g$ there exists at most finitely many closed points x' in M_g such that $\pi_1(x') \simeq \pi_1(x)$.

One could ask the preceding question more generally, without the condition that the point x be closed. However, the condition that x be closed is essential in the proof of our results.

Acknowledgments. Parts of this paper were written while Saïdi was a post-doc in the Graduiertenkolleg of the Mathematics Institute, University of Bonn, from September 1998 to June 1999. He would like very much to thank the members of the Mathematics Institute for their hospitality and the very good working conditions. The manuscript was completed during the fall 1999 at the Mathematical Sciences Research Institute in Berkeley. The authors would like to express their gratitude for the support from MSRI and the wonderful working conditions there.

1. Preliminaries and Notations

1.1. The sheaf of locally exact differentials in characteristic p > 0 and the associated theta divisor. We recall here the definition of the sheaf of locally exact differentials associated to an algebraic curve in positive characteristic and its associated theta divisor, mainly following Raynaud (see [Ra-1], 4). Let X be a proper smooth and connected algebraic curve of genus $g_X := g \geq 2$, over an algebraically closed field k of characteristic $p > 0$. Consider the Cartesian diagram

$$\begin{array}{ccc} X^1 & \longrightarrow & X \\ \downarrow & & \downarrow \\ \mathrm{Spec}\, k & \xrightarrow{\ F\ } & \mathrm{Spec}\, k \end{array}$$

where F denotes the absolute Frobenius morphism. The projection $X^1 \to X$ is a scheme isomorphism, in particular X^1 is a smooth and proper curve of genus g. The absolute Frobenius morphism $F : X \to X$ induces in a canonical way a morphism $\pi : X \to X^1$ called the *relative Frobenius* which is a radicial morphism of k-curves of degree p. The canonical differential $\pi_* d : \pi_* \mathcal{O}_X \to \pi_* \Omega^1_X$ is a morphism of $\mathcal{O}_{X^1}$-modules. Its image $B_X := B := \mathrm{Im}(\pi_* d)$ is the *sheaf of locally exact differentials*. One has the exact sequence

$$0 \to \mathcal{O}_{X^1} \to \pi_* \mathcal{O}_X \to B \to 0,$$

and B is a vector bundle on X^1 of rank $p - 1$. Let $c : \pi_*(\Omega^1_X) \to \Omega^1_{X^1}$ be the *Cartier operator*; this is a morphism of $\mathcal{O}_{X^1}$-modules. The kernel $\ker(c)$ of c is equal to B, and the following sequence of $\mathcal{O}_{X^1}$-modules is exact (see [Se], 10):

$$0 \to B \to \pi_*(\Omega^1_X) \to \Omega^1_{X^1} \to 0$$

Let L be a *universal Poincaré bundle* on $X^1 \times_k J^1$ where $J^1 := \mathrm{Pic}^0(X^1)$ is the Jacobian of X^1. This is a line bundle such that its restriction to $X^1 \times \{a\}$ for any $a \in J^1(k)$ is isomorphic to the invertible sheaf $\mathcal{L}_a$ which is the image of a under the natural isomorphism $J^1(k) \simeq \mathrm{Pic}^0(X^1)$. Let $h : X^1 \times J^1 \to X^1$ and $f : X^1 \times J^1 \to J^1$ be the canonical projections. As $R^i f_*(h^*B \otimes L) = 0$ for $i \geq 2$, the total direct image $Rf_*(h^*B \otimes L)$ of $(h^*B \otimes L)$ by f can be realized by a complex $u : \mathcal{M}^0 \to \mathcal{M}^1$ of length 1, where $\mathcal{M}^0$ and $\mathcal{M}^1$ are vector bundles on J^1, $\ker u = R^0 f_*(h^*B \otimes L)$, and $\mathrm{coker}\, u = R^1 f_*(h^*B \otimes L)$. Moreover as the Euler-Poincaré characteristic $\chi(h^*B \otimes L) = 0$, the vector bundles $\mathcal{M}^0$ and $\mathcal{M}^1$ have the same rank. In [Ra-1], théorème 4.1.1, it has been proved that the determinant $\det u$ of u is not identically zero on J^1, hence one can consider the divisor $\theta := \theta_X$ on J^1, which is the positive Cartier divisor locally generated by $\det u$, it is the *theta divisor* associated to the vector bundle B (note that the definition of θ_X is independant on the above chosen complex u). By definition a point $a \in J^1(k)$ lies on the support of θ if and only if $H^0(X^1, B \otimes \mathcal{L}_a) \neq 0$.

1.2. p-Rank of cyclic étale covers of degree prime to p. We use the same notations as in 1.1. The *p-rank* r_X of X is the dimension of the maximal subspace of $H^1(X, \mathcal{O}_X)$ on which the absolute Frobenius F acts bijectively. By duality it is also the dimension of the maximal subspace of $H^0(X, \Omega^1_X)$ on which the Cartier operator c is bijective (see [Se-1], 10). The p-rank r_X of X is also the rank of the maximal pro-p-quotient $\pi_1^p(X)$ of the fundamental group $\pi_1(X)$ of X, which is a free pro-p-group (see [Sh]).

The relative Frobenius morphism $\pi : X \to X^1$ induces a "canonical" isomorphism $\pi_1(X) \to \pi_1(X^1)$ between fundamental groups (see [SGA-1]). In particular for any positive integer n which is prime to p one has a one to one correspondence between μ_n-torsors of X^1 and μ_n-torsors of X. More precisely the canonical homomorphism $H^1_{\mathrm{et}}(X^1, \mu_n) \to H^1_{\mathrm{et}}(X, \mu_n)$ induced by π is an isomorphism. Consider a μ_n-torsor $f : Y \to X$ with Y connected. By Kummer theory f is given by an invertible sheaf $\mathcal{L}$ of order n on X, and $Y := \mathrm{Spec}(\oplus_{i=0}^{n-1} \mathcal{L}^{\otimes i})$. Thus there exists an invertible sheaf $\mathcal{L}^1$ on X^1 of order n, such that if $f' : Y^1 \to X^1$ is the associated μ_n-torsor we have a Cartesian diagram

$$
\begin{array}{ccc}
Y & \xrightarrow{\;f\;} & X \\
{\scriptstyle \pi'}\downarrow & & \downarrow{\scriptstyle \pi} \\
Y^1 & \xrightarrow{\;f'\;} & X^1
\end{array}
$$

Let J_Y (resp. J_X) denote the Jacobian variety of Y (resp. the Jacobian of X). The morphism $f : Y \to X$ induces a homomorphism $f^* : J_X \to J_Y$ between Jacobians. Let $J^{\text{new}} := J_{Y/X}$ denote the quotient of J_Y by the image $f^*(J_X)$ of J_X, that is the *new part* of the Jacobian J_Y of Y with respect to the morphism f.

1.3. Definition. The μ_n-torsor $f : Y \to X$ is said to be *new-ordinary* if the new part J^{new} of the Jacobian of Y with respect to the morphism f is an ordinary abelian variety.

Since the dimension of the abelian variety J^{new} is $h = g_Y - g_X$, it follows that J^{new} is ordinary if the étale part of the kernel of the multiplication by p in J^{new} has order p^h. This is also equivalent to the fact that the absolute Frobenius F acts bijectively on $H^1(J^{\text{new}}, \mathcal{O}_{J^{\text{new}}})$. One has $H^1(J_Y, \mathcal{O}_{J_Y}) \simeq H^1(Y, \mathcal{O}_Y)$, and $H^1(Y, \mathcal{O}_Y) = H^1(X, f_* \mathcal{O}_Y) = H^1(X, \oplus_{i=0}^{n-1} \mathcal{L}^{\otimes i})$. Moreover $H^1(J^{\text{new}}, \mathcal{O}_{J^{\text{new}}}) \simeq H^1(X, \oplus_{i=1}^{n-1}(\mathcal{L})^{\otimes i})$ and these identifications are compatible with the action of Frobenius. Hence the kernel of Frobenius on $H^1(J^{\text{new}}, \mathcal{O}_{J^{\text{new}}})$ is isomorphic to the kernel of Frobenius acting on $H^1(X, \oplus_{i=1}^{n-1} \mathcal{L}^{\otimes i})$. On the other hand as f' is étale $(f')^*(B_X) = B_Y$, thus also $(f')_*(B_Y) = B_X \otimes (f')_*(\mathcal{O}_{Y^1}) = \oplus_{i=0}^{n-1}(B_X \otimes (\mathcal{L}^1)^{\otimes i})$. Now by duality, the kernel of the Frobenius acting on $H^1(X^1, \oplus_{i=1}^{n-1} \mathcal{L}^{1 \otimes i})$ is isomorphic to the kernel of the Cartier operator on $H^0(X^1, \pi_* \Omega_X^1 \otimes (\oplus_{i=1}^{n-1}(\mathcal{L}^1)^{\otimes i}))$, which is $\oplus_{i=1}^{n-1} H^0(X^1, B_X \otimes (\mathcal{L}^1)^{\otimes i})$. Thus the above μ_n-torsor $f : Y \to X$ is new-ordinary if and only if $H^0(X^1, B \otimes (\mathcal{L}^1)^{\otimes i}) = 0$ for $i \in \{1, \ldots, n-1\}$, which is also equivalent to the fact that the subgroup $\langle \mathcal{L}^1 \rangle$ generated by $\mathcal{L}^1$ in J^1 intersects the support of the theta divisor θ_X associated to B_X at most at the zero point 0_{J^1} of J^1.

2. μ_n-Torsors of Curves over Finite Fields and Ordinariness

In this section we consider curves over the algebraic closure $\bar{\mathbb{F}}_p$ of the prime field $\mathbb{F}_p$. We establish that after finite étale covers the theta divisor associated to the sheaf B of locally exact differentials contains infinitely many torsion points of order prime to p. This indeed gives information on the fundamental group of these curves.

PROPOSITION 2.1. *Let A be an abelian variety of dimension ≥ 2 over $\bar{\mathbb{F}}_p$, and let Y be a closed sub-variety of A of dimension ≥ 1. Assume either A is a simple abelian variety, or $Y(\bar{\mathbb{F}}_p)$ contains the zero point 0_A of A. Then $Y(\bar{\mathbb{F}}_p)$ contains an infinity of torsion points of pairwise prime order.*

PROOF. First note that the abelian group $A(\bar{\mathbb{F}}_p) = A(\bar{\mathbb{F}}_p)^{\text{tor}}$ is torsion. We will use the following result from [An-In]:

PROPOSITION. *Let C be a proper smooth and connected curve over $\bar{\mathbb{F}}_p$ of genus $g \geq 1$, and let $J := \text{Pic}^0(C)$ be its Jacobian. Let $\phi : C \to J$ be the embedding of C in J associated to a point $x_0 \in C(\bar{\mathbb{F}}_p)$. For any integer m denote by $_m J(\bar{\mathbb{F}}_p)$ the m-primary part of the torsion group $J(\bar{\mathbb{F}}_p)$ (i.e., $_m J(\bar{\mathbb{F}}_p) := \oplus_l(_l J(\bar{\mathbb{F}}_p))$ the* *sum being taken over all primes ℓ dividing m), and let $\lambda : J(\bar{\mathbb{F}}_p) \to_m J(\bar{\mathbb{F}}_p)$ be the canonical projection. Then the map $\lambda \circ \phi : C(\bar{\mathbb{F}}_p) \to_m J(\bar{\mathbb{F}}_p)$ is surjective.*

This was proved in [An-In] only in the case where $m = l$ is a prime number, but it is easy to check that the proof there works also in the case of any positive integer m. It follows immediately from the above result that $C(\bar{\mathbb{F}}_p)$ contains infinitely many points which have pairwise prime orders, in particular it contains infinitely many points of order prime to p. Indeed, If $\{x_1, \ldots, x_n\}$ are finitely many points of $C(\bar{\mathbb{F}}_p)$, r is the least common multiple of the orders of the points $\{x_1, \ldots, x_n\}$, and if $s > 1$ is an integer which is relatively prime to r, and $x \neq 0$ is an s-torsion point on J, then by the above result one can find a point on $C(\bar{\mathbb{F}}_p)$ whose s primary part equal x and whose r-primary part equals 0, in particular such a point has an order which is prime to r. $\square$

For the proof of 2.1, let $y \in Y(\bar{\mathbb{F}}_p)$ be a closed point in Y and let C be an irreducible sub-scheme of Y of dimension 1 which contains y. We endow C with its reduced structure. Let $\tilde{C}$ be the normalization of C which is a smooth and connected curve of genus ≥ 1, and let $\tilde{J}$ be its Jacobian. One has a commutative diagram:

$$\begin{array}{ccc} \tilde{J} & \xrightarrow{f} & A \\ {\scriptstyle\tilde{\phi}}\uparrow & & \uparrow{\scriptstyle i} \\ \tilde{C} & \xrightarrow{\tilde{f}} & C \end{array}$$

where $\tilde{f}$ is the normalization morphism, $\tilde{\phi}$ is the embedding of $\tilde{C}$ in its Jacobian associated to a point $\tilde{y}$ above y, and f is the morphism induced by the universal property of $\tilde{J}$, which is a composition of a homomorphism g and a translation τ_y by the point y. If y is a point of order prime to p then the image via f of the points of order prime to p on $\tilde{\phi}(\tilde{C})$ (which exists and are an infinity by the above result) yields infinitely many points in $C(\bar{\mathbb{F}}_p)$ which have pairwise prime orders. Moreover if $0_A \in Y(\bar{\mathbb{F}}_p)$ and one takes $y = 0_A$, then with the same notations as above, the images via f of the points of $\tilde{\phi}(\tilde{C})$ having pairwise prime orders yield infinitely many points on C having pairwise prime orders. Assume now that A is a simple abelian variety. Then the above homomorphism g is necessarily surjective, in particular there exists x in $\tilde{J}$ such that $g(x) = y$, and $C = g(\tau_x(\tilde{C}))$, where τ_x denotes the translation by x inside $\tilde{J}$. On the other hand it is easy to see, using the above result in [An-In] in the same way that was used above, that $\tau_x(\tilde{C})$ also contains infinitely many points which have pairwise prime orders in $\tilde{J}$ hence the result in this case.

PROPOSITION/DEFINITION 2.2. *With the same hypothesis as in Proposition 2.1 let Y_i be an irreducible component of Y which has dimension ≥ 1, and denote by $A(\bar{\mathbb{F}}_p)^{(p')}$ the prime to p-part of the torsion group $A(\bar{\mathbb{F}}_p)$. Then:*

(1) *either $Y_i(\bar{\mathbb{F}}_p) \cap A(\bar{\mathbb{F}}_p)^{(p')}$ is Zariski dense in Y_i, in which case we call Y_i an abelian like sub-variety of A, or*

(2) $Y_i(\bar{\mathbb{F}}_p) \cap A(\bar{\mathbb{F}}_p)^{(p')}$ is empty in which case Y_i must be a translate of an abelian like sub-variety of A by a point which necessarily has order divisible by p.

PROOF. After eventually a translation we can assume that Y_i contains the zero point of A and then we can assume by 2.1 that $Y_i(\bar{\mathbb{F}}_p) \cap A(\bar{\mathbb{F}}_p)^{(p')}$ is non empty. Assume that the closure Z_i of $Y_i(\bar{\mathbb{F}}_p) \cap A(\bar{\mathbb{F}}_p)^{(p')}$ is distinct from Y_i. Let x be a point in $Y_i(\bar{\mathbb{F}}_p) \cap A(\bar{\mathbb{F}}_p)^{(p')}$ and $y \in Y_i(\bar{\mathbb{F}}_p)$, but y is not contained in Z_i. Then one can find a curve C which contains both y and x (see [Mu], lemma on p. 56). It follows then from the same argument used in the proof of 2.1 that C contains infinitely many points of order prime to p, hence $Y_i - Z_i$ contains such a point which contradicts the fact that $Z_i \neq Y_i$.   □

Here is an immediate consequence of these propositions:

PROPOSITION 2.3. Let X be a proper smooth and connected curve over $\bar{\mathbb{F}}_p$. Let θ_X be the theta divisor associated to the sheaf B_X of locally exact differentials on X (see Section 1.1). Assume: either the Jacobian J of X is a simple abelian variety, or that the curve X is not ordinary which is equivalent to the fact that $0 \in \theta_X(\bar{\mathbb{F}}_p)$. Then $\theta_X(\bar{\mathbb{F}}_p)$ contains infinitely many torsion points of the Jacobian J^1 of X^1 having pairwise prime orders. In general, if $\theta_X(\bar{\mathbb{F}}_p)$ contains a torsion point of order prime to p, then $\theta_X(\bar{\mathbb{F}}_p)$ contains infinitely many torsion points of order prime to p. In both cases θ_X has an irreducible component which is an abelian like sub-variety of J^1.

In the general case where the conditions of 2.3 are not satisfied one has the following.

PROPOSITION 2.4. Let X be a proper smooth and connected curve over $\bar{\mathbb{F}}_p$. Then there exists an étale Galois cover $Y \to X$ with Galois group G of order prime to p such that the theta divisor θ_Y associated to the sheaf of locally exact differentials on Y contains infinitely many $\bar{\mathbb{F}}_p$-torsion points of pairwise prime order.

PROOF. By a result of Raynaud (see [Ra-2]) there exists an étale Galois cover $Y \to X$ with Galois group G of order prime to p such that Y is not ordinary. In particular the theta divisor θ_Y associated to the sheaf of locally exact differentials on Y contains the zero point of J^1_Y. Hence the result follows from 2.3.   □

3. On the Theta Divisor θ of Curves with Simple Jacobians

THEOREM 3.1. Let A be a **simple** abelian variety of dimension $g \geq 2$ over an algebraically closed field K of characteristic $p > 0$. Assume that A is not defined over a finite field, and that the p-rank of A equals g or $g - 1$. Let D be a closed sub-variety of codimension ≥ 1 of A. Then $D(K)$ contains at most finitely many torsion points of $A(K)^{\text{tor}}$ of order prime to p.

PROOF. Since A is simple, any K-homomorphism from A to an abelian variety is either trivial or an isogeny. In particular, the $\bar{\mathbb{F}}_p$-trace of A is either trivial or isogenous to A in which case the kernel of such an isogeny is automatically defined over a $\bar{\mathbb{F}}_p$ because of the condition on the p-rank of A (see [Oo], 3.4), hence the $\bar{\mathbb{F}}_p$-trace of A equals 0 necessarily, since A is not defined over a finite field by assumption. Let $D^{(p')}$ be the closure in A of the intersection of D with the prime to p-part of the torsion group $J(K)^{\mathrm{tor}}$. By the results of Hrushovski on the analog of the 'Mordell-Lang conjecture over function fields in positive characteristic $D^{(p')}$ is a finite union $\cup_i (a_i + A_i)$ of translates of abelian subvarieties A_i of A (see [Hr], Corollary 1.2). As A is simple, $\dim A_i = 0$ and hence $D^{(p')}$ consists of at most finitely many points.   □

COROLLARY 3.2. *Let X be a proper smooth and connected curve of genus $g \geq 2$ over an algebraically closed field K of characteristic $p > 0$. Assume that X is not defined over a finite field. Let θ_X be the theta divisor associated to the sheaf of locally exact differentials B_X on X^1 (see Section 1.1). Assume that the Jacobian J of X is a simple abelian variety and that the p-rank of X equals g or $g - 1$. Then $\theta_X(K)$ contains at most finitely many torsion points of order prime to p.*

PROOF. Since X is not defined over a finite field this is also the case for its Jacobian J By Torelli's theorem [We]; hence 3.2 follows from 3.1.   □

4. Proof of Theorem A, Theorem B, and Corollary

We reformulate the assertions of the theorems as follows:

Let x, y be points of M_g with x a specialization of y. Thus the local ring $\mathcal{O}_{M_g,x}$ of the point x contains a prime ideal $\mathcal{P}_y$ corresponding to y, and $\mathcal{O}_{M_g,y}$ is the localization of $\mathcal{O}_{M_g,x}$ at $\mathcal{P}_y$. Let K be an algebraic closure of $\kappa(y)$. Then there exits a valuation ring R of K dominating the factor ring $\mathcal{O}_{M_g,x}/\mathcal{P}_y$ inside $\kappa(y) \subset K$, such that the residue field of R is an algebraic closure κ of $\kappa(x)$. Thus $\bar{y} = \mathrm{Spec}\, K$ is the generic point, and $\bar{x} = \mathrm{Spec}\, \kappa$ is the closed point of $\mathrm{Spec}\, R$. We choose a smooth projective curve $f : X \to \mathrm{Spec}\, R$ so that we have a morphism $g : \mathrm{Spec}\, R \to M_g$ such that the induced morphisms $\bar{y} \to M_g$ and $\bar{x} \to M_g$ define the generic fiber $X_{\bar{y}} \to \mathrm{Spec}\, K$, respectively the special fiber $X_{\bar{x}} \to \mathrm{Spec}\, \kappa$ as points in $M_g(K)$, respectively $M_g(\kappa)$. We can identify $\pi_1(X_\kappa)$ with $\pi_1(x)$, and $\pi_1(X_K)$ with $\pi_1(y)$ respectively, in such a way that the Grothendieck's specialization homomorphism $\pi_1(X_K) \to \pi_1(X_\kappa)$ is exactly the specialization homomorphism $\mathrm{Sp} : \pi_1(y) \to \pi_1(x)$.

Now we suppose that the points y and x are of a special nature, as in Theorem A and/or Theorem B. This means in particular, that y might be the generic point η of M_g, and x is a point s in $S^{\mathrm{a.s.}}$ or $S_{\geq 2g-1}^{\mathrm{a.s.}}$. Assuming that $\mathrm{Sp}_{y,x}$ is an isomorphism, we will get a contradiction by showing that the morphism $g : \mathrm{Spec}\, R \to M_g$ is constant.

Concerning Theorem A. In the above notations, let $x = s$ and $y = \eta$, thus κ is an algebraic closure of the finite field $\kappa(s)$, and K is the algebraic closure of $\kappa(\eta)$. We denote by $J_s = J_\kappa$ the Jacobian of $X_s := X_\kappa$, respectively by $J_\eta = J_K$ the Jacobian of $X_\eta := X_K$. Further let θ_s, respectively θ_η be the theta divisor in $(J_s)^1$ associated to the sheaf of locally exact differentials on X_s, respectively the theta divisor in $(J_{\bar\eta})^1$ associated to the sheaf of locally exact differentials on X_η. It follows from 2.3 that θ_s contains infinitely many torsion points of order prime to p. Let $\mathcal{L}$ be an invertible sheaf of order n prime to p on $X \to S$. Let $\mathcal{L}_\eta$, respectively $\mathcal{L}_s$ be the restriction of $\mathcal{L}$ to X_η, respectively its restriction to X_s. The assumption that $\mathrm{Sp} : \pi_1(X_\eta) \to \pi_1(X_s)$ is an isomorphism implies in particular: The μ_n-torsor associated to $\mathcal{L}_\eta$ is new-ordinary (in the sense of 1.3) if and only if the μ_n-torsor associated to $\mathcal{L}_s$ is new ordinary. In other words: The subgroup $\langle \mathcal{L}_\eta^1 \rangle$ generated by $\mathcal{L}_\eta^1$ intersects the theta divisor θ_η at a non zero point if and only if the subgroup $\langle \mathcal{L}_s^1 \rangle$ generated by $\mathcal{L}_s^1$ intersects the theta divisor θ_s at a non zero point. Hence we deduce from Proposition 2.3, it follows that θ_η contains infinitely many torsion points of J_η of order prime to p. On the other hand, it is well known that all cyclic étale covers $Y \to X_\eta$ of degree n prime to p (and even without this condition) are new-ordinary (see [Na], for instance). This means that the theta divisor θ_η contains no torsion point of order prime to p. Thus a contradiction in this case.

Concerning Theorem B. One proceeds as above, but without using the assumption that y is the generic point of M_g. In the above notations we then have: Let $J \to \mathrm{Spec}\, R$ be the Jacobian of the projective smooth curve $X \to \mathrm{Spec}\, R$. Thus $J \to \mathrm{Spec}\, R$ is an abelian scheme over $\mathrm{Spec}\, R$, and $J_s = J \times_R \kappa$ is the special fiber of J, and $J_y = J \times_R K$ is the generic fiber of J. Since J_s is a simple abelian variety (by the hypothesis on s), it follows that its generic fiber J_y is simple too. Since f is non iso-trivial, it follows that $X_y := X_K$ is not defined over a finite field. Hence Corollary 3.2 implies that the theta-divisor θ_y of X_y^1 is such that $\theta_y(\lambda)$ contains at most finitely many torsion points of order prime to p. This is a contradiction, so Sp cannot be an isomorphism in this case.

We next prove the second assertion of Theorem B. Let $x \in S_{\geq g-1}^{\mathrm{a.s.}}$ be a closed point of M_g. By contradiction, suppose that there exists infinitely many points $x' \in S_{\geq g-1}^{\mathrm{a.s.}}$ such that $\pi_1(x) \simeq \pi_1(x')$. Let S_x denote the subset of those points, and $\overline{S}_x$ be the closure of S_x in M_g. Then $\overline{S}_x$ is a closed sub-scheme of M_g of dimension $d \geq 1$. Let z be a point of $\overline{S}_x$ which is not a closed point. By hypothesis there exists a point $x' \in S_x$ such that z specializes in x', and hence there exists a continuous surjective homomorphism $\mathrm{Sp} : \pi_1(z) \to \pi_1(x')$. In particular one has an inclusion of sets $\pi_A(x') \subset \pi_A(z)$. On the other hand it is well known that every finite group $G \in \pi_A(z)$ belongs to π_A in an open neighborhood of z (see [St]), and as each such a neighborhood contains a point of S_x one deduces in fact that one has an equality $\pi_A(x') = \pi_A(z)$, and the above homomorphism $\mathrm{Sp} : \pi_1(z) \to \pi_1(x')$ is an isomorphism (this follows from the

Hopfian property for finitely generated profinite groups; see [Fr-Ja], Prop. 15.4). But this can not be the case by the first half of Theorem B since $x' \in S^{\text{a.s.}}_{\geq g-1}$.

Concerning the Introduction's Corollary. We finally come to the proof of the Corollary. First, the fact that the subset $S^{\text{a.s.}}$ of closed points with absolutely simple Jacobian has positive Dirichlet density implies in particular that $S^{\text{a.s.}} \cap U$ is dense in U for every open (nonempty) subset U of M_g (see [Se-2]). Further, since the Jacobian of the generic curve C_η is ordinary, it follows that every curve C_x with $\pi_1(x) \cong \pi_1(\eta)$ is ordinary too. Thus we have: If π_1 is constant on U, then $S^{\text{a.s.}} \cap U = S^{\text{a.s.}}_{\geq g} \cap U$ is dense in U, in particular infinite. This in turn is a contradiction by the second part of Theorem B.

References

[An-In] G. W. Anderson and R. Indik, "On primes of degree one in function fields", *Proc. Amer. Math. Soc.* **94** (1985), 31–32.

[Bo-Lu-Ra] S. Bosch, W. Lütkebohmert and M. Raynaud, *Néron Models*, Ergebnisse der Mathematik (3. Folge) **21**, Springer, Berlin, 1990.

[Ch-Oo] C. L. Chai and F. Oort, "A note on the existence of absolutely simple jacobians", *Jour. Pure Appl. Alg.* **155**:2–3 (2001), 115–120.

[D-M] P. Deligne and D. Mumford, "The irreducibility of the space of curves with a given genus", *Publ. Math. IHES* **36** (1969), 75–110.

[EGA] A. Grothendieck and J. Dieudonné, "Élements de la géométrie algébrique, Chap. 2", *Pub. Math. IHES* **8** (1961).

[Fr-Ja] M. Fried and M. Jarden, *Field Arithmetic*, Ergebnisse der Mathematik (3. Folge) **11**, Springer, Berlin 1986.

[Hr] E. Hrushovski, "The Mordell-Lang conjecture for function fields", *J. Amer. Math. Soc.* **9**:3 (1996), 667–690.

[Mu] D. Mumford, *Abelian Varieties*, Oxford Univ. Press, Oxford, 1970.

[Na] S. Nakajima, "On generalized Hasse–Witt invariants of an algebraic curve", *Adv. Stud. Pure Math.* **12** (1987), 69–88.

[Oo] F. Oort, "The isogeny class of a CM-type abelian variety is defined over a finite extension of the prime field", *J. Pure Appl. Alg.* **3** (1973), 399–408.

[Ra-1] M. Raynaud, "Section des fibrés vectoriel sur une courbe", *Bull. Soc. Math. France*, **110** (1982), 103–125.

[Ra-2] M. Raynaud, "Revêtements des courbes en caractéristique $p > 0$ et ordinarité", *Comp. Mathematica* **123** (2000), 73–88.

[SGA-1] A. Grothendieck, *Revêtements étales et groupe fondamental*, Lecture Notes in Math. **224**, Springer, Berlin, 1971.

[Se-1] J.-P. Serre, "Sur la topologie des variétés algébriques en caractériqstique $p > 0$", pp. 24–53 in *Symposium Internacional de Topologia Algebraica* (Mexico, 1958), Universidad Nacional Autonoma de Mexico and UNESCO, Mexico City, 1958.

[Se-2] J.-P. Serre, "Zeta and L-functions", pp. 82–92 in *Arithmetical Algebraic geometry*, Harper and Row, New York, 1965.

[Sh] I. Shafarevich, "On p-extensions", *Mat. Sb. Nov. Ser.* **20** (62) (1947), 351–363; *AMS Transl. Series 2*, **4** (1965), 59–72.

[St] K. Stevenson, "Quotients of the fundamental group of algebraic curves in positive characteristics", *J. Alg.* **182** (1996), 770–804.

[We] A. Weil, "Zum Beweis des Torellischen Satzes", in *Collected papers*, Volume II, (1951–1964).

FLORIAN POP
MATHEMATISCHES INSTITUT
UNIVERSITÄT BONN
BERINGSTRASSE 6
53115 BONN
GERMANY
pop@math.uni-bonn.de

MOHAMED SAïDI
DEPARTMENT OF MATHEMATICAL SCIENCES
UNIVERSITY OF DURHAM
SCIENCE LABORATORIES
SOUTH ROAD
DURHAM DH1 3LE
UNITED KINGDOM
mohamed.saidi@durham.ac.uk

Galois Groups and Fundamental Groups
MSRI Publications
Volume 41, 2003

Topics Surrounding the Anabelian Geometry of Hyperbolic Curves

SHINICHI MOCHIZUKI

CONTENTS

Introduction

We give an exposition of various ideas and results related to the fundamental results of [Tama1-2], [Mzk1-2] concerning *Grothendieck's Conjecture of Anabelian Geometry* (which we refer to as the "Grothendieck Conjecture" for short; see [Mzk2], Introduction, for a brief introduction to this conjecture). Many of these ideas existed prior to the publication of [Tama1-2], [Mzk1-2], but were not discussed in these papers because of their rather elementary nature and secondary importance (by comparison to the main results of these papers). Nevertheless, it is the hope of the author that the reader will find this article useful as a supplement to [Tama1-2], [Mzk1-2]. In particular, we hope that *the discussion of this article will serve to clarify the meaning and motivation behind the main result of* [Mzk2].

Our main results are the following:

(1) In Section 1, we take the reverse point of view to the usual one (i.e., that the Grothendieck Conjecture should be regarded as a sort of (anabelian) Tate Conjecture) and show that in a certain case, *the Tate Conjecture may be regarded as a sort of Grothendieck Conjecture* (see Theorem 1.1, Corollary 1.2). In particular, Corollary 1.2 is interesting in that *it allows one to express the fundamental phenomenon involved in the Tate and Grothendieck Conjectures using elementary language that can, in principle, be understood even by high school students* (see the Introduction to Section 1; the Remarks following Corollary 1.2).

(2) In Section 2, we show how the main result of [Mzk2] gives rise to a purely *algebro-geometric* corollary (i.e., one which has nothing to do with Galois groups, arithmetic considerations, etc.) in characteristic 0 (see Corollary 2.1). Moreover, we give a partial generalization of this result to positive characteristic (see Theorem 2.2).

(3) In Section 3, we discuss *real analogues of anabelian geometry.* Not surprisingly, the real case is substantially easier than the case where the base field is p-adic or a number field. Thus, we are able to prove much stronger results in the real case than in the p-adic or number field cases (see Theorem 3.6, Corollaries 3.7, 3.8, 3.10, 3.11, 3.13, 3.14, 3.15). In particular, we are able to prove various *real analogues of the so-called Section Conjecture of anabelian geometry* (which has not been proven, at the time of writing, for any varieties over p-adic or number fields) — see [Groth], p. 289, (2); [NTM], § 1.2, (GC3), for a discussion of the Section Conjecture. Also, we note that the real case is interesting relative to the *analogy* between the differential geometry that occurs in the real case and certain aspects of the p-adic case (see [Mzk4], Introduction, § 0.10; the Introduction to Section 3 of this article). It was this analogy that led the author to the proof of the main result of [Mzk2].

(4) In Section 4, we show that a certain *isomorphism version* (see Theorem 4.12) of the main result of [Mzk2] can be proven over *"generalized sub-p-adic fields"* (see Definition 4.11), which form a somewhat larger class of fields than the class of "sub-p-adic fields" dealt with in [Mzk2]. This result is interesting in that it is *reminiscent of the main results of* [Tama2]*, as well as of the rigidity theorem of Mostow–Prasad for hyperbolic manifolds of real dimension* 3 (see the Remarks following the proof of Theorem 4.12).

Although we believe the results of Section 4 to be essentially new, we make no claim of essential originality relative to the results of Sections 1–3, which may be proven using well-known standard techniques. Nevertheless, we believe that it is likely that, even with respect to Sections 1–3, *the point of view of the discussion is likely to be new* (and of interest relative to understanding the main result of [Mzk2]).

Finally, before beginning our exposition, *we pause to review the main result of* [Mzk2] (which is the central result to which the ideas of the present article are related). To do this, we must introduce some notation. Let Σ be a nonempty set of *prime numbers*. If K is a field, denote its *absolute Galois group* $\operatorname{Gal}(\bar{K}/K)$ (where $\bar{K}$ is some algebraic closure of K) by Γ_K. If X is a geometrically connected K-scheme, recall that its algebraic fundamental group $\pi_1(X)$ (for some choice of base-point) fits into a natural exact sequence

$$1 \to \pi_1(X \otimes_K \bar{K}) \to \pi(X) \to \Gamma_K \to 1.$$

Denote by Δ_X the *maximal pro-Σ quotient of* $\pi_1(X \otimes_K \bar{K})$ (i.e., the inverse limit of those finite quotients whose orders are products of primes contained in Σ). The profinite group Δ_X is often referred to as the *(pro-Σ) geometric fundamental group* of X. Note that since the kernel of $\pi_1(X \otimes_K \bar{K}) \to \Delta_X$ is a *characteristic subgroup* of $\pi_1(X \otimes_K \bar{K})$, it follows that it is normal *inside* $\pi_1(X)$. Denote the quotient of $\pi_1(X)$ by this normal subgroup by Π_X. The profinite group Π_X is often referred to as the *(pro-Σ) arithmetic fundamental group* of X. (When it is necessary to specify the set of primes Σ, we will write Δ_X^Σ, Π_X^Σ.) Thus, we have a natural exact sequence

$$1 \to \Delta_X \to \Pi_X \to \Gamma_K \to 1.$$

In [Mzk2] we proved the following result:

THEOREM A. *Let K be a **sub-p-adic field** (i.e., a field isomorphic to a subfield of a finitely generated field extension of $\mathbb{Q}_p$), where $p \in \Sigma$. Let X_K be a smooth variety over K, and Y_K a hyperbolic curve over K. Let $\operatorname{Hom}_K^{\mathrm{dom}}(X_K, Y_K)$ be the set of dominant K-morphisms from X_K to Y_K. Let $\operatorname{Hom}_{\Gamma_K}^{\mathrm{open}}(\Pi_X, \Pi_Y)$ be the set of open, continuous group homomorphisms $\Pi_X \to \Pi_Y$ over Γ_K, considered up to composition with an inner automorphism arising from Δ_Y. Then the natural map*

$$\operatorname{Hom}_K^{\mathrm{dom}}(X_K, Y_K) \to \operatorname{Hom}_{\Gamma_K}^{\mathrm{open}}(\Pi_X, \Pi_Y)$$

is bijective.

REMARK. Theorem A as stated above is a formal consequence of "Theorem A" of [Mzk2]. In [Mzk2], only the cases of $\Sigma = \{p\}$, and Σ equal to the set of all prime numbers are discussed, but it is easy to see that the case of arbitrary Σ containing p may be derived from the case $\Sigma = \{p\}$ by precisely the same argument as that used in [Mzk2] (see [Mzk2], the Remark following Theorem 16.5) to derive the case of Σ equal to the set of all prime numbers from the case of $\Sigma = \{p\}$.

Acknowledgements. The author would like to thank A. Tamagawa and T. Tsuji for stimulating discussions of the various topics presented in this article. In particular, the author would especially like to express his gratitude A. Tamagawa for discussions concerning removing the hypothesis of "monodromy type" from Theorem 4.12 in an earlier version of this article. Also, the author greatly appreciates the advice given to him by Y. Ihara (orally) concerning the Remark following Corollary 1.2, and by M. Seppala and C. McMullen (by email) concerning the Teichmüller theory used in Section 3.

1. The Tate Conjecture as a Sort of Grothendieck Conjecture

In this section, we attempt to present what might be referred to as the most fundamental "prototype result" among the family of results (including the Tate and Grothendieck Conjectures) that states that maps between varieties are "essentially equivalent" to maps between arithmetic fundamental groups. The result given below, especially in the form Corollary 1.2, is interesting in that *it allows one to express the fundamental phenomenon involved using elementary language that can, in principle, be understood even by high school students* (see the Remark following Corollary 1.2). In particular, it does not require a knowledge of the notion of a Galois group or any another advanced notions, hence provides a convincing example of how advanced mathematics can be applied to prove results which can be stated in simple terms. Also, it may be useful for explaining to mathematicians in other fields (who may not be familiar with Galois groups or other notions used in arithmetic geometry) *the essence of the Tate and Grothendieck Conjectures.* Another interesting feature of Corollary 1.2 is that *it shows how the Tate conjecture may be thought of as being of the "same genre" as the Grothendieck Conjecture in that it expresses how the isomorphism class of a curve (in this case, an elliptic curve) may be recovered from Galois-theoretic information.*

1.1. The Tate conjecture for non-CM elliptic curves. Let K be a *number field* (i.e., a finite extension of $\mathbb{Q}$). If E is an *elliptic curve* over K, and N is a natural number, write

$$K(E[N])$$

for the minimal finite extension field of K over which all of the N-*torsion points are defined.* Note that the extension $K(E[N])$ will always be *Galois.* Then we have the following elementary consequence of the "Tate Conjecture for abelian varieties over number fields" proven in [Falt]:

THEOREM 1.1. *Let K be a number field. Let E_1 and E_2 be elliptic curves over K such that neither E_1 nor E_2 admits complex multiplication over $\overline{\mathbb{Q}}$. Then E_1 and E_2 are isomorphic as elliptic curves over K if and only if $K(E_1[N]) = K(E_2[N])$ for all natural numbers N.*

REMARK. The equality $K(E_1[N]) = K(E_2[N])$ is to be understood in the sense of subfields of some fixed algebraic closure of K. The substance of this expression is independent of the choice of algebraic closure precisely because both fields in question are *Galois* extensions of K.

PROOF. If $E_1 \cong E_2$ over K, then it is clear that $K(E_1[N]) = K(E_2[N])$ for all natural numbers N. Thus, assume that $K(E_1[N]) = K(E_2[N])$ for all natural numbers N, and prove that $E_1 \cong E_2$ over K. In this proof, we use the notation and results of Section 1.2 below. Since we assume that $K(E_1[N]) = K(E_2[N])$, we denote this field by $K[N]$. Also, if p is a prime number, then we write $K[p^\infty]$ for the union of the $K[p^n]$, as n ranges over the positive integers. Finally, for $n \geq 0$, we denote the Galois group $\mathrm{Gal}(K[p^\infty]/K[p^n])$ by $\Gamma[p^n]$; the center of $\Gamma[p^n]$ by $Z\Gamma[p^n]$; and the quotient $\Gamma[p^n]/Z\Gamma[p^n]$ by $P\Gamma[p^n]$.

Let p be a prime number. Then by the semisimplicity of the Tate module, together with the Tate conjecture (both proven in general in [Falt]; see also [Ser2], IV), the fact that neither E_1 nor E_2 admits complex multiplication over $\overline{\mathbb{Q}}$ implies that there exists an integer $n \geq 1$ such that the Galois representation on the p-power torsion points of E_1 (respectively, E_2) induces an *isomorphism* $\beta_1 : \Gamma[p^n] \cong \mathrm{GL}_2^{[n]}(\mathbb{Z}_p)$ (respectively, $\beta_2 : \Gamma[p^n] \cong \mathrm{GL}_2^{[n]}(\mathbb{Z}_p)$), where $\mathrm{GL}_2^{[n]}(\mathbb{Z}_p) \subseteq \mathrm{GL}_2(\mathbb{Z}_p)$ is the subgroup of matrices that are $\equiv 1$ modulo p^n. Since the kernel of $\mathrm{GL}_2^{[n]}(\mathbb{Z}_p) \to \mathrm{PGL}_2^{[n]}(\mathbb{Z}_p)$ is easily seen to be equal to the center of $\mathrm{GL}_2^{[n]}(\mathbb{Z}_p)$, it thus follows that β_1, β_2 induce isomorphisms

$$\alpha_1 : P\Gamma[p^n] \cong \mathrm{PGL}_2^{[n]}(\mathbb{Z}_p), \quad \alpha_2 : P\Gamma[p^n] \cong \mathrm{PGL}_2^{[n]}(\mathbb{Z}_p).$$

Thus, in particular, by Lemma 1.3 of Section 1.2 below, we obtain that (after possibly increasing n) the automorphism $\alpha \overset{\mathrm{def}}{=} \alpha_1 \circ \alpha_2^{-1}$ of $\mathrm{PGL}_2^{[n]}(\mathbb{Z}_p)$ is defined by conjugation by an element of $\mathrm{PGL}_2(\mathbb{Z}_p)$. In particular, we obtain that there exists a $\mathbb{Z}_p$-linear isomorphism

$$\psi : T_p(E_1) \cong T_p(E_2)$$

between the p-adic Tate modules of E_1 and E_2 with the property that for $\sigma \in \Gamma[p^n]$, we have $\psi(\sigma(t)) = \lambda_\sigma \sigma(\psi(t))$ ($\forall t \in T_p(E_1)$), for some $\lambda_\sigma \in \mathbb{Z}_p^\times$ which is independent of t. On the other hand, since the determinant of ψ is clearly compatible with the Galois actions on both sides (given by the cyclotomic character), it thus follows (by taking determinants of both sides of the equation $\psi(\sigma(t)) = \lambda_\sigma \sigma(\psi(t))$) that $\lambda_\sigma^2 = 1$. Since the correspondence $\sigma \mapsto \lambda_\sigma$ is clearly a homomorphism (hence a character of order 2), we conclude:

(∗) *There exists a finite extension K' of K over which the $\mathrm{Gal}(\overline{K}/K')$-modules $T_p(E_1)$ and $T_p(E_2)$ become isomorphic.*

(Here, K' is the extension of $K[p^n]$ (of degree ≤ 2) defined by the kernel of $\sigma \mapsto \lambda_\sigma$. In fact, if $p > 2$, then this extension is trivial (since $\Gamma[p^n]$ is a pro-p-group).) Thus, by the Tate Conjecture proven in [Falt], we obtain that $\mathrm{Hom}_{K'}(E_1, E_2) \otimes_{\mathbb{Z}}$

$\mathbb{Z}_p$ contains an element that induces an isomorphism on p-adic Tate modules. On the other hand, since $H_{K'} \overset{\text{def}}{=} \text{Hom}_{K'}(E_1, E_2)$ (the module of homomorphisms $(E_1)_{K'} \to (E_2)_{K'}$ over K') is a finitely generated free $\mathbb{Z}$-module of rank ≤ 1 (since E_1, E_2 do not have complex multiplication over $\overline{\mathbb{Q}}$), we thus obtain that $H_{K'}$ is a free $\mathbb{Z}$-module of rank 1. Let $\varepsilon \in H_{K'}$ be a generator of $H_{K'}$. Then ε necessarily corresponds to an isogeny $E_1 \to E_2$ that induces an isomorphism on p-power torsion points.

Now write $H_{\overline{K}} \overset{\text{def}}{=} \text{Hom}_{\overline{K}}(E_1, E_2)$. Then the above argument shows that $H_{\overline{K}}$ is a free $\mathbb{Z}$-module of rank 1 with a generator ε that induces an isomorphism on p-power torsion points for every prime number p. But this implies that $\varepsilon : (E_1)_{\overline{K}} \to (E_2)_{\overline{K}}$ is an *isomorphism*, i.e., that E_1 and E_2 become *isomorphic over $\overline{K}$*.

Thus, it remains to check that E_1 and E_2 are, in fact, isomorphic over K. Let $p \geq 5$ be a prime number which is sufficiently large that: (i) K is absolutely unramified at p; (ii) the Galois representations on the p-power torsion points of E_1 and E_2 induce isomorphisms

$$\beta_1 : \Gamma[p^0] \cong \text{GL}_2(\mathbb{Z}_p); \quad \beta_2 : \Gamma[p^0] \cong \text{GL}_2(\mathbb{Z}_p)$$

(the existence of such p follows from the "modulo l versions" (for large l) of the semisimplicity of the Tate module, together with the Tate conjecture in [Mord], VIII, § 5 ; see also [Ser2], IV). Now we would like to consider the extent to which the automorphism $\beta \overset{\text{def}}{=} \beta_1 \circ \beta_2^{-1}$ of $\text{GL}_2(\mathbb{Z}_p)$ is defined by conjugation by an element of $\text{GL}_2(\mathbb{Z}_p)$. Note that by what we did above, we know that the morphism induced by β on $\text{PGL}_2^{[n]}(\mathbb{Z}_p)$ (for some large n) is given by conjugation by some element $A \in \text{GL}_2(\mathbb{Z}_p)$. Let $\gamma : \text{GL}_2(\mathbb{Z}_p) \to \text{GL}_2(\mathbb{Z}_p)$ be the automorphism of $\text{GL}_2(\mathbb{Z}_p)$ obtained by composing β with the automorphism given by conjugation by A^{-1}. Thus, γ induces the identity on $\text{PGL}_2^{[n]}(\mathbb{Z}_p)$. But this implies (by Lemma 1.4 below) that γ induces the identity on $\text{PGL}_2(\mathbb{Z}_p)$. In particular, it follows that there exists a homomorphism $\lambda : \text{GL}_2(\mathbb{Z}_p) \to \mathbb{Z}_p^\times$ such that $\gamma(\sigma) = \lambda(\sigma) \cdot \sigma$ ($\forall \sigma \in \text{GL}_2(\mathbb{Z}_p)$). Next, recall that since $p \geq 5$, the topological group $SL_2(\mathbb{Z}_p)$ has *no abelian quotients* (an easy exercise). Thus, λ factors through the determinant map $\text{GL}_2(\mathbb{Z}_p) \to \mathbb{Z}_p^\times$. Moreover, (see the argument at the beginning of the proof involving arbitrary p) since the composites of β_1, β_2 with the determinant map are given by the cyclotomic character, we obtain that $\lambda^2 = 1$. In particular, we obtain that λ is trivial on the index 2 subgroup of $\text{GL}_2(\mathbb{Z}_p)$ of elements whose determinant is a square. Put another way, if we write K_p for the quadratic extension of K determined by composing the cyclotomic character $\text{Gal}(\overline{K}/K) \to \mathbb{Z}_p^\times$ (which is surjective since K is absolutely unramified at p) with the unique surjection $\mathbb{Z}_p^\times \twoheadrightarrow \mathbb{Z}/2\mathbb{Z}$, then over K_p, the Tate modules $T_p(E_1)$, $T_p(E_2)$ become isomorphic as Galois modules, which implies that $\text{Hom}_{K_p}(E_1, E_2) \neq 0$. But this implies that E_1 and E_2 become isomorphic over K_p.

On the other hand, for distinct primes p, p' as above, K_p, $K_{p'}$ form *linearly disjoint quadratic extensions* of K (as can be seen by considering the ramification at p, p'). Thus, the fact that both $\text{Gal}(\bar{K}/K_p)$ and $\text{Gal}(\bar{K}/K_{p'})$ act trivially on $\text{Hom}_{\bar{K}}(E_1, E_2)$ implies that $\text{Gal}(\bar{K}/K)$ acts trivially on $\text{Hom}_{\bar{K}}(E_1, E_2)$, so $E_1 \cong E_2$ over K, as desired. □

REMARK. The above proof benefited from discussions with A. Tamagawa and T. Tsuji.

REMARK. In the preceding proof (see also the arguments of Section 1.2 below), we use in an *essential* way the *strong rigidity properties* of the *simple p-adic Lie group* $\text{PGL}_2(\mathbb{Z}_p)$. Such rigidity properties are not shared by abelian Lie groups such as $\mathbb{Z}_p$; this is why it was necessary to assume in Theorem 1.1 that the elliptic curves in question do not *admit complex multiplication*.

COROLLARY 1.2. *There is a finite set $\mathfrak{CM} \subseteq \mathbb{Z}$ such that if E_1 and E_2 are arbitrary elliptic curves over $\mathbb{Q}$ whose j-invariants $j(E_1)$, $j(E_2)$ do not belong to $\mathfrak{CM}$, then E_1 and E_2 are isomorphic as elliptic curves over $\mathbb{Q}$ if and only if $\mathbb{Q}(E_1[N]) = \mathbb{Q}(E_2[N])$ for all natural numbers N.*

PROOF. In light of Theorem 1.1, it suffices to show that there are only finitely many possibilities (all of which are integral — see, e.g., [Shi], p. 108, Theorem 4.4) for the j-invariant of an elliptic curve over $\mathbb{Q}$ which has complex multiplication over $\bar{\mathbb{Q}}$. But this follows from the finiteness of the number of imaginary quadratic extensions of $\mathbb{Q}$ with class number one (see, e.g., [Stk]), together with the theory of [Shi] (see [Shi], p. 123, Theorem 5.7, (i), (ii)). (Note that we also use here the elementary facts that: (i) the class group of any order surjects onto the class group of the maximal order; (ii) in a given imaginary quadratic extension of $\mathbb{Q}$, there are only finitely many orders with trivial class group.) □

REMARK. According to an (apparently) unpublished manuscript of J.-P. Serre ([Ser3]) whose existence was made known to the author by Y. Ihara, the set $\mathfrak{CM}$ of Corollary 1.2, i.e., the list of rational j-invariants of elliptic curves with complex multiplication, is as follows:

$$d = 1, \mathfrak{f} = 1 \Longrightarrow j = j(i) = 2^6 \cdot 3^3$$
$$d = 1, \mathfrak{f} = 2 \Longrightarrow j = j(2i) = (2 \cdot 3 \cdot 11)^3$$
$$d = 2, \mathfrak{f} = 1 \Longrightarrow j = j(\sqrt{-2}) = (2^2 \cdot 5)^3 \quad ([\text{Weber}], \text{p.}\,721)$$
$$d = 3, \mathfrak{f} = 1 \Longrightarrow j = j(\tfrac{-1+\sqrt{-3}}{2}) = 0$$
$$d = 3, \mathfrak{f} = 2 \Longrightarrow j = j(\sqrt{-3}) = 2^4 \cdot 3^3 \cdot 5^3 \quad ([\text{Weber}], \text{p.}\,721)$$
$$d = 3, \mathfrak{f} = 3 \Longrightarrow j = j(\tfrac{-1+3\sqrt{-3}}{2}) = -3 \cdot 2^{15} \cdot 5^3 \quad ([\text{Weber}], \text{p.}\,462)$$
$$d = 7, \mathfrak{f} = 1 \Longrightarrow j = j(\tfrac{-1+\sqrt{-7}}{2}) = -3^3 \cdot 5^3 \quad ([\text{Weber}], \text{p.}\,460)$$
$$d = 7, \mathfrak{f} = 2 \Longrightarrow j = j(\sqrt{-7}) = (3 \cdot 5 \cdot 17)^3 \quad ([\text{Weber}], \text{p.}\,475)$$
$$d = 11, \mathfrak{f} = 1 \Longrightarrow j = j(\tfrac{-1+\sqrt{-11}}{2}) = -2^{15} \quad ([\text{Weber}], \text{p.}\,462)$$

$$d = 19, \mathfrak{f} = 1 \Longrightarrow j = -(2^5 \cdot 3)^3 \quad ([\text{Weber}], \, \text{p.}\, 462)$$

$$d = 43, \mathfrak{f} = 1 \Longrightarrow j = -(2^6 \cdot 3 \cdot 5)^3 \quad ([\text{Weber}], \, \text{p.}\, 462)$$

$$d = 67, \mathfrak{f} = 1 \Longrightarrow j = -(2^5 \cdot 3 \cdot 5 \cdot 11)^3 \quad ([\text{Weber}], \, \text{p.}\, 462)$$

$$d = 163, \mathfrak{f} = 1 \Longrightarrow j = -(2^6 \cdot 3 \cdot 5 \cdot 23 \cdot 29)^3 \quad ([\text{Weber}], \, \text{p.}\, 462)$$

Here $\mathbb{Q}(\sqrt{-d})$ is the imaginary quadratic extension of $\mathbb{Q}$ containing the order in question, $\mathfrak{f}$ is the conductor of the order, and the reference given in parentheses is for the values of the invariants "f" and "f_1" of [Weber], which are related to the j-invariant as follows: $j = (f^{24} - 16)^3/f^{24} = (f_1^{24} + 16)^3/f_1^{24}$.

REMARK. Thus, if one defines elliptic curves over $\mathbb{Q}$ using cubic equations, constructs the group law on elliptic curves by considering the intersection of the cubic with various lines, and interprets the notion of isomorphism of elliptic curves (over $\mathbb{Q}$) to mean "being defined by the same cubic equation, up to coordinate transformations," then Corollary 1.2 may be expressed as follows:

> Except for the case of finitely many exceptional j-invariants, two elliptic curves E_1, E_2 over $\mathbb{Q}$ are isomorphic if and only if for each natural number N, the coordinates ($\in \mathbb{C}$) necessary to define the N-torsion points of E_1 generate the same "subfield of $\mathbb{C}$" — i.e., "collection of complex numbers closed under addition, subtraction, multiplication, and division" — as the coordinates necessary to define the N-torsion points of E_2.

(Here, of course, the j-invariant is defined as a polynomial in the coefficients of the cubic.) In this form, the essential phenomenon at issue in the Tate or Grothendieck Conjectures may be understood even by high school students or mathematicians unfamiliar with Galois theory.

1.2. Some pro-p group theory. Let $n \geq 1$ be an integer. In this section we denote by PGL_2 the algebraic group (defined over $\mathbb{Z}$) obtained by forming the quotient of GL_2 by $\mathbb{G}_m$ (where $\mathbb{G}_m \hookrightarrow \mathrm{GL}_2$ is the standard embedding by scalars), and by

$$\mathrm{PGL}_2^{[n]}(\mathbb{Z}_p) \subseteq \mathrm{PGL}_2(\mathbb{Z}_p)$$

the subgroup of elements which are $\equiv 1$ modulo p^n. Write $\mathrm{pgl}_2(\mathbb{Z}_p)$ for the quotient of the Lie algebra $M_2(\mathbb{Z}_p)$ (of 2 by 2 matrices with $\mathbb{Z}_p$ coefficients) by the scalars $\mathbb{Z}_p \subseteq M_2(\mathbb{Z}_p)$. Thus, $\mathrm{pgl}_2(\mathbb{Z}_p) \subseteq \mathrm{pgl}_2(\mathbb{Z}_p) \otimes_{\mathbb{Z}_p} \mathbb{Q}_p = \mathrm{pgl}_2(\mathbb{Q}_p)$. Write $\mathrm{pgl}_2^{[n]}(\mathbb{Z}_p) \subseteq \mathrm{pgl}_2(\mathbb{Z}_p)$ for the submodule which is the image of matrices in $M_2(\mathbb{Z}_p)$ which are $\equiv 0$ modulo p^n. Thus, for n sufficiently large, $\mathrm{pgl}_2^{[n]}(\mathbb{Z}_p)$ maps bijectively onto $\mathrm{PGL}_2^{[n]}(\mathbb{Z}_p)$ via the exponential map (see [Ser1], Chapter V, § 7).

LEMMA 1.3. *Let $\alpha : \mathrm{PGL}_2^{[n]}(\mathbb{Z}_p) \to \mathrm{PGL}_2^{[n]}(\mathbb{Z}_p)$ be an automorphism of the profinite topological group $\mathrm{PGL}_2^{[n]}(\mathbb{Z}_p)$ such that $\alpha(\mathrm{PGL}_2^{[m]}(\mathbb{Z}_p)) = \mathrm{PGL}_2^{[m]}(\mathbb{Z}_p)$ for all $m \geq n$. Then there exists an element $A \in \mathrm{PGL}_2(\mathbb{Z}_p)$ such that for some $m \geq n$, the restriction $\alpha|_{\mathrm{PGL}_2^{[m]}(\mathbb{Z}_p)}$ is given by conjugation by A.*

PROOF. Write

$$\mathcal{A} : \mathrm{pgl}_2(\mathbb{Q}_p) \to \mathrm{pgl}_2(\mathbb{Q}_p)$$

for the morphism on Lie algebras induced by α. By [Ser1], Chapter V, § 7, 9, after possibly replacing n by a larger n, we may assume that α is the homomorphism obtained by exponentiating $\mathcal{A}$. Moreover, by the well-known theory of the Lie algebra $\mathrm{pgl}_2(\mathbb{Q}_p)$, it follows that $\mathcal{A}$ may be obtained by conjugating by some $A' \in \mathrm{PGL}_2(\mathbb{Q}_p)$. (Indeed, this may be proven by noting that $\mathcal{A}$ induces an automorphism of the "variety of Borel subalgebras of $\mathrm{pgl}_2(\mathbb{Q}_p)$." Since this variety is simply $\mathbb{P}^1_{\mathbb{Q}_p}$, we thus get an automorphism of $\mathbb{P}^1_{\mathbb{Q}_p}$, hence an element of $\mathrm{PGL}_2(\mathbb{Q}_p)$, as desired.) On the other hand, it follows immediately from the structure theory of finitely generated $\mathbb{Z}_p$-modules that A' may be written as a product

$$A' = C_1 \cdot A'' \cdot C_2,$$

where $C_1, C_2 \in \mathrm{PGL}_2(\mathbb{Z}_p)$, and A'' is defined by a matrix of the form

$$\begin{pmatrix} \lambda_1 & 0 \\ 0 & \lambda_2 \end{pmatrix},$$

where $\lambda_1, \lambda_2 \in \mathbb{Q}_p^{\times}$.

Now observe that the fact that $\mathcal{A}$ arises from an automorphism of $\mathrm{PGL}_2^{[n]}(\mathbb{Z}_p)$ implies that $\mathcal{A}$ induces an automorphism of $\mathrm{pgl}_2^{[n]}(\mathbb{Z}_p)$ (see the discussion at the beginning of this section). Since conjugation by C_1 and C_2 clearly induces automorphisms of $\mathrm{pgl}_2^{[n]}(\mathbb{Z}_p)$, it thus follows that conjugation by A'' induces an automorphism of $\mathrm{pgl}_2^{[n]}(\mathbb{Z}_p)$. Now, by considering, for instance, upper triangular matrices with zeroes along the diagonal, one sees that A'' can only induce an automorphism of $\mathrm{pgl}_2^{[n]}(\mathbb{Z}_p)$ if $\lambda_1 = \lambda_2 \cdot u$, where $u \in \mathbb{Z}_p^{\times}$. Let $A \stackrel{\mathrm{def}}{=} \lambda_1^{-1} \cdot A'$. Then clearly $A \in \mathrm{PGL}_2(\mathbb{Z}_p)$, and conjugation by A induces $\mathcal{A}$. Thus, by using the exponential map, we obtain that for some $m \geq n$, the restriction $\alpha|_{\mathrm{PGL}_2^{[m]}(\mathbb{Z}_p)}$ is given by conjugation by A, as desired. $\square$

The following lemma was pointed out to the author by A. Tamagawa:

LEMMA 1.4. *Let $\alpha : \mathrm{PGL}_2(\mathbb{Z}_p) \to \mathrm{PGL}_2(\mathbb{Z}_p)$ be an automorphism of the profinite topological group $\mathrm{PGL}_2(\mathbb{Z}_p)$ such that for some integer $m \geq 1$, the restriction $\alpha|_{\mathrm{PGL}_2^{[m]}(\mathbb{Z}_p)}$ is the identity. Then α itself is the identity.*

PROOF. First let us show that α is the identity on the image in $\mathrm{PGL}_2(\mathbb{Z}_p)$ of matrices of the form $\begin{pmatrix} 1 & \lambda \\ 0 & 1 \end{pmatrix}$, where $\lambda \in \mathbb{Z}_p$. For $m \geq 0$ an integer, write $U_m \subseteq \mathrm{PGL}_2(\mathbb{Z}_p)$ for the subgroup of images in $\mathrm{PGL}_2(\mathbb{Z}_p)$ of matrices of the form $\begin{pmatrix} 1 & \lambda \\ 0 & 1 \end{pmatrix}$, where $\lambda \in p^m \cdot \mathbb{Z}_p$. Since, by hypothesis, α preserves U_m for some m, it follows that α preserves the *centralizer* $Z(U_m)$ of U_m in $\mathrm{PGL}_2(\mathbb{Z}_p)$. On the other hand, one checks easily that $Z(U_m) = U_0$. Thus, α preserves U_0, i.e., induces an automorphism of the topological group $U_0 \cong \mathbb{Z}_p$ which is the identity on $p^m \cdot \mathbb{Z}_p$. Since $\mathbb{Z}_p$ is torsion free, it thus follows that α is the identity on U_0, as desired. Moreover, let us observe that since conjugation commutes with the operation of taking centralizers, one sees immediately that the above argument implies also that α *is the identity on all conjugates of U_0 in* $\mathrm{PGL}_2(\mathbb{Z}_p)$.

Next, observe that α is the identity on the subgroup $B \subseteq \mathrm{PGL}_2(\mathbb{Z}_p)$ consisting of images of matrices of the form

$$\begin{pmatrix} \mu_1 & \lambda \\ 0 & \mu_2 \end{pmatrix}$$

(where $\lambda \in \mathbb{Z}_p$, $\mu_1, \mu_2 \in \mathbb{Z}_p^{\times}$). Indeed, since B is generated by U_0 and the subgroup $T \subseteq \mathrm{PGL}_2(\mathbb{Z}_p)$ of images of matrices of the form $\begin{pmatrix} \mu & 0 \\ 0 & 1 \end{pmatrix}$, it suffices to see that α is the identity on T. But T acts faithfully by conjugation on U_0, and α is the identity on U_0. This implies that α is the identity on T, hence on B. Moreover, as in the previous paragraph, this argument implies that α *is the identity on all conjugates of B in* $\mathrm{PGL}_2(\mathbb{Z}_p)$. Since $\mathrm{PGL}_2(\mathbb{Z}_p)$ is generated by the union of the conjugates of B, it thus follows that α is the identity on $\mathrm{PGL}_2(\mathbb{Z}_p)$. $\qquad\square$

2. Hyperbolic Curves As Their Own "Anabelian Albanese Varieties"

In this section, we present an application (Corollary 2.1) of the main theorem of [Mzk2] which is interesting in that it is *purely algebro-geometric*, i.e., it makes no mention of Galois actions or other arithmetic phenomena.

2.1. A corollary of the Main Theorem of [Mzk2]. We fix a nonempty set of *prime numbers* Σ, and use the notation of the discussion of Theorem A in the Introduction. Now Theorem A has the following immediate consequence:

COROLLARY 2.1. *Let K be a field of characteristic 0. Let C be a hyperbolic curve over K, and let $\psi : X \to Y$ be a morphism of (geometrically integral) smooth varieties over K which induces an isomorphism $\Delta_X \cong \Delta_Y$. Write "$\mathrm{Hom}_K^{\mathrm{dom}}(-, C)$" for the set of dominant K-morphisms from "$-$" to C. Then the natural morphism of sets*

$$\mathrm{Hom}_K^{\mathrm{dom}}(Y, C) \to \mathrm{Hom}_K^{\mathrm{dom}}(X, C)$$

induced by $\psi : X \to Y$ is a bijection.

PROOF. By a standard technique involving the use of subfields of K which are finitely generated over $\mathbb{Q}$, we reduce immediately to the case where K is finitely generated over $\mathbb{Q}$. (We recall for the convenience of the reader that the essence of this technique lies in the fact that since we are working with K-schemes of finite type, all schemes and morphisms between schemes are defined by *finitely many polynomials* with coefficients in K, hence may be defined over any subfield of K that contains these coefficients — of which there are only finitely many!)

Next, observe that since the morphism $\psi : X \to Y$ induces an isomorphism between the respective geometric fundamental groups, it follows from the exact sequences reviewed in the Introduction that it induces an isomorphism $\Pi_X \cong \Pi_Y$. By Theorem A of the Introduction, it thus follows that the morphism of sets under consideration — i.e., $\operatorname{Hom}_K^{\mathrm{dom}}(Y, C) \to \operatorname{Hom}_K^{\mathrm{dom}}(X, C)$ — is naturally isomorphic to the morphism of sets given by

$$\operatorname{Hom}_{\Gamma_K}^{\mathrm{open}}(\Pi_Y, \Pi_C) \to \operatorname{Hom}_{\Gamma_K}^{\mathrm{open}}(\Pi_X, \Pi_C)$$

which is bijective.   □

REMARK. As stated above, Corollary 2.1 is interesting in that it is a *purely algebro-geometric application* of Theorem A, i.e., it makes no mention of Galois actions or other arithmetic phenomena. The observation that Corollary 2.1 holds first arose in discussions between the author and A. Tamagawa. Typical examples of morphisms $\psi : X \to Y$ as in Corollary 2.1 are:

(1) the case where $X \to Y$ is a fiber bundle in, say, the étale topology, with proper, simply connected fibers;

(2) the case where $Y \subseteq \mathbb{P}_k^n$ is a closed subvariety of dimension ≥ 3 in some projective space, and X is obtained by intersecting Y with a hyperplane in $\mathbb{P}_k^n$.

In these cases, the fact that the resulting morphism on geometric fundamental groups is an isomorphism follows from the long exact homotopy sequence of a fiber bundle in the first case (see [SGA1], X, Corollary 1.4), and Lefshetz-type theorems (see [SGA2], XII, Corollary 3.5) in the second case. Since this consequence of Theorem A (i.e., Corollary 2.1) is purely algebro-geometric, *it is natural to ask if one can give a purely algebro-geometric proof of Corollary 2.1.* In Section 2.2 below, we give a partial answer to this question.

2.2. A partial generalization to finite characteristic. Let k be an *algebraically closed field.* Let C be a *proper hyperbolic curve* over k. Suppose that we are also given a *connected, smooth closed subvariety*

$$Y \subseteq \mathbb{P}_k^n$$

of projective space, of dimension ≥ 3, together with a *hyperplane* $H \subseteq \mathbb{P}_k^n$ such that the scheme-theoretic intersection $X \overset{\mathrm{def}}{=} H \bigcap Y$ is still smooth. Note that X is necessarily *connected* (see [SGA2], XII, Corollary 3.5) and of dimension ≥ 2.

If k is of *characteristic $p > 0$*, and S is a k-scheme, then let us write $\Phi_S : S \to S$ for the *Frobenius morphism* on S (given by raising regular functions on S to the power p). If k is of characteristic 0, then we make the convention that $\Phi_S : S \to S$ denotes the *identity morphism*. If T is a k-schemes, we define

$$\operatorname{Hom}^{\Phi}(T, C)$$

to be the inductive limit of the system

$$\operatorname{Hom}_k^{\mathrm{dom}}(T, C) \to \operatorname{Hom}_k^{\mathrm{dom}}(T, C) \to \cdots \to \operatorname{Hom}_k^{\mathrm{dom}}(T, C) \to \cdots,$$

where the arrows are those induced by applying the functor $\mathrm{Hom}_k^{\mathrm{dom}}(-, C)$ to the morphism Φ_T. Thus, in particular, if k is of characteristic 0, then $\mathrm{Hom}^\Phi(T, C) = \mathrm{Hom}_k^{\mathrm{dom}}(T, C)$.

Now we have the following partial generalization of Corollary 2.1 of Section 2.1 to the case of varieties over a field of arbitrary characteristic:

THEOREM 2.2. *Let k, C, X, and Y be as above. Then the natural morphism*

$$\mathrm{Hom}^\Phi(Y, C) \to \mathrm{Hom}^\Phi(X, C)$$

induced by the inclusion $X \hookrightarrow Y$ is a bijection.

PROOF. Denote by A_X, A_Y, and A_C the *Albanese varieties* of X, Y, and C, respectively. We refer to [Lang], Chapter II, § 3, for basic facts concerning Albanese varieties. Thus, the inclusion $X \hookrightarrow Y$ induces a morphism $A_X \to A_Y$. I *claim* that this morphism is a *purely inseparable isogeny*. Indeed, by various well-known Leftshetz theorem-type results (see, [SGA2], XII, Corollary 3.5), the inclusion $X \hookrightarrow Y$ induces an isomorphism $\pi_1(X) \cong \pi_1(Y)$; since (by the universal property of the Albanese variety as the "minimal abelian variety to which the original variety maps") we have surjections $\pi_1(X) \twoheadrightarrow \pi_1(A_X)$, $\pi_1(Y) \twoheadrightarrow \pi_1(A_Y)$, we thus obtain that $\pi_1(A_X) \twoheadrightarrow \pi_1(A_Y)$ is a surjection. Moreover, since $X \to A_X$, $Y \to A_Y$ induce isomorphisms on the respectively étale first cohomology groups with $\mathbb{Z}_l$-coefficients (where l is prime to the characteristic of k), we thus obtain that $\pi_1(A_X) \twoheadrightarrow \pi_1(A_Y)$ is a *surjection which is an isomorphism on the respective maximal pro-l quotients*. Now it follows from the elementary theory of abelian varieties that this implies that $A_X \to A_Y$ is a *isogeny of degree a power of p*. Finally, applying *again* the fact that $\pi_1(A_X) \twoheadrightarrow \pi_1(A_Y)$ is *surjective* (i.e., even on maximal pro-p quotients), we conclude (again from the elementary theory of abelian varieties) that this isogeny has *trivial étale part*, hence is *purely inseparable*, as desired. Note that since $A_X \to A_Y$ is an isogeny, it follows in particular that it is *faithfully flat*.

Now let $\gamma_X : X \to C$ be a *dominant k-morphism*. Write $\alpha_X : A_X \to A_C$ for the induced morphism on Albanese varieties. If γ_X arises from some $\gamma_Y : Y \to C$, then this γ_Y is *unique*. Indeed, γ_Y is determined by its associated α_Y, and the composite of α_Y with $A_X \to A_Y$ is given by α_X (which is uniquely determined by γ_X). Thus, the fact that α_Y is uniquely determined follows from the fact that $A_X \to A_Y$ is faithfully flat. This completes the proof of the claim, and hence of the injectivity portion of the bijectivity assertion in Theorem 2.2.

Now suppose that γ_X is *arbitrary* (i.e., does not necessarily arise from some γ_Y). The surjectivity portion of the bijectivity assertion in Theorem 2.2 amounts to showing that, up to replacing γ_X by the composite of γ_X with some power of Φ_X, γ_X necessarily arises from some $\gamma_Y : Y \to C$. Now although $\alpha_X : A_X \to A_C$ itself might not factor through A_Y, since $A_X \to A_Y$ is *purely inseparable*, it follows that the composite of α_X with some power of Φ_{A_X} will factor through A_Y. Thus, if we replace γ_X by the composite of γ_X with some power of Φ_X, then

α_X will factor (uniquely) through A_Y. Denote this morphism by $\alpha_Y : A_Y \to A_C$. Thus, in order to complete the proof of surjectivity, it suffices to show:

The restriction $\alpha_Y|_Y$ of α_Y to Y (relative to the natural morphism $Y \to A_Y$) maps into the subvariety $C \subseteq A_C$.

Before continuing, we make some observations:

(1) The assertion (∗) for characteristic zero k follows immediately from the assertion (∗) for k of finite characteristic. Indeed, this follows via the usual argument of replacing k first by a finitely generated $\mathbb{Z}$-algebra, and then reducing modulo various primes. Thus, in the following, we assume the k is of characteristic $p > 0$.

(2) The assertion (∗) will follow if we can show that the restriction $\alpha_Y|_{\hat{Y}}$ (where we write $\hat{Y}$ for the *completion* of Y along X) maps into $C \subseteq A_C$.

Now we show that (up to possibly composing γ_X again with a power of Frobenius), γ_X *extends to* $\hat{Y}$. If $\mathcal{I}$ is the sheaf of ideals on Y that defines the closed subscheme $X \subseteq Y$, then let us write $Y_n \stackrel{\text{def}}{=} V(\mathcal{I}^n) \subseteq Y$ for the *n-th infinitesimal neighborhood* of X in Y, and $\mathcal{J} \stackrel{\text{def}}{=} \mathcal{I}|_X \cong \mathcal{O}_X(-1)$. Write $\mathcal{T}$ for the pull-back of the tangent bundle of C to X via γ_X. Since $\mathcal{T}^{-1}$ is generated by global sections, it thus follows that $\mathcal{T}^{-1} \otimes \mathcal{J}^{-1}$ is *ample*, hence, by *Serre duality* (see, e.g., [Harts], Chapter III, Theorem 7.6), together with the fact that $\dim(X) \geq 2$, that there exists a natural number N such that

$$H^1(X, \mathcal{T}^{\otimes p^N} \otimes \mathcal{J}^{\otimes p^N}) = 0.$$

Note that this implies that for all $n \geq p^N$, we have:

$$H^1(X, \mathcal{T}^{\otimes p^N} \otimes \mathcal{J}^{\otimes n}) = 0.$$

(Indeed, it suffices to assume that $n > p^N$. Then since $\mathcal{J}^{-1} \cong \mathcal{O}_X(1)$ is *very ample*, it follows that there exists a section $s \in \Gamma(X, \mathcal{O}_X(1))$ whose zero locus $Z \stackrel{\text{def}}{=} V(s) \subseteq X$ is smooth of dimension ≥ 1. Thus, s defines an exact sequence

$$0 \to \mathcal{T}^{\otimes p^N} \otimes \mathcal{J}^{\otimes n} \to \mathcal{T}^{\otimes p^N} \otimes \mathcal{J}^{\otimes n-1} \to \mathcal{T}^{\otimes p^N} \otimes \mathcal{J}^{\otimes n-1}|_Z \to 0,$$

whose associated long exact cohomology sequence yields

$$H^0(Z, \mathcal{T}^{\otimes p^N} \otimes \mathcal{J}^{\otimes n-1}|_Z) \to H^1(X, \mathcal{T}^{\otimes p^N} \otimes \mathcal{J}^{\otimes n}) \to H^1(X, \mathcal{T}^{\otimes p^N} \otimes \mathcal{J}^{\otimes n-1})$$

But $H^0(Z, \mathcal{T}^{\otimes p^N} \otimes \mathcal{J}^{\otimes n-1}|_Z) = 0$ since $\mathcal{T}^{\otimes p^N} \otimes \mathcal{J}^{\otimes n-1}|_Z$ is the inverse of an ample line bundle on a smooth scheme of dimension ≥ 1, while $H^1(X, \mathcal{T}^{\otimes p^N} \otimes \mathcal{J}^{\otimes n-1}) = 0$ by the induction hypothesis.)

Next, observe that $\Phi^N_{Y_{p^N}} : Y_{p^N} \to Y_{p^N}$ factors through X (since $\Phi^N_{Y_{p^N}}$ is induced by raising functions to the p^N-th power). Thus, if we compose $\gamma_X : X \to C$ with Φ^N_X, we see that this composite extends to a morphism $Y_{p^N} \to C$. Moreover, since the pull-back to X via this composite of the tangent bundle on

C is given by $\mathcal{T}^{\otimes p^N}$, it follows that the obstruction to extending this composite to Y_{n+1} for $n \geq p^N$ is given by an element of the cohomology group

$$H^1(X, \mathcal{T}^{\otimes p^N} \otimes \mathcal{J}^{\otimes n}),$$

which (by the above discussion concerning cohomology groups) is zero. Thus, in summary, if we replace the given γ_X by its composite with Φ_X^N, the resulting γ_X extends to a morphism $\hat{Y} \to C$. This completes the proof of $(*)$, and hence of the entire proof of Theorem 2.2. □

REMARK. The above proof benefited from discussions with A. Tamagawa.

REMARK. The other case discussed in the remark at the end of Section 2.1, i.e., the case of a fiber bundle with proper, simply connected fibers also admits a purely algebro-geometric proof: namely, it follows immediately from the theory of Albanese varieties that there do not exist any nonconstant morphisms from a simply connected smooth proper variety to an abelian variety.

REMARK. The role played by the Albanese variety in the proof of Theorem 2.2 given above suggests that the property proven in Corollary 2.1 and Theorem 2.2 might be thought of as asserting that a hyperbolic curve is, so to speak, *its own "anabelian Albanese variety."* This is the reason for the title of Section 2.

3. Discrete Real Anabelian Geometry

The original motivation for the p-adic result of [Mzk2] came from *the (differential) geometry of the upper half-plane uniformization* of a hyperbolic curve. This point of view — and, especially, the related idea that Kähler geometry at archimedean primes should be regarded as analogous to Frobenius actions at p-adic primes — is discussed in detail in [Mzk4], Introduction (especially Section 0.10; see also the Introduction of [Mzk3]). In the present section, we attempt to make this motivation more rigorous by presenting the *real analogues of various theorems/conjectures* of anabelian geometry. The substantive mathematics here — i.e., essentially the geometry of the Siegel upper half-plane and Teichmüller space — is not new, but has been well-known to topologists, Teichmüller theorists, and symmetric domain theorists for some time. What is (perhaps) new is the formulation or point of view presented here, namely, that these geometric facts should be regarded as real analogues of Grothendieck's conjectured anabelian geometry.

3.1. Real complex manifolds. We begin with the following purely analytic definition: Let X be a *complex manifold* and ι an *antiholomorphic involution* (i.e., automorphism of order 2) of X.

DEFINITION 3.1. A pair such as (X, ι) will be referred to as a *real complex manifold*. If X has the structure of an abelian variety whose origin is fixed by ι, then (X, ι) will be referred to as a *real abelian variety*. If $\dim_{\mathbb{C}}(X) = 1$, then

(X, ι) will be referred to as a *real Riemann surface*. A real Riemann surface (X, ι) will be called *hyperbolic* if the universal covering space of X is isomorphic (as a Riemann surface) to the upper half-plane $\mathfrak{H} \overset{\text{def}}{=} \{z \in \mathbb{C} \mid \operatorname{Im}(z) > 0\}$.

REMARK. If $X_{\mathbb{R}}$ is a *smooth algebraic variety over* $\mathbb{R}$, then $X_{\mathbb{R}}(\mathbb{C})$ equipped with the antiholomorphic involution defined by complex conjugation defines a *real complex manifold* (X, ι). Moreover, one checks easily that $X_{\mathbb{R}}$ is uniquely determined by (X, ι). Conversely, any real complex manifold (X, ι) such that X is *projective* arises from a unique algebraic variety $X_{\mathbb{R}}$ over $\mathbb{R}$. Indeed, this follows easily from "Chow's Theorem" (that any projective complex manifold is necessarily algebraic) and the (related) fact that any holomorphic isomorphism between projective algebraic varieties (in this case, the given X and its complex conjugate) is necessarily algebraic. Thus, in summary, one motivating reason for the introduction of Definition 3.1 is that *it allows one to describe the notion of a (proper, smooth) algebraic variety over* $\mathbb{R}$ *entirely in terms of complex manifolds and analytic maps.*

REMARK. In the case of one complex dimension, one does not even need to assume projectivity: That is, any real Riemann surface (X, ι) such that X is algebraic arises from a unique algebraic curve $X_{\mathbb{R}}$ over $\mathbb{R}$. Indeed, this follows easily by observing that any holomorphic isomorphism between Riemann surfaces associated to complex algebraic curves is necessarily algebraic. (This may be proven by noting that any such isomorphism extends naturally to the "one-point compactifications" of the Riemann surfaces (which have natural algebraic structures), hence is necessarily algebraizable.) It is not clear to the author whether or not this can be generalized to higher dimensions.

In the following, we shall consider various groups G with natural augmentations $G \to \operatorname{Gal}(\mathbb{C}/\mathbb{R})$. In this sort of situation, we shall denote the inverse image of the identity element (respectively, the complex conjugation element) in $\operatorname{Gal}(\mathbb{C}/\mathbb{R})$ by G^{+} (respectively, G^{-}).

If X is a *complex manifold*, we shall denote by

$$\operatorname{Aut}(X) \to \operatorname{Gal}(\mathbb{C}/\mathbb{R})$$

the group of automorphisms of X which are *either holomorphic or antiholomorphic*, equipped with its natural augmentation (which sends holomorphic (respectively, antiholomorphic) automorphisms to the identity (respectively, complex conjugation element) in $\operatorname{Gal}(\mathbb{C}/\mathbb{R})$). Thus,

$$\operatorname{Aut}^{+}(X), \ \operatorname{Aut}^{-}(X) \subseteq \operatorname{Aut}(X)$$

denote the subsets of holomorphic and antiholomorphic automorphisms, respectively. In many cases, X will come equipped with a *natural Riemannian metric which is preserved by* $\operatorname{Aut}(X)$. The principal examples of this situation are:

EXAMPLE 3.2 (THE SIEGEL UPPER HALF-PLANE). Let $g \geq 1$ be an integer. The *Siegel upper half-plane* $\mathfrak{H}_g$ is the set

$$\mathfrak{H}_g \stackrel{\text{def}}{=} \{Z \in M_g(\mathbb{C}) \mid Z = {}^tZ; \operatorname{Im}(Z) > 0\},$$

where ^{t} denotes the transpose matrix, and > 0 means positive definite. (Thus, $\mathfrak{H}_1$ is the usual upper half-plane $\mathfrak{H}$.) We shall regard $\mathfrak{H}_g$ as a *complex manifold* (equipped with the obvious complex structure). Set

$$J_g \stackrel{\text{def}}{=} \begin{pmatrix} 0 & I_g \\ -I_g & 0 \end{pmatrix} \in M_{2g}(\mathbb{R}),$$

where $I_g \in M_g(\mathbb{R})$ is the identity matrix. Write

$$\operatorname{GSp}_{2g} \stackrel{\text{def}}{=} \{M \in M_{2g}(\mathbb{R}) \mid M \cdot J \cdot {}^tM = \eta \cdot J, \ \eta \in \mathbb{R}^\times\}$$

for the group of *symplectic similitudes*. Thus, we have a natural character

$$\chi : \operatorname{GSp}_{2g} \to \operatorname{Gal}(\mathbb{C}/\mathbb{R})$$

that maps an $M \in \operatorname{GSp}_{2g}$ to the sign of η (where η is as in the above definition of GSp_{2g}). In particular, χ defines $\operatorname{GSp}_{2g}^+$, $\operatorname{GSp}_{2g}^-$. Then we have a natural homomorphism

$$\phi : \operatorname{GSp}_{2g} \to \operatorname{Aut}(\mathfrak{H}_g)$$

given by letting $M = \begin{pmatrix} A & B \\ C & D \end{pmatrix} \in \operatorname{GSp}_{2g}$ act on $Z \in \mathfrak{H}_g$ by

$$Z \mapsto (AZ^{\chi(M)} + B)(CZ^{\chi(M)} + D)^{-1}.$$

Thus, ϕ is compatible with the augmentations to $\operatorname{Gal}(\mathbb{C}/\mathbb{R})$. Now it is clear that the kernel of ϕ is given by the scalars $\mathbb{R}^\times \subseteq \operatorname{GSp}_{2g}$. In fact, ϕ is *surjective*. Indeed, this is well-known when $+$'s are added to both sides (i.e., for holomorphic automorphisms — see, e.g., [Maass], § 4, Theorem 2). On the other hand, since ϕ is compatible with the augmentations to $\operatorname{Gal}(\mathbb{C}/\mathbb{R})$, the surjectivity of ϕ thus follows from the "5-Lemma." Thus, in summary, we have a natural isomorphism

$$\operatorname{GSp}_{2g}/\mathbb{R}^\times \cong \operatorname{Aut}(\mathfrak{H}_g)$$

Moreover, the space $\mathfrak{H}_g$ admits a natural *Riemannian metric*. Relative to this metric, any two points Z_1, Z_2 of $\mathfrak{H}_g$ can be joined by a *unique geodesic* (see [Maass], § 3, Theorem). Moreover, this Riemannian metric is *preserved by the action of* GSp_{2g} *on* $\mathfrak{H}_g$. (Indeed, this follows from [Maass], § 4, Theorem 1, in the holomorphic case. As for the antiholomorphic case, it suffices to check that the metric is preserved by a single antiholomorphic map. But this is clear from [Maass], § 4, Theorem 1, for the map $Z \mapsto -\bar{Z}$.)

EXAMPLE 3.3 (TEICHMÜLLER SPACE). Let $g, r \geq 0$ be integers such that $2g - 2 + r > 0$. Denote by $T_{g,r}$ the *Teichmüller space of genus g Riemann surfaces with r marked points*. Thus, $T_{g,r}$ has a natural structure of complex manifold. Moreover, $T_{g,r}$ is equipped with a natural Kähler metric, called the *Weil–Petersson metric*, whose associated Riemannian metric has the property that *any two points $t_1, t_2 \in T_{g,r}$ may be joined by a unique geodesic* (see [Wolp], §5.1).

Write

$$\mathrm{Mod}_{g,r}$$

for the *full modular group*, i.e., the group of homotopy classes of homeomorphisms of a topological surface of type (g, r) onto itself. Note that $\mathrm{Mod}_{g,r}$ is equipped with an augmentation $\mathrm{Mod}_{g,r} \to \mathrm{Gal}(\mathbb{C}/\mathbb{R})$ given by considering whether or not the homeomorphism preserves the orientation of the surface. The quotient $T_{g,r}/\mathrm{Mod}_{g,r}^{+}$ (in the sense of stacks) may be identified with the moduli stack $\mathcal{M}_{g,r}$ of hyperbolic curves of type (g, r) over $\mathbb{C}$, and the Weil–Petersson metric *descends to $\mathcal{M}_{g,r}$*. Moreover, the Riemannian metric arising from the Weil–Petersson metric on $\mathcal{M}_{g,r}$ is *preserved by complex conjugation*. Indeed, this follows easily, for instance, from the definition of the Weil–Petersson metric in terms of integration of the square of the absolute value of a quadratic differential (on the Riemann surface in question) divided by the $(1,1)$-form given by the Poincaré metric (on the Riemann surface in question) — see, e.g, [Wolp], § 1.4.

If (g, r) is not *exceptional* (i.e., not equal to the cases $(0, 3)$, $(0, 4)$, $(1, 1)$, $(1, 2)$, or $(2, 0)$), then it is known (by a theorem of Royden — see, e.g., [Gard], §9.2, Theorem 2) that one has a natural isomorphism

$$\mathrm{Mod}_{g,r} \cong \mathrm{Aut}(T_{g,r}),$$

which is compatible with the natural augmentations to $\mathrm{Gal}(\mathbb{C}/\mathbb{R})$. Now I claim that (at least if (g, r) is nonexceptional, then) $\mathrm{Aut}(T_{g,r})$ *preserves (the Riemannian metric arising from) the Weil–Petersson metric*. Indeed, since $T_{g,r}/\mathrm{Mod}_{g,r}^{+} = \mathcal{M}_{g,r}$, and the Weil–Petersson metric descends to $\mathcal{M}_{g,r}$, it thus follows that $\mathrm{Mod}_{g,r}^{+}$ preserves the Weil–Petersson metric. Thus, the claim follows from the fact (observed above) that (the Riemannian metric arising from) the Weil–Petersson metric on $\mathcal{M}_{g,r}$ is preserved by complex conjugation.

We now return to our discussion of an arbitrary *real complex manifold* (X, ι). By analogy with the case when (X, ι) arises from a real algebraic variety (see the Remark following Definition 3.1), we will refer to the fixed point locus of ι as the *real locus of* (X, ι), and use the notation

$$X(\mathbb{R})$$

for this locus. Observe that $X(\mathbb{R})$ is necessarily a *real analytic submanifold* of X of real dimension equal to the complex dimension of X. (Indeed, this follows immediately by considering the local structure of ι at a point $x \in X(\mathbb{R})$.)

Moreover, at any $x \in X(\mathbb{R})$, the involution ι induces a *semi-linear* (i.e., with respect to complex conjugation) *automorphism* ι_x of order 2 of the complex vector space $T_x(X)$ (i.e., the tangent space to the complex manifold X at x). That is to say, ι_x defines a *real structure* $T_x(X)_{\mathbb{R}} \subseteq T_x(X)_{\mathbb{R}} \otimes_{\mathbb{R}} \mathbb{C} = T_x(X)$ on $T_x(X)$. Put another way, this real structure $T_x(X)_{\mathbb{R}}$ is simply the tangent space to the real analytic submanifold $X(\mathbb{R}) \subseteq X$.

Since ι acts *without fixed points* on $X \backslash X(\mathbb{R})$, it follows that the quotient of $X \backslash X(\mathbb{R})$ by the action of ι defines a real analytic manifold over which $X \backslash X(\mathbb{R})$ forms an unramified double cover. In the following, in order to analyze the action of ι on all of X, we would like to consider the quotient of X by the action of ι *in the sense of real analytic stacks*. Denote this quotient by X^ι. Thus, we have an unramified double cover

$$X \to X^\iota$$

which extends the cover discussed above over $X \backslash X(\mathbb{R})$.

The Galois group of this double cover (which is isomorphic to $\mathbb{Z}/2\mathbb{Z}$) may be identified with the Galois group $\mathrm{Gal}(\mathbb{C}/\mathbb{R})$. Thus, this double cover induces a short exact sequence of fundamental groups

$$1 \to \pi_1(X) \to \pi_1(X^\iota) \to \mathrm{Gal}(\mathbb{C}/\mathbb{R}) \to 1$$

where we omit base-points, since they are inessential to the following discussion. (Here, by "π_1" we mean the usual (discrete) topological fundamental group in the sense of algebraic topology.)

Now write $\tilde{X} \to X$ for the *universal covering space* of X. Thus, $\tilde{X}$ also has a natural structure of complex manifold, and ι induces an antiholomorphic automorphism $\tilde{\iota}$ (not necessarily of order 2!) of $\tilde{X}$, which is uniquely determined up to composition with the covering transformations of $\tilde{X} \to X$. Since $\tilde{X}$ is also the universal cover of the real analytic stack X^ι, it thus follows that by considering the covering transformations of the covering $\tilde{X} \to X^\iota$, we get a natural homomorphism

$$\pi_1(X^\iota) \to \mathrm{Aut}(\tilde{X})$$

which is compatible with the natural projections of both sides to $\mathrm{Gal}(\mathbb{C}/\mathbb{R})$.

Thus, if, for instance, (X, ι) is a *hyperbolic real Riemann surface*, then by Example 3.2, there is a natural isomorphism $\mathrm{Aut}(\tilde{X}) \cong \mathrm{PGL}_2(\mathbb{R}) = \mathrm{GSp}_2/\mathbb{R}^\times$ (well-defined up to conjugation by an element of $\mathrm{PGL}_2^+(\mathbb{R})$). Thus, we obtain a natural representation

$$\rho_X : \pi_1(X^\iota) \to \mathrm{PGL}_2(\mathbb{R})$$

which is compatible with the natural projections of both sides to $\mathrm{Gal}(\mathbb{C}/\mathbb{R})$.

DEFINITION 3.4.   Let (X, ι) be a hyperbolic real Riemann surface. Then the representation

$$\rho_X : \pi_1(X^\iota) \to \mathrm{PGL}_2(\mathbb{R})$$

just constructed (which is defined up to composition with conjugation by an element of $\mathrm{PGL}_2^+(\mathbb{R})$) will be referred to as the *canonical representation of* (X, ι).

REMARK. The point of view of Definition 3.4 is discussed in [Mzk3], § 1, "Real Curves," although the formulation presented there is somewhat less elegant.

3.2. Fixed points of antiholomorphic involutions. Let T be a (nonempty) *complex manifold* which is also equipped with a *smooth Riemannian metric*. Assume also that the Riemannian metric on T satisfies the following property:

(∗) *For any two distinct points* $t_1, t_2 \in T$, *there exists a unique geodesic joining* t_1 *and* t_2.

Then we have the following result, which is fundamental to the theory of the present Section 3:

LEMMA 3.5.   *Let* T *be a (nonempty)* **complex manifold** *equipped with a* **smooth Riemannian metric** *satisfying the condition* (∗). *Let* $\iota_T : T \to T$ *be an antiholomorphic involution of* T *which preserves this Riemannian metric. Then the fixed point set* $F_{\iota_T} \stackrel{\text{def}}{=} \{t \in T \mid \iota_T(t) = t\}$ *of* ι_T *is a* **nonempty, connected** *real analytic submanifold of* T *of real dimension equal to the complex dimension of* T.

PROOF.   By the discussion of Section 3.1, it follows that it suffices to prove that ι_T is nonempty and connected. First, we prove nonemptiness. Let $t_1 \in T$ be any point of T, and set $t_2 \stackrel{\text{def}}{=} \iota_T(t_1)$. If $t_1 = t_2$, then $t_1 \in F_{\iota_T}$, so we are done. If $t_1 \neq t_2$, then let γ be the *unique geodesic* joining t_1, t_2. Then since the subset $\{t_1, t_2\}$ is preserved by ι_T, it follows that γ *is also preserved by* ι_T. Thus, it follows in particular that the *midpoint* t of γ is preserved by ι_T, i.e., that $t \in F_{\iota_T}$, so F_{ι_T} is nonempty as desired. Connectedness follows similarly: If $t_1, t_2 \in F_{\iota_T}$, then the unique geodesic γ joining t_1, t_2 is also clearly fixed by ι_T, i.e., $\gamma \subseteq F_{\iota_T}$, so F_{ι_T} is pathwise connected. $\square$

REMARK. The idea for this proof (using the Weil–Petersson metric in the case of Teichmüller space) is essentially due to Wolpert ([Wolp]), and was related to the author by C. McMullen. We remark that this idea has been used to give a solution of the *Nielsen Realization Problem* (see the Introduction of [Wolp]). It is easiest to see what is going on by thinking about what happens in the case when $T = \mathfrak{H}$ (the upper half-plane) equipped with the *Poincaré metric* $\frac{dx^2 + dy^2}{y^2}$. Also, we remark that in the case when $T = \mathbb{P}^1_{\mathbb{C}}$, both the hypothesis and the conclusion of Lemma 3.5 are *false*! (That is, the hypothesis is false because there will always exist "conjugate points," and the conclusion is false because it is easy to construct examples of antiholomorphic involutions without fixed points.)

Now assume that (X, ι) is any *real complex manifold* equipped with a smooth Riemannian metric (i.e., X is equipped with a smooth Riemannian metric preserved by ι) such that the induced Riemannian metric on the universal cover

$T \stackrel{\text{def}}{=} \tilde{X}$ satisfies $(*)$. Let $Y \subseteq X(\mathbb{R})$ be a *connected component of the real analytic manifold* $X(\mathbb{R})$. Then since ι acts trivially on Y, the quotient of Y by the action of ι forms a real analytic stack Y^ι whose associated coarse space is Y itself, and which fits into a commutative diagram:

$$\begin{array}{ccc} Y & \to & Y^\iota \\ \downarrow & & \downarrow \\ X & \to & X^\iota \end{array}$$

Moreover, the mapping $Y^\iota \to Y$ (where we think of Y as the coarse space associated to the stack Y^ι) defines a splitting of the exact sequence

$$1 \to \pi_1(Y) \to \pi_1(Y^\iota) \to \text{Gal}(\mathbb{C}/\mathbb{R}) \to 1,$$

hence a homomorphism $\text{Gal}(\mathbb{C}/\mathbb{R}) \to \pi_1(Y^\iota)$. If we compose this homomorphism with the natural homomorphism $\pi_1(Y^\iota) \to \pi_1(X^\iota)$, then we get a morphism

$$\alpha_Y : \text{Gal}(\mathbb{C}/\mathbb{R}) \to \pi_1(X^\iota)$$

naturally associated to Y, which is well-defined up to composition with an inner autormorphism of $\pi_1(X)$. In particular, the image of complex conjugation under α_Y defines a conjugacy class of involutions ι_Y of $\pi_1(X^\iota)$. Thus, to summarize, *we have associated to each connected component* $Y \subseteq X(\mathbb{R})$ *of the real locus of* (X, ι) *a conjugacy class of involutions* ι_Y *in* $\pi_1(X^\iota)$.

Now we have the following immediate consequence of Lemma 3.5:

THEOREM 3.6 (GENERAL DISCRETE REAL SECTION CONJECTURE). *Let* (X, ι) *be a* **real complex manifold** *equipped with a smooth Riemannian metric (i.e.,* X *is equipped with a smooth Riemannian metric preserved by* ι*) such that the induced Riemannian metric on the universal cover* $\tilde{X}$ *satisfies* $(*)$. *Then the correspondence* $Y \mapsto \iota_Y$ *defines a bijection*

$$\pi_0(X(\mathbb{R})) \cong \text{Hom}_{\text{Gal}(\mathbb{C}/\mathbb{R})}(\text{Gal}(\mathbb{C}/\mathbb{R}), \pi_1(X^\iota))$$

from the set of connected components of the real locus $X(\mathbb{R})$ *to the set of conjugacy classes of sections of* $\pi_1(X^\iota) \to \text{Gal}(\mathbb{C}/\mathbb{R})$ *(or, equivalently, involutions in* $\pi_1(X^\iota)$*). Moreover, the centralizer of an involution* $\iota_Y \in \pi_1(X^\iota)$ *is the image of* $\pi_1(Y^\iota)$ *in* $\pi_1(X^\iota)$.

PROOF. Indeed, let $\iota_T \in \pi_1(X^\iota)$ be an involution. Then ι_T may be thought of as an antiholomorphic involution of $T \stackrel{\text{def}}{=} \tilde{X}$. By Lemma 3.5, the fixed point locus F_{ι_T} of ι_T is nonempty and connected. Thus, F_{ι_T} maps into some connected component $Y \subseteq X(\mathbb{R})$. (In fact, the morphism $F_{\iota_T} \to Y$ is a covering map.) By *functoriality* (consider the map of triples $(T, \iota_T, F_{\iota_T}) \to (X, \iota, Y)$!), it follows that $\iota_Y = \iota_T$. Thus, every involution in $\pi_1(X^\iota)$ arises as some ι_Y. Next, let us show uniqueness. If ι_T arises from two distinct $Y_1, Y_2 \subseteq X(\mathbb{R})$, then it would follow that the fixed point locus F_{ι_T} contains at least two distinct connected components (corresponding to Y_1, Y_2), thus contradicting Lemma 3.5. Finally, if $\alpha \in \pi_1(X^\iota)$ commutes with ι_Y, then α preserves F_{ι_Y}, hence induces an automorphism of F_{ι_Y} over Y^ι. But since $F_{\iota_Y} \to Y^\iota$ is a covering map, this implies that α is in the image of $\pi_1(Y^\iota)$ in $\pi_1(X^\iota)$. This completes the proof. $\square$

REMARK. Thus, Theorem 3.6 is a sort of analogue of the so-called "Section Conjecture" of anabelian geometry for the *discrete fundamental groups of real complex manifolds*—see [Groth], p. 289, (2); [NTM], §1.2, (GC3), for more on the Section Conjecture.

REMARK. Theorem 3.6 generalizes immediately to the case where X is a *complex analytic stack*. In this case, "Y^ι" is to be understood to be the real analytic stack whose stack structure is inherited from that of the real analytic stack X^ι. We leave the routine details to the reader.

3.3. Hyperbolic curves and their moduli. By the discussion of Examples 3.2 (in the case of $\mathfrak{H}$), 3.3, in Section 3.1, together with Theorem 3.6 of Section 3.2, we obtain:

COROLLARY 3.7 (DISCRETE REAL SECTION CONJECTURE FOR HYPERBOLIC REAL RIEMANN SURFACES). *Let (X, ι) be a **hyperbolic real Riemann surface**. Then the correspondence $Y \mapsto \iota_Y$ of Section 3.2 defines a bijection*

$$\pi_0(X(\mathbb{R})) \cong \mathrm{Hom}_{\mathrm{Gal}(\mathbb{C}/\mathbb{R})}(\mathrm{Gal}(\mathbb{C}/\mathbb{R}), \pi_1(X^\iota))$$

from the set of connected components of the real locus $X(\mathbb{R})$ to the set of conjugacy classes of sections of $\pi_1(X^\iota) \to \mathrm{Gal}(\mathbb{C}/\mathbb{R})$ (or, equivalently, involutions in $\pi_1(X^\iota)$).

REMARK. Some readers may find it strange that there is no discussion of "*tangential sections*" (at the "points at infinity" of X) in Corollary 3.7. The reason for this is that in the present "real context," where we only consider *connected components* of the set of real points, every tangential section arising from a real point at infinity may be obtained as a limit of a sequence of real points that are not at infinity (and, which, moreover, may be chosen to lie in the same connected components of the real locus), hence is "automatically included" in the connected component containing those real points.

COROLLARY 3.8 (DISCRETE REAL SECTION CONJECTURE FOR MODULI OF HYPERBOLIC CURVES). *Let $g, r \geq 0$ be integers such that $2g - 2 + r > 0$. Write $(\mathcal{M}_{g,r}, \iota_\mathcal{M})$ for the moduli stack of complex hyperbolic curves of type (g, r), equipped with its natural antiholomorphic involution (arising from the structure of $\mathcal{M}_{g,r}$ as an algebraic stack defined over $\mathbb{R}$). If (X, ι) arises from a real hyperbolic curve of type (g, r), then the exact sequence*

$$1 \to \pi_1(X) \to \pi_1(X^\iota) \to \mathrm{Gal}(\mathbb{C}/\mathbb{R}) \to 1$$

defines a homomorphism

$$\alpha_{(X,\iota)} : \mathrm{Gal}(\mathbb{C}/\mathbb{R}) \to \pi_1(\mathcal{M}_{g,r}^{\iota_\mathcal{M}}) = \mathrm{Mod}_{g,r} \subseteq \mathrm{Out}(\pi_1(X))$$

(*where* "Out$(-)$" *denotes the group of outer automorphisms of the group in paren-theses*). *This correspondence* $(X, \iota) \mapsto \alpha_{(X,\iota)}$ *defines a bijection*

$$\pi_0(\mathcal{M}_{g,r}(\mathbb{R})) \cong \mathrm{Hom}_{\mathrm{Gal}(\mathbb{C}/\mathbb{R})}(\mathrm{Gal}(\mathbb{C}/\mathbb{R}), \pi_1(\mathcal{M}_{g,r}^{\iota \mathcal{M}}))$$

from the set of connected components of $\mathcal{M}_{g,r}(\mathbb{R})$ *to the set of conjugacy classes of sections of* $\pi_1(\mathcal{M}_{g,r}^{\iota \mathcal{M}}) \to \mathrm{Gal}(\mathbb{C}/\mathbb{R})$, *or, equivalently, involutions in* $\pi_1(\mathcal{M}_{g,r}^{\iota \mathcal{M}})$. *Moreover, the centralizer of an involution* $\iota_Y \in \pi_1(\mathcal{M}_{g,r}^{\iota \mathcal{M}})$ *is the image of* $\pi_1(Y^{\iota})$ *in* $\pi_1(\mathcal{M}_{g,r}^{\iota \mathcal{M}})$.

REMARK. The *injectivity portion* of the bijection of Corollary 3.8, together with the determination of the centralizer of an involution (the final sentence in the statement of Corollary 3.8), may be regarded as the discrete real analogue of the so-called "Strong Isomorphism Version of the Grothendieck Conjecture." (For the convenience of the reader, we recall that the "Strong Isomorphism Version of the Grothendieck Conjecture" is the statement of Theorem A in the Intro-duction, except with K-morphism (respectively, homomorphism) replaced by K-isomorphism (respectively, isomorphism).)

REMARK. The author was informed by M. Seppala that results similar to Corol-lary 3.8 have been obtained in [AG].

3.4. Abelian varieties and their moduli.

LEMMA 3.9. *Let* (X, ι) *be a real complex manifold such that* X *is an abelian variety over* $\mathbb{C}$. *Then there exists a translation-invariant Riemannian metric on* X *which is preserved by* ι.

PROOF. By the Remark following Definition 3.1, (X, ι) arises from a projec-tive algebraic variety $X_{\mathbb{R}}$ over $\mathbb{R}$. Write X^c for the complex conjugate of the complex manifold X (i.e., X^c and X have the same underlying real analytic manifold, but holomorphic functions on X^c are antiholomorphic functions on X). Since X is an abelian variety over $\mathbb{C}$, it follows that X^c is also an abelian variety over $\mathbb{C}$. Thus, the holomorphic isomorphism $\iota : X \cong X^c$ is the compos-ite of an isomorphism of abelian varieties (i.e., one which preserves the group structures) with a translation. In particular, it follows that ι *preserves the in-variant differentials* $V \stackrel{\mathrm{def}}{=} \Gamma(X, \Omega_X)$ *on* X. Thus, ι induces a semi-linear (with respect to complex conjugation) automorphism of V, i.e., ι induces a *real struc-ture* $V_{\mathbb{R}} \subseteq V_{\mathbb{R}} \times_{\mathbb{R}} \mathbb{C} = V$ on V. Then any inner product on the real vector space $V_{\mathbb{R}}$ induces an ι-invariant inner product on the underlying real vector space of V which, in turn, induces a translation-invariant Riemannian metric on X which is preserved by ι, as desired. $\qquad \square$

REMARK. Any Riemannian metric on $\tilde{X}$ arising from a Riemannian metric as in the conclusion of Lemma 3.9 induces a geometry on $\tilde{X}$ which is isomorphic to *Euclidean space*, hence enjoys the property that any two points are joined by a unique geodesic.

Now if we apply Theorem 3.6 using Lemma 3.9, Example 3.2, we obtain:

COROLLARY 3.10 (DISCRETE REAL SECTION CONJECTURE FOR REAL ABELIAN VARIETIES). *Let (X, ι) be a **real abelian variety**. Then the correspondence $Y \mapsto \iota_Y$ of Section 3.2 defines a bijection*

$$\pi_0(X(\mathbb{R})) \cong \mathrm{Hom}_{\mathrm{Gal}(\mathbb{C}/\mathbb{R})}(\mathrm{Gal}(\mathbb{C}/\mathbb{R}), \pi_1(X^\iota))$$

from the set of connected components of the real locus $X(\mathbb{R})$ to the set of conjugacy classes of sections of $\pi_1(X^\iota) \to \mathrm{Gal}(\mathbb{C}/\mathbb{R})$ (or, equivalently, involutions in $\pi_1(X^\iota)$).

COROLLARY 3.11 (DISCRETE REAL SECTION CONJECTURE FOR MODULI OF ABELIAN VARIETIES). *Let $g \geq 1$ be a positive integer. Write $(\mathcal{A}_g, \iota_\mathcal{A})$ for the moduli stack of principally polarized abelian varieties of dimension g, equipped with its natural antiholomorphic involution (arising from the structure of $\mathcal{A}_g$ as an algebraic stack defined over $\mathbb{R}$). If (X, ι) is a real abelian variety of dimension g, then the exact sequence*

$$1 \to \pi_1(X) \to \pi_1(X^\iota) \to \mathrm{Gal}(\mathbb{C}/\mathbb{R}) \to 1$$

defines a homomorphism $\alpha_{(X,\iota)} : \mathrm{Gal}(\mathbb{C}/\mathbb{R}) \to \pi_1(\mathcal{A}_g^{\iota_\mathcal{A}}) \cong \mathrm{GSp}(\pi_1(X))$ (where "GSp" denotes the automorphisms that preserve, up to a constant multiple, the symplectic form defined by the polarization). This correspondence $(X, \iota) \mapsto \alpha_{(X,\iota)}$ defines a bijection

$$\pi_0(\mathcal{A}_g(\mathbb{R})) \cong \mathrm{Hom}_{\mathrm{Gal}(\mathbb{C}/\mathbb{R})}(\mathrm{Gal}(\mathbb{C}/\mathbb{R}), \pi_1(\mathcal{A}_g^{\iota_\mathcal{A}}))$$

from the set of connected components of $\mathcal{A}_g(\mathbb{R})$ to the set of conjugacy classes of sections of $\pi_1(\mathcal{A}_g^{\iota_\mathcal{A}}) \to \mathrm{Gal}(\mathbb{C}/\mathbb{R})$ (or, equivalently, involutions in $\pi_1(\mathcal{A}_g^{\iota_\mathcal{A}})$). Moreover, the centralizer of an involution $\iota_Y \in \pi_1(\mathcal{A}_g^{\iota_\mathcal{A}})$ is the image of $\pi_1(Y^\iota)$ in $\pi_1(\mathcal{A}_g^{\iota_\mathcal{A}})$.

PROOF. The bijectivity of the natural morphism

$$\pi_1(\mathcal{A}_g^{\iota_\mathcal{A}}) \to \mathrm{GSp}(\pi_1(X))$$

follows from the fact that it is compatible with the projections on both sides to $\mathrm{Gal}(\mathbb{C}/\mathbb{R})$ (where the projection $\mathrm{GSp}(\pi_1(X)) \to \mathbb{Z}^\times = \mathrm{Gal}(\mathbb{C}/\mathbb{R})$ is given by looking at the constant multiple to which the symplectic form arising from the polarization is mapped), together with the well-known bijectivity of this morphism on the "+" portions of both sides. $\square$

3.5. Profinite real anabelian geometry. So far we have considered the real analogue of Grothendieck's anabelian geometry given by using the *discrete* fundamental groups of varieties. Another "real analogue" of anabelian geometry is that given by using the *profinite* fundamental groups. Just as in the discrete, the fundamental result was an existence theorem for real points in the presence of involutions (i.e., Lemma 3.5), *in the profinite case, the fundamental existence is given by the following theorem of Cox* (see [Frdl], Corollary 11.3):

LEMMA 3.12. *Let X be a connected real algebraic variety. Then $X(\mathbb{R}) \neq \varnothing$ if and only if $H^i_{\mathrm{et}}(X, \mathbb{Z}/2\mathbb{Z}) \neq 0$ (where "H^i_{et}" denotes étale cohomology) for infinitely many i.*

REMARK. In particular, if the complex manifold $X(\mathbb{C})$ is a "$K(\pi, 1)$" space (i.e., its universal cover is contractible), and, moreover, its fundamental group $\pi_1(X(\mathbb{C}))$ is *good* (i.e., the cohomology of $\pi_1(X(\mathbb{C}))$ with coefficients in any finite $\pi_1(X(\mathbb{C}))$-module is isomorphic (via the natural morphism) to the cohomology of the profinite completion of $\pi_1(X(\mathbb{C}))$ with coefficients in that module), then we obtain:

(∗) $X(\mathbb{R}) \neq \varnothing$ *if and only if* $H^i_{\mathrm{et}}(\pi_1^{\mathrm{alg}}(X), \mathbb{Z}/2\mathbb{Z}) \neq 0$ *for infinitely many integers i.*

(Here $\pi_1^{\mathrm{alg}}(X)$ denotes the algebraic fundamental group of the scheme X.) Also, if the projection $\pi_1^{\mathrm{alg}}(X) \to \mathrm{Gal}(\mathbb{C}/\mathbb{R})$ possesses a splitting, then the fact that $H^i_{\mathrm{et}}(\mathrm{Gal}(\mathbb{C}/\mathbb{R}), \mathbb{Z}/2\mathbb{Z}) \neq 0$ for infinitely many i implies that

$$H^i_{\mathrm{et}}(\pi_1^{\mathrm{alg}}(X), \mathbb{Z}/2\mathbb{Z}) \neq 0$$

for infinitely many i.

Since hyperbolic curves and abelian varieties satisfy the conditions of the preceding remark, we obtain:

COROLLARY 3.13 (PROFINITE REAL SECTION CONJECTURE FOR REAL HYPERBOLIC CURVES). *Let X be a **hyperbolic curve** over $\mathbb{R}$. Then the profinite version of the correspondence $Y \mapsto \iota_Y$ of Section 3.2 defines a bijection*

$$\pi_0(X(\mathbb{R})) \cong \mathrm{Hom}_{\mathrm{Gal}(\mathbb{C}/\mathbb{R})}(\mathrm{Gal}(\mathbb{C}/\mathbb{R}), \pi_1^{\mathrm{alg}}(X))$$

from the set of connected components of the real locus $X(\mathbb{R})$ to the set of conjugacy classes of sections of $\pi_1^{\mathrm{alg}}(X) \to \mathrm{Gal}(\mathbb{C}/\mathbb{R})$ (or, equivalently, involutions in $\pi_1^{\mathrm{alg}}(X)$).

PROOF. *Surjectivity* follows from the above Remark, using the technique of [Tama1]: Namely, given a section $\alpha : \mathrm{Gal}(\mathbb{C}/\mathbb{R}) \to \pi_1^{\mathrm{alg}}(X)$ of $\pi_1^{\mathrm{alg}}(X) \to \mathrm{Gal}(\mathbb{C}/\mathbb{R})$, the family of open subgroups of $\pi_1^{\mathrm{alg}}(X)$ containing $\mathrm{Im}(\alpha)$ defines a system of coverings $\{X_i \to X\}$ (as i varies over the elements of some index set I) such that (by the above Remark) each $X_i(\mathbb{R}) \neq 0$. Since each $X_i(\mathbb{R})$ has only finitely many connected components, it thus follows that there exists a compatible system (indexed by I) of connected components of $X_i(\mathbb{R})$. But this amounts to the assertion that α arises from some connected component of $X(\mathbb{R})$, as desired (see [Tama1], Corollary 2.10).

Injectivity follows from the fact that involutions arising from distinct connected components define distinct elements of $H^1(\pi_1^{\mathrm{alg}}(X), \mathbb{Z}/2\mathbb{Z})$ — see [Schd], § 20, Propositions 20.1.8, 20.1.12. □

REMARK. To the author's knowledge, the first announcement in the literature of a result such as Corollary 3.13 (in the proper case) appears in a manuscript of Huisman ([Huis]). (In fact, [Huis] also treats the one-dimensional case of Corollary 3.14 below.) Unfortunately, however, the author was not able to follow the portion of Huisman's proof ([Huis], Lemma 5.7) that corresponds to the application of Cox's theorem (as in the Remark following Lemma 3.12).

COROLLARY 3.14 (PROFINITE REAL SECTION CONJECTURE FOR REAL ABELIAN VARIETIES). *Let X be an **abelian variety** over $\mathbb{R}$. Then the profinite version of the correspondence $Y \mapsto \iota_Y$ of Section 3.2 defines a bijection*

$$\pi_0(X(\mathbb{R})) \cong \mathrm{Hom}_{\mathrm{Gal}(\mathbb{C}/\mathbb{R})}(\mathrm{Gal}(\mathbb{C}/\mathbb{R}), \pi_1^{\mathrm{alg}}(X))$$

from the set of connected components of the real locus $X(\mathbb{R})$ to the set of conjugacy classes of sections of $\pi_1^{\mathrm{alg}}(X) \to \mathrm{Gal}(\mathbb{C}/\mathbb{R})$ (or, equivalently, involutions in $\pi_1^{\mathrm{alg}}(X)$).

PROOF. *Surjectivity* follows as in the proof of Corollary 3.13. *Injectivity* follows, for instance, from the discrete result (Corollary 3.10), together with the injectivity of the natural morphism

$$H^1(\mathrm{Gal}(\mathbb{C}/\mathbb{R}), \pi_1(X(\mathbb{C}))) \to H^1(\mathrm{Gal}(\mathbb{C}/\mathbb{R}), \pi_1^{\mathrm{alg}}(X \otimes_\mathbb{R} \mathbb{C}))$$

— itself a consequence of the fact that $\pi_1^{\mathrm{alg}}(X \otimes_\mathbb{R} \mathbb{C}) = \pi_1(X(\mathbb{C})) \otimes_\mathbb{Z} \hat{\mathbb{Z}}$, where $\hat{\mathbb{Z}}$ is the profinite completion of $\mathbb{Z}$ (hence a faithfully flat $\mathbb{Z}$-module). □

As for the case of moduli, the above argument breaks down in most cases since it is either false that or unknown whether or not the fundamental group of the corresponding moduli stacks is *good*. More precisely, $\pi_1(\mathcal{A}_g) = \mathrm{Sp}(2g, \mathbb{Z})$ is known *not to be good* if $g \geq 2$ (see Lemma 3.16 below). (If $g = 1$, then the "profinite real section conjecture" for $\mathcal{A}_g$ is essentially contained in Corollary 3.13 above.) On the other hand, to the author's knowledge, *it is not known whether or not $\pi_1(\mathcal{M}_{g,r})$ is good if $g > 2$. If $g \leq 2$*, then, up to passing to finite étale coverings, $\mathcal{M}_{g,r}$ may be written as a successive extension of smooth families of hyperbolic curves, hence has a good fundamental group. Thus, we obtain:

COROLLARY 3.15 (PROFINITE REAL SECTION CONJECTURE FOR MODULI OF HYPERBOLIC CURVES OF GENUS ≤ 2). *Let $g, r \geq 0$ be integers such that $2g - 2 + r > 0$, $g \leq 2$. Write $(\mathcal{M}_{g,r})_\mathbb{R}$ for the moduli stack of complex hyperbolic curves of type (g, r) over $\mathbb{R}$. If X is a real hyperbolic curve of type (g, r), then X defines a section $\alpha_{(X,\iota)} : \mathrm{Gal}(\mathbb{C}/\mathbb{R}) \to \pi_1^{\mathrm{alg}}((\mathcal{M}_{g,r})_\mathbb{R})$. This correspondence $(X, \iota) \mapsto \alpha_{(X,\iota)}$ defines a bijection*

$$\pi_0((\mathcal{M}_{g,r})_\mathbb{R}(\mathbb{R})) \cong \mathrm{Hom}_{\mathrm{Gal}(\mathbb{C}/\mathbb{R})}(\mathrm{Gal}(\mathbb{C}/\mathbb{R}), \pi_1^{\mathrm{alg}}((\mathcal{M}_{g,r})_\mathbb{R}))$$

from the set of connected components of $(\mathcal{M}_{g,r})_{\mathbb{R}}(\mathbb{R})$ *to the set of conjugacy classes of sections of* $\pi_1^{\mathrm{alg}}((\mathcal{M}_{g,r})_{\mathbb{R}}) \to \mathrm{Gal}(\mathbb{C}/\mathbb{R})$ *(or, equivalently, conjugacy classes of involutions in* $\pi_1^{\mathrm{alg}}((\mathcal{M}_{g,r})_{\mathbb{R}})$). *Moreover, the centralizer of an involution* $\iota_Y \in \pi_1^{\mathrm{alg}}((\mathcal{M}_{g,r})_{\mathbb{R}})$ *is the image of the profinite completion of* $\pi_1(Y^{\iota})$ *in* $\pi_1^{\mathrm{alg}}((\mathcal{M}_{g,r})_{\mathbb{R}})$.

PROOF. Since (as just remarked) the fundamental groups involved are *good*, *surjectivity* follows as in Corollaries 3.13, 3.14.

As for *injectivity*, we reason as follows. Given two real hyperbolic curves X, Y of the same type (g, r) which induce the same section α (up to conjugacy) of $\pi_1^{\mathrm{alg}}((\mathcal{M}_{g,r})_{\mathbb{R}}) \to \mathrm{Gal}(\mathbb{C}/\mathbb{R})$, we must show that they belong to the same connected component of $(\mathcal{M}_{g,r})_{\mathbb{R}}(\mathbb{R})$. First, observe that since $[\mathbb{C} : \mathbb{R}] = 2$, it follows that the marked points of X and Y over $\mathbb{C}$ consist of: (i.) points defined over $\mathbb{R}$; (ii) complex conjugate pairs. Moreover, the combinatorial data of which points are defined over $\mathbb{R}$ and which points are conjugate pairs is clearly determined by the section α. Thus, there exists an "ordering of connected components of the divisor of marked points over $\mathbb{R}$" of X, Y, which is compatible with α. Write

$$\mathcal{N} \to (\mathcal{M}_{g,r})_{\mathbb{R}}$$

for the finite étale covering defined by the moduli stack (over $\mathbb{R}$) of hyperbolic curves equipped with such an ordering. Note, in particular, that the injectivity assertion under consideration for $(\mathcal{M}_{g,r})_{\mathbb{R}}$ follows formally from the corresponding injectivity assertion for $\mathcal{N}$. Moreover, $\mathcal{N}$ may be written as a *"successive extension"*

$$\mathcal{N} = \mathcal{N}_r \to \mathcal{N}_{r-1} \to \cdots \to \mathcal{N}_1 \to \mathcal{N}_0$$

of smooth families (i.e., the $\mathcal{N}_{j+1} \to \mathcal{N}_j$) of either *hyperbolic curves* (where we include curves which are stacks — see the remark in parentheses following the list below) or *surfaces* (of a special type, to be described below) *over the stack* $\mathcal{N}_0$, where $\mathcal{N}_0$ may be described as follows:

(1) If $g = 0$, then $\mathcal{N}_0$ is the moduli stack of 4-pointed curves of genus 0, equipped with an "ordering type" $\mathcal{T}$, where $\mathcal{T}$ is one of the following: a total ordering of the four points; a total ordering of two points, plus a pair of conjugate points; a total ordering of two pairs of conjugate points. In each of these three cases, one sees that $\mathcal{N}_0$ is a *hyperbolic curve* over $\mathbb{R}$, so we conclude the corresponding injectivity assertion for $\mathcal{N}_0$ from Corollary 3.13.

(2) If $g = 1$, then $\mathcal{N}_0$ is either the moduli stack of 1-pointed curves of genus 1 (which is a hyperbolic curve, so we may conclude the corresponding injectivity assertion for $\mathcal{N}_0$ from Corollary 3.13), or $\mathcal{N}_0$ is the moduli stack of 2-pointed curves of genus 1, where the two points are unordered. In the latter case, by considering the "group of automorphisms of the underlying genus 1 curve which preserve the invariant differentials," we get a morphism $\mathcal{N}_0 \to (\mathcal{M}_{1,1})_{\mathbb{R}}$, which is a smooth family whose fiber over the elliptic curve E is the stack

given by forming the quotient of $E\backslash\{0_E\}$ (where 0_E is the origin of E) by the action of ± 1. (Indeed, this fiber parametrizes the "difference" of the two unordered points.) In particular, $(\mathcal{M}_{1,1})_{\mathbb{R}}$, as well as these fibers over $(\mathcal{M}_{1,1})_{\mathbb{R}}$ are hyperbolic curves, so the corresponding injectivity assertion for $\mathcal{N}_0$ follows from Corollary 3.13.

(3) If $g = 2$, then $\mathcal{N}_0$ is the moduli stack of 0-pointed curves of genus 2. Moreover, by using the well-known morphism $(\mathcal{M}_{2,0})_{\mathbb{R}} \to (\mathcal{M}_{0,6})_{\mathbb{R}}$ (given by considering the ramification points of the canonical double covering of the projective line associated to a curve of genus 2), this case may be reduced to the case $g = 0$, which has already been dealt with.

(We remark here that even though some of the hyperbolic curves appearing above are in fact stacks, by passing to appropriate finite étale coverings which are still defined over $\mathbb{R}$ and for which the real points in question lift to real points of the covering, injectivity for such "stack-curves" follows from injectivity for usual curves as proven in Corollary 3.13.) Finally, the surfaces that may appear as fibers in the families $\mathcal{N}_{j+1} \to \mathcal{N}_j$ appearing above are of the following type: If C is a hyperbolic curve over $\mathbb{R}$, write $\Delta_C \subseteq C \times_{\mathbb{R}} C$ for the diagonal. Then the surfaces in question are of the form $\{(C \times_{\mathbb{R}} C)\backslash\Delta_C\}/\mathfrak{S}_2$ (where $\mathfrak{S}_2$ is the symmetric group on two letters permuting the two factors of C, and we note that this quotient is the same whether taken in the sense of schemes or of stacks). Now by passing (as in the one-dimensional case) to appropriate finite étale coverings of these surfaces which are still defined over $\mathbb{R}$ and for which the real points in question lift to real points of the covering, the corresponding injectivity assertion for such surfaces follows from injectivity for surfaces that may be written as a smooth family of hyperbolic curves parametrized by a hyperbolic curve, hence is a consequence of Corollary 3.13. Thus, by *"dévissage"* we conclude the desired *injectivity* for $(\mathcal{M}_{g,r})_{\mathbb{R}}$.

Before proceeding, we *observe* that the above argument shows that the bijectivity assertion of Corollary 3.15 also holds for any finite étale covering of $\mathcal{M}_{g,r}$ which is defined over $\mathbb{R}$.

The final statement on centralizers may be proven as follows: Given an involution ι_Y, write $\mathcal{M}_Y \to (\mathcal{M}_{g,r})_{\mathbb{R}}$ for the pro-covering defined by the subgroup generated by ι_Y in $\pi_1^{\mathrm{alg}}((\mathcal{M}_{g,r})_{\mathbb{R}})$. Then the statement on centralizers follows from the fact that the conjugates of ι_Y in $\pi_1^{\mathrm{alg}}((\mathcal{M}_{g,r})_{\mathbb{R}})$ are in bijective correspondence with the connected components of the inverse images of Y in $\mathcal{M}_Y$ (where we note that this bijective correspondence follows from the observation of the preceding paragraph). $\square$

REMARK. In many respects the profinite theory is more difficult and less elegant than the discrete theory, where everything follows easily from the very general Lemma 3.5. It is thus the feeling of the author that *the discrete theory provides a more natural real analogue of anabelian geometry than the profinite theory.*

LEMMA 3.16. *Let* $g \geq 2$, *and let* $H \subseteq \mathrm{Sp}(2g, \mathbb{Z})$ *be a subgroup of finite index. Then there exists a subgroup* $H' \subseteq H$ *which is normal and of finite index in* $\mathrm{Sp}(2g, \mathbb{Z})$ *such that the cohomological dimension of the profinite completion of* H' *is* $> \dim_{\mathbb{R}}(\mathcal{A}_g) = 2 \cdot \dim_{\mathbb{C}}(\mathcal{A}_g) = g(g+1)$. (*This estimate holds even if one restricts to* H'-*modules of order equal to a power of p, for any fixed prime number p.*) *In particular, if* $g \geq 2$, *then* $\mathrm{Sp}(2g, \mathbb{Z})$ *is* **not good**.

PROOF. First, note that if $\mathrm{Sp}(2g, \mathbb{Z})$ is good, then so is any subgroup H of finite index. But there exist H such that if we write $\mathcal{A}_{H'} \to \mathcal{A}_g$ for the finite étale covering defined by a finite index subgroup $H' \subseteq H$, then $\mathcal{A}_{H'}$ is a *complex manifold* (i.e., not just a stack). The cohomology of H' is then given by the cohomology of $\mathcal{A}_{H'}$. Moreover, the cohomological dimension of $\mathcal{A}_{H'}$ is $= \dim_{\mathbb{R}}(\mathcal{A}_{H'}) = \dim_{\mathbb{R}}(\mathcal{A}_g)$. Thus, if the cohomological dimension of the profinite completion of H' is $> \dim_{\mathbb{R}}(\mathcal{A}_g)$, it follows that the cohomology of H' and of its profinite completion (with coefficients in a finite module) are not isomorphic in general, i.e., that H' is not good. But this implies that $\mathrm{Sp}(2g, \mathbb{Z})$ is not good, as desired.

Next, assume that we are given H as in the statement of Lemma 3.16, and prove the existence of an H' as stated. First, observe that since the *congruence subgroup problem* has been resolved affirmatively for $\mathrm{Sp}(2g, \mathbb{Z})$ (see [BMS]), it follows that

$$\mathrm{Sp}(2g, \mathbb{Z})^{\wedge} = \mathrm{Sp}(2g, \hat{\mathbb{Z}}) = \prod_p \mathrm{Sp}(2g, \mathbb{Z}_p)$$

(where the "$\wedge$" denotes the profinite completion, and the product is taken over all prime numbers p). Thus, it follows that the cohomological dimension of $\mathrm{Sp}(2g, \mathbb{Z})^{\wedge}$ is $\geq$ the cohomological dimension of $\mathrm{Sp}(2g, \mathbb{Z}_p)$ for any prime p. In particular, in order to complete the proof of Lemma 3.16, *it suffices to show that* $\mathrm{Sp}(2g, \mathbb{Z}_p)$ *admits a collection of arbitrarily small normal open subgroups whose p-cohomological dimension is* $> g(g+1)$.

But this follows from the theory of [Laz]: Indeed, by [Laz], V, §2.2.8, it follows that that the p-cohomological dimension of any "p-valuable group" is equal to the "rank" r of the group. Here, a *p-valuable group* (see [Laz], III, §2.1.2) is a topogical group with a filtration satisfying certain properties. In the present context, the topological group $\mathrm{Sp}^{[n]}(2g, \mathbb{Z}_p)$ (i.e., symplectic matrices which are $\equiv$ to the identity matrix modulo p^n), equipped with the filtration defined by the $\mathrm{Sp}^{[m]}(2g, \mathbb{Z}_p)$ for $m \geq n$, will satisfy these properties. Moreover, the *rank* r of a p-valuable group (see [Laz], III, §2.1.1, §2.1.3) is the $\mathbb{Q}_p$-dimension of the Lie algebra $\mathrm{sp}(2g, \mathbb{Q}_p)$ of $\mathrm{Sp}(2g, \mathbb{Z}_p)$. Thus, in this case,

$$r = \dim_{\mathbb{Q}_p}(\mathrm{sp}(2g, \mathbb{Q}_p)) = \dim_{\mathbb{R}}(\mathrm{sp}(2g, \mathbb{R}))$$
$$= \dim_{\mathbb{R}}(\mathrm{Sp}(2g, \mathbb{R})) > \dim_{\mathbb{R}}(\mathfrak{H}_g) = \dim_{\mathbb{R}}(\mathcal{A}_g)$$

(where $\mathfrak{H}_g$ is the *Siegel upper half-plane*— see Example 3.2). Indeed, the inequality here follows from the fact that $\mathrm{Sp}(2g, \mathbb{R})$ acts *transitively* on $\mathfrak{H}_g$, with *positive dimensional* isotropy subgroups. This completes the proof.   □

4. Complements to the p-adic Theory

In this section, we present certain complements to the p-adic theory of [Mzk2] which allow us to prove a certain *isomorphism version* of Theorem A of [Mzk2] (see Section 0 of the present article) over a somewhat larger class of fields K than was treated in [Mzk2]. This larger class of fields—which we refer to as *generalized sub-p-adic*—consists of those fields which may be embedded as subfields of a finitely generated extension of the quotient field of $W(\bar{\mathbb{F}}_p)$ (the ring of Witt vectors with coefficients in the algebraic closure of $\mathbb{F}_p$, for some prime number p).

4.1. Good Chern classes. In this section, we work over a base field K, which we assume (for simplicity, although it is not absolutely necessary for much of what we shall do) to be *of characteristic* 0. Let X_K be a *smooth, geometrically connected variety* over K.

If p is a prime number, and $n \geq 1$ an integer, then we may consider the *Kummer sequence on X_K*, i.e., the exact sequence of sheaves on $(X_K)_{\mathrm{et}}$ (i.e., the étale site of X_K) given by

$$0 \to (\mathbb{Z}/p^n\mathbb{Z})(1) \to \mathbb{G}_m \to \mathbb{G}_m \to 0$$

(where the (1) is a "Tate twist," and the morphism from $\mathbb{G}_m$ to $\mathbb{G}_m$ is given by raising to the p^n-th power.) The connecting morphism induced on étale cohomology by the Kummer sequence then gives us a morphism

$$\delta_{p,n} : H^1_{\mathrm{et}}(X_K, \mathbb{G}_m) \to H^2_{\mathrm{et}}(X_K, (\mathbb{Z}/p^n\mathbb{Z})(1))$$

Now suppose that $\mathcal{L}$ is a *line bundle on X_K*. Then applying $\delta_{p,n}$ to $\mathcal{L} \in H^1_{\mathrm{et}}(X_K, \mathbb{G}_m)$ gives us a compatible system of classes

$$\delta_{p,n}(\mathcal{L}) \in H^2_{\mathrm{et}}(X_K, (\mathbb{Z}/p^n\mathbb{Z})(1)),$$

and hence (by letting p, n vary) a class $c_1(\mathcal{L}) \in H^2_{\mathrm{et}}(X_K, \hat{\mathbb{Z}}(1))$.

DEFINITION 4.1. We shall refer to $c_1(\mathcal{L}) \in H^2_{\mathrm{et}}(X_K, \hat{\mathbb{Z}}(1))$ as the *(profinite, étale-theoretic) first Chern class of $\mathcal{L}$.* If $N \geq 1$ is an integer, then we shall refer to $c_1(\mathcal{L}) \mod N \in H^2_{\mathrm{et}}(X_K, (\mathbb{Z}/N\mathbb{Z})(1))$ as the *(étale-theoretic) first Chern class of $\mathcal{L}$ modulo N.*

Next, write

$$\pi_1(X_K)$$

for the (*algebraic*) *fundamental group* of X_K (where we omit the base-point since it will not be explicitly necessary in our discussion). Also, assume that we are given a *quotient*

$$\pi_1(X_K) \twoheadrightarrow Q$$

(where Q is profinite, and the surjection is continuous). Then we make the following *crucial definition*:

DEFINITION 4.2. Let $N \geq 1$ be an integer. For $i, j \in \mathbb{Z}$, a cohomology class $\eta \in H^i_{\text{et}}(X_K, (\mathbb{Z}/N\mathbb{Z})(j))$ will be called *good* if there exists a (nonempty) finite étale covering $Y \to X_K$ such that $\eta|_Y \in H^i_{\text{et}}(Y, (\mathbb{Z}/N\mathbb{Z})(j))$ is zero.

Next, suppose that $\pi_1(X_K) \twoheadrightarrow Q$ is a surjection such that the composite of the natural surjection $\pi_1(X_K) \twoheadrightarrow \Gamma_K$ with the *cyclotomic character* $\Gamma_K \to (\mathbb{Z}/N\mathbb{Z})^\times$ *factors through* Q. Then we shall say that η is *Q-good* if this covering $Y \to X_K$ may be chosen to arise from a quotient of $\pi_1(X_K)$ that factors through $\pi_1(X_K) \twoheadrightarrow Q$. If $\mathcal{L}$ is a line bundle on X_K, then we will say that *its Chern class is good* (*respectively, Q-good*) *modulo N* if the Chern class of $\mathcal{L}$ modulo N in $H^2_{\text{et}}(X_K, (\mathbb{Z}/N\mathbb{Z})(1))$ is good (respectively, Q-good).

Recall that a discrete group Γ is said to be *good* if the cohomology of Γ with coefficients in any finite Γ-module is isomorphic (via the natural morphism) to the cohomology of the profinite completion of Γ with coefficients in that module. Then the justification for the terminology of Definition 4.2 is the following:

LEMMA 4.3. *Suppose that K is a subfield of $\mathbb{C}$ (the complex number field); that the topological space $\mathcal{X} \overset{\text{def}}{=} X_{\mathbb{C}}(\mathbb{C})$ is a "$K(\pi, 1)$" space (i.e., its universal cover is contractible); and that the topological fundamental group $\pi_1^{\text{top}}(\mathcal{X})$ is good. Then it follows that all cohomology classes $\eta \in H^i_{\text{et}}(X_K, (\mathbb{Z}/N\mathbb{Z})(j))$ are good.*

PROOF. Write $X_{\mathbb{C}} \overset{\text{def}}{=} X_K \otimes_K \mathbb{C}$, $X_{\bar{K}} \overset{\text{def}}{=} X_K \otimes_K \bar{K}$. Since finite étale coverings of $X_{\mathbb{C}}$ are always defined over a finite extension of K, and (by well-known elementary properties of étale cohomology) the natural morphism

$$H^i_{\text{et}}(X_{\bar{K}}, (\mathbb{Z}/N\mathbb{Z})(j)) \to H^i_{\text{et}}(X_{\mathbb{C}}, (\mathbb{Z}/N\mathbb{Z})(j))$$

is an isomorphism, one sees immediately that it suffices to prove Lemma 4.3 when $K = \mathbb{C}$, $j = 0$. But then

$$H^i_{\text{et}}(X_{\mathbb{C}}, \mathbb{Z}/N\mathbb{Z}) \cong H^i_{\text{sing}}(X_{\mathbb{C}}, \mathbb{Z}/N\mathbb{Z}) \cong H^i(\pi_1^{\text{top}}(\mathcal{X}), \mathbb{Z}/N\mathbb{Z})$$

(where the second isomorphism (between singular and group cohomology) follows from the fact that $\mathcal{X}$ is a "$K(\pi, 1)$" space). Thus, the fact that η vanishes upon restriction to a (nonempty) finite étale covering follows from the fact that $\pi_1^{\text{top}}(\mathcal{X})$ is assumed to be good. $\square$

REMARK. Thus, under the hypotheses of Lemma 4.3, *every* cohomology class is good. In general, however, we would like to work with varieties X_K that do *not* satisfy the hypotheses of Lemma 4.3, but which nonetheless have the property that *the cohomology classes that we are interested in are good.*

Then the image of B lies in the diagonal

$$\overline{\mathcal{H}}/\mathrm{PSL}(2,\mathbb{Z}) \to \overline{\mathcal{H}}/\mathrm{PSL}(2,\mathbb{Z}) \times \overline{\mathcal{H}}/\mathrm{PSL}(2,\mathbb{Z}).$$

The preimage of the diagonal in $\overline{\mathcal{H}}/\Gamma \times \overline{\mathcal{H}}/\Gamma'$ decomposes into the union

$$\cup_{g\in\mathrm{PSL}(2,\mathbb{Z})}\overline{\mathcal{H}}/g\Gamma g^{-1} \cap \Gamma'.$$

Since B is irreducible, it dominates exactly one of such curves. If for all g the group $g\Gamma g^{-1}\cap\Gamma' \neq \Gamma$ then Γ is not minimal. Finally, the index of Γ in $\mathrm{PSL}(2,\mathbb{Z})$ is bounded from above by the degree of j.

The elliptic fibration $\mathcal{E} \to B$ defines a rational function on B: the j-function. There is a relationship between the j-function and local (resp. global) monodromies. By Lemma 2.3 above, j determines the monodromy invariant $\rho_{\mathcal{E}}^{c}$ (modulo conjugation in $\mathrm{PSL}(2,\mathbb{Z})$).

Now consider the local situation: the restriction of j to the disc Δ_b is analytically equivalent to $j(b) + z^k$ if $j(b)$ is finite or z^{-k} if $j(b)$ is infinite ($k \in \mathbb{N}$). Here z is a local parameter. There are certain compatibility conditions between the local monodromy ρ_b and k. Kodaira classifies all pairs (ρ_b, k) which occur (see [2]). The types are labeled by I_n, II, III, IV, and I_n^*, II*, III*, IV*). The local monodromy ρ_b around fibers of type I_n is unipotent. The local monodromy around the fibers of type II, III and IV is finite. For $*$-fibers the local monodromy is multiplied by -1 (I_0 is nonsingular, with trivial monodromy).

THEOREM 2.4. *The pair (ρ_b, k) from Kodaira's list defines a unique (in the analytic category) relatively minimal Jacobian fibration over Δ_b. Any two Jacobian elliptic fibrations over an analytic disc Δ_b with the same (ρ_b, k) are fiberwise birationally isomorphic.*

THEOREM 2.5. *For any nonconstant map $j : \mathbb{P}^1 \to \mathbb{P}^1$ there exists an elliptic fibration $\mathcal{E} \to \mathbb{P}^1$ with j-map j.*

If $\mathcal{E}$ and $\mathcal{E}'$ are elliptic fibrations over $\mathbb{P}^1$ such that $j = j'$ then there exists a function $\chi : \mathbb{P}^1 \to \mathbb{Z}/2 = \pm1$ of finite support such that $\prod_{b\in\mathbb{P}^1} \chi(b) = 1$ and $\rho_b = \chi(b)\rho_b'$ for all $b \in \mathbb{P}^1$ (here ρ_b, resp. ρ_b' are the local monodromies for j, resp. j'). Conversely, for every such function χ there exists an elliptic fibration $\mathcal{E}'$ such that $\rho_b = \chi(b)\rho_b'$ (for all $b \in \mathbb{P}^1$) and $j' = j$.

REMARK 2.6. The theorem says that if $\rho_{\mathcal{E}'}^c = \rho_{\mathcal{E}}^c$, then

$$\rho_{\mathcal{E}} = \chi \cdot \rho_{\mathcal{E}'}.$$

In general, there are exactly $2^{g(B)+k-1}$ different liftings of the standard generators of $\pi_1(B^0)$ to $\mathrm{SL}(2,\mathbb{Z})$, (see part (a) of Theorem 11.1 p. 160 in [2]).

We are interested in classifying global monodromies in some restricted class of surfaces, for example rational elliptic or elliptic K3 surfaces. For each of these classes the degree of j is bounded. This implies a bound on the index of the global monodromy group $\tilde{\Gamma}$ in $\mathrm{SL}(2,\mathbb{Z})$. Only a finite number of possible global monodromy groups $\tilde{\Gamma}$ and only few homological invariants can occur if we fix the image of $\tilde{\Gamma}$ in $\mathrm{PSL}(2, \mathbb{Z})$. Elliptic surfaces with the same j-invariant but different homological invariants are scattered through different topological classes. Our point of departure was that Kodaira's theory does not provide a sufficiently simple combinatorial control over the topology of the resulting surfaces. In the following sections we give some technical improvements of Kodaira's theory which lead to an effective algorithm.

3. j-Modular Curves

Let $\mathcal{E} \to B$ be an elliptic fibration as above. By Lemma 2.3 j-map decomposes as a product $j = j_\Gamma \circ j_\mathcal{E}$ where $j_\mathcal{E} : B \to \overline{\mathcal{H}}/\Gamma$ is a natural lifting of j onto the modular curve $M_\Gamma = \overline{\mathcal{H}}/\Gamma$ corresponding to Γ and

$$j_\Gamma : \overline{\mathcal{H}}/\Gamma \to \overline{\mathcal{H}}/\mathrm{PSL}(2, \mathbb{Z}) = \mathbb{P}^1. \qquad (3\text{--}1)$$

The above decomposition shows that $\deg(j) = \deg(j_\mathcal{E}) \cdot \deg(j_\Gamma)$. In particular, for any non-isotrivial elliptic surface the group Γ is a subgroup of finite index in $\mathrm{PSL}(2, \mathbb{Z})$.

DEFINITION 3.1. We call the pair (M_Γ, j_Γ) the j-modular curve corresponding to the monodromy group Γ.

REMARK 3.2. Usual modular curves are j-modular. A j-modular curve is simply any curve defined over a number field together with a special rational function on it (this follows from the theorem of Belyi [3], see 3.8). There is a countable number of such functions for each curve.

Let us give a combinatorial description of j-modular curves. They correspond to special triangulations of Riemann surfaces.

DEFINITION 3.3. Let R be an oriented Riemann surface. A triangulation $\tau(R) = (\tau_0, \tau_1, \tau_2)$ of R is a decomposition of R into a finite union of open 2-cells τ_2 and a connected graph τ_1 with vertices τ_0 such that the complement $\tau_1 \setminus \tau_0$ is a disjoint union of open segments and the closure of any open 2-cell is isomorphic to the image of a triangle under a simplicial map.

The number of edges originating in a vertex x is called the valence at x and is denoted by $v(x)$.

DEFINITION 3.4. A j-triangulation of R is a triangulation together with a coloring of vertices in three colors A, B and I such that

(i) The colors of any two adjacent vertices are different.
(ii) There are 2 or 6 edges at vertices of color A and 2 or 4 edges at vertices of color B.

We will refer to vertices of color A (resp. B) with valence j as A_j (resp. B_j) vertices. If we delete the I-vertices from τ_0 and all edges AI and BI from τ_1 then the remaining connected 3-valent graph on R with A- and B-ends is called the j-graph associated to the j-triangulation. The complement to this graph is a disjoint union of a finite number of cells (neighborhoods of I-vertices). It might look as in Figure 2, where we use a small circle to indicate an A-

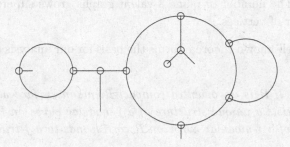

Figure 2.

vertex. The B-vertices are placed on the edges between two A-vertices. A "loose" end represents a B-vertex. A j-graph is called saturated if all A-vertices are A_6-vertices. Saturated graphs can be considered as arising from generalized triangulations of $\mathbb{P}^1$. An arbitrary graph can be obtained from a saturated graph by addition of trees.

Figure 3 shows saturated graphs with $a_6 = 4$.

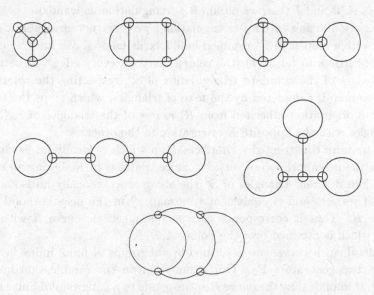

Figure 3.

A j-triangulation on R can be reconstructed from a j-graph by placing one I-vertex into each connected component of R minus the j-graph and by connecting (cyclically) the I-vertex with vertices on the boundary of the corresponding connected component. The valences of A-vertices in an j-graph are 1 or 3, the valences of B-vertices are 1 or 2 and vertices of the same color are not connected by an edge.

REMARK 3.5. The number of plane 3-valent graphs grows superexeponentially with the number of vertices.

The following well-known theorem forms the basis for our analysis of monodromy groups.

THEOREM 3.6. *If R is an oriented compact Riemann surface with a j-triangulation, there exists a unique structure of a j-modular curve on R. Conversely, every structure of a j-modular curve on R corresponds to a j-triangulation.*

PROOF. We first show how j_Γ defines a triangulation of M_Γ. The map $j : \overline{\mathcal{H}} \to \overline{\mathcal{H}}/\mathrm{PSL}(2,\mathbb{Z}) = \mathbb{P}^1$ is ramified over three points $0 = A, 1 = B, \infty = I$. The ramification index at 0 is equal to 3, the ramification index at 1 is 2 and the ramification index at ∞ is infinite. Similar result is true for

$$j_\Gamma : \overline{\mathcal{H}}/\Gamma = M_\Gamma \to \overline{\mathcal{H}}/\mathrm{PSL}(2,\mathbb{Z}) = \mathbb{P}^1.$$

Consider the standard triangulation $\tau_{st}(\mathbb{S}^2)$ of the sphere $\mathbb{S}^2 = \mathbb{P}^1$ into a union of two triangles with vertices $0, 1$ and ∞. The preimage of this triangulation provides a triangulation of M_Γ. If we color the preimages of the corresponding vertices in A, B and I then we obtain a j-triangulation as wanted.

Conversely, starting with a j-triangulation τ we construct an algebraic curve R together with a map $R \to \mathbb{S}^2$ ramified in $0, 1, \infty$ as follows. We have a map from the set of vertices to (A, B, I) (the color). Further, every edge will be mapped into the edges of the standard triangulation of $\mathbb{S}^2$, respecting the colors of the ends. This map is completed by the map of triangles, which maps the triangles ABI (with orientation inherited from R) to one of the triangles of $\tau_{st}(\mathbb{S}^2)$ and the triangles with the opposite R-orientation to the other.

Thus we have constructed a simplicial map which is locally an isomorphism except in the neighborhood of vertices. Since triangles in R sharing an edge are mapped into different triangles of $\mathbb{S}^2$ the above map is locally an isomorphism outside of vertices and is equivalent to a map z^n in the neighborhood of each vertex in R. Thus it corresponds to a unique algebraic curve R with a map $R \to \mathbb{P}^1$ which is ramified over the points A, B, I.

In general, such curves are described by subgroups of finite index in the free group on two generators $\mathbf{F}_2$. Our assumption on the ramification indices at points A, B implies that the curve R corresponds to a subgroup of finite index in the quotient $\mathbb{Z}/2 * \mathbb{Z}/3$ of $\mathbf{F}_2$. The group $\mathbb{Z}/2 * \mathbb{Z}/3$ equals $\mathrm{PSL}(2,\mathbb{Z})$. Thus local monodromy groups over A-vertices can be either 1 or $\mathbb{Z}/3$ and over B either 1 or $\mathbb{Z}/2$. This finishes the proof of the theorem. $\square$

COROLLARY 3.7. *The number of triangles in any j-triangulation is equal to $2\deg(j_\Gamma)$. Moreover, $2\deg(j_\Gamma) = \sum_i v(i)$, where the summation is over all vertices i with color I.*

REMARK 3.8. It follows from Belyi's theorem that every arithmetic curve (an algebraic curve defined over a number field) can be realized as a j-modular curve. Moreover, the corresponding triangulation of the curve is a barycentric subdivision of an arbitrary triangulation the underlying Riemann surface. In this case, Γ is torsion free and a subgroup of $\mathbf{F}_2 \subset \mathrm{PSL}(2,\mathbb{Z})$. The corresponding j-graph is a trivalent graph *without ends*. These type of j-graphs are called *saturated*. They correspond to a relatively small fraction of possible monodromy groups.

Many properties of Γ as a subgroup of $\mathrm{SL}(2,\mathbb{Z})$ can be easily recovered from the j-triangulation. For example, there is a bijection between the set of B_2-vertices and conjugacy classes of subgroups of order 2 in Γ. Similarly, there is a bijection between A_2-vertices and conjugacy classes of subgroups of order 3 in Γ. Finally, there is a bijection between the I-vertices and conjugacy classes of unipotent subgroups in $\Gamma \subset \mathrm{PSL}(2,\mathbb{Z})$. The generator of the unipotent subgroup is given by

$$\begin{pmatrix} 1 & v(i)/2 \\ 0 & 1 \end{pmatrix},$$

where $v(i)$ is the valence of the corresponding I-vertex i.

4. j-Modular Surfaces

In this section we study Jacobian elliptic surfaces such that the map $j_\mathcal{E}$ has degree 1. Here $\tilde{\Gamma} \subset \mathrm{SL}(2,\mathbb{Z})$ is the global monodromy group of the elliptic fibration $\mathcal{E}$. We call such surfaces j-modular surfaces and denote them by $S_{\tilde{\Gamma}}$.

Consider the j-modular curve M_Γ where Γ is the image of $\tilde{\Gamma}$ in $\mathrm{PSL}(2,\mathbb{Z})$ under the natural projection. We want to solve the following problem: describe all surfaces $S_{\tilde{\Gamma}}$ together with the structure of a Jacobian elliptic fibration over the j-modular curve M_Γ such that the monodromy group $\tilde{\Gamma}$ surjects onto Γ. We want to give a complete answer to this question using the j-triangulation of M_Γ.

We have an exact sequence

$$0 \to \mathbb{Z}/2 \to \mathrm{SL}(2,\mathbb{Z}) \to \mathrm{PSL}(2,\mathbb{Z}) \to 1 \qquad (4\text{--}1)$$

which induces a sequence

$$0 \to \mathbb{Z}/2 \to \Gamma' \to \Gamma \to 1, \qquad (4\text{--}2)$$

where $\Gamma' \subset \mathrm{SL}(2,\mathbb{Z})$.

LEMMA 4.1. *If Γ does not contain elements of order 2 then the exact sequence (4-2) splits. Equivalently, the j-triangulation of M_Γ does not contain B_2-vertices.*

PROOF. The group $\mathrm{PSL}(2,\mathbb{Z}) = \mathbb{Z}/2 * \mathbb{Z}/3$. Any subgroup of finite index is a finite free product of groups isomorphic to $\mathbb{Z}, \mathbb{Z}/2, \mathbb{Z}/3$. Assuming that Γ has no elements of order 2 we have a representation of Γ as a free product of groups $\mathbb{Z}, \mathbb{Z}/3$. If we lift the generators of these free generating subgroups to elements of the same order in Γ' we obtain a subgroup of Γ' which projects isomorphically onto Γ, in other words, a splitting of the exact sequence 4-2.   □

REMARK 4.2. All splittings differ by $\mathbb{Z}/2$-characters of Γ $(H^1(\Gamma, \mathbb{Z}/2))$ and the one we obtain may not be the best (this will be specified in Section 5). Namely, the preimages of unipotent generators can be products of unipotent elements by the central element in $\mathrm{SL}(2,\mathbb{Z})$. There may be no natural splitting.

We have to consider three cases:

Case 1: $\tilde{\Gamma} \simeq \Gamma$. There are finitely many such $\tilde{\Gamma}$ and they differ by a character of Γ. For each such character there exists a unique (up to birational morphisms) j-modular $S_{\tilde{\Gamma}}$. Indeed, take the quotient $V^o \to \mathcal{H}/\Gamma$ of the universal elliptic curve $\mathcal{E}^u \to \mathcal{H}$ by $\tilde{\Gamma}$. It has the structure of a fibration with a section and with generic fibers smooth elliptic curves. The monodromy of this fibration (over the open curve $B = \mathcal{H}/\Gamma$) is $\tilde{\Gamma} \simeq \Gamma$. Compactify V^o keeping the structure of an elliptic fibration (over $M_\Gamma = \bar{\mathcal{H}}/\Gamma$) and the zero section as above. It is clear that this construction is birationally universal. Indeed, if there is a Jacobian elliptic fibration V' with the given monodromy group $\tilde{\Gamma}$ then there is a rational fiberwise map $V \to V'$ which is regular on the grouplike parts of V and V'.

Case 2: There exists a lifting $\Gamma' \simeq \Gamma$ but $\Gamma' \not\simeq \tilde{\Gamma}$. The corresponding surfaces $S_{\tilde{\Gamma}}$ are obtained from surfaces in Case 1 by an even number of twists. Thus the set of such surfaces is parametrized by a symmetric power of $\mathbb{P}^1$ (modulo the action of the finite group of automorphisms of the embedded j-graph).

Case 3: The general case. Consider the universal elliptic curve $\mathcal{E}^u \to \mathcal{H}$ given as a quotient of $\mathbb{C} \times \mathcal{H}$ by $\mathbb{Z}e_1 \oplus \mathbb{Z}e_2$. The action of $\mathbb{Z}e_1 \oplus \mathbb{Z}e_2$ on $\mathbb{C} \times \lambda$ is given by

$$e_1(z, \lambda) = (z+1, \lambda),$$
$$e_2(z, \lambda) = (z+\lambda, \lambda)$$

(here $(z, \lambda) \in \mathbb{C} \times \mathcal{H}$). The group $\mathrm{SL}(2,\mathbb{Z})$ acts on $\mathcal{E}^u$, stabilizing the section $(0, \lambda)$. Consider the quotient of the universal elliptic curve $\mathcal{E}^u \to \mathcal{H}$ by Γ'. We get an open surface V' admitting a fibration (with a section) over the open curve $B' = \mathcal{H}/\Gamma'$, whose generic fiber is a smooth rational curve. The map $\mathcal{E}^u \to V'$ is ramified over a divisor D which has at least two horizontal components: D_0 (which is a smooth zero-section of $V' \to B'$) and D_1 which projects to B' with degree 3 and is smooth and unramified over B' in the complement of singular fibers. Denote by V^o the open surface obtained by removing from V' the singular fibers. The surface V^o is fibered over an open curve B^o with fibers $\mathbb{P}^1$. The intersection of the divisor D with each fiber consists of exactly 4 points and D is unramified over B^o.

We want to define a double covering of V^o which is ramified on every component of D. There is a correspondence between such double coverings and special characters

$$\chi \in \text{Hom}(\pi_1(V^o \setminus D), \mathbb{Z}/2).$$

The group $\pi_1(V^o \setminus D)$ has a quotient which is a central $\mathbb{Z}/2$-extension of the free group $\pi_1(B^o)$. This extension has a section (since the fibration $V^o \to B^o$ has a section) and therefore it splits into a product $\mathbb{Z}/2 \times \pi_1(B^o)$. A character χ defining a double cover of $V^o \setminus D$ is a character which is induced from $\mathbb{Z}/2 \times \pi_1(B^o)$ and which is an isomorphism on the central subgroup $\mathbb{Z}/2$ in $\mathbb{Z}/2 \times \pi_1(B^o)$.

In other words, the restriction of χ to the subgroup $\pi_1(\mathbb{P}^1 \setminus 4 \text{ points})$ (for every fiber $\mathbb{P}^1$ of the fibration $V^o \to B^o$) is equal to the standard character of $\mathbf{F}_3$ (realized as $\pi_1(\mathbb{P}^1 \setminus 4 \text{ points})$) which sends the standard generators of $\mathbf{F}_3$ into the non-zero element of $\mathbb{Z}/2$.

We summarize this in the diagram:

$$
\begin{array}{ccc}
\mathbf{F}_3 & \longrightarrow & \pi_1(V^o \setminus D) \\
\downarrow & & \downarrow \\
\mathbb{Z}/2 \quad \times & \pi_1(B^o) & \to \mathbb{Z}_2 \\
\uparrow & \downarrow & \\
\text{Ker}(\chi) & \longrightarrow \quad \Gamma' &
\end{array}
$$

The group $\text{Ker}(\chi)$ is a subgroup of $\mathbb{Z}/2 \times \pi_1(B^o)$ of index 2 and it is isomorphic to $\pi_1(B^o)$. This induces a map $\text{Ker}(\chi) \to \Gamma'$.

The character χ defines a double cover $W^o(\chi)$ of V^o. The preimage of every fiber $\mathbb{P}^1$ of $V^o \to B^o$ is an elliptic curve realized as a standard double cover of this $\mathbb{P}^1$. Thus we obtain an open surface $W^o(\chi)$ with a structure of an elliptic fibration over B^o. All fibers are smooth. The monodromy $\tilde{\Gamma}$ of this elliptic fibration coincides with the image of $\text{Ker}(\chi)$ in Γ'. If $\tilde{\Gamma}$ is not equal to the whole of Γ' then the sequence 4–2 splits. This also means that the character χ is induced from Γ'.

The character χ completely defines the local monodromy around the points in $M_\Gamma \setminus B^o$. Now we compactify V^o keeping the structure of an elliptic fibration over M_Γ and keeping the zero section. Locally, in the neighborhood of $b \in M_\Gamma$ corresponding to singular fibers our elliptic fibration is birationally isomorphic to a standard fibration from the Kodaira list. The corresponding birational isomorphism is biregular on the complement to the singular fiber. The zero section is preserved under this birational isomorphism. Now we can modify our initial fibration via this fiberwise transformation along neighborhoods of singular fibers. The resulting surface V is smooth and it admits a structure of a Jacobian elliptic fibration with the same monodromy group $\tilde{\Gamma}$.

Now consider the diagram

$$
\begin{array}{ccc}
\mathcal{E} & & S_{\tilde{\Gamma}} \\
\downarrow & & \downarrow \\
B & \to M_\Gamma & \to \mathbb{P}^1
\end{array}
$$

Let $B^o = B \setminus j^{-1}\{0, 1, \infty\}$ and $M_\Gamma^o = M_\Gamma \setminus j_\Gamma^{-1}\{0, 1, \infty\}$ (the points deleted from M_Γ are the A, B and I -vertices of the j-triangulation). There is a natural map $\pi_1(B^o) \to \pi_1(M_\Gamma^o)$ and a commutative diagram of monodromy homomorphisms:

$$
\begin{array}{ccc}
\pi_1(B^o) & \longrightarrow & \Gamma' \\
\downarrow & & \downarrow \\
\pi_1(M_\Gamma^o) & \longrightarrow & \Gamma
\end{array}
$$

and a monodromy homomorphism $\pi_1(M_\Gamma^o) \to \Gamma'$, compatible with the projection $\Gamma' \to \Gamma$.

Finally, we want to compare the lifting of the elliptic fibration $S_{\tilde{\Gamma}}$ to B and $\mathcal{E}$.

Case A: $\tilde{\Gamma} \simeq \Gamma$. Then $\mathcal{E}$ is (fiberwise birationally) isomorphic to one of the j-modular surfaces constructed in Case 1 above, namely to the surface corresponding to the section $\Gamma \to \mathrm{SL}(2, \mathbb{Z})$.

Case B: The general case. Then $\mathcal{E}$ is (fiberwise birationally) obtained as a composition of a pullback of a corresponding $S_{\tilde{\Gamma}}$ to B followed by an even number of twists.

5. The Topological Type of j-Modular Surfaces

In this section we determine j-modular surfaces of smallest possible $\chi(\mathcal{E})$ among all $S_{\tilde{\Gamma}}$ with fixed Γ. For simplicity we assume that $B = \mathbb{P}^1$. Similar techniques work for any base M_Γ.

Jacobian elliptic fibrations over $\mathbb{P}^1$ arise in families, defined (in Weierstrass form) as follows. Denote by $U_0 = \mathbb{A}^1$ a chart of $\mathbb{P}^1$ obtained by deleting $(0 : 1)$ and by $U_\infty = \mathbb{A}^1$ the chart obtained by deleting $(1 : 0)$. On U_0 we use the coordinate t and on U_∞ the coordinate $s = 1/t$. Consider a hypersurface in $\mathbb{P}^2 \times U_0$ given by

$$
zy^2 = x^3 + p_0(t)xz^2 + q_0(t)z^3
$$

where p (resp. q) is a polynomial of degree $4r$ (resp. $6r$). In U_∞ the equation is similar, with $p_\infty(s) = p_0(1/s)s^{4r}$ and $q_\infty(s) = q_0(1/s)s^{6r}$. We get elliptic fibrations over U_0, U_∞ which we can glue to an elliptic surface $\mathcal{E} \to B$. The j-function (on U_0) is given by $p_0(t)^3/(4p_0(t)^3 + 27q_0(t)^2)$. The obtained fibration can be singular in fibers corresponding to $b \in B$ where $4p_0(t)^3 + 27q_0(t)^2 = 0$ and the singularities can be resolved by a sequence of blow-ups. The outcome is a (unique) smooth relatively minimal Jacobian elliptic fibration. Thus we get a family $\mathcal{F}_r$ of such elliptic fibrations. Notice that $12r = \chi(\mathcal{E})$. Conversely, a simply connected, compact, minimal Jacobian elliptic fibration with $\chi(\mathcal{E}) = 12r$ belongs to $\mathcal{F}_r$. The family $\mathcal{F}_r$ is parametrized by the coefficients of p_0, q_0 (subject to certain constrains): it is a smooth irreducible variety. Every Jacobian elliptic fibration is birational to a minimal elliptic fibration and the j-map for both fibrations is the same.

On the other hand, the Euler characteristic $\chi(\mathcal{E})$ can be computed as a sum of contributions from singular fibers:

	χ_b		χ_b
I_0		I_0^*	6
I_n	n	I_n^*	$n+6$
II	2	IV*	8
III	3	III*	9
IV	4	II*	10

Here I_0 is a smooth fiber, I_n is a multiplicative fiber with n-irreducible components. The types II, III and IV correspond to the case of potentially good reduction. More precisely, the neighborhood of such a fiber is a (desingularization of a) quotient of a local fibration with smooth fibers by an automorphism of finite order. The corresponding order is 4 for the case III and 3 in the cases II, IV. The fibers of type I_0^*, (resp. I_n^*, II*, III*, IV*) are obtained from fibers I_0 (resp. I_n, IV, III, II) by twisting (changing the local automorphism by the involution $x \mapsto -x$ in the local group structure of the fibration). We shall call them *-*fibers* in the sequel.

REMARK 5.1. The local invariant χ_b has a monodromy interpretation. Any element of a local monodromy at $b \in B^s$ has a minimal representation as a product of elements conjugated to $\left(\begin{smallmatrix} 1 & 1 \\ 0 & 1 \end{smallmatrix}\right)$ in $\mathrm{SL}(2, \mathbb{Z})$. The length of this representation equals χ_b. This explains the equality $\chi_{b^*} = \chi_b + 6$: the element $\left(\begin{smallmatrix} -1 & 0 \\ 0 & -1 \end{smallmatrix}\right) \in \mathrm{SL}_2(\mathbb{Z})$ is a product of 6 elements conjugated to $\left(\begin{smallmatrix} 1 & 1 \\ 0 & 1 \end{smallmatrix}\right)$ (elementary Dehn twists).

PROPOSITION 5.2. *Fix Γ and assume that the number of vertices v_0 in the associated j-modular graph is divisible by 4. Then there exists a unique (up to birational transformations) j-modular surface with (minimal)*

$$\chi(S_{\tilde{\Gamma}}) = 3v_0.$$

If v_0 is not divisible by 4 then the set of j-modular surfaces with (minimal)

$$\chi(S_{\tilde{\Gamma}}) = 3(v_0 + 2)$$

forms a 1-parameter family.

PROOF. First of all observe that v_0 is always even. Let $S_{\tilde{\Gamma}} \to M_\Gamma$ be a j-modular surface with given Γ and singular fibers exactly over the vertices of the j-modular triangulation. Then

$$\chi(S_{\tilde{\Gamma}}) = \sum_b \chi_b \geq 2a_2 + 3b_2 + \sum_I \chi_b.$$

At the same time

$$2 \sum_I \chi_b = 2(3a_6 + a_2)$$

equals the number of triangles in the j-modular triangulation: every triangle contains exactly one A-vertex, A_2-vertices are contained in two triangles and A_6-vertices in 6 triangles. Therefore,

$$\chi(S_{\tilde{\Gamma}}) \geq 3a_6 + a_2 + 2a_2 + 3b_2 = 3v_0.$$

The difference $\chi(S_{\tilde{\Gamma}}) - 3v_0$ is equal to 6 times the number of $*$-fibers. Twisting an even number of $*$-fibers we can diminish this difference — either to zero (when v_0 is divisible by 4) or to 6 (otherwise). In the latter case the $*$-fiber can be chosen freely on B^o. □

References

[1] J. Alexander, *Note on Riemann spaces*, Bull. AMS, **26**, 370–372, (1920).

[2] W. Barth, C. Peters, A. van de Ven, *Compact complex surfaces*, Ergebnisse der Mathematik und ihrer Grenzgebiete, Springer-Verlag, Berlin-New York, (1984).

[3] V. G. Belyi, *On Galois extensions of the maximal cyclotomic field*, Izv. AN USSR, **43:2**, 269–276, (1979).

[4] F. Bogomolov, T. Petrov, Yu. Tschinkel, *Rationality of moduli of elliptic fibrations with fixed monodromy*, Geom. and Funct. Analysis, **12, no. 6**, 1105-1160, (2002).

[5] R. Friedman, J. Morgan, *Smooth Four-Manifolds and Complex Surfaces*, Ergebnisse der Math. und ihrer Grenzgebiete, vol. 27, Springer-Verlag, (1994).

[6] V. A. Iskovskikh, I. R. Shafarevich, *Algebraic surfaces*, Current problems in mathematics. Fundamental directions, Vol. 35 (Russian), 131–263, 272, VINITI, Moscow, (1989).

[7] K. Kodaira, *On compact analytic surfaces II, III*, Ann. of Math. (2) **77** (1963), 563–626; ibid. **78** (1963), 1–40.

[8] A. Lubotzky, *Subgroup growth*, Proceedings of the International Congress of Mathematicians, (Zürich, 1994), 309–317, Birkhäuser, Basel, (1995).

[9] *The Grothendieck Theory of Dessins d'Enfants*, ed. L. Schneps, LMS Lecture Notes, vol. 200, Cambridge Univ. Press, (1994).

[10] *Geometric Galois Actions*, ed. L. Schneps and P. Lochak, LMS Lecture Notes, vol. 242, Cambridge Univ. Press, (1997).

[11] G. Shabat, *On the classification of plane trees by their Galois orbits*, The Grothendieck theory of dessins d'enfants (Luminy, 1993), 169–177, London Math. Soc. Lecture Note Ser., 200, Cambridge Univ. Press, Cambridge, (1994).

[12] G. G. Shabat, V. Voevodsky, *Drawing curves over number fields*, Grothendieck Festschrift, vol. 3, 199–227, (1990).

[13] I. R. Shafarevich, *Algebraic surfaces*, Proc. Steklov Inst. Math., **75**, III-VI, 52-84, 162-182, (1965).

FEDOR BOGOMOLOV
COURANT INSTITUTE, NEW YORK UNIVERSITY
251 MERCER ST.
NEW YORK, NY 10012
UNITED STATES
   bogomolo@cims.nyu.edu

YURI TSCHINKEL
DEPARTMENT OF MATHEMATICS
PRINCETON UNIVERSITY
PRINCETON, NJ 08544
UNITED STATES
   ytschink@math.princeton.edu

Galois Groups and Fundamental Groups
MSRI Publications
Volume 41, 2003

Tannakian Fundamental Groups
Associated to Galois Groups

RICHARD HAIN AND MAKOTO MATSUMOTO

ABSTRACT. The goal of this paper is to give background and motivation for several conjectures of Deligne and Goncharov concerning the action of the absolute Galois group on the fundamental group of the thrice punctured line, and to sketch solutions, complete and partial, of several of them. A major ingredient in these is the theory of weighted completion of profinite groups. An exposition of weighted completion from the point of view of tannakian categories is included.

1. Introduction

The goal of this paper is to provide background, heuristics and motivation for several conjectures of Deligne [14, 8.2, p. 163], [14, 8.9.5, p. 168] and [27, p. 300] and Goncharov [19, Conj. 2.1], presumably along the lines used to arrive at them. A complete proof of the third of these conjectures, and partial solutions of the remaining three are given in [23].[1] A second goal of this paper is to show that the weighted completion of a profinite group, developed in [23], and a key ingredient in the proofs referred to above, can be defined as the tannakian fundamental group[2] of certain categories of modules of the group. This should help clarify the role of weighted completion in [23].

Hain was supported in part by grants from the National Science Foundation. Matsumoto was supported in part by a Mombusho Grant and also by MSRI during a visit in the fall of 1999.

[1]After writing that paper, we learned from Goncharov that proofs of ℓ-adic versions of [14, 8.2, p. 163] and [14, 8.9.5, p. 168] had previously been given in the unpublished manuscript [2] of Beilinson and Deligne.

[2]A tannakian category $\mathcal{T}$ with fiber functor ω is equivalent to the category of representations of the automorphism group of ω. We shall refer to this proalgebraic group as the *tannakian fundamental group* of $\mathcal{T}$ with respect to the base point ω. Basic material on tannakian categories, such as their definition, can be found in [13].

2. Motivic Cohomology

It is believed that there is a universal cohomology theory, called *motivic cohomology*. It should be defined for all schemes X. It is indexed by two integers m and n. The coefficient ring Λ is typically $\mathbb{Z}$, $\mathbb{Z}/N$, $\mathbb{Z}_\ell$, $\mathbb{Q}$ or $\mathbb{Q}_\ell$; the corresponding motivic cohomology group is denoted

$$H^m_{\mathrm{mot}}(X, \Lambda(n)).$$

There should be cup products

$$H^{m_1}_{\mathrm{mot}}(X, \Lambda(n_1)) \otimes H^{m_2}_{\mathrm{mot}}(X, \Lambda(n_2)) \to H^{m_1+m_2}_{\mathrm{mot}}(X, \Lambda(n_1 + n_2)). \qquad (2\text{--}1)$$

Motivic cohomology should have the following universal mapping property: if $H^\bullet_{\mathcal{C}}(\ , \Lambda(\ ))$ is any Bloch-Ogus cohomology theory [9] (such as étale cohomology, Deligne cohomology, Betti (i.e., singular) cohomology, de Rham cohomology, and crystalline cohomology) there should be a unique natural transformation

$$H^m_{\mathrm{mot}}(\ , \Lambda(n)) \to H^m_{\mathcal{C}}(\ , \Lambda(n))$$

compatible with products and Chern classes

$$c_n : K_m(X) \to H^{2n-m}_{\mathcal{C}}(X, \Lambda(n)),$$

where $K_\bullet$ denotes Quillen's algebraic K-group functor, [39].

Beilinson's definition. Beilinson [1] observed that the motivic cohomology with $\mathbb{Q}$ coefficients of a large class of schemes could be defined in terms of Quillen's algebraic K-theory [39].

Suppose that X is the spectrum of the ring of S-integers in a number field or a smooth scheme over a perfect field. Denote the algebraic K-theory of X by $K_\bullet(X)$. As in the case of topological K-theory, there are Adams operations ([25], [31])

$$\psi^k : K_\bullet(X) \to K_\bullet(X),$$

defined for all $k \in \mathbb{Z}_+$. They can be simultaneously diagonalized:

$$K_m(X) \otimes \mathbb{Q} = \bigoplus_{n \in \mathbb{Z}} K_m(X)^{(n)}$$

where ψ^k acts as k^n on $K_m(X)^{(n)}$.

DEFINITION 2.1 (BEILINSON). For a ring Λ containing $\mathbb{Q}$, define the motivic cohomology groups of X by

$$H^m_{\mathrm{mot}}(X, \Lambda(n)) = K_{2n-m}(X)^{(n)} \otimes_{\mathbb{Q}} \Lambda.$$

The ring structure of $K_\bullet(X)$ induces a cup product (2–1) as

$$\psi^k(xy) = \psi^k(x)\psi^k(y) \quad x, y \in K_\bullet(X).$$

Motivation for Beilinson's definition comes from topological K-theory and can be found in the introduction of [4].

If X is smooth, then it follows from a result of Grothendieck (see [12]) that

$$H^{2n}_{\mathrm{mot}}(X, \mathbb{Q}(n)) \cong CH^n(X) \otimes \mathbb{Q}.$$

In the next section, we present the well-known computation of the motivic cohomology of the ring of S-integers in a number field.

PROPOSITION 2.2. *There are Chern classes*

$$c^{\mathrm{mot}}_j : K_m(X) \to H^{2j-m}_{\mathrm{mot}}(X, \mathbb{Q}(j))$$

such that for each Bloch-Ogus cohomology theory $H^\bullet_C(\ , \Lambda(\ ))$, where Λ contains $\mathbb{Q}$, there is a natural transformation

$$H^\bullet_{\mathrm{mot}}(\ , \Lambda(\ )) \to H^\bullet_C(\ , \Lambda(\ ))$$

that is compatible with Chern classes.

PROOF. The basic tool needed to construct the natural transformations to other cohomology theories is the theory of Chern classes

$$c_j : K_m(X) \to H^{2j-m}_C(X, \mathbb{Z}(j))$$

constructed by Beilinson [1] and Gillet [17] for a very large set of cohomology theories $H^\bullet_C$ that includes all Bloch-Ogus cohomology theories. These give rise to the Chern character maps

$$ch : K_m(X) \to \prod_{j \geq 0} H^{2j-m}_C(X, \mathbb{Q}(j)).$$

The degree j part of this

$$ch_j : K_m(X) \to H^{2j-m}_C(X, \mathbb{Q}(j))$$

is a homogeneous polynomial of degree j in the Chern classes, just as in the topological case. The key point is the compatibility with the Adams operations which implies that the restriction of ch_k to $K_m(X)^{(j)}$ vanishes unless $k = j$. It follows that ch_j factors through the projection onto $K_m(X)^{(j)}$:

$$
\begin{array}{ccc}
K_m(X) & \xrightarrow{\ ch_j\ } & H^{2j-m}_C(X, \mathbb{Q}(j)) \\
\text{proj} \downarrow & \nearrow & \\
K_m(X)^{(j)} & &
\end{array}
$$

Thus the Chern character induces a natural transformation

$$H_{\mathrm{mot}}^m(X, \mathbb{Q}(n)) \to H_{\mathcal{C}}^m(X, \mathbb{Q}(n)).$$

It is a ring homomorphism as the Chern character is. Define

$$ch_j^{\mathrm{mot}} : K_m(X) \to H_{\mathrm{mot}}^{2j-m}(X, \mathbb{Q}(j))$$

to be the projection $K_m(X) \to K_m(X)^{(j)}$. From this, one can inductively construct Chern classes $c_j^{\mathrm{mot}} : K_m(X) \to H_{\mathrm{mot}}^{2j-m}(X, \mathbb{Q}(j))$. Compatibility with the Chern classes $c_j : K_m(X) \to H_{\mathcal{C}}^{2j-m}(X, \Lambda(j))$ is automatic and guarantees the uniqueness of natural transformation $H_{\mathrm{mot}}^\bullet \to H_{\mathcal{C}}^\bullet$.   □

The quest for cochains. Beilinson's definition raises many questions and problems such as:

(i) How does one define motivic cohomology with *integral* coefficients?
(ii) Find natural cochain complexes (a.k.a., motivic complexes) whose homology groups are motivic cohomology.
(iii) Compute motivic cohomology groups.

Bloch's higher Chow groups [5] provide an integral version of motivic cohomology as well as a chain complex whose homology is motivic cohomology. (See also [6] and [33].) One difficulty with this approach is that, being based on algebraic cycles and rational equivalence, it is difficult to compute with.

More fundamentally, one would also like motivic cohomology groups to be the ext or hyper-ext groups associated to a suitable category of motives. In the ideal case, this category will be tannakian after tensoring all its objects with $\mathbb{Q}$, so that the category of $\mathbb{Q}$-motives will be equivalent to the category of representations of a proalgebraic group defined over $\mathbb{Q}$. These goals have been achieved to some degree. For all fields k, Voevodsky [44] and Levine [34] have each constructed a triangulated tensor category of "mixed motives over k". For each scheme X, smooth and quasi-projective over k, there is an object $M(X)$ in this category such that $\mathrm{Ext}^\bullet(\mathbb{Z}(-n), M(X))$ is isomorphic to the integral motivic cohomology groups of X (i.e., Bloch's higher Chow groups). However, the categories obtained from the categories of Levine and Voevodsky by tensoring their objects with $\mathbb{Q}$ are not tannakian.

One can also propose that there should be a tannakian category of *mixed Tate motives* over a field k. The motivic cohomology of k should be an ext in this category. In the case where k is a number field (or any field satisfying Beilinson-Soulé vanishing), Levine [32] has constructed such a tannakian category of mixed Tate motives. Goncharov [18, p. 611] later proved a result similar to Levine's and proved, in addition, that the bounded derived category of this tannakian category of mixed Tate motives is equivalent to the full subcategory of mixed Tate motives of the category of mixed motives over k.

An older and less fundamental approach to constructing categories of motives, proposed by Deligne [14] and Jannsen [29], is to view them as "compatible systems of realizations". These also form a tannakian category. We shall take this approach in this paper as it is more accessible and is more consistent with our point of view.

3. The Motivic Cohomology of the Spectrum of a Ring of S-integers

Basic results of Quillen [37] and Borel [10] give the computation of the motivic cohomology of the spectra of rings of S-integers in number fields. Suppose that F is a number field with ring of integers $\mathcal{O}_F$ and that S is a finite subset of $\operatorname{Spec} \mathcal{O}_F$. Set $X_{F,S} = \operatorname{Spec} \mathcal{O}_F - S$. Set

$$
d_n = \operatorname{ord}_{s=1-n} \zeta_F(s) = \begin{cases} r_1 + r_2 - 1 & \text{when } n = 1, \\ r_1 + r_2 & \text{when } n \text{ is odd and } n > 1, \\ r_2 & \text{when } n \text{ is even,} \end{cases} \tag{3-1}
$$

where $\zeta_F(s)$ denotes the Dedekind zeta function of F and r_1, and r_2 denote the number of real and complex places of F, respectively.

THEOREM 3.1. *For all n and m, $H^m_{\mathrm{mot}}(X_{F,S}, \mathbb{Q}(n))$ is a finite dimensional rational vector space whose dimension is given by*

$$
\dim H^m_{\mathrm{mot}}(X_{F,S}, \mathbb{Q}(n)) = \begin{cases} d_1 + \#S & \text{when } m = n = 1, \\ d_n & \text{when } m = 1 \text{ and } n > 1, \\ 0 & \text{otherwise.} \end{cases}
$$

PROOF. First suppose that S is empty. Quillen [37] showed that each K-group $K_m(X_{F,S})$ is a finitely generated abelian group. It follows that each of the groups $H^j_{\mathrm{mot}}(X_{F,S}, \mathbb{Q}(n))$ is finite dimensional. The rank of $K_0(X_{F,S})$ is 1 and the rank of $K_1(X_{F,S})$ is $r_1 + r_2 - 1$ by the Dirichlet Unit Theorem. The ranks of the remaining $K_m(X_{F,S})$ were computed by Borel [10]. It is zero when m is even and > 0, and d_n when $m = 2n - 1 > 1$. It is easy to see that

$$
H^0_{\mathrm{mot}}(X_{F,S}, \mathbb{Q}(0)) = K_0(X_{F,S}) \otimes \mathbb{Q} \cong \mathbb{Q}.
$$

Borel [11] constructed regulator mappings

$$
K_{2n-1}(X_{F,S}) \to \mathbb{R}^{d_n}, \quad n > 0,
$$

and showed that each is injective mod torsion. Beilinson [1] showed that Borel's regulator is a non-zero rational multiple of the regulator mapping

$$
ch_n : K_{2n-1}(X_{F,S}) \to H^1_{\mathcal{D}}(X_{F,S}, \mathbb{R}(n)) \cong \mathbb{R}^{d_n}
$$

to Deligne cohomology. The properties of the Chern character and Borel's injectivity together imply that

$$
H^1_{\mathrm{mot}}(X_{F,S}, \mathbb{Q}(n)) = K_{2n-1}(X_{F,S})^{(n)} = K_{2n-1}(X_{F,S}) \otimes \mathbb{Q}
$$

and that $H^m_{\mathrm{mot}}(X_{F,S}, \mathbb{Q}(n))$ vanishes when $m > 1$, and when $m = 0$ and $n \neq 0$. The result when S is non-empty follows from this using the localization sequence [39], and the fact, due to Quillen [38], that the K-groups of finite fields are torsion groups in positive degree. Together these imply that each prime removed adds one to the rank of K_1 and does not change the rank of any other K-group. $\square$

Denote the Galois group of the maximal algebraic extension of F, unramified outside S, by $G_{F,S}$. In this paper, a finite dimensional $G_{F,S}$-module means a finite dimensional $\mathbb{Q}_\ell$-vector space with continuous $G_{F,S}$-action. Denote the category of $\mathbb{Q}$ mixed Hodge structures by $\mathcal{H}$. Denote the ext functor in the category of finite dimensional $G_{F,S}$-modules by $\mathrm{Ext}_{G_{F,S}}$, and the ext functor in $\mathcal{H}$ by $\mathrm{Ext}_{\mathcal{H}}$. The results on regulators of Borel [11] and Soulé [41] can be stated as follows.

THEOREM 3.2. *The natural transformation from motivic to étale cohomology induces isomorphisms*

$$H^1_{\mathrm{mot}}(X_{F,S}, \mathbb{Q}_\ell(n)) \cong H^1_{\text{ét}}(X_{F,S}, \mathbb{Q}_\ell(n)) \cong \mathrm{Ext}^1_{G_{F,S}}(\mathbb{Q}_\ell, \mathbb{Q}_\ell(n)).$$

for all $n \geq 1$. The natural transformation from motivic to Deligne cohomology induces injections

$$H^1_{\mathrm{mot}}(X_{F,S}, \mathbb{Q}(n)) \hookrightarrow H^1_{\mathcal{D}}(X_{F,S}, \mathbb{Q}(n)) := \left[\bigoplus_{\nu : F \hookrightarrow \mathbb{C}} \mathrm{Ext}^1_{\mathcal{H}}(\mathbb{Q}, \mathbb{Q}(n)) \right]^{\mathrm{Gal}(\mathbb{C}/\mathbb{R})}. \square$$

Thus each element x of $K_{2n-1}(X_{F,S})$ determines an extension

$$0 \to \mathbb{Q}_\ell(n) \to E_{\ell,x} \to \mathbb{Q}_\ell(0) \to 0$$

of ℓ-adic local systems over $X_{F,S}$ and a $\mathrm{Gal}(\mathbb{C}/\mathbb{R})$-equivariant extension

$$0 \to \mathbb{Q}(n) \to E_{\mathrm{Hodge},x} \to \mathbb{Q}(0) \to 0$$

of mixed Hodge structures over $X_{F,S} \otimes \mathbb{C}$. One can think of these as the étale and Hodge realizations of $x \in K_{2n-1}(X_{F,S})$.

4. Mixed Tate Motives

As mentioned earlier, one approach to motivic cohomology is to postulate that to each sufficiently nice scheme X (say, smooth and quasi-projective over a field, or regular over a ring $\mathcal{O}_{F,S}$ of S-integers in a number field) one can associate a category $\mathcal{T}(X)$ of *mixed Tate motives over X*. This should satisfy the following conjectural properties.

(i) $\mathcal{T}(X)$ is a (neutral) tannakian category over $\mathbb{Q}$ with a fiber functor $\omega : \mathcal{T}(X) \to \mathrm{Vec}_{\mathbb{Q}}$ to the category of finite dimensional rational vector spaces.

(ii) Each object M of $\mathcal{T}(X)$ has an increasing filtration called the weight filtration

$$\cdots \subseteq W_{m-1}M \subseteq W_m M \subseteq W_{m+1}M \subseteq \cdots,$$

whose intersection is 0 and whose union is M. Morphisms of $\mathcal{T}(X)$ should be strictly compatible with the weight filtration — that is, the functor

$$\mathrm{Gr}^W_\bullet : M \mapsto \bigoplus_m \mathrm{Gr}^W_m M := \bigoplus_m W_m M / W_{m-1} M$$

to graded objects in $\mathcal{T}(X)$ should be an exact tensor functor.

(iii) $\mathcal{T}(X)$ contains "the Tate motive $\mathbb{Q}(1)$" over $\mathrm{Spec}\,R$ where R is the base ring (here either a field or $\mathcal{O}_{F,S}$). This can be considered as the dual of the local system $R^1 f_*(\mathbb{Q})$ over $\mathrm{Spec}\,R$, where f is the structure morphism of the multiplicative group $\mathbb{G}_m$, i.e., $f : \mathbb{G}_m \otimes R \to \mathrm{Spec}\,R$. Put $\mathbb{Q}(n) := \mathbb{Q}(1)^{\otimes n}$ for $n \in \mathbb{Z}$. (Negative tensor powers are defined by duality.)

(iv) There should be *realization functors* to various categories such as ℓ-adic étale local systems over $X[1/\ell] := X \otimes_R R[1/\ell]$ (where R is the base ring and ℓ does not divide the characteristic of R), variations of mixed Hodge structure over X, etc. These functors should be faithful, exact tensor functors. These functors are related by natural comparison transformations. The Betti, de Rham, ℓ-adic and crystalline realizations of $\mathbb{Q}(1)$ should be the Betti, de Rham, ℓ-adic and crystalline versions of $H_1(\mathbb{G}_m)$.

(v) For each object M, $\mathrm{Gr}^W_{2m+1} M$ is trivial and $\mathrm{Gr}^W_{2m} M$ is isomorphic to the direct sum of a finite number of copies of $\mathbb{Q}(-m)$.

The last property characterizes *mixed Tate* motives among mixed motives. The category $\mathcal{T}(X)$, being tannakian, is equivalent to the category of finite dimensional representations of a proalgebraic $\mathbb{Q}$-group $\pi_1(\mathcal{T}(X), \omega)$, which represents the tensor automorphism group of the fiber functor ω. We denote it simply by $\pi_1(\mathcal{T}(X))$, if the selection of ω does not matter.

There are several approaches to constructing the category $\mathcal{T}(X)$, at least when X is the spectrum of a field or $X = X_{F,S}$, such as those of Bloch-Kriz [8], Levine [32], and Goncharov[18].

We follow Deligne [14] and Jannsen [29], who define a motive to be a "compatible set of realizations" of "geometric origin." This is a tannakian category. Deligne does not define what it means to be of geometric origin, but wants it to be broad enough to include those compatible realizations that occur in the unipotent completion of fundamental groups of varieties in addition to subquotients of cohomology groups. We refer the reader to Section 1 of Deligne's paper [14] for the definition of compatible set of realizations. One example is $\mathbb{Q}(1)$, defined as $H_1(\mathbb{G}_{m/\mathbb{Z}})$, another is the extension E_x of $\mathbb{Q}(0)$ by $\mathbb{Q}(n)$ coming from $x \in K_{2n-1}(X_{F,S})$ described in the previous section.

The hope is that

$$H^m_{\mathrm{mot}}(X, \mathbb{Q}(n)) \cong \mathrm{Ext}^m_{\mathcal{T}(X)}(\mathbb{Q}(0), \mathbb{Q}(n))$$

holds when X is the spectrum of a field or $X_{F,S}$.[3] This covers the cases of interest for us. In general, one expects that motivic cohomology groups of X can be computed as hyper-exts:

$$H^m_{\mathrm{mot}}(X, \mathbb{Q}(n)) \cong \mathbb{H}^m(X, \mathcal{E}xt^\bullet_{\mathcal{T}}(\mathbb{Q}(0), \mathbb{Q}(n))).$$

Deligne's conjecture (Conjecture 5.5) will be a consequence of:

POSTULATE 4.1. If $X = X_{F,S}$, there is a category of mixed Tate motives $\mathcal{T}(X)$ over X with the above mentioned properties. It has the property that there is a natural isomorphism

$$H^m_{\mathrm{mot}}(X, \mathbb{Q}(n)) \cong \mathrm{Ext}^m_{\mathcal{T}(X)}(\mathbb{Q}(0), \mathbb{Q}(n)),$$

which is compatible with Chern maps.

Examples of mixed Tate motives over $\mathrm{Spec}\,\mathbb{Z}$. One of the main points of [14] is to show that the unipotent completion of the fundamental group of $\mathbb{P}^1 - \{0, 1, \infty\}$ is an example of a mixed Tate motive (actually a pro-mixed Tate motive), smooth over $\mathrm{Spec}\,\mathbb{Z}$.

As base point, take $\overrightarrow{01}$, the tangent vector of $\mathbb{P}^1$ based at 0 that corresponds to $\partial/\partial t$, where t is the natural parameter on $\mathbb{P}^1 - \{0, 1, \infty\}$. Deligne [14] shows that the unipotent completion of $\pi_1(\mathbb{P}^1 - \{0, 1, \infty\}, \overrightarrow{01})$ is a mixed Tate motive over $\mathrm{Spec}\,\mathbb{Z}$ by exhibiting compatible Betti, étale, de Rham and crystalline realizations of it. It is smooth over $\mathrm{Spec}\,\mathbb{Z}$ essentially because the pair $(\mathbb{P}^1 - \{0, 1, \infty\}, \overrightarrow{01})$ has everywhere good reduction.

There is an interesting relation to classical polylogarithms which was discovered by Deligne (cf. [14], [3], [20]). There is a polylog local system P, which is a motivic local system over $\mathbb{P}^1_{\mathbb{Z}} - \{0, 1, \infty\}$ in the point of view of compatible realizations. Its Hodge-de Rham realization is a variation of mixed Hodge structure over the complex points of $\mathbb{P}^1 - \{0, 1, \infty\}$ whose periods are given by $\log x$ and the classical polylogarithms: $\mathrm{Li}_1(x) = -\log(1 - x)$, $\mathrm{Li}_2(x)$ (Euler's dilogarithm), $\mathrm{Li}_3(x)$, and so on. Here $\mathrm{Li}_n(x)$ is the multivalued holomorphic function on $\mathbb{P}^1(\mathbb{C}) - \{0, 1, \infty\}$ whose principal branch is given by

$$\mathrm{Li}_n(x) = \sum_{k=1}^{\infty} \frac{x^k}{k^n}$$

in the unit disk.[4]

The fiber $P_{\overrightarrow{01}}$ of P over the base point $\overrightarrow{01}$ is a mixed Tate motive over $\mathrm{Spec}\,\mathbb{Z}$ and has periods the values of the Riemann zeta function at integers $n > 1$. In

---

[3]If this is true, then $H^m_{\mathrm{mot}}(\mathrm{Spec}\,F, \mathbb{Q}(n))$ will vanish when $n < 0$ and $m = 0$, and when $n \leq 0$ and $m > 0$. This vanishing is a conjecture of Beilinson and Soulé. It is known for number fields, for example.

[4]This goes back to various letters of Deligne. Accounts can be found, for example, in [3] and [20].

fact, $P_{\overrightarrow{01}}$ is an extension

$$0 \to \bigoplus_{n \geq 1} \mathbb{Q}(n) \to P_{\overrightarrow{01}} \to \mathbb{Q}(0) \to 0$$

and thus determines a class

$$(e_n)_n \in \bigoplus_{n \geq 1} \mathrm{Ext}^1_{\mathcal{H}}(\mathbb{Q}(0), \mathbb{Q}(n)).$$

The class e_n is trivial when $n = 1$ and is the coset of $\zeta(n)$ in

$$\mathbb{C}/(2\pi i)^n \mathbb{Q} \cong \mathrm{Ext}^1_{\mathcal{H}}(\mathbb{Q}(0), \mathbb{Q}(n))$$

when $n > 1$. Since $\zeta(2n)$ is a rational multiple of π^{2n}, each e_{2n} is trivial.

Deligne computes the ℓ-adic realization of $P_{\overrightarrow{01}}$ in [14] and shows that the polylogarithm motive is a canonical quotient of the enveloping algebra of the Lie algebra of the unipotent completion of $\pi_1(\mathbb{P}^1 - \{0, 1, \infty\}, \overrightarrow{01})$. (See also [3] and [20].)

5. The Motivic Lie Algebra of $X_{F,S}$ and Deligne's Conjectures

Assume that X is as in Section 4, and that there is a category of mixed Tate motives $\mathcal{T}(X)$ with properties (i)–(v) in Section 4. Since $\mathcal{T}(X)$ is tannakian, it is determined by its tannakian fundamental group $\pi_1(\mathcal{T}(X))$, which is an extension of $\mathbb{G}_m$ by a prounipotent $\mathbb{Q}$-group

$$1 \to \mathcal{U}_X \to \pi_1(\mathcal{T}(X)) \to \mathbb{G}_m \to 1 \tag{5–1}$$

as we shall now explain.

The category of pure Tate motives is the tannakian subcategory of $\mathcal{T}(X)$ generated by $\mathbb{Q}(1)$. By the faithfulness of realization functors, it is equivalent to the category of finite dimensional graded $\mathbb{Q}$-vector spaces, and hence to the category of finite dimensional representations of $\mathbb{G}_m$; $\mathbb{Q}(n)$ corresponds to the nth power of the standard representation. This induces a group homomorphism between the tannakian fundamental groups

$$\pi_1(\mathcal{T}(X)) \to \pi_1(\text{pure Tate motives}) \cong \mathbb{G}_m.$$

Since the category of pure Tate motives is a full subcategory and every subobject of a pure Tate motive is pure, this morphism is surjective (cf. [13, Proposition 2.21a]), and the properties of the weight filtration imply the unipotence of its kernel $\mathcal{U}_X$, thus we have (5–1). The Lie algebra $\mathfrak{t}_X$ of $\pi_1(\mathcal{T}(X))$ is an extension

$$0 \to \mathfrak{u}_X \to \mathfrak{t}_X \to \mathbb{Q} \to 0$$

where $\mathfrak{u}_X$ is pronilpotent. This $\mathfrak{t}_X$ is called the motivic Lie algebra of X. We shall see that the knowledge of the cohomologies of $\mathfrak{u}_X$ (as $\mathbb{G}_m$-modules) is equivalent to the knowledge of the extension groups $\text{Ext}^{\bullet}_{\mathcal{T}(X)}(\mathbb{Q}(0), \mathbb{Q}(m))$ for all m in the next section.

5.1. Extension groups in a tannakian category.

We start with a general setting. Let K be a field of characteristic zero. Let $\mathcal{G}$ be a proalgebraic group (in this paper a proalgebraic group means an affine proalgebraic group) over K, or equivalently, an affine group scheme over K (cf. [13]). A $\mathcal{G}$-module V is a (possibly infinite dimensional) K-vector space with algebraic $\mathcal{G}$-action (cf. [30]). The category of $\mathcal{G}$-modules is abelian with enough injectives, and hence we have the cohomology groups

$$H^m(\mathcal{G}, V) := \text{Ext}^m_{\mathcal{G}}(K, V)$$

defined as the extension groups, where K denotes the trivial representation. The right hand side has an interpretation as Yoneda's extension groups, i.e., as the set of equivalence classes of m-step extensions (see [45]). Since each $\mathcal{G}$-module is locally finite [30, 2.13], every m-step extension representing an element of $\text{Ext}^m_{\mathcal{G}}(K, V)$ can be replaced by an equivalent extension consisting of finite dimensional modules when V is finite dimensional. Thus, the right hand side does not change when the category of $\mathcal{G}$-modules is replaced by the category of finite dimensional $\mathcal{G}$-modules.

Let $\mathcal{T}$ be a neutral tannakian category over K with a fiber functor ω, and let $\mathcal{G}$ be its tannakian fundamental group with base point ω. Since $\mathcal{T}$ is isomorphic to the category of finite dimensional $\mathcal{G}$-modules, we have the following.

LEMMA 5.1. *Let $\mathcal{T}$ be a neutral tannakian category and $\mathcal{G}$ be its tannakian fundamental group. Then, for any object V, we have*

$$\text{Ext}^m_{\mathcal{T}}(K, V) \cong H^m(\mathcal{G}, V). \qquad \square$$

Suppose that $\mathcal{G}$ is an extension

$$1 \to \mathcal{U} \to \mathcal{G} \to R \to 1$$

of proalgebraic groups over K. Then, for any $\mathcal{G}$-module V, we have the Lyndon-Hochschild-Serre spectral sequence (cf. [30, 6.6 Proposition]):

$$E_2^{s,t} = H^s(R, H^t(\mathcal{U}, V)) \Rightarrow H^{s+t}(\mathcal{G}, V).$$

If R is a reductive algebraic group, then every R-module is completely reducible. Consequently, $H^s(R, V)$ vanishes for $s \geq 1$ for all V, and

$$H^m(\mathcal{G}, V) \cong H^0(R, H^m(\mathcal{U}, V)).$$

If, in addition, the action of $\mathcal{G}$ on V factors through R, then one has an R-module isomorphism

$$H^m(\mathcal{U}, V) \cong H^m(\mathcal{U}, K) \otimes V.$$

Moreover, if we assume that $\mathcal{U}$ is prounipotent, then its Lie algebra $\mathfrak{u}$ is a projective limit

$$\mathfrak{u} \cong \varprojlim_{\mathfrak{n}} \mathfrak{u}/\mathfrak{n}$$

of finite dimensional nilpotent Lie algebras. It has a topology as a projective limit, where each $\mathfrak{u}/\mathfrak{n}$ is viewed as a discrete topological space.

Let V be a continuous $\mathfrak{u}$-module over K. The continuous cohomology $H^m_{\text{cts}}(\mathfrak{u}, V)$ is defined as the extension group $\text{Ext}^m(K, V)$ in the category of continuous $\mathfrak{u}$-modules. We denote $H^m_{\text{cts}}(\mathfrak{u}, K)$ by $H^m_{\text{cts}}(\mathfrak{u})$. It is easy to show that

$$H^m_{\text{cts}}(\mathfrak{u}) \cong \varinjlim_{\mathfrak{n}} H^m(\mathfrak{u}/\mathfrak{n}),$$

where $H^m(\mathfrak{u}/\mathfrak{n})$ can be computed as the cohomology of the complex of cochains

$$\text{Hom}(\Lambda^{\bullet}(\mathfrak{u}/\mathfrak{n}), K).$$

The following is standard.

PROPOSITION 5.2. *Let $\mathfrak{u}$ be a pronilpotent Lie algebra, and let $H_1(\mathfrak{u})$ denote the abelianization of $\mathfrak{u}$. Then*

$$H_1(\mathfrak{u}) \cong \text{Hom}(H^1_{\text{cts}}(\mathfrak{u}), K).$$

If $H^2(\mathfrak{u}) = 0$, then $\mathfrak{u}$ is free.

It is also well known that the category of $\mathcal{U}$-modules is equivalent to the category of continuous $\mathfrak{u}$-modules. Hence we have

$$H^m(\mathcal{U}, K) \cong H^m_{\text{cts}}(\mathfrak{u}).$$

Putting this together, we have the following.

THEOREM 5.3. *Suppose that $1 \to \mathcal{U} \to \mathcal{G} \to R \to 1$ is a short exact sequence of pro-algebraic groups over a field K of characteristic zero. Assume that R is a reductive algebraic group, and that $\mathcal{U}$ is a prounipotent group. Let $\mathfrak{u}$ be the Lie algebra of $\mathcal{U}$. If V is an R-module, considered as a $\mathcal{G}$-module, then*

$$H^m(\mathcal{G}, V) \cong (H^m_{\text{cts}}(\mathfrak{u}) \otimes V)^R.$$

Consequently, we have the R-module isomorphism

$$H^m_{\text{cts}}(\mathfrak{u}) \cong \bigoplus_{\alpha} (H^m(\mathcal{G}, V_{\alpha}) \otimes V^*_{\alpha}),$$

where $\{V_{\alpha}\}$ is a set of representatives of the isomorphism classes of irreducible R-modules, and $(\ )^*$ denotes $\text{Hom}(\ , K)$.

5.2. Deligne's conjecture. By applying Theorem 5.3 to (5–1), we have

$$\text{Ext}^m_{\mathcal{T}(X)}(\mathbb{Q}(0), \mathbb{Q}(n)) \cong [H^m_{\text{cts}}(\mathfrak{u}_X) \otimes \mathbb{Q}(n)]^{\mathbb{G}_m}$$

and $\mathbb{G}_m$-module isomorphisms

$$H^m_{\text{cts}}(\mathfrak{u}_X) \cong \bigoplus_{n \in \mathbb{Z}} \text{Ext}^m_{\mathcal{T}(X)}(\mathbb{Q}(0), \mathbb{Q}(n)) \otimes \mathbb{Q}(-n), \qquad (5\text{–}2)$$

where $\mathfrak{u}_X$ is the Lie algebra of $\mathcal{U}_X$, the prounipotent radical of $\pi_1(\mathcal{T}(X), \omega)$. By a weight argument, each extension on the right hand side vanishes if $n \leq m - 1$. Postulate 4.1 says that these Ext groups should be the motivic cohomology groups of X, and Theorem 3.1 says that they should be isomorphic to the Adams eigenspaces of the K-groups of X:

PROPOSITION 5.4. *Assume the existence of a category $\mathcal{T}(X_{F,S})$ of mixed Tate motives over $X_{F,S}$ with properties (i)–(v) as in Section 4. Suppose that Postulate 4.1 holds for all $n \geq 1$. Let $\mathcal{U}_{X_{F,S}}$ be the unipotent radical of $\pi_1(\mathcal{T}(X_{F,S}), \omega)$, and $\mathfrak{u}_{X_{F,S}}$ be its Lie algebra. Then there is a natural $\mathbb{G}_m$-module isomorphism*

$$H^1_{\text{cts}}(\mathfrak{u}_{X_{F,S}}) \cong \bigoplus_{n \geq 1} K_{2n-1}(X_{F,S}) \otimes_{\mathbb{Z}} \mathbb{Q}(-n),$$

and $H^m_{\text{cts}}(\mathfrak{u}_{X_{F,S}}) = 0$ whenever $m \geq 2$. Moreover, the exactness of $\text{Gr}^W_\bullet$ implies that

$$H_1(\text{Gr}^W_\bullet \mathfrak{u}_{X_{F,S}}) = \bigoplus_{n \geq 1} K_{2n-1}(X_{F,S})^* \otimes \mathbb{Q}(n)$$

and that

$$H^m(\text{Gr}^W_\bullet \mathfrak{u}_{X_{F,S}}) = 0 \text{ when } m > 1.$$

It follows from this that $\text{Gr}^W_\bullet \mathfrak{u}_{X_{F,S}}$ is isomorphic to the free Lie algebra generated by $H_1(\text{Gr}^W_\bullet \mathfrak{u}_{X_{F,S}})$.

Let us assume that there is a category $\mathcal{T}(X_{F,S})$ satisfying (i)–(v) in Section 4. Then Postulate 4.1 is equivalent to the following conjecture of Deligne:

CONJECTURE 5.5 (DELIGNE). (i) [14, 8.2.1] For the category $\mathcal{T}(X_{F,S})$ of mixed Tate motives smooth over $X_{F,S}$ one has a natural isomorphism

$$\text{Ext}^1_{\mathcal{T}(X_{F,S})}(\mathbb{Q}(0), \mathbb{Q}(n)) \cong K_{2n-1}(X_{F,S}) \otimes \mathbb{Q} \text{ for all } n,$$

which is compatible with the Chern mappings.

(ii) [14, 8.9.5] The group $\pi_1(\mathcal{T}(X_{F,S}))$ is an extension of $\mathbb{G}_m$ by a free prounipotent group.

Note that by Definition 2.1, Theorem 3.1 and the isomorphism (5–2), (i) is equivalent to Postulate 4.1 for $m = 1$, and that (ii) is equivalent to Postulate 4.1 for $m \geq 2$.

Consequences of Deligne's conjecture. Deligne's conjecture suggests restrictions on the action of Galois groups on pro-ℓ completions of fundamental groups of curves. Here is a sketch of how this should work.

As in the beginning of Section 4, there should be a Betti realization functor

$$\mathrm{real}_B : \mathcal{T}(X_{F,S}) \to \{\mathbb{Q}\text{-vector spaces}\}$$

to the category of $\mathbb{Q}$-vector spaces, and an ℓ-adic realization functor

$$\mathrm{real}_\ell : \mathcal{T}(X_{F,S}) \to \{\ell\text{-adic } G_F\text{-modules}\},$$

to the category of the $\mathbb{Q}_\ell$-vector spaces with a continuous G_F-action. The Galois modules should be unramified outside $S \cup [\ell]$, where $[\ell]$ denotes the set of primes of F over ℓ. We choose real_B as our fiber functor ω. Let ω_ℓ denote the functor real_ℓ which forgets the G_F-action. Conjecturally, there is a comparison isomorphism

$$\omega \otimes \mathbb{Q}_\ell \cong \omega_\ell,$$

so we shall identify these two. Define $\mathcal{T}(X_{F,S}) \otimes \mathbb{Q}_\ell$ to be the tannakian category whose objects are the same as those of $\mathcal{T}(X_{F,S})$ and whose hom-sets are those of $\mathcal{T}(X_{F,S})$ tensored with $\mathbb{Q}_\ell$. The ℓ-adic realization functor induces a functor

$$\mathrm{real}_\ell : \mathcal{T}(X_{F,S}) \otimes \mathbb{Q}_\ell \to \{\ell\text{-adic } G_F\text{-modules}\} \qquad (5\text{--}3)$$

(by an abuse of notation we denote it by real_ℓ again), and by forgetting the Galois action a fiber functor $\omega_\ell : \mathcal{T}(X_{F,S}) \otimes \mathbb{Q}_\ell \to \mathrm{Vec}_{\mathbb{Q}_\ell}$ (under a similar abuse of notation). Through the comparison isomorphism, it is easy to show that

$$\pi_1(\mathcal{T}(X_{F,S}) \otimes \mathbb{Q}_\ell, \omega_\ell) \cong \pi_1(\mathcal{T}(X_{F,S}), \omega) \otimes \mathbb{Q}_\ell.$$

The following is closely related to the Tate conjecture on Galois modules.

POSTULATE 5.6. The realization functor real_ℓ in (5–3) is fully faithful, and its image is closed under taking subobjects.

The first condition is that every Galois compatible morphism comes from a morphism of motives up to extension of scalars, and the second condition is that every Galois submodule arises as an ℓ-adic realization of a motive. We shall see that this postulate follows from Deligne's Conjecture 5.5 and our Theorem 9.2 (see Corollary 9.4).

Every element of G_F gives an automorphism of the forgetful fiber functor of the category of G_F-modules (i.e. forgetting the Galois action), and hence an automorphism of ω_ℓ. Thus we have a homomorphism

$$G_F \to \pi_1(\mathcal{T}(X_{F,S}) \otimes \mathbb{Q}_\ell, \omega_\ell)(\mathbb{Q}_\ell) \cong \pi_1(\mathcal{T}(X_{F,S}), \omega)(\mathbb{Q}_\ell). \qquad (5\text{--}4)$$

In addition, the G_F-action on the ℓ-adic realization of any (pro)object M of $\mathcal{T}(X_{F,S})$ factors through $\pi_1(\mathcal{T}(X_{F,S})) \otimes \mathbb{Q}_\ell$ via the morphism (5–4).

PROPOSITION 5.7. *Postulate 5.6 is equivalent to the statement that the above morphism (5–4) has Zariski dense image.*

PROOF. Let $\mathcal{G}$ denote the tannakian fundamental group of the category of finite dimensional ℓ-adic G_F-modules. By [13, Prop. 2.21a], the conditions in Postulate 5.6 are equivalent to the surjectivity of $\mathcal{G} \to \pi_1(\mathcal{T}(X_{F,S}) \otimes \mathbb{Q}_\ell, \omega_\ell)$. It is a general fact that the image of $G_F \to \mathcal{G}(\mathbb{Q}_\ell)$ is Zariski dense.   □

Assuming Postulate 5.6, the Zariski density of the image of (5–4) implies that for any object M of $\mathcal{T}(X_{F,S})$, the Zariski closure of the image of G_F in $\mathrm{Aut}(M)$ should be a quotient of $\pi_1(\mathcal{T}(X_{F,S})) \otimes \mathbb{Q}_\ell$. We can define a filtration $J_M^\bullet$ on G_F (which depends on M) by

$$J_M^m G_F := \text{ the inverse image of } W_m \mathrm{Aut}\, M.$$

The image of the Galois group $G_{F(\mu_{\ell^\infty})}$ of $F(\mu_{\ell^\infty})$ in $\pi_1(\mathcal{T}(X_{F,S}), \omega_\ell)$ will lie in its prounipotent radical and should be Zariski dense in it. The exactness of $\mathrm{Gr}_\bullet^W$ will then imply that

$$\left(\bigoplus_{m<0} \mathrm{Gr}_{J_M}^m G_F \right) \otimes_{\mathbb{Z}_\ell} \mathbb{Q}_\ell$$

(a Lie algebra) is a quotient of $\mathrm{Gr}_\bullet^W \mathfrak{u}_{X_{F,S}}$, and hence generated by

$$\bigoplus_{m\geq 1} \mathrm{Hom}(K_{2m-1}(X_{F,S}), \mathbb{Q}_\ell(m)).$$

For example, the pronilpotent Lie algebra $\mathfrak{p}$ of the unipotent completion of $\pi_1(\mathbb{P}^1 - \{0,1,\infty\}, \overrightarrow{01})$ should be a pro-object of $\mathcal{T}(\mathrm{Spec}\,\mathbb{Z})$. One should therefore expect that the graded Lie algebra

$$\left(\bigoplus_{m<0} \mathrm{Gr}_{J_{\mathfrak{p}}}^m G_{\mathbb{Q}} \right) \otimes_{\mathbb{Z}_\ell} \mathbb{Q}_\ell$$

is generated by elements $z_3, z_5, z_7, \ldots$, where z_m has weight $-2m$.

Following Ihara [26], we define

$$I_\ell^m G_{\mathbb{Q}} = \ker\{G_{\mathbb{Q}} \to \mathrm{Out}(\pi_1^{(\ell)}(\mathbb{P}^1(\mathbb{C}) - \{0,1,\infty\}, \overrightarrow{01})/L^{m+1})\}$$

where L^m denotes the mth term of the lower central series of the pro-ℓ completion of $\pi_1(\mathbb{P}^1(\mathbb{C}) - \{0,1,\infty\}, \overrightarrow{01})$. This is related to the filtration $J_{\mathfrak{p}}^\bullet$ by

$$I_\ell^m G_{\mathbb{Q}} = J_{\mathfrak{p}}^{-2m} G_{\mathbb{Q}} = J_{\mathfrak{p}}^{-2m+1} G_{\mathbb{Q}}.$$

Making this substitution, we are led to the following conjecture, stated by Ihara in [27, p. 300] and which he attributes to Deligne.

CONJECTURE 5.8 (DELIGNE). The Lie algebra

$$\left[\bigoplus_{m>0} \mathrm{Gr}_{I_\ell}^m G_{\mathbb{Q}} \right] \otimes \mathbb{Q}_\ell$$

is generated by generators $s_3, s_5, s_7, \ldots$, where $s_m \in \mathrm{Gr}_{I_\ell}^m G_{\mathbb{Q}}$.

Deligne also asked whether this Lie algebra is free. A related conjecture of Goncharov [19, Conj. 2.1], stated below, and the questions of Drinfeld [15] can be 'derived' from Deligne's Conjecture 5.5 in a similar fashion. The freeness questions are more optimistic and are equivalent to the statement that the representation of the motivic Galois group $\pi_1(\mathcal{T}(\mathrm{Spec}\,\mathbb{Z}))$ in the automorphisms of the ℓ-adic unipotent completion of the fundamental group of $\mathbb{P}^1 - \{0,1,\infty\}$ is faithful. The computational results [27], [35] and [43] give support to the belief that this Lie algebra is free. Indeed, these computations show that $\mathrm{Gr}_{I_\ell}^{>0} G_\mathbb{Q}$ is free up to $\mathrm{Gr}_{I_\ell}^{12} G_\mathbb{Q}$.

CONJECTURE 5.9 (GONCHAROV). The Lie algebra of the Zariski closure of the Galois group of $\mathbb{Q}(\mu_{\ell^\infty})$ in the automorphism group of the ℓ-adic unipotent completion of $\pi_1(\mathbb{P}^1 - \{0,1,\infty\}, \overrightarrow{01})$ is a prounipotent Lie algebra freely generated by elements $z_3, z_5, z_7, \ldots$, where z_m has weight $-2m$.

Deligne's Conjecture 5.8 above and the generation part of Goncharov's conjecture are proved in [23]. A brief sketch of their proofs is given in Section 10. Modulo technical details, the main point is the computation of the tannakian fundamental group of the candidate for $\mathcal{T}(X_{F,S}) \otimes \mathbb{Q}_\ell$ given in the next section.

Polylogarithms revisited. Assuming the existence of $\mathfrak{t}_{\mathrm{Spec}\,\mathbb{Z}}$ (i.e. the Lie algebra of $\pi_1(\mathcal{T}(\mathrm{Spec}\,\mathbb{Z}))$), we can give another interpretation of the fiber $P_{\overrightarrow{01}}$ of the polylogarithm local system. Being a motive over $\mathrm{Spec}\,\mathbb{Z}$, it is a $\mathfrak{t}_{\mathrm{Spec}\,\mathbb{Z}}$-module. Note that since

$$W_{-1} P_{\overrightarrow{01}} = \bigoplus_{n \geq 1} \mathbb{Q}(n),$$

a direct sum of Tate motives (no nontrivial extensions), the restriction of the $\mathfrak{t}_{\mathrm{Spec}\,\mathbb{Z}}$-action on $W_{-1} P_{\overrightarrow{01}}$ to $\mathfrak{u}_{\mathrm{Spec}\,\mathbb{Z}}$ is trivial and $[\mathfrak{u}_{\mathrm{Spec}\,\mathbb{Z}}, \mathfrak{u}_{\mathrm{Spec}\,\mathbb{Z}}]$ annihilates $P_{\overrightarrow{01}}$. Since $P_{\overrightarrow{01}}$ is an extension of $\mathbb{Q}(0)$ by $W_{-1} P_{\overrightarrow{01}}$, this implies that there is a homomorphism

$$\psi : \left(\mathfrak{t}_{\mathrm{Spec}\,\mathbb{Z}}/[\mathfrak{u}_{\mathrm{Spec}\,\mathbb{Z}}, \mathfrak{u}_{\mathrm{Spec}\,\mathbb{Z}}]\right) \otimes \mathbb{Q}(0) \longrightarrow P_{\overrightarrow{01}}$$

such that the diagram

$$
\begin{array}{ccc}
\mathfrak{t}_{\mathrm{Spec}\,\mathbb{Z}} \otimes P_{\overrightarrow{01}} & \xrightarrow{\text{action}} & P_{\overrightarrow{01}} \\
\Big\downarrow{\scriptstyle\text{quotient}} & \nearrow{\scriptstyle\psi} & \\
\mathfrak{t}_{\mathrm{Spec}\,\mathbb{Z}}/[\mathfrak{u}_{\mathrm{Spec}\,\mathbb{Z}}, \mathfrak{u}_{\mathrm{Spec}\,\mathbb{Z}}] & &
\end{array}
$$

commutes. By comparing graded quotients, it follows that ψ is an isomorphism

$$P_{\overrightarrow{01}} \cong \mathfrak{t}_{\mathrm{Spec}\,\mathbb{Z}}/[\mathfrak{u}_{\mathrm{Spec}\,\mathbb{Z}}, \mathfrak{u}_{\mathrm{Spec}\,\mathbb{Z}}]$$

of motives over $\mathrm{Spec}\,\mathbb{Z}$.

6. ℓ-adic Mixed Tate Modules over $X_{F,S}$

In this section, we describe a candidate for the category of ℓ-adic realizations of objects and morphisms of $\mathcal{T}(X_{F,S})$. This is essentially the category constructed by Deligne and Beilinson in their unpublished manuscript [2]. It is purely Galois-representation theoretic, and requires no postulates. For technical reasons, we assume that S contains $[\ell]$, the set of primes over ℓ. This condition will be removed in Section 11. By a finite dimensional $G_{F,S}$-module, we shall mean a finite dimensional $\mathbb{Q}_\ell$-vector space on which $G_{F,S}$ acts continuously.

We define the category $\mathcal{T}_\ell(X_{F,S})$ of ℓ-adic mixed Tate modules which are smooth over $X_{F,S}$ to be the category whose objects are finite dimensional $G_{F,S}$-modules M that are equipped with a *weight filtration*

$$\cdots \subseteq W_{m-1}M \subseteq W_m M \subseteq W_{m+1}M \subseteq \cdots$$

of M by $G_{F,S}$-submodules. The weight filtration satisfies:

(i) all odd weight graded quotients of M vanish: $\mathrm{Gr}^W_{2m+1} M = 0$;

(ii) $G_{F,S}$ acts on its $2m$th graded quotient $\mathrm{Gr}^W_{2m} M$ via the $(-m)$th power of the cyclotomic character,

(iii) the intersection of the $W_m M$ is trivial and their union is all' of M.

Morphisms are $\mathbb{Q}_\ell$-linear, $G_{F,S}$-equivariant mappings. These will necessarily preserve the weight filtration, so that $\mathcal{T}_\ell(X_{F,S})$ is a full subcategory of the category of $G_{F,S}$-modules.

The category $\mathcal{T}_\ell(X_{F,S})$ is a tannakian category over $\mathbb{Q}_\ell$ with a fiber functor ω' that takes an object to its underlying $\mathbb{Q}_\ell$-vector space. We shall denote the tannakian fundamental group of this category by $\mathcal{A}^\ell_{F,S} := \pi_1(\mathcal{T}_\ell(X_{F,S}), \omega')$. Every element of $G_{F,S}$ acts on ω', which induces a natural, continuous homomorphism

$$\rho : G_{F,S} \to \mathcal{A}^\ell_{F,S}(\mathbb{Q}_\ell).$$

This has Zariski-dense image as $\mathcal{T}_\ell(X_{F,S})$ is a full subcategory of the category of $G_{F,S}$-modules, closed under taking subobjects (cf. [13, Proposition 2.21a]).

Relation to mixed Tate motives over $X_{F,S}$. As explained in Section 4, the existence of a category $\mathcal{T}(X_{F,S})$ of mixed Tate motives over $X_{F,S}$ satisfying (i)–(v) in Section 4 implies the existence of an ℓ-adic realization functor

$$\mathrm{real}_\ell : \mathcal{T}(X_{F,S}) \otimes \mathbb{Q}_\ell \to \mathcal{T}_\ell(X_{F,S}).$$

This will induce a morphism of tannakian fundamental groups

$$\mathcal{A}^\ell_{F,S} = \pi_1(\mathcal{T}_\ell(X_{F,S}), \omega') \to \pi_1(\mathcal{T}(X_{F,S}), \omega) \otimes \mathbb{Q}_\ell.$$

The main result of [23] may be regarded as saying that $\mathcal{A}^\ell_{F,S} = \pi_1(\mathcal{T}_\ell(X_{F,S}), \omega')$ is isomorphic to the conjectured value of the $\mathbb{Q}_\ell$-form $\pi_1(\mathcal{T}(X_{F,S}), \omega) \otimes \mathbb{Q}_\ell$ of the motivic fundamental group of $X_{F,S}$. We shall explain this in Section 9.

It is interesting to note that we have not restricted to $G_{F,S}$-modules of geometric origin as Deligne would like to. So one consequence of our result is that, if Deligne's Conjecture 5.5 is true, then all objects and morphisms of $\mathcal{T}(X_{F,S})$ will be of geometric origin.

7. Weighted Completion of Profinite Groups

In this and the subsequent two sections we will sketch how to compute the tannakian fundamental group $\pi_1(\mathcal{T}_\ell(X_{F,S}), \omega')$ of the category of ℓ-adic mixed Tate modules smooth over $X_{F,S}$, which was defined in Section 6. It is convenient to work in greater generality.[5]

Suppose that R is a reductive algebraic group over $\mathbb{Q}_\ell$ and that $w : \mathbb{G}_m \to R$ is a central cocharacter — that is, its image is contained in the center of R. It is best to imagine that w is nontrivial as the theory of weighted completion is uninteresting if w is trivial.

Suppose that Γ is a profinite group and that a homomorphism $\rho : \Gamma \to R(\mathbb{Q}_\ell)$ has Zariski dense image and is continuous where we view $R(\mathbb{Q}_\ell)$ as an ℓ-adic Lie group.

By a *weighted* Γ-module with respect to ρ and w we shall mean a finite dimensional $\mathbb{Q}_\ell$-vector space with continuous Γ-action together with a weight filtration

$$\cdots \subseteq W_{m-1}M \subseteq W_m M \subseteq W_{m+1}M \subseteq \cdots$$

by Γ-invariant subspaces. These should satisfy:

(i) the intersection of the $W_m M$ is 0 and their union is M,
(ii) for each m, the representation $\Gamma \to \operatorname{Aut} \operatorname{Gr}_m^W M$ should factor through ρ and a homomorphism $\phi_m : R \to \operatorname{Aut} \operatorname{Gr}_m^W M$,
(iii) $\operatorname{Gr}_m^W M$ has weight m when viewed as a $\mathbb{G}_m$-module via

$$\mathbb{G}_m \xrightarrow{\ w\ } R \xrightarrow{\ \phi_m\ } \operatorname{Aut} \operatorname{Gr}_m^W M.$$

That is, $\mathbb{G}_m$ acts on $\operatorname{Gr}_m^W M$ via the mth power of the standard character.

The category of weighted Γ-modules consists of the Γ-equivariant morphisms between weighted Γ-modules. These morphisms automatically preserve the weight filtration and are strict with respect to it; that is, the functor $\operatorname{Gr}_\bullet^W$ is exact.

One can show that the category of weighted Γ-modules is tannakian, with fiber functor ω' given by forgetting the Γ-action.

DEFINITION 7.1. The weighted completion of Γ with respect to $\rho : \Gamma \to R(\mathbb{Q}_\ell)$ and $w : \mathbb{G}_m \to R$ is the tannakian fundamental group of the category of weighted Γ-modules with respect to ρ and w.

---

[5]We may generalize further: weighted completion and its properties in this section are unchanged even if we replace $\mathbb{Q}_\ell$ by an arbitrary topological field of characteristic zero and Γ by an arbitrary topological group.

Denote the weighted completion of Γ with respect to ρ and w by $\mathcal{G}$. There is a natural homomorphism $\Gamma \to \mathcal{G}(\mathbb{Q}_\ell)$ which has Zariski dense image as we shall see below.

This definition differs from the one given in [23, Section 5], but is easily seen to be equivalent to it. (See below.) In particular, we can apply it when:

- Γ is $G_{F,S}$,
- R is $\mathbb{G}_m$ and $w : \mathbb{G}_m \to \mathbb{G}_m$ takes x to x^{-2},
- ρ is the composite of the ℓ-adic cyclotomic character $\chi_\ell : G_{F,S} \to \mathbb{Z}_\ell^\times$ with the inclusion $\mathbb{Z}_\ell^\times \hookrightarrow \mathbb{Q}_\ell^\times$.

In this case, the category of weighted Γ-modules is nothing but the category of mixed Tate modules $\mathcal{T}_\ell(X_{F,S})$. Recall that we denote the corresponding weighted completion by $\mathcal{A}_{F,S}^\ell := \pi_1(\mathcal{T}_\ell(X_{F,S}), \omega')$.

Equivalence of definitions. Here we show that the definition of weighted completion given in [23] agrees with the one given here.

Suppose that G is a linear algebraic group over $\mathbb{Q}_\ell$ which is an extension

$$1 \to U \to G \to R \to 1$$

of R by a unipotent group U. Note that $H_1(U)$ is naturally an R-module, and therefore a $\mathbb{G}_m$-module via the given central cocharacter $w : \mathbb{G}_m \to R$. We can decompose $H_1(U)$ as a $\mathbb{G}_m$-module:

$$H_1(U) = \bigoplus_{n \in \mathbb{Z}} H_1(U)_n$$

where $\mathbb{G}_m$ acts on $H_1(U)_n$ via the nth power of the standard character. We say that G is a *negatively weighted extension of R* if $H_1(U)_n$ vanishes whenever $n \geq 0$.

Given a continuous homomorphism $\rho : \Gamma \to R(\mathbb{Q}_\ell)$ with Zariski dense image, we can form a category of pairs $(\tilde{\rho}, G)$, where G is a negatively weighted extension of R and $\tilde{\rho} : \Gamma \to G(\mathbb{Q}_\ell)$ is a continuous homomorphism that lifts ρ. Morphisms in this category are given by homomorphisms between the Gs that respect the projection to R and the lifts $\tilde{\rho}$ of ρ. The objects of this category, where $\tilde{\rho}$ is Zariski dense, form an inverse system. Their inverse limit is an extension

$$1 \to \mathcal{U} \to \mathcal{G} \to R \to 1$$

of R by a prounipotent group. There is a natural homomorphism $\hat{\rho} : \Gamma \to \mathcal{G}(\mathbb{Q}_\ell)$, which is continuous in a natural sense. It has the following universal mapping property: if $\tilde{\rho} : \Gamma \to G(\mathbb{Q}_\ell)$ is an object of this category, then there is a unique homomorphism $\phi : \mathcal{G} \to G$ that commutes with the projections to R and with the homomorphisms $\tilde{\rho} : \Gamma \to G(\mathbb{Q}_\ell)$ and $\hat{\rho} : \Gamma \to \mathcal{G}(\mathbb{Q}_\ell)$. In [23], the weighted completion is defined to be this inverse limit. The equivalence of the two definitions follows from the following result.

PROPOSITION 7.2. *The inverse limit defined above is naturally isomorphic to the weighted completion of* Γ *relative to* ρ *and* w.

PROOF. Denote the inverse limit by $\mathcal{G}$ and by $\mathcal{M} = \mathcal{M}(\rho, w)$ the category of weighted Γ-modules with respect to $\rho : \Gamma \to R(\mathbb{Q}_\ell)$ and w. We will show that $\mathcal{M}$ is the category of finite dimensional $\mathcal{G}$-modules, from which the result follows.

Suppose that M is an object of $\mathcal{M}$. Then the Zariski closure of Γ in $\operatorname{Aut} M$ is an extension

$$1 \to U \to G \to R' \to 1$$

of a quotient of R by a unipotent group. Here R' is the Zariski closure of the image of Γ in $\operatorname{Aut} \operatorname{Gr}^W_\bullet M$. Because the action of Γ on each weight graded quotient factors through ρ, and because $\mathbb{G}_m$ acts on the mth weight graded quotient of M with weight m, it follows that this is a negatively weighted extension of R'. Pulling back this extension along the projection $R \to R'$, we obtain a negatively weighted extension

$$1 \to U \to \widetilde{G} \to R \to 1$$

of R and a continuous homomorphism $\Gamma \to \widetilde{G}(\mathbb{Q}_\ell)$ that lifts both ρ and the homomorphism $\Gamma \to R'(\mathbb{Q}_\ell)$. By the universal mapping property of $\mathcal{G}$, there is a natural homomorphism $\mathcal{G} \to \widetilde{G}$ compatible with the projections to R and the homomorphisms from Γ to $\mathcal{G}(\mathbb{Q}_\ell)$ and $\widetilde{G}(\mathbb{Q}_\ell)$. Thus every object of $\mathcal{M}$ is naturally a $\mathcal{G}$-module. It is also easy to see that every morphism of $\mathcal{M}$ is $\mathcal{G}$-equivariant.

Conversely, suppose that M is a finite dimensional $\mathcal{G}$-module. Composing with the natural homomorphism $\hat{\rho} : \Gamma \to \mathcal{G}(\mathbb{Q}_\ell)$ gives M the structure of a Γ-module. In [23, Sect. 4], it is proven that every $\mathcal{G}$-module has a natural weight filtration with the property that the action of $\mathcal{G}$ on each weight graded quotient factors through the projection $\mathcal{G} \to R$ and that $\mathbb{G}_m$ acts with weight m on the mth weight graded quotient. It follows that M is naturally an object of $\mathcal{M}$. Since $\mathcal{G}$-equivariant mappings are naturally Γ-equivariant, this proves that $\mathcal{M}$ is naturally the category of finite dimensional $\mathcal{G}$-modules, which completes the proof. $\square$

8. Computation of Weighted Completions

Suppose that Γ, R, $\rho : \Gamma \to R(\mathbb{Q}_\ell)$ and $w : \mathbb{G}_m \to R$ are as above. The weighted completion $\mathcal{G}$ of Γ is controlled by the low-dimensional cohomology groups $H^\bullet_{\mathrm{cts}}(\Gamma, V)$ of Γ with coefficients in certain irreducible representations V of R. If one knows these cohomology groups, as we do in the case of $G_{F,S}$, one can sometimes determine the structure of the weighted completion. These cohomological results are stated in this section.

The weighted completion of Γ with respect to ρ and w is an extension

$$1 \to \mathcal{U} \to \mathcal{G} \to R \to 1$$

where $\mathcal{U}$ is prounipotent. Now we are in the situation of Theorem 5.3. Denote the Lie algebra of $\mathcal{U}$ by $\mathfrak{u}$. Since $\mathfrak{u}$ is a $\mathcal{G}$-module by the adjoint action, the natural weight filtration on $\mathfrak{u}$ induces one on $H_{\mathrm{cts}}^{\bullet}(\mathfrak{u})$. By looking at cochains, it is not difficult to see that if $\mathfrak{u} = W_{-N}\mathfrak{u}$ for some $N > 0$, then

$$W_n H_{\mathrm{cts}}^m(\mathfrak{u}) = 0 \text{ if } n < Nm. \tag{8–1}$$

Let V_α be an irreducible R-module. Since w is central in R, the $\mathbb{G}_m$-action commutes with the R-action, so Schur's Lemma implies that there is an integer $n(\alpha)$ such that $\mathbb{G}_m$ acts on V_α via the $n(\alpha)$th power of the standard character. This is the weight of V_α as a $\mathcal{G}$-module. Now (8–1) and Theorem 5.3 imply

$$H^m(\mathcal{G}, V_\alpha) = 0$$

if $n(\alpha) > -Nm$. Note that always $\mathfrak{u} = W_{-1}\mathfrak{u}$.

Suppose that V is an ℓ-adic Γ-module, i.e., a $\mathbb{Q}_\ell$-vector space with continuous Γ-action. We shall need the continuous cohomology $H_{\mathrm{cts}}^{\bullet}(\Gamma, V)$, which is defined as the cohomology of a suitable complex of continuous cochains as in [42, Sect. 2]. A $\mathcal{G}$-module V can be considered as an ℓ-adic Γ-module through $\hat{\rho} : \Gamma \to \mathcal{G}(\mathbb{Q}_\ell)$. There is a natural group homomorphism

$$\Phi^m : H^m(\mathcal{G}, V) \to H_{\mathrm{cts}}^m(\Gamma, V)$$

for each $m \geq 0$.

Let $\{V_\alpha\}_\alpha$ be as in Theorem 5.3. These are considered as Γ-modules via ρ. The following theorem is our basic tool for computing $\mathfrak{u}$ when the appropriate continuous cohomology groups $H_{\mathrm{cts}}^i(\Gamma, V_\alpha)$ are known for $i = 1, 2$.

THEOREM 8.1. *For $m = 1, 2$, the mappings Φ^m defined above satisfy:*

(i) $\Phi^1 : H^1(\mathcal{G}, V_\alpha) \to H_{\mathrm{cts}}^1(\Gamma, V_\alpha)$ *is an isomorphism if $n(\alpha) < 0$;*
(ii) $\Phi^2 : H^2(\mathcal{G}, V_\alpha) \to H_{\mathrm{cts}}^2(\Gamma, V_\alpha)$ *is injective.*

This and Theorem 5.3 imply the following, by using (8–1) and the comment following it.

COROLLARY 8.2. (i) *There is a natural R-equivariant isomorphism*

$$H_{\mathrm{cts}}^1(\mathfrak{u}) \cong \bigoplus_{\{\alpha : n(\alpha) \leq -1\}} H_{\mathrm{cts}}^1(\Gamma, V_\alpha) \otimes V_\alpha^*.$$

(ii) *If N is an integer such that $H_{\mathrm{cts}}^1(\Gamma, V_\alpha) = 0$ for $0 > n(\alpha) > -N$, then there is a natural R-equivariant inclusion*

$$\Phi : H_{\mathrm{cts}}^2(\mathfrak{u}) \hookrightarrow \bigoplus_{\{\alpha : n(\alpha) \leq -2N\}} H_{\mathrm{cts}}^2(\Gamma, V_\alpha) \otimes V_\alpha^*.$$

This is proved in [23]. Below we shall give another more categorical proof, similar to that in [2].

COROLLARY 8.3. (i) *If $H_{\mathrm{cts}}^1(\Gamma, V_\alpha) = 0$ whenever $n(\alpha) < 0$, then $\mathfrak{u} = 0$.*

(ii) *Let N be as in Corollary 8.2. If $H^2_{\text{cts}}(\Gamma, V_\alpha) = 0$ whenever $n(\alpha) \leq -2N$, then $\mathfrak{u}$ is free as a pronilpotent Lie algebra.*   □

In the proof of Theorem 8.1 we shall use Yoneda extensions. Let V be a finite dimensional ℓ-adic Γ-module. For each $m \geq 0$, define

$$\text{Ext}^m_\Gamma(\mathbb{Q}_\ell, V)$$

to be the m-th Yoneda extension group in the category of finite dimensional ℓ-adic Γ-modules, where $\mathbb{Q}_\ell$ denotes the trivial Γ-module. For each $m \geq 1$, there is a natural homomorphism

$$\text{Ext}^m_\Gamma(\mathbb{Q}_\ell, V) \to H^m_{\text{cts}}(\Gamma, V),$$

which, by Theorem A.6, is an isomorphism when $m = 1$, and injective when $m = 2$.

There is an exact functor from the category of weighted $\mathcal{G}$-modules to the category of ℓ-adic Γ-modules. It induces morphisms between the extension groups, and hence homomorphisms

$$\Psi^m : H^m(\mathcal{G}, V) \to \text{Ext}^m_\Gamma(\mathbb{Q}_\ell, V), \qquad m \geq 0.$$

The homomorphisms Φ^m above factor through these:

$$H^m(\mathcal{G}, V) \xrightarrow[\Psi^m]{} \text{Ext}^m_\Gamma(\mathbb{Q}_\ell, V) \longrightarrow H^m_{\text{cts}}(\Gamma, V).$$

with Φ^m labeling the composite arrow from $H^m(\mathcal{G},V)$ to $H^m_{\text{cts}}(\Gamma,V)$.

In fact, this is one of several equivalent ways to define the natural mappings Φ^m.

PROOF OF THEOREM 8.1. In view of Theorem A.6, it suffices to prove that Ψ^1 is an isomorphism, and that Ψ^2 is injective.

Since the functor from the category of weighted Γ-modules to the category of Γ-modules is fully faithful, a 1-step extension of weighted Γ-modules splits if it splits as an extension of Γ-modules. This establishes the injectivity of Ψ^1.

To prove surjectivity of Ψ^1, we define a natural weight filtration on each Γ-module extension E of $\mathbb{Q}_\ell$ by V_α. Simply set $W_0 E = E$ and $W_{n(\alpha)} E = V_\alpha$. Since $n(\alpha) < 0$, this makes E a weighted Γ-module.

To prove that Ψ^2 is injective, we need to show that if a 2-step extension

$$1 \to V_\alpha \to E_2 \to E_1 \to \mathbb{Q}_\ell \to 1 \tag{8--2}$$

lies in the trivial class of extensions of Γ-modules, then it also lies in the trivial class of extensions of *weighted* Γ-modules.

If $n(\alpha) \geq -1$, then $H^2(\mathcal{G}, V_\alpha) = 0$, and there is nothing to prove. Thus we may assume $n(\alpha) \leq -2$. Since W_m is an exact functor, we may apply W_0 to (8--2) to obtain another 2-step extension, without changing the extension class. Then, taking Gr_0^W, we have a short exact sequence

$$0 \to \text{Gr}_0^W E_2 \to \text{Gr}_0^W E_1 \to \mathbb{Q}_\ell \to 0$$

of R-modules. Since R is reductive, this has a splitting $\mathbb{Q}_\ell \hookrightarrow \mathrm{Gr}_0^W E_1$. Taking the inverse images of this copy of $\mathbb{Q}_\ell$ along $E_2 \to E_1 \to \mathrm{Gr}_0^W E_1$ in E_2 and in E_1, we obtain a 2-step extension

$$0 \to V_\alpha \to E_2' \to E_1' \to \mathbb{Q}_\ell \to 0$$

equivalent to (8–2) satisfying $W_{-1} E_2' = E_2'$ and $W_0 E_1' = E_1'$. Using the dual argument, we may assume that (8–2) satisfies $W_0 E_1 = E_1$, $W_{-1} E_2 = E_2$, $W_{n(\alpha)-1} E_2 = 0$, and $W_{n(\alpha)} E_1 = 0$.

By Yoneda's characterization [45, p. 575] of trivial m-step extensions, the extension (8–2) represents the trivial 2-step extension class as Γ-modules if and only if there is a Γ-module E and exact sequences

$$0 \to E_2 \to E \to \mathbb{Q}_\ell \to 0 \text{ and } 0 \to V_\alpha \to E \to E_1 \to 0$$

which are compatible with the existing mappings $V_\alpha \hookrightarrow E_2$ and $E_1 \twoheadrightarrow \mathbb{Q}_\ell$. To establish the injectivity of Ψ^2, it suffices to prove that E is a weighted Γ-module. But E has the weight structure $W_0 E = E$ and $W_{-1} E = E_2$. This completes the proof of Theorem 8.1.   □

EXAMPLE 8.4. Suppose that $\Gamma = \mathbb{Z}_\ell^\times$, that $R = \mathbb{G}_{m/\mathbb{Q}_\ell}$ and that $\rho : \mathbb{Z}_\ell^\times \hookrightarrow \mathbb{G}_m(\mathbb{Q}_\ell) = \mathbb{Q}_\ell^\times$ is the natural inclusion. Take w to be the inverse of the square of the standard character. (With this choice, representation theoretic weights coincide with the weights from Hodge and Galois theory.) In this example we compute the weighted completion of $\mathbb{Z}_\ell^\times$ with respect to ρ and w. Note that

$$H^1_{\mathrm{cts}}(\mathbb{Z}_\ell^\times, \mathbb{Q}_\ell(n)) = 0,$$

for all non-zero $n \in \mathbb{Z}$, where $\mathbb{Q}_\ell(n)$ denotes the nth power of the standard representation of $\mathbb{G}_m$. It has weight $-2n$ under the central cocharacter.

Corollary 8.3 tells us that the unipotent radical $\mathcal{U}$ of the weighted completion of $\mathbb{Z}_\ell^\times$ is trivial, so that the weighted completion of $\mathbb{Z}_\ell^\times$ with respect to ρ is just $\rho : \mathbb{Z}_\ell^\times \to \mathbb{G}_m(\mathbb{Q}_\ell)$. More generally, if Γ is an open subgroup of $\mathbb{Z}_\ell^\times$, then the weighted completion of Γ, relative to the restriction $\Gamma \to \mathbb{Q}_\ell^\times$ of the homomorphism ρ above and the same w, is simply $\mathbb{G}_m$.

EXAMPLE 8.5. Let $\mathcal{M}_g$ be the moduli stack of genus g curves over $\mathrm{Spec}\,\mathbb{Z}$. Suppose that there is a $\mathbb{Z}[1/\ell]$-section $x : \mathrm{Spec}\,\mathbb{Z}[1/\ell] \to \mathcal{M}_g$. We allow tangential sections, and then such x exist for all g.

Let $\bar{x} : \mathrm{Spec}\,\overline{\mathbb{Q}} \to \mathrm{Spec}\,\mathbb{Z}[1/\ell] \to \mathcal{M}_g$ be a geometric point on the generic point of x. Let $C_{\bar{x}}$ be the curve corresponding to $\bar{x}$.

There is a short exact sequence of algebraic fundamental groups

$$1 \to \pi_1(\mathcal{M}_g \otimes \overline{\mathbb{Q}}, \bar{x}) \to \pi_1(\mathcal{M}_g \otimes \mathbb{Q}, \bar{x}) \to G_\mathbb{Q} \to 1, \tag{8-3}$$

where the left group is isomorphic to the profinite completion $\hat{\Gamma}_g$ of the mapping class group Γ_g of a genus g surface. We fix such an isomorphism. We have the natural representation

$$\pi_1(\mathcal{M}_g \otimes \mathbb{Q}, \bar{x}) \to \operatorname{Aut} H^1_{\text{ét}}(C_{\bar{x}}, \mathbb{Q}_\ell). \tag{8-4}$$

It is known that the image of (8–4) is isomorphic to $\operatorname{GSp}_g(\mathbb{Z}_\ell)$, where GSp_g denotes the group of symplectic similitudes of a symplectic module of rank $2g$.

By considering the action of the mapping class group on the $\mathbb{Z}/\ell\mathbb{Z}$ homology of the surface, we obtain a natural representation $\hat{\Gamma}_g \to \operatorname{Sp}_g(\mathbb{Z}/\ell)$. Let Γ^ℓ_g be the largest quotient of $\hat{\Gamma}_g$ that also maps to $\operatorname{Sp}_g(\mathbb{Z}/\ell)$ and such that the kernel of the induced mapping $\Gamma^\ell_g \to \operatorname{Sp}_g(\mathbb{Z}/\ell)$ is a pro-ℓ group.

One can construct a quotient

$$1 \to \Gamma^\ell_g \to \Gamma^{\text{arith},\ell}_g \to G_{\mathbb{Q},\{\ell\}} \to 1$$

of the short exact sequence (8–3) such that the homomorphism (8–4) induces a homomorphism

$$\rho : \Gamma^{\text{arith},\ell}_g \to \operatorname{GSp}_g(\mathbb{Q}_\ell)$$

from (8–4).

Define the central cocharacter $\omega : \mathbb{G}_m \to \operatorname{GSp}_g$ by $x \to x^{-1} I_{2g}$. In [24] we show that the weighted completion $\mathcal{G}^{\text{arith},\ell}_g$ of $\Gamma^{\text{arith},\ell}_g$ is an extension

$$\mathcal{G}_g \otimes \mathbb{Q}_\ell \to \mathcal{G}^{\text{arith},\ell}_g \to \mathcal{A}^\ell_{\mathbb{Q},\{\ell\}} \to 1,$$

where $\mathcal{G}_g$ is the completion of Γ_g relative to the standard homomorphism $\rho : \Gamma_g \to \operatorname{Sp}_g(\mathbb{Q})$, which is studied in [21] and for which a presentation is given in [22]. We expect that the left homomorphism is injective.

9. Computation of $\mathcal{A}^\ell_{F,S}$

In this section, we compute $\mathcal{A}^\ell_{F,S}$, the tannakian fundamental group of the category of ℓ-adic mixed Tate modules over $X_{F,S}$. An equivalent computation was done by Beilinson and Deligne in [2]. We shall need the following result of Soulé [41] when ℓ is odd. The case $\ell = 2$ follows from [40]. Recall that d_n is defined in (3–1).

THEOREM 9.1 (SOULÉ [41]). *With notation as above,*

$$K_{2n-1}(X_{F,S}) \otimes \mathbb{Q}_\ell \cong H^1_{\text{cts}}(G_{F,S}, \mathbb{Q}_\ell(n))$$

and hence

$$\dim_{\mathbb{Q}_\ell} H^1_{\text{cts}}(G_{F,S}, \mathbb{Q}_\ell(n)) = \begin{cases} d_1 + \#S & \text{when } n = 1, \\ d_n & \text{when } n > 1. \end{cases}$$

In addition, $H^2_{\text{cts}}(G_{F,S}, \mathbb{Q}_\ell(n))$ vanishes for all $n \geq 2$.

Denote the unipotent radical of $\mathcal{A}_{F,S}^\ell$ by $\mathcal{K}_{F,S}^\ell$. We have the exact sequence

$$1 \to \mathcal{K}_{F,S}^\ell \to \mathcal{A}_{F,S}^\ell \to \mathbb{G}_m \to 1,$$

and the corresponding exact sequence of Lie algebras

$$0 \to \mathfrak{k}_{F,S}^\ell \to \mathfrak{a}_{F,S}^\ell \to \mathbb{Q}_\ell \to 0.$$

The Lie algebra $\mathfrak{a}_{F,S}^\ell$, being the Lie algebra of a weighted completion, has a natural weight filtration. Note that since w is the inverse of the square of the standard character, all weights are even. Thus the weight filtration of $\mathfrak{a}_{F,S}^\ell$ satisfies

$$\mathfrak{a}_{F,S}^\ell = W_0 \mathfrak{a}_{F,S}^\ell, \quad \mathfrak{k}_{F,S}^\ell = W_{-2} \mathfrak{a}_{F,S}^\ell \text{ and } \mathrm{Gr}_{2n+1}^W \mathfrak{a}_{F,S}^\ell = 0 \text{ for all } n \in \mathbb{Z},$$

and we may take $N = 2$ in Corollary 8.2. The basic structure of $\mathcal{A}_{F,S}^\ell$ now follows from Corollary 8.2, Corollary 8.3 and Soulé's computation above.

THEOREM 9.2 (HAIN-MATSUMOTO [23]). *The Lie algebra* $\mathrm{Gr}_\bullet^W \mathfrak{k}_{F,S}^\ell$ *is a free Lie algebra and there is a natural* $\mathbb{G}_m$-*equivariant isomorphism*

$$H_{\mathrm{cts}}^1(\mathfrak{k}_{F,S}^\ell) \cong \bigoplus_{n=1}^\infty H_{\mathrm{cts}}^1(G_{F,S}, \mathbb{Q}_\ell(n)) \otimes \mathbb{Q}_\ell(-n) \cong \mathbb{Q}_\ell(-1)^{d_1 + \#S} \oplus \bigoplus_{n>1} \mathbb{Q}_\ell(-n)^{d_n},$$

where d_n *is defined in* (3–1). *Any lift of a graded basis of* $H_1(\mathrm{Gr}_\bullet^W \mathfrak{k}_{F,S}^\ell)$ *to a graded set of elements of* $\mathrm{Gr}_\bullet^W \mathfrak{k}_{F,S}^\ell$ *freely generates* $\mathrm{Gr}_\bullet^W \mathfrak{k}_{F,S}^\ell$.

As a corollary of the proof, we have:

COROLLARY 9.3. *There are natural isomorphisms*

$$\mathrm{Ext}_{\mathcal{T}_\ell(X_{F,S})}^m(\mathbb{Q}_\ell, \mathbb{Q}_\ell(n)) \cong \begin{cases} \mathbb{Q}_\ell & \text{when } m = n = 0, \\ H_{\mathrm{cts}}^1(G_{F,S}, \mathbb{Q}_\ell(n)) & \text{when } m = 1 \text{ and } n > 0, \\ 0 & \text{otherwise.} \end{cases}$$

Consequently, for all $n \in \mathbb{Z}$, *there are natural isomorphisms*

$$\mathrm{Ext}_{\mathcal{T}_\ell(X_{F,S})}^1(\mathbb{Q}_\ell, \mathbb{Q}_\ell(n)) \cong K_{2n-1}(\mathrm{Spec}\, \mathcal{O}_{F,S}) \otimes \mathbb{Q}_\ell.$$

COROLLARY 9.4. *Suppose that there is a category of mixed Tate motives* $\mathcal{T}(X_{F,S})$ *with properties* (i)–(v) *as in Section 4. If Deligne's conjecture 5.5 is true, then, the image of the* ℓ-*adic realization functor* real$_\ell$ *in* (5–3) *is equivalent to the category of weighted* $G_{F,S}$-*modules. In particular, Postulate 5.6 follows.*

PROOF. Deligne's conjecture 5.5 implies that $\pi_1(\mathcal{T}(X_{F,S}))$, the tannakian fundamental group, is an extension

$$1 \to \mathcal{U}_{X_{F,S}} \to \pi_1(\mathcal{T}(X_{F,S})) \to \mathbb{G}_m \to 1,$$

where $\mathcal{U}_{X_{F,S}}$ is a free prounipotent group generated by $K_{2n-1}(X_{F,S})^*$. This and Theorem 9.2 show that the natural map $\pi_1(\mathcal{T}_\ell(X_{F,S})) \to \pi_1(\mathcal{T}(X_{F,S})) \otimes \mathbb{Q}_\ell$ is an isomorphism, and it follows that

$$\mathrm{real}_\ell : \mathcal{T}(X_{F,S}) \otimes \mathbb{Q}_\ell \to \ell\text{-adic } G_F\text{-modules}$$

is fully faithful and its image is equivalent to the category of weighted $G_{F,S}$-modules. $\qquad\square$

Note that these theorems can be generalized to the case where S may not contain all the primes above ℓ, see Section 11.

Another example. Suppose that S is a finite set of rational primes containing ℓ. Suppose that F is a finite Galois extension of $\mathbb{Q}$ with Galois group G, which is unramified outside S. Define

$$\rho : G_{\mathbb{Q},S} \to \mathbb{G}_m(\mathbb{Q}_\ell) \times G$$

by

$$\rho(\sigma) = (\chi_\ell(\sigma), f(\sigma))$$

where $f : G_{\mathbb{Q},S} \to G$ is the quotient homomorphism and χ_ℓ is the ℓ-adic cyclotomic character. Define

$$w : \mathbb{G}_m \to \mathbb{G}_m \times G$$

by $w : x \mapsto (x^{-2}, 1)$. It is a central cocharacter. Denote the weighted completion of $G_{\mathbb{Q},S}$ with respect to ρ and w by $\mathcal{G}_{\mathbb{Q},S}$.

Denote the set of primes in $\mathcal{O}_F$ that lie over $S \subset \mathrm{Spec}\,\mathbb{Z}$ by T.

PROPOSITION 9.5. *There is a natural inclusion* $\iota : \mathcal{A}^\ell_{F,T} \to \mathcal{G}_{\mathbb{Q},S}$ *and an exact sequence*

$$1 \longrightarrow \mathcal{A}^\ell_{F,T} \overset{\iota}{\longrightarrow} \mathcal{G}_{\mathbb{Q},S} \longrightarrow G \longrightarrow 1.$$

PROOF. If $\{V_\alpha\}$ is a set of representatives of the isomorphism classes of irreducible representations of G, then $\{\mathbb{Q}_\ell(m) \boxtimes V_\alpha\}$ is a set of representatives of the isomorphism classes of irreducible representations of $\mathbb{G}_m \times G$, where $W \boxtimes V$ denotes the exterior tensor product of a representation W of $\mathbb{G}_m$ and V of G. Consider the restriction mapping

$$\phi : H^i_{\mathrm{cts}}(G_{\mathbb{Q},S}, \mathbb{Q}_\ell(m) \boxtimes V_\alpha) \to H^i_{\mathrm{cts}}(G_{F,T}, \mathbb{Q}_\ell(m) \boxtimes V_\alpha)^G$$

and the transfer mapping [42]

$$\psi : H^i_{\mathrm{cts}}(G_{F,T}, \mathbb{Q}_\ell(m) \boxtimes V_\alpha)^G \to H^i_{\mathrm{cts}}(G_{\mathbb{Q},S}, \mathbb{Q}_\ell(m) \boxtimes V_\alpha).$$

A direct computation on cocycles shows that $\phi \circ \psi$ and $\psi \circ \phi$ are both multiplication by the order of G, and are thus isomorphisms.

Therefore $H^i_{\mathrm{cts}}(G_{\mathbb{Q},S}, \mathbb{Q}_\ell(m) \boxtimes V_\alpha)$ vanishes if V_α is nontrivial, and it equals $H^i_{\mathrm{cts}}(G_{F,T}, \mathbb{Q}_\ell(m))$ if V_α is trivial. This shows that the unipotent radical of the completion $\mathcal{G}_{\mathbb{Q},S}$ is isomorphic to that of $\mathcal{A}^\ell_{F,T}$.

By functoriality of weighted completion, we have a homomorphism $\mathcal{A}^\ell_{F,T} \to \mathcal{G}_{\mathbb{Q},S}$ which induces the isomorphism on the unipotent radical. The statement follows. $\square$

10. Applications to Galois Actions on Fundamental Groups

Let G_ℓ denote $G_{\mathbb{Q},\{\ell\}}$. In this section, we sketch how our computation of the weighted completion of G_ℓ can be used to prove Deligne's Conjecture 5.8 about the action of the absolute Galois group $G_\mathbb{Q}$ on the pro-ℓ completion of the fundamental group of $\mathbb{P}^1(\mathbb{C}) - \{0, 1, \infty\}$. Modulo a few technical details, which are addressed in [23], the proof proceeds along the expected lines suggested in Section 5.2 given the computation of $\mathcal{A}^\ell_{F,S}$.

We begin in a more general setting. Suppose that F is a number field and that X is a variety over F. Set $\overline{X} = X \otimes \overline{\mathbb{Q}}$ and denote the absolute Galois group of F by G_F. Suppose that the étale cohomology group $H^1_{\text{ét}}(\overline{X}, \mathbb{Q}_\ell(1))$ is a trivial G_F-module. Let S be a set of finite primes of F, containing those above ℓ. Suppose that X has a model $\mathcal{X}$ over $\operatorname{Spec}\mathcal{O}_{F,S}$ which has a base point section $x : \operatorname{Spec}\mathcal{O}_{F,S} \to \mathcal{X}$ (possibly tangential) such that $(\mathcal{X}, x)$ has good reduction outside S.[6] Then the G_F-action on the pro-ℓ fundamental group $\pi^\ell_1(\overline{X}, x)$ factors through $G_{F,S}$.

Denote the ℓ-adic unipotent completion of $\pi^\ell_1(\overline{X}, x)$ by $\mathcal{P}$ (see [23, Appendix A]) and its Lie algebra by $\mathfrak{p}$. The lower central series filtration of $\mathfrak{p}$ gives it the structure of a pro-object of the category $\mathcal{T}_\ell(X_{F,S})$ of ℓ-adic mixed Tate modules over $X_{F,S}$. It follows that the G_F-action on $\mathcal{P}$ induces a homomorphism

$$\mathcal{A}^\ell_{F,S} \to \operatorname{Aut}\mathcal{P} \cong \operatorname{Aut}\mathfrak{p}$$

and that the action of G_F on $\mathcal{P}$ factors through the composition of this with the natural homomorphism $G_F \to G_{F,S} \to \mathcal{A}^\ell_{F,S}(\mathbb{Q}_\ell)$. One can show (see [23, Sect. 8]) that the image of $G_{F(\mu_{\ell^\infty})}$ in $\mathcal{A}^\ell_{F,S}$ lies in and is Zariski dense in $\mathcal{K}^\ell_{F,S}$.

For the rest of this section, we consider the case where $X = \mathbb{P}^1 - \{0, 1, \infty\}$, $F = \mathbb{Q}$, $S = \{\ell\}$ and x is the tangential base point $\overrightarrow{01}$. Goncharov's conjecture [19, Conj. 2.1] (cf. the generation part of Conjecture 5.9) follows immediately, since $\mathfrak{k}_{\mathbb{Q},\{\ell\}}$ is generated by $z_1, z_3, z_5, \ldots$, where z_j has weight $-2j$. The image of z_1 can be shown to be trivial.

We are now ready to give a brief sketch of the proof of Conjecture 5.8. One can define a filtration $\mathcal{I}^\bullet_\ell$ on $G_\mathbb{Q}$ similar to $I^\bullet_\ell$ using the lower central series $L^\bullet\mathcal{P}$ of $\mathcal{P}$ instead:

$$\mathcal{I}^m_\ell G_\mathbb{Q} = \ker\{G_\mathbb{Q} \to \operatorname{Out}\mathcal{P}/L^{m+1}\mathcal{P}\}$$

---

[6]What we mean here is that there is a scheme $\widetilde{\mathcal{X}}$, proper over $\operatorname{Spec}\mathcal{O}_{F,S}$, and a divisor D in $\widetilde{\mathcal{X}}$ which is relatively normal crossing over $\operatorname{Spec}\mathcal{O}_{F,S}$ such that $\mathcal{X} = \widetilde{\mathcal{X}} - D$, and D does not intersect with x. In the tangential case, the tangent vector should be non-zero over each point of $\operatorname{Spec}\mathcal{O}_{F,S}$.

where $L^m \mathcal{P}$ is the mth term of its lower central series. The lower central series of $\mathcal{P}$ is related to its weight filtration by

$$W_{-2m}\mathcal{P} = L^m \mathcal{P}, \quad \mathrm{Gr}^W_{2m+1}\mathcal{P} = 0.$$

There is a natural isomorphism (see [23, Sect. 10])

$$[\mathrm{Gr}^m_{\mathcal{I}_\ell} G_\mathbb{Q}] \otimes \mathbb{Q}_\ell \cong [\mathrm{Gr}^m_{\mathcal{I}_\ell} G_\mathbb{Q}] \otimes \mathbb{Q}_\ell.$$

Thus it suffices to prove that $[\mathrm{Gr}^m_{\mathcal{I}_\ell} G_\mathbb{Q}] \otimes \mathbb{Q}_\ell$ is generated by elements $s_3, s_5, s_7, \ldots$, where s_j has weight $-2j$.

As above, the homomorphism $G_\mathbb{Q} \to \mathrm{Out}\,\mathcal{P}$ factors through the sequence

$$G_\mathbb{Q} \to G_{\mathbb{Q},\{\ell\}} \to \mathcal{A}_{\mathbb{Q},\{\ell\}} \to \mathrm{Out}\,\mathcal{P}$$

of natural homomorphisms. A key point ([23, Sect. 8]) is that the image of $I_\ell^{-1} G_\mathbb{Q}$ in $\mathcal{K}^\ell_{\mathbb{Q},\ell}$ is Zariski dense. This and the strictness can be used to establish isomorphisms

$$Gr^m_{\mathcal{I}_\ell} G_\mathbb{Q} \otimes \mathbb{Q}_\ell \cong \mathrm{Gr}^W_{-2m}(\mathrm{im}\{\mathfrak{k}^\ell_{\mathbb{Q},\ell} \to \mathrm{OutDer}\,\mathfrak{p}\})$$
$$\cong \mathrm{im}\{\mathrm{Gr}^W_{-2m}\,\mathfrak{k}^\ell_{\mathbb{Q},\ell} \to \mathrm{Gr}^W_{-2m}\,\mathrm{OutDer}\,\mathfrak{p}\}$$

for each $m > 0$.

Theorem 9.2 implies that $\mathrm{Gr}^W_\bullet\,\mathfrak{k}^\ell_{\mathbb{Q},\ell}$ is freely generated by $\sigma_1, \sigma_3, \sigma_5, \ldots$ where $\sigma_{2i+1} \in \mathrm{Gr}^W_{-2(2i+1)}\,\mathfrak{k}^\ell_{\mathbb{Q},\ell}$. It is easy to show that the image of σ_1 vanishes in $\mathrm{Gr}^W_\bullet\,\mathrm{OutDer}\,\mathfrak{p}$. It follows that the image of $\mathrm{Gr}^W_\bullet\,\mathfrak{k}^\ell_{\mathbb{Q},\ell}$ is generated by the images of $\sigma_3, \sigma_5, \sigma_7, \ldots$, which completes the proof.

REMARK 10.1. Ihara proves the openness of the group generated by σ_{2i+1} in a suitable Galois group, see [28]. He also establishes the non-vanishing of the images of the σ_{2i+1} and some of their brackets in [27].

11. When ℓ is not contained in S

Let $[\ell]$ denote the set of all primes above ℓ in $\mathcal{O}_F$. In this section, we generalize the definition of the category $\mathcal{T}_\ell(X_{F,S})$ of ℓ-adic mixed Tate modules smooth over $X_{F,S} = \mathrm{Spec}\,\mathcal{O}_F - S$ (see Section 6) to the case where S does not necessarily contain $[\ell]$.

For this, we define the category $\mathcal{T}_\ell(X_{F,S})$ of ℓ-adic mixed Tate modules over $X_{F,S}$ to be the full subcategory of $\mathcal{T}_\ell(X_{F,S\cup[\ell]})$ (defined in Section 6) consisting of the Galois modules which are crystalline at every prime $\mathfrak{p} \in [\ell] - S$. (Recall that an ℓ-adic G_F-module M is crystalline at a prime $\mathfrak{p}$ of F if it is crystalline as $G_{F_\mathfrak{p}}$-module, where $F_\mathfrak{p}$ is the completion of F at $\mathfrak{p}$ and $G_{F_\mathfrak{p}}$ is identified with the decomposition group of G_F at $\mathfrak{p}$, see [16; 7] for crystalline representations.)

It is known that the crystalline property is closed under tensor products, direct sums, duals, and subquotients [16], so that $\mathcal{T}_\ell(X_{F,S})$ is a tannakian category. Denote its tannakian fundamental group by $\mathcal{A}^\ell_{F,S}$. We have a short exact sequence

$$1 \to \mathcal{K}^{\ell}_{F,S} \to \mathcal{A}^{\ell}_{F,S} \to \mathbb{G}_m \to 1,$$

and the corresponding exact sequence of Lie algebras

$$0 \to \mathfrak{k}^{\ell}_{F,S} \to \mathfrak{a}^{\ell}_{F,S} \to \mathbb{Q}_{\ell} \to 0.$$

Let V be a $G_{F,S}$-module. The *finite part* of the first degree Galois cohomology $H^1_{\mathrm{cts}f}(G_{F,S}, V) \subset H^1_{\mathrm{cts}}(G_{F,S}, V)$ is defined in [7, (3.7.2)]. This corresponds to those extensions of $\mathbb{Q}_{\ell}$ by V as $G_{F,S}$-modules, which are crystalline at every prime in $[\ell]$ outside S. By a remark on p. 354 in [7], $H^1_{\mathrm{cts}f}(G_{F,S}, \mathbb{Q}_{\ell}) = (\mathcal{O}_{F,S}^{\times}) \otimes_{\mathbb{Z}_{\ell}} \mathbb{Q}_{\ell}$, so its dimension is $d_1 + \#S = r_1 + r_2 + \#S - 1$. Theorem 9.2 is generalized as follows, by replacing H^1_{cts} with $H^1_{\mathrm{cts}f}$ [23]. We shall give a categorical proof below.

THEOREM 11.1. *The Lie algebra* $\mathrm{Gr}^W_{\bullet} \mathfrak{k}^{\ell}_{F,S}$ *is a free Lie algebra and there is a natural* $\mathbb{G}_m$-*equivariant isomorphism*

$$H^1_{\mathrm{cts}}(\mathfrak{k}^{\ell}_{F,S}) \cong \bigoplus_{n=1}^{\infty} H^1_{\mathrm{cts}f}(G_{F,S}, \mathbb{Q}_{\ell}(n)) \otimes \mathbb{Q}_{\ell}(-n) \cong \mathbb{Q}_{\ell}(-1)^{d_1 + \#S} \oplus \bigoplus_{n>1} \mathbb{Q}_{\ell}(-n)^{d_n},$$

where d_n *is defined in* (3–1). *Any lift of a graded basis of* $H_1(\mathrm{Gr}^W_{\bullet} \mathfrak{k}^{\ell}_{F,S})$ *to a graded set of elements of* $\mathrm{Gr}^W_{\bullet} \mathfrak{k}^{\ell}_{F,S}$ *freely generates* $\mathrm{Gr}^W_{\bullet} \mathfrak{k}^{\ell}_{F,S}$.

COROLLARY 11.2. *There are natural isomorphisms*

$$\mathrm{Ext}^m_{\mathcal{T}_{\ell}(X_{F,S})}(\mathbb{Q}_{\ell}, \mathbb{Q}_{\ell}(n)) \cong \begin{cases} \mathbb{Q}_{\ell} & \text{when } m = n = 0, \\ H^1_{\mathrm{cts}f}(G_{F,S}, \mathbb{Q}_{\ell}(n)) & \text{when } m = 1 \text{ and } n > 0, \\ 0 & \text{otherwise.} \end{cases}$$

Consequently, for all $n \in \mathbb{Z}$, *there are natural isomorphisms*

$$\mathrm{Ext}^1_{\mathcal{T}_{\ell}(X_{F,S})}(\mathbb{Q}_{\ell}, \mathbb{Q}_{\ell}(n)) \cong K_{2n-1}(\mathrm{Spec}\, \mathcal{O}_{F,S}) \otimes \mathbb{Q}_{\ell}. \qquad \square$$

This shows that $\mathcal{T}_{\ell}(X_{F,S})$ has all the properties of the category $\mathcal{T}(X_{F,S}) \otimes \mathbb{Q}_{\ell}$, where $\mathcal{T}(X_{F,S})$ is the category whose existence is conjectured by Deligne. In particular, $\mathrm{Gr}^W_{\bullet} \mathfrak{k}^{\ell}_{\mathbb{Q},\varnothing}$ is free with generators $\sigma_3, \sigma_5, \ldots$.

PROOF OF THEOREM 11.1. It suffices to show that the natural mapping

$$\Phi^1 : H^1(\mathcal{A}^{\ell}_{F,S}, \mathbb{Q}_{\ell}(n)) \to H^1_{\mathrm{cts}f}(G_{F,S}, \mathbb{Q}_{\ell}(n))$$

is an isomorphism when $n \geq 1$ and that the natural mapping

$$\Phi^2 : H^2(\mathcal{A}^{\ell}_{F,S}, \mathbb{Q}_{\ell}(n)) \to H^2_{\mathrm{cts}}(G_{F,S}, \mathbb{Q}_{\ell}(n))$$

is injective when $n \geq 2$. The proof is similar to that of Theorem 8.1. To show that Φ^1 is an isomorphism, it suffices to show that an extension E of $\mathbb{Q}_{\ell}$ by $\mathbb{Q}_{\ell}(n)$ corresponding to an element of $H^1_{\mathrm{cts}f}(G_{F,S}, \mathbb{Q}_{\ell}(n))$ is crystalline, which is well-known. So the first assertion follows.

We now consider the case of Φ^2. Set $V_{\alpha} = \mathbb{Q}_{\ell}(n)$. We may assume $n \geq 2$. It suffices to show that E in the proof of Theorem 8.1 is crystalline provided

E_1 and E_2 are crystalline. But this follows from the next result, which will be proved below.

PROPOSITION 11.3. *Let*

$$0 \longrightarrow V \longrightarrow U \longrightarrow \mathbb{Q}_\ell(1)^n \longrightarrow 0$$

be a short exact sequence of crystalline ℓ-adic representations of $G_{F_\mathfrak{p}}$. Assume that V is a successive extension of direct sums of a finite number of copies of $\mathbb{Q}_\ell(r)$ with $r \geq 2$. Then, for any extension

$$0 \longrightarrow U \longrightarrow E \longrightarrow \mathbb{Q}_\ell \longrightarrow 0$$

of ℓ-adic representations of $G_{F_\mathfrak{p}}$, E is crystalline if and only if its pushout by the surjection $U \to \mathbb{Q}_\ell(1)^n$ is crystalline.

Let U be E_2 as in the proof of Theorem 5.3. Since $W_{-2}E_2 = E_2$, U is an extension of $\mathbb{Q}_\ell(1)^n$ for some n. Since $m \geq 2$, the pushout of E along $U \to \mathbb{Q}_\ell(1)^n$ is a quotient of E_1, and hence is crystalline. Thus the proposition says that E is crystalline, which completes the proof of Theorem 11.1. $\square$

Proposition 11.3 follows from the following two lemmas.

LEMMA 11.4. *Let*

$$0 \to V_1 \to V_2 \to V_3 \to 0$$

be a short exact sequence of crystalline ℓ-adic representations of $G_{F_\mathfrak{p}}$. Then we have a long exact sequence

$$0 \to H^0(G_{F_\mathfrak{p}}, V_1) \to H^0(G_{F_\mathfrak{p}}, V_2) \to H^0(G_{F_\mathfrak{p}}, V_3)$$
$$\to H^1_{\mathrm{cts}f}(G_{F_\mathfrak{p}}, V_1) \to H^1_{\mathrm{cts}f}(G_{F_\mathfrak{p}}, V_2) \to H^1_{\mathrm{cts}f}(G_{F_\mathfrak{p}}, V_3) \to 0. \quad \square$$

This follows from [7, Cor. 3.8.4].

LEMMA 11.5. *Let V be a crystalline ℓ-adic representation of $G_{F_\mathfrak{p}}$. If V is a successive extension of $\mathbb{Q}_\ell(r)$ $(r \geq 2)$, then $H^1_{\mathrm{cts}f}(G_{F_\mathfrak{p}}, V) = H^1_{\mathrm{cts}}(G_{F_\mathfrak{p}}, V)$.*

PROOF. The proof is by induction on the dimension of V. In the case $\dim(V) = 1$, this is well-known (loc. cit. Example 3.9). Assume $\dim V = n \geq 2$ and the claim is true for $n - 1$. By assumption, there exists an exact sequence of ℓ-adic representations of $G_{F_\mathfrak{p}}$:

$$0 \longrightarrow V' \longrightarrow V \longrightarrow \mathbb{Q}_\ell(r) \longrightarrow 0$$

for some integer $r \geq 2$ such that V' satisfies the assumption of the lemma. By Lemma 11.4, we have the following commutative diagram whose two rows are exact:

$$0 \longrightarrow H^1_{\mathrm{cts}f}(G_{F_\mathfrak{p}}, V') \longrightarrow H^1_{\mathrm{cts}f}(G_{F_\mathfrak{p}}, V) \longrightarrow H^1_{\mathrm{cts}f}(G_{F_\mathfrak{p}}, \mathbb{Q}_\ell(r)) \longrightarrow 0$$

$$0 \longrightarrow H^1_{\mathrm{cts}}(G_{F_\mathfrak{p}}, V') \longrightarrow H^1_{\mathrm{cts}}(G_{F_\mathfrak{p}}, V) \longrightarrow H^1_{\mathrm{cts}}(G_{F_\mathfrak{p}}, \mathbb{Q}_\ell(r))$$

The right vertical arrow is an isomorphism and the left one is also an isomorphism by the induction hypothesis. Hence the middle one is also an isomorphism. $\square$

PROOF OF PROPOSITION 11.3. By Lemma 11.4, we have the following commutative diagram whose two rows are exact:

$$0 \longrightarrow H^1_{\mathrm{cts}f}(G_{F_\mathfrak{p}}, V) \longrightarrow H^1_{\mathrm{cts}f}(G_{F_\mathfrak{p}}, U) \longrightarrow H^1_{\mathrm{cts}f}(G_{F_\mathfrak{p}}, \mathbb{Q}_\ell(1)^n) \longrightarrow 0$$

$$0 \longrightarrow H^1_{\mathrm{cts}}(G_{F_\mathfrak{p}}, V) \longrightarrow H^1_{\mathrm{cts}}(G_{F_\mathfrak{p}}, U) \longrightarrow H^1_{\mathrm{cts}}(G_{F_\mathfrak{p}}, \mathbb{Q}_\ell(1)^n)$$

and the left vertical arrow is an isomorphism by Lemma 11.5. Hence the right square is cartesian. $\square$

Appendix: Continuous Cohomology and Yoneda Extensions

In this appendix we prove a result about the relation between continuous cohomology and Yoneda extension groups in low degrees. It is surely well known, but we know of no reference.

Suppose that K is a topological field, and Γ a topological group. A *continuous* Γ-*module* is a Γ-module V, where V is a finite dimensional K-vector space. The action $\Gamma \to \mathrm{GL}(V)$ is required to be continuous, where $\mathrm{GL}(V)$ is given the topology induced from that of K.

Denote by $\mathcal{C}(\Gamma, K)$ the category of finite dimensional continuous Γ-modules. Since any K-linear morphism between finite dimensional vector spaces is continuous, this is an abelian category. For continuous Γ-modules A and B, define $\mathrm{Ext}^\bullet_\Gamma(A, B)$ to be the graded group of Yoneda extensions of B by A in the category $\mathcal{C}(\Gamma, K)$.

For a continuous Γ-module A, one also has the continuous cohomology groups $H^\bullet_{\mathrm{cts}}(\Gamma, A)$ defined by Tate [42], which are defined using the complex of continuous cochains.

THEOREM A.6. *If A is a continuous Γ-module, there is a natural isomorphism* $\mathrm{Ext}^1_\Gamma(K, A) \cong H^1_{\mathrm{cts}}(\Gamma, A)$ *and a natural injection* $\mathrm{Ext}^2_\Gamma(K, A) \hookrightarrow H^2_{\mathrm{cts}}(\Gamma, A)$.

PROOF. It is well known that an extension $0 \to A \to E \to K \to 0$ in $\mathcal{C}(\Gamma, K)$ gives a continuous cocycle $f : \Gamma \to A$ by choosing a lift $e \in E$ of $1 \in K$ and defining $f(\sigma) = \sigma(e) - e$. Conversely, for a given continuous cocycle f, we may define continuous Γ-action on $A \oplus K$ by $\sigma : (a, k) \mapsto (\sigma(a) + kf(\sigma), k)$. These are mutually inverse, which establishes the first claim.

To prove the second claim, we first define a K-linear mapping

$$\varphi : \mathrm{Ext}^2_\Gamma(K, A) \to H^2_{\mathrm{cts}}(\Gamma, A)$$

as follows. For $c \in \mathrm{Ext}^2_\Gamma(K, A)$, choose a 2-fold extension $0 \to A \to E_2 \to E_1 \to K \to 0$ that represents it. By [45], c is the image under the connecting homomorphism

$$\delta : \mathrm{Ext}^1_\Gamma(K, E_2/A) \to \mathrm{Ext}^2_\Gamma(K, A)$$

of the class $\tilde{c}$ of the extension $0 \to E_2/A \to E_1 \to K \to 0$.

We shall construct φ so that the diagram

$$
\begin{array}{ccccc}
\mathrm{Ext}^1_\Gamma(K, E_2) & \longrightarrow & \mathrm{Ext}^1_\Gamma(K, E_2/A) & \overset{\delta}{\longrightarrow} & \mathrm{Ext}^2_\Gamma(K, A) \\
\downarrow{\simeq} & & \psi\downarrow{\simeq} & & \downarrow{\varphi} \\
H^1_{\mathrm{cts}}(\Gamma, E_2) & \longrightarrow & H^1_{\mathrm{cts}}(\Gamma, E_2/A) & \overset{\delta_{\mathrm{cts}}}{\longrightarrow} & H^2_{\mathrm{cts}}(\Gamma, A)
\end{array}
$$

commutes, where the rows are parts of the standard long exact sequences constructed in [45] and [42, Sect. 2]. Define $\varphi(c)$ to be $\delta_{\mathrm{cts}}(\psi(\tilde{c}))$.

To prove $\varphi(c)$ is well-defined, it suffices to show that two 2-fold extensions that fit into a commutative diagram

$$
\begin{array}{ccccccccc}
0 & \longrightarrow & A & \longrightarrow & E'_2 & \longrightarrow & E'_1 & \longrightarrow & K & \longrightarrow & 0 \\
& & \| & & \downarrow & & \downarrow & & \| & & \\
0 & \longrightarrow & A & \longrightarrow & E_2 & \longrightarrow & E_1 & \longrightarrow & K & \longrightarrow & 0
\end{array}
$$

give a same element of $H^2_{\mathrm{cts}}(\Gamma, A)$. But this follows from the functoriality of the connecting homomorphism for $H^\bullet_{\mathrm{cts}}$, i.e., the commutativity of

$$
\begin{array}{ccc}
H^1_{\mathrm{cts}}(\Gamma, E'_2/A) & \longrightarrow & H^2_{\mathrm{cts}}(\Gamma, A) \\
\downarrow & & \| \\
H^1_{\mathrm{cts}}(\Gamma, E_2/A) & \longrightarrow & H^2_{\mathrm{cts}}(\Gamma, A).
\end{array}
$$

The K-linearity of φ is easily checked. Finally, the injectivity of φ follows from the fact that for each extension as above, φ is injective on the image of the connecting homomorphism $\delta : \mathrm{Ext}^1_\Gamma(K, E_2/A) \to \mathrm{Ext}^2_\Gamma(K, A)$. $\quad\square$

Note that one may define

$$\mathrm{Ext}^m_\Gamma(K, A) \to H^m_{\mathrm{cts}}(G, A)$$

by induction on m in the same way.

Acknowledgements

We would like to thank Marc Levine for clarifying several points about motivic cohomology and Owen Patashnick for his helpful comments on the manuscript. We are indebted to Kazuya Kato and Akio Tamagawa for pointing out a subtlety regarding continuous cohomology related to Theorem A.6, and to Takeshi Tsuji for the proof of Proposition 11.3. We would also like to thank Sasha Goncharov for pointing out the existence and relevance of the unpublished manuscript [2] of Beilinson and Deligne, and Romyar Sharifi for correspondence of Galois cohomology when $\ell = 2$. Finally, we would like to thank the referee for doing a very thorough job and for many useful comments.

References

[1] A. Beilinson: *Higher regulators and values of L-functions* (Russian), Current problems in mathematics, Vol. 24 (1984), 181–238.

[2] A. Beilinson, P. Deligne: *Motivic Polylogarithms and Zagier's Conjecture*, unpublished manuscript, 1992.

[3] A. Beilinson, P. Deligne: *Interprétation motivique de la conjecture de Zagier reliant polylogarithmes et régulateurs*, Motives (Seattle, WA, 1991), Proc. Sympos. Pure Math., 55, Part 2, Amer. Math. Soc., Providence, RI, 1994, 97–121.

[4] A. Beilinson, R. MacPherson, V. Schechtman: *Notes on motivic cohomology*, Duke Math. J. 54 (1987), 679–710.

[5] S. Bloch: *Algebraic cycles and higher K-theory*, Adv. in Math. 61 (1986), 267–304.

[6] S. Bloch: *The moving lemma for higher Chow groups*, J. Algebraic Geom. 3 (1994), 537–568.

[7] S. Bloch, K. Kato: *L-Functions and Tamagawa Numbers of Motives*, The Grothendieck Festschrift Volume I, Progress in Math. Vol.86, Birkhäuser, (1990), 333-400.

[8] S. Bloch, I. Kriz: *Mixed Tate motives*, Ann. of Math. 140 (1994), 557–605.

[9] S. Bloch, A. Ogus: *Gersten's conjecture and the homology of schemes*, Ann. Sci. École Norm. Sup. (4) (1974), 181–201.

[10] A. Borel: *Stable real cohomology of arithmetic groups*, Ann. Sci. École Norm. Sup. (4) 7 (1974), 235–272 (1975).

[11] A. Borel: *Cohomologie de* SL_n *et valeurs de fonctions zêta aux points entiers*, Ann. Scuola Norm. Sup. Pisa Cl. Sci. (4) 4 (1977), 613–636.

[12] A. Borel, J.-P. Serre: *Le théorème de Riemann-Roch*, Bull. Soc. Math. France 86 1958, 97–136.

[13] P. Deligne and J. Milne: *Tannakian categories*, in Hodge Cycles, Motives, and Shimura Varieties, (P. Deligne, J. Milne, A. Ogus, K.-Y. Shih editors), Lecture Notes in Mathematics 900, Springer-Verlag, 1982.

[14] P. Deligne: *Le groupe fondamental de la droite projective moins trois points*, in Galois groups over Q (Berkeley, CA, 1987), 79–297, Math. Sci. Res. Inst. Publ., 16, Springer, New York-Berlin, 1989.

[15] V. Drinfeld: *On quasitriangular quasi-Hopf algebras and on a group that is closely connected with* Gal $\overline{\mathbb{Q}}/\mathbb{Q}$), (Russian) Algebra i Analiz 2 (1990), 149–181; translation in Leningrad Math. J. 2 (1991), no. 4, 829–860.

[16] J.M. Fontaine: *Sur certains types de représentations p-adiques du groupe de Galois d'un corps local: construction d'un anneau de Barsotti-Tate*, Ann. of Math. 115 (1982), 529–577.

[17] H. Gillet: *Riemann-Roch theorems for higher algebraic K-theory*, Adv. in Math. 40 (1981), 203–289.

[18] A. Goncharov: *Volumes of hyperbolic manifolds and mixed Tate motives*, J. Amer. Math. Soc. 12 (1999), 569–618.

[19] A. Goncharov: *Multiple ζ-values, Galois groups, and geometry of modular varieties*, European Congress of Mathematics, Vol. I (Barcelona, 2000), 361–392, Progr. Math., 201, Birkhäuser, 2001.

[20] R. Hain: *Classical polylogarithms*, Motives (Seattle, WA, 1991), Proc. Sympos. Pure Math., 55, Part 2, Amer. Math. Soc., Providence, RI, 1994, 3–42.

[21] R. Hain: *Completions of mapping class groups and the cycle $C - C^-$*, in *Mapping Class Groups and Moduli Spaces of Riemann Surfaces*, Contemp. Math. 150 (1993), 75–105.

[22] R. Hain: *Infinitesimal presentations of the Torelli groups*, J. Amer. Math. Soc. 10 (1997), 597–651.

[23] R. Hain, M. Matsumoto: *Weighted Completion of Galois Groups and Galois Actions on Fundamental Groups*, Compositio Math., to appear, math.AG/0006158.

[24] R. Hain, M. Matsumoto: *Completions of Arithmetic Mapping Class Groups*, in preparation.

[25] H. Hiller: *λ-rings and algebraic K-theory*, J. Pure Appl. Algebra 20 (1981), 241–266.

[26] Y. Ihara: *Profinite braid groups, Galois representations and complex multiplications*, Ann. of Math., 123 (1986), 43–106.

[27] Y. Ihara: *The Galois representation arising from* $\mathbb{P}^1 - \{0, 1, \infty\}$ *and Tate twists of even degree*, in *Galois groups over* $\mathbb{Q}$, Publ. MSRI, No. 16 (1989), Springer-Verlag, 299–313.

[28] Y. Ihara: *Some arithmetic aspects of Galois actions in the pro-p fundamental group of* $\mathbb{P}^1 - \{0, 1, \infty\}$, Arithmetic fundamental groups and noncommutative algebra (Berkeley, CA, 1999), 247–273, Proc. Sympos. Pure Math., 70, Amer. Math. Soc., 2002.

[29] U. Jannsen: *Mixed motives and algebraic K-theory*, Lecture Notes in Mathematics, 1400, Springer-Verlag, Berlin, 1990.

[30] J. Jantzen: *Representations of Algebraic Groups*, Pure and Applied Mathematics Vol.131, Academic Press, 1987.

[31] C. Kratzer: *λ-structure en K-théorie algébrique*, Comment. Math. Helv. 55 (1980), 233–254.

[32] M. Levine: *Tate motives and the vanishing conjectures for algebraic K-theory*, Algebraic K-theory and algebraic topology (Lake Louise, AB, 1991), NATO Adv.

Sci. Inst. Ser. C Math. Phys. Sci., 407, Kluwer Acad. Publ., Dordrecht, 1993, 167–188.

[33] M. Levine: *Bloch's higher Chow groups revisited*, K-theory (Strasbourg, 1992), Astérisque No. 226, (1994), 10, 235–320.

[34] M. Levine: *Mixed motives*, Mathematical Surveys and Monographs, 57, Amer. Math. Soc., Providence, RI, 1998.

[35] M. Matsumoto: *On the Galois image in derivation of π_1 of the projective line minus three points*, in "Recent developments in the inverse Galois problem (Seattle, WA, 1993)," Contemp. Math. 186 (1995), 201–213.

[36] J. Milnor: *Introduction to algebraic K-theory*, Annals of Mathematics Studies, No. 72. Princeton University Press, Princeton, N.J.; University of Tokyo Press, Tokyo, 1971.

[37] D. Quillen: *Finite generation of the groups K_i of rings of algebraic integers*, Algebraic K-theory, I: Higher K-theories (Proc. Conf., Battelle Memorial Inst., Seattle, Wash., 1972), pp. 179–198. Lecture Notes in Math., Vol. 341, Springer, Berlin, 1973.

[38] D. Quillen: *On the cohomology and K-theory of the general linear groups over a finite field*, Ann. of Math. 96 (1972), 552–586.

[39] D. Quillen: *Higher algebraic K-theory, I*, Algebraic K-theory, I: Higher K-theories (Proc. Conf., Battelle Memorial Inst., Seattle, Wash., 1972), pp. 85–147. Lecture Notes in Math., Vol. 341, Springer, Berlin 1973.

[40] J. Rognes and C. Weibel: *Two-primary algebraic K-theory of rings of integers in number fields. (Appendix A by Manfred Kolster)*, J. Amer. Math. Soc., 13 (2000), 1–54.

[41] C. Soulé: *On higher p-adic regulators*, Lecture Notes in Math. 854 (1981), 372–401.

[42] J. Tate: *Relations between K_2 and Galois Cohomology*, Invent. Math., 30 (1976), 257–274.

[43] H. Tsunogai: *On ranks of the stable derivation algebra and Deligne's problem*, Proc. Japan Academy Ser. A 73 (1997), 29–31.

[44] V. Voevodsky, A. Suslin, E. Friedlander: *Cycles, transfers, and motivic homology theories*, Annals of Mathematics Studies, 143, Princeton University Press, 2000.

[45] N. Yoneda: *On Ext and exact sequences*, J. Fac. Sci. Univ. Tokyo Sect. I 8 (1960), 507–576.

RICHARD HAIN
DEPARTMENT OF MATHEMATICS
DUKE UNIVERSITY
DURHAM, NC 27708-0320
hain@math.duke.edu

MAKOTO MATSUMOTO
DEPARTMENT OF MATHEMATICS
FACULTY OF SCIENCE
HIROSHIMA UNIVERSITY
HIGASHI-HIROSHIMA, 739-8526 JAPAN
m-mat@math.sci.hiroshima-u.ac.jp

Galois Groups and Fundamental Groups
MSRI Publications
Volume 41, 2003

Special Loci in Moduli Spaces of Curves

LEILA SCHNEPS

ABSTRACT. Let S be a topological surface of genus g with n marked points, and let φ be a finite-order element of the mapping class group of S. We study the *special locus* associated to φ in the moduli space $\mathcal{M}(S)$ of Riemann surfaces of topological type (g, n); this is the set of points in $\mathcal{M}(S)$ corresponding to Riemann surfaces admitting φ as an automorphism. Another definition of the special locus is that it is the image on $\mathcal{M}(S)$ of the points in the Teichmüller space $\mathcal{T}(S)$ fixed by φ under the natural action of the mapping class group on $\mathcal{T}(S)$. We completely describe all special loci in the moduli spaces of small type $(0, 4)$, $(0, 5)$, $(1, 1)$ and $(1, 2)$, and also of the general genus zero spaces $(0, n)$, including determining their fields of moduli. Then, based on results of Harvey et al., we show how the (normalization of the) special locus of φ in $\mathcal{M}(S)$ provides a finite covering of the moduli space of the topological quotient S/φ, and give conditions on φ for this covering to be as close as possible to an isomorphism. Finally, we translate these results in terms of the mapping class groups and show that when the conditions on φ are satisfied, we obtain a homomorphism between mapping class groups which has geometric and arithmetic significance, and that in genus zero, these two conditions are always satisfied. We end with two explicit examples of such homomorphisms, one in genus zero and one in genus one.

CONTENTS

Mathematics Subject Classification: 14H10, 14H37, 14H30.

Keywords: Moduli spaces of curves, automorphisms of curves, coverings and fundamental groups.

1. Introduction

1.1. Overview. Let $\mathcal{M}_{g,n}$ denote the moduli space of Riemann surfaces of genus g with n ordered marked points. By permuting the marked points on the Riemann surfaces, the permutation group S_n acts naturally on this space; the moduli space $\mathcal{M}_{g,[n]} = \mathcal{M}_{g,n}/S_n$ classifies the Riemann surfaces of genus g with n unordered marked points. It is sometimes useful to consider 'partially ordered' moduli spaces, i.e., quotients of $\mathcal{M}_{g,n}$ by subgroups of S_n.

The main goal of this article is to study *special loci* on moduli spaces. Topologically, the moduli spaces of curves are *orbifolds*; in fact, the moduli space $\mathcal{M}_{g,n}$ (resp. $\mathcal{M}_{g,[n]}$) is a quotient of a contractible space of complex dimension $3g - 3 + n$, the Teichmüller space $\mathcal{T}_{g,n}$, by the action of a discrete group called the *mapping class group* $\Gamma_{g,n}$ (resp. $\Gamma_{g,[n]}$). If S denotes a topological surface of genus g with n marked points, then $\Gamma_{g,n}$ (resp. $\Gamma_{g,[n]}$) is exactly the group of orientation-preserving diffeomorphisms fixing (resp. permuting) the marked points of S, up to those isotopic to the identity. The mapping class groups act properly discontinuously on the Teichmüller space, but not always freely; some points of Teichmüller space have isotropy groups of finite order inside the mapping class group, and conversely, every finite-order subgroup of the mapping class group fixes some point on Teichmüller space. The quotient of a simply connected space by a group acting in this way is called a topological *orbifold*, and the groups themselves are called *orbifold fundamental groups* (see [HN] for an introduction to these groups). The images in moduli space of the points with nontrivial isotropy in Teichmüller space are called *special orbifold points*. If φ is an element of finite order in the mapping class group, then we consider the set of points in Teichmüller space fixed by φ; the image of this set in the quotient moduli space is called the *special locus of φ*. This article is essentially devoted to studying these special loci, and the morphisms between moduli spaces and the corresponding homomorphisms between their fundamental groups which can be deduced from them. The main observations are as follows.

Harvey showed that in a suitable quotient of $\mathcal{M}_{g,n}$, the normalization of the special locus of a finite-order element φ in $\Gamma_{g,[n]}$ naturally gives a finite covering of the moduli space of the topological quotient of S by the action of a finite-order diffeomorphism lifting the diffeomorphism class φ (see Section 4.1). We give two conditions on φ that ensure that each irreducible component of the special locus of φ is actually isomorphic to the moduli space of S/φ up to a trivial orbifold structure. We then translate these results in terms of fundamental groups and show that when the two corresponding conditions are satisfied, we obtain interesting *special homomorphisms* between mapping class groups. Finally, we show that when S is of genus 0, these two conditions are always fulfilled.

1.2. Outline of the article. In Section 2, we recall the basic facts about Teichmüller space and moduli spaces of curves (see [M] for a beautiful introduction) and mapping class groups. We recall that the mapping class group of type (g, n) has three different descriptions, namely as the group of diffeomorphisms of a topological surface S of type (g, n) up to isotopy, the orbifold fundamental group of the moduli space of type (g, n) (Section 2.1) and the group of special outer automorphisms of $\pi_1(S)$ (Section 2.2). Then we concentrate on the genus zero moduli spaces and mapping class groups, and give several propositions showing how to pass explicitly between these three descriptions (Section 2.3).

Section 3 is devoted to explicitly examining the details of the structure of the genus zero ordered and unordered moduli spaces. Working over the complex numbers (i.e., topologically rather than algebraically), we review well-known features such as their mapping class groups, their stable compactifications, their topological tangential base points, their orbifold structures, paths given by standard Dehn twists, their fundamental groups, the points of special automorphism group on the ordered moduli space, and the special loci on the unordered spaces. We first consider the one-dimensional spaces $(0, 4)$ (Section 3.1) and $(1, 1)$ (Section 3.2), then the two-dimensional spaces $(0, 5)$ (Section 3.3) and $(1, 2)$ (Section 3.4). Finally, in Section 3.5, we give an explicit description of the special loci in genus zero moduli spaces with any number of marked points and determine their fields of definition.

In Section 4, we continue to investigate special loci in moduli spaces associated to finite cyclic subgroups $\langle \varphi \rangle$ in the mapping class groups. In Section 4.1, we compare a special locus in the moduli space of a topological surface S of type (g, n) to the moduli space of the topological quotient S/φ, showing (based on a theorem due to Harvey et al.) that the normalization of the first provides a finite covering of the second. We give two conditions on φ that ensure that each irreducible component of this finite covering is actually an isomorphism, up to a trivial orbifold structure due to the automorphism associated to φ at every point of the special locus. In Section 4.2, we translate the finite covering into a homomorphism of the associated fundamental groups, and translate the two conditions on φ into *splitting* and *surjectivity* conditions on this homomorphism. When these two conditions are fulfilled, we show that we obtain new and interesting *special homomorphisms* between mapping class groups, which have geometric and arithmetic significance. In Section 4.3, we prove that the two conditions are satisfied whenever φ is a finite-order element of a genus zero mapping class group $\Gamma_{0,[n]}$. This means that *all special loci in genus zero moduli space corresponding to cyclic subgroups of the mapping class groups are themselves moduli spaces*. The last two sections are devoted to two examples of the splitting and surjectivity conditions and explicit determination of the corresponding special homomorphisms.

1.3. Connections with Galois Theory. In this section, we give a very brief sketch of the connection between the special loci in moduli space and the associated special homomorphisms between mapping class groups, and the wider world of Galois and Grothendieck–Teichmüller theory. Although we do not return to this topic within the paper, it should provide an understanding of the motivation behind the results.

The arithmetic significance of homomorphisms between mapping class groups coming from topological manipulations leading to geometric morphisms between the moduli spaces is the following. The moduli spaces themselves are defined over $\mathbb{Q}$. Let S and T be topological surfaces and let $\mathcal{M}(S)$ and $\mathcal{M}(T)$ denote their associated moduli spaces (with or without allowing permutation of the marked points). If we have a geometric morphism $f : \mathcal{M}(S) \to \mathcal{M}(T)$ which is defined over $\mathbb{Q}$, then up to inner automorphisms, the following diagram commutes for all $\sigma \in \mathrm{Gal}(\overline{\mathbb{Q}}/\mathbb{Q})$:

$$
\begin{array}{ccc}
\widehat{\pi}_1(\mathcal{M}(S)) & \xrightarrow{\ f_*\ } & \widehat{\pi}_1(\mathcal{M}(T)) \\
\Big\downarrow{\scriptstyle \sigma} & & \Big\downarrow{\scriptstyle \sigma} \\
\widehat{\pi}_1(\mathcal{M}(S)) & \xrightarrow{\ f_*\ } & \widehat{\pi}_1(\mathcal{M}(T)),
\end{array}
\qquad (1.3.1)
$$

where the $\widehat{\pi}_1$ are the algebraic (profinite) fundamental groups of the moduli spaces, which are equipped with a canonical outer $\mathrm{Gal}(\overline{\mathbb{Q}}/\mathbb{Q})$-action.

Let the *tower* $\mathcal{T}$ consist of the groups $\widehat{\pi}_1(\mathcal{M}(S))$ for all topological types S, equipped with homomorphisms f_* coming from geometric morphisms f between the moduli spaces which are defined over $\mathbb{Q}$. We can define the *special outer automorphism group of* $\mathcal{T}$ to be the collection of tuples $(\phi_S)_S$ where ϕ_S is a special outer automorphism of $\widehat{\pi}_1(\mathcal{M}(S))$, i.e., an outer automorphism preserving conjugacy classes of inertia generators, and such that the tuple makes all diagrams (1.3.1) commute for every f_* in the tower. It is then clear that $\mathrm{Gal}(\overline{\mathbb{Q}}/\mathbb{Q})$ injects into the automorphism group of $\mathcal{T}$, and it is an open question (due to Grothendieck) whether this automorphism group, of which various versions have been explicitly computed, according to the precise collection of homomorphisms f_* which are included in the tower, and which is generically known as the Grothendieck–Teichmüller group, may actually be equal to $\mathrm{Gal}(\overline{\mathbb{Q}}/\mathbb{Q})$. The more $\mathbb{Q}$-homomorphisms are included in the tower $\mathcal{T}$, the closer the corresponding automorphism group (Grothendieck–Teichmüller group) will be to $\mathrm{Gal}(\overline{\mathbb{Q}}/\mathbb{Q})$. Until now, the Teichmüller tower has been equipped with homomorphisms coming from morphisms between moduli spaces coming from erasing marked points on the topological surfaces, and including subsurfaces into surfaces by cutting along simple closed loops. As a natural sequel to the present article, the author hopes to compute the new Grothendieck–Teichmüller group associated to the tower equipped with the *special homomorphisms* as well as these.

2. Moduli Spaces of Surves

2.1. Teichmüller space and the mapping class group. Let S be a topological surface of genus g equipped with n ordered marked points $x_1, \ldots, x_n$ (we say that S is of type (g, n)). A Riemann surface X of genus g with n ordered marked points $y_1, \ldots, y_n$ is said to be *marked* if we choose a diffeomorphism $\Phi : S \to X$ such that $\Phi(x_i) = y_i$ for $1 \leq i \leq n$. Two marked Riemann surfaces X (with marked points $y_1, \ldots, y_n$ and marking Φ) and X' (with marked points $y_1', \ldots, y_n'$ and marking Φ') are said to be *isomorphic* if there exists an isomorphism $\alpha : X \to X'$ of Riemann surfaces with $\alpha(y_i) = y_i'$ for $1 \leq i \leq n$, and a diffeomorphism $h : S \to S$ with $h(x_i) = x_i$ for $1 \leq i \leq n$, which is isotopic to the identity, such that the following diagram commutes:

$$\begin{array}{ccc} S & \xrightarrow{\Phi} & X \\ {\scriptstyle h}\big\downarrow & & \big\downarrow{\scriptstyle \alpha} \\ S & \xrightarrow{\Phi'} & X'. \end{array} \qquad (2.1.1)$$

The *Teichmüller space* $\mathcal{T}_{g,n}$ is the set of isomorphism classes of marked Riemann surfaces of type (g, n). It is well-known that the Teichmüller space forms a contractible complex analytic space of dimension $3g - 3 + n$.

Let S be a topological surface of type (g, n) as above, and set

$$\Gamma_{g,[n]} = \mathrm{Diff}^+([S])/\mathrm{Diff}^0(S),$$

where $\mathrm{Diff}^+([S])$ denotes the group of orientation-preserving diffeomorphisms of S which permute the marked points, and $\mathrm{Diff}^0(S)$ is the group of those which are isotopic to the identity. The group $\Gamma_{g,[n]}$ is known as the *full mapping class group*. We also define the *pure mapping class group*, or *pure subgroup of the full mapping class group*, by setting

$$\Gamma_{g,n} = \mathrm{Diff}^+(S)/\mathrm{Diff}^0(S),$$

where $\mathrm{Diff}^+(S)$ is the subgroup of $\mathrm{Diff}^+([S])$ consisting of diffeomorphisms which fix each marked point.

The mapping class group $\Gamma_{g,[n]}$ acts on the Teichmüller space $\mathcal{T}_{g,n}$, as follows. To begin with, if $\psi \in \Gamma_{g,[n]}$, let ψ' denote a lifting of ψ to a diffeomorphism of S. Then ψ' maps the marked Riemann surface (Φ, X) to $(\Phi \circ \psi', X)$. Now we show explicitly that elements of $\Gamma_{g,[n]}$ act on isomorphism classes of marked Riemann surfaces. First, we show that if (Φ, X) and (Φ', X') are isomorphic marked Riemann surfaces and ψ' is a diffeomorphism of S, then the images $(\Phi\psi', X)$ and $(\Phi'\psi', X')$ are isomorphic. To see this, let α be as in (2.1.1). Then the diagram

$$
\begin{array}{ccc}
S \xrightarrow{\psi'} S \xrightarrow{\Phi} X \\
{\scriptstyle \psi'^{-1}h\psi'}\downarrow \quad\quad h\downarrow \quad\quad \downarrow{\scriptstyle \alpha} \\
S \xrightarrow{\psi'} S \xrightarrow{\Phi'} X'
\end{array}
$$

commutes, i.e.

$$
\begin{array}{ccc}
S \xrightarrow{\Phi\psi'} X \\
{\scriptstyle \psi'^{-1}h\psi'}\downarrow \quad\quad \downarrow{\scriptstyle \alpha} \\
S \xrightarrow{\Phi'\psi'} X'
\end{array}
$$

commutes, and $\psi'^{-1}h\psi'$ is isotopic to the identity since the diffeomorphisms isotopic to the identity form a normal subgroup of the group of diffeomorphisms. Next, we show that two equivalent diffeomorphisms of S take a marked Riemann surface (Φ, X) to two isomorphic marked Riemann surfaces. Let h be a diffeomorphism of S which is isotopic to the identity; then $(\Phi \circ \psi', X)$ is isomorphic to $(\Phi \circ \psi' \circ h, X)$, since we have

$$
\begin{array}{ccc}
S \xrightarrow{\Phi\circ\psi'} X \\
{\scriptstyle h^{-1}}\downarrow \quad\quad \downarrow{\scriptstyle \mathrm{id}} \\
S \xrightarrow{\Phi\circ\psi'\circ h} X.
\end{array}
$$

This shows that the mapping class group (equivalence classes of diffeomorphisms of S) acts on the Teichmüller space $\mathcal{T}_{g,n}$ (isomorphism classes of marked Riemann surfaces).

It is well-known that this action of $\Gamma_{g,[n]}$ on $\mathcal{T}_{g,n}$ is *properly discontinuous*. This means that for any compact subset K of $\mathcal{T}_{g,n}$, there are at most finitely many elements γ in the mapping class group such that $\gamma(K) \cap K \neq \varnothing$. Note that this fact implies that the stabilizer in the mapping class group of any point $x \in \mathcal{T}_{g,n}$ is a finite group, since if K is a small compact neighborhood of x, then $\gamma(K) \cap K \neq \varnothing$ for every γ in the stabilizer of x.

The *unordered moduli space* $\mathcal{M}_{g,[n]}$ is realized as the quotient of the Teichmüller space $\mathcal{T}_{g,n}$ by the action of the mapping class group $\Gamma_{g,[n]}$. This is tantamount to forgetting the marking, so the points of $\mathcal{M}_{g,[n]}$ correspond to isomorphism classes of Riemann surfaces of genus g with n unordered marked points. Similarly, the *ordered moduli space* $\mathcal{M}_{g,n}$ is the quotient of $\mathcal{T}_{g,n}$ by the pure subgroup $\Gamma_{g,n}$ of $\Gamma_{g,[n]}$, and its points correspond to isomorphism classes of Riemann surfaces of genus g with n ordered marked points. Because the Teichmüller space is topologically just a ball, and the moduli space is the quotient of Teichmüller space by the proper discontinuous action of a discrete group, the moduli spaces are *topological orbifolds* (if the group acted *freely*, as is actually the case for the pure genus zero groups $\Gamma_{0,n}$, or whenever n is sufficiently large with respect to g, they would be simply *ordinary topological manifolds*). When an orbifold arises in this manner, as a quotient of a simply-connected topological space by a discrete group acting properly discontinuously, it is called a *good orbifold*, and the discrete group is called the *orbifold fundamental group*. Thus we have

$$\Gamma_{g,[n]} = \pi_1^{\text{orbifold}}(\mathcal{M}_{g,[n]}) \text{ and } \Gamma_{g,n} = \pi_1^{\text{orbifold}}(\mathcal{M}_{g,n}). \qquad (2.1.2)$$

The key to studying such orbifolds is the study of the (finite) isotropy subgroups of the fundamental group, i.e., the subgroups which fix points of the simply connected space.

In the case of the moduli space, these isotropy subgroups have a particular geometric meaning. Namely, the isotropy subgroup of a point of moduli space (i.e., an isomorphism class of Riemann surfaces) inside the mapping class group is exactly the automorphism group of the Riemann surface associated to the point. The main focus of this article is the set of elements φ of finite order inside the mapping class groups (particularly in genus zero), and the corresponding *special loci*, i.e., the set of points on the moduli space having an automorphism which, topologically, corresponds to φ.

It is easy to give an explicit description of the genus zero moduli spaces. Indeed, an isomorphism class of spheres with n ordered marked points is an orbit of n-tuples of points up to the action of $\text{PSL}_2(\mathbb{C})$. This means that we can choose a unique representative of each class with the first three points fixed at three given values, usually taken to be 0, 1 and ∞. Thus, points of the ordered moduli space $\mathcal{M}_{0,n}$ are in bijection with n-tuples $(0, 1, \infty, x_4, \ldots, x_n)$ where the x_i are distinct from 0, 1, ∞ and each other, which gives

$$\mathcal{M}_{0,n} \simeq \left(\mathbb{P}^1 - \{0, 1, \infty\}\right)^{n-3} - \Delta \qquad (2.1.3)$$

where Δ denotes the multidiagonal of points with $x_i = x_j$. The unordered moduli space is the quotient of this space by the action of S_n.

2.2. A second definition of the mapping class group. As before, let S be a topological surface of type (g, n), and consider its fundamental group given by generators and relations as

$$\pi_{g,n} = \pi_1(S) = \left\langle a_1, b_1, \ldots, a_g, b_g, c_1, \ldots, c_n \mid \prod_{i=1}^{g}(a_i, b_i)c_1 \cdots c_n = 1 \right\rangle. \quad (2.2.1)$$

Here, the loops $c_1, \ldots, c_n$ correspond to loops around each of the ordered marked points of S, and they are the generators of the inertia subgroups in $\pi_1(S)$.

For an element $c \in \pi_1(S)$, let $\langle c \rangle$ denote the conjugacy class of c. Recall that outer automorphisms act on conjugacy classes. Define the group of "inertia-preserving" outer automorphisms $\text{Out}^*([\pi_{g,n}])$ to be the group

$$\{\psi \in \text{Out}(\pi_{g,n}) \mid \exists \sigma \in S_n \text{ such that } \psi(\langle c_i \rangle) = \langle c_{\sigma(i)} \rangle \text{ for } 1 \leq i \leq n\}.$$

The notation with square brackets [   ] indicates that the inertia subgroups can be permuted, mimicking the notation $\Gamma_{g,[n]}$ when the marked points can be permuted. We have a natural homomorphism

$$\mathrm{Out}^*([\pi_{g,n}]) \to S_n$$

$$\psi \mapsto \sigma,$$

and we let $\mathrm{Out}^*(\pi_{g,n})$ be the kernel of this homomorphism; thus it is the group of *pure* automorphisms ψ, i.e., automorphisms such that $\psi(\langle c_i \rangle) = \langle c_i \rangle$ for $1 \le i \le n$.

This definition affords a new, and very useful, definition of the mapping class group, as attested in the following well-known theorem.

THEOREM 2.2.1. *Let S be a topological surface of type (g, n). Then*

$$\mathrm{Out}^*([\pi_{g,n}]) \simeq \Gamma_{g,[n]} \quad and \quad \mathrm{Out}^*(\pi_{g,n}) \simeq \Gamma_{g,n}.$$

This is a classical result (see [Mac]), so we do not give the complete proof here. We simply indicate the general wherefore of it, by giving an explicit description of the homomorphism

$$\Gamma_{g,[n]} \to \mathrm{Out}^*([\pi_{g,n}]).$$

It is easy to see that an element $\psi \in \Gamma_{g,n}$ can be lifted to a diffeomorphism ψ' which fixes the base point of $\pi_{g,n}$ on $S - \{x_1, \ldots, x_n\}$, and which thus induces an automorphism of $\pi_{g,n}$ simply by acting on S. If ψ'' is another lifting of ψ also fixing the base point, then $\psi'' \cdot \psi'^{-1}$ is isotopic to the identity and fixes the base point of $\pi_{g,n}$, so it acts by an inner automorphism on $\pi_{g,n}$. Thus one obtains a well-defined map from $\Gamma_{g,n}$ to $\mathrm{Out}(\pi_{g,n})$, whose image actually lies in $\mathrm{Out}^*([\pi_{g,n}])$ since each c_i can be represented by a loop $\tilde{c}_i$ around x_i which consists of a path γ_i from the base point nearly to x_i, followed by a tiny circle around x_i and then γ_i^{-1}; a diffeomorphism necessarily maps this loop $\tilde{c}_i$ to a conjugate of a power of $\tilde{c}_j$. It is straightforward to check that we obtain a group homomorphism.

2.3. Genus zero mapping class groups. In this section, we consider a certain generating system for the mapping class groups $\Gamma_{g,[n]}$ and $\Gamma_{g,n}$, by elements known as *Dehn twists*. Then we consider the case $g = 0$ and show how the Dehn twists allow us to make the isomorphisms

$$\Gamma_{0,[n]} \simeq \pi_1^{\mathrm{orbifold}}(\mathcal{M}_{0,[n]}) \quad and \quad \Gamma_{0,n} \simeq \mathrm{Out}^*(\pi_{0,n})$$

explicit, and to show that the genus zero mapping class groups are closely related to the Artin braid groups. Recall that for $n \ge 2$, the *Artin braid group* B_n is generated by $n - 1$ generators denoted $\sigma_1, \ldots, \sigma_{n-1}$ subject to the relations

$$\sigma_i \sigma_{i+1} \sigma_i = \sigma_{i+1} \sigma_i \sigma_{i+1}$$

and

$$\sigma_i \sigma_j = \sigma_j \sigma_i \quad \text{if} \quad |i - j| \ge 2.$$

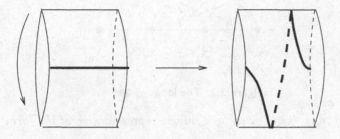

Figure 2.1. A Dehn twist.

Set $y_1 = z_1 = 1$, and for $2 \leq i \leq n$, set $y_i = \sigma_{i-1} \cdots \sigma_1 \cdot \sigma_1 \cdots \sigma_{i-1}$ and $z_i = (\sigma_1 \ldots \sigma_{i-1})^i$. It is known that the center of the group B_n is cyclic, generated by z_n. For $1 \leq i < j \leq n$, we write $x_{ij} = \sigma_{j-1} \cdots \sigma_{i+1} \sigma_i \sigma_{i+1}^{-1} \cdots \sigma_{j-1}^{-1}$; in particular $x_{i,i+1} = \sigma_i^2$ for $1 \leq i < n$. These elements generate the subgroup of *pure* braids, i.e., braids each of whose strands ends up in the same position it started from.

Definition. Let S be a topological surface of type (g, n) and let γ be a simple closed loop on S passing through either zero or two marked points. We define a certain diffeomorphism of S associated to the loop γ, in the following way. First, cut out a neighborhood of γ in S; the neighborhood of a simple closed curve has the form of a cylinder with the curve itself as a sort of "belt". Parametrize the cylinder by parameters (y, θ) with $y \in [-1, 1]$ and $\theta \in [0, 2\pi)$, in such a way that $y = 0$ corresponds to the simple closed loop γ, such that if the loop passes through two marked points, they lie at $(0, 0)$ and $(0, \pi)$. Then, define the Dehn twist diffeomorphism on this cylinder by $(y, \theta) \mapsto (y, \theta + \pi(y + 1))$. This diffeomorphism acts like the identity on the boundaries of the cylinder, and we extend it to the whole of S by the identity.

All of the following propositions are standard results (see [B]).

PROPOSITION 2.3.1 (DEHN). *The pure mapping class groups $\Gamma_{g,n}$ are generated by Dehn twists along simple closed loops passing through 0 marked points, and the full mapping class groups are generated by Dehn twists along simple closed loops passing through 0 or 2 marked points.*

PROPOSITION 2.3.2. *Let S be a sphere with marked points $x_1, \ldots, x_n$ with $n \geq 5$, which can be considered (topologically) as lying on a line. For $1 \leq i \leq n-1$, let $\sigma_i \in \Gamma_{0,[n]}$ denote the Dehn twist along the simple closed loop γ_i passing through the neighboring points x_i and x_{i+1}. Then $\Gamma_{0,[n]}$ is generated by $\sigma_1, \ldots, \sigma_{n-1}$.*

PROPOSITION 2.3.3. *Let $n \geq 5$, and recall from (2.1.3) that*

$$\mathcal{M}_{0,n} \simeq \left(\mathbb{P}^1 - \{0, 1, \infty\}\right)^{n-3} - \Delta,$$

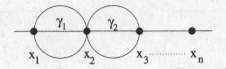

Figure 2.2. The loops γ_1 and γ_2.

each point of $\mathcal{M}_{0,n}$ being given by a unique representative of the form

$$(x_1 = 0, x_2 = 1, x_3 = \infty, x_4, \ldots, x_n).$$

Fix a base point $X = (0, 1, \infty, x_4, \ldots, x_n)$ such that $x_4 < x_5 < \ldots < x_n < 0$ are ordered real numbers. The Dehn twists σ_i based at X correspond to paths (not loops) on $\mathcal{M}_{0,n}$, since they permute the marked points; these paths become loops on $\mathcal{M}_{0,[n]}$. On $\mathcal{M}_{0,n}$, these paths can be explicitly parametrized by

$$\sigma_i \mapsto \Big(x_1, \ldots, x_{i-1}, f_i(t), g_{i+1}(t), x_{i+2}, \ldots, x_n\Big),$$

with $t \in [0, 1]$, where (because $x_3 = \infty$) we have

$$f_i(t) = \begin{cases} \dfrac{x_i - x_{i+1}}{2} e^{\pi i t} + \dfrac{x_i + x_{i+1}}{2} & \text{for } i \neq 2, 3, \\[2mm] 1 - i\Big(\dfrac{t}{1-t}\Big) & \text{for } i = 2, \\[2mm] x_4 - i\Big(\dfrac{1-t}{t}\Big) & \text{for } i = 3, \end{cases}$$

and

$$g_{i+1}(t) = \begin{cases} \dfrac{x_{i+1} - x_i}{2} e^{\pi i t} + \dfrac{x_i + x_{i+1}}{2} & \text{for } i \neq 3, 4, \\[2mm] 1 + i\Big(\dfrac{1-t}{t}\Big) & \text{for } i = 3, \\[2mm] x_4 + i\Big(\dfrac{t}{1-t}\Big) & \text{for } i = 4. \end{cases}$$

PROPOSITION 2.3.4. *Let $c_1, \ldots, c_n$ with $c_1 \ldots c_n = 1$ be standard generators of $\pi_{0,n}$ as in (2.2.1). The isomorphism $\Gamma_{0,[n]} \simeq \mathrm{Out}^*([\pi_{0,n}])$ of Theorem 2.2.1 of $\pi_{0,n}$ associates the following automorphism of $\pi_{0,n}$ to the Dehn twist $\sigma_i \in \Gamma_{0,[n]}$ for $1 \leq i \leq n - 1$:*

$$c_i \mapsto c_{i+1}, \quad c_{i+1} = c_{i+1}^{-1} c_i c_{i+1}, \quad c_j \mapsto c_j \text{ for } j \neq i, i+1.$$

Finally, the following proposition relates the genus zero mapping class group $\Gamma_{0,[n]}$ with the generators $\sigma_1, \ldots, \sigma_{n-1}$ to the Artin braid group B_n.

PROPOSITION 2.3.5. *The genus zero mapping class group $\Gamma_{0,[n]}$ (resp. the pure subgroup $\Gamma_{0,n}$) is isomorphic to the quotient of the Artin braid group B_n (resp. the pure Artin braid group K_n) by the relations $z_n = 1$ and $y_n = 1$, where z_n and y_n are as at the beginning of this section.*

3. Geometry and Special Loci of Small Moduli Spaces

In this section, we turn our attention to the explicit example of moduli spaces in dimension 1 and 2, as well as the higher-dimensional genus zero moduli spaces. For these spaces, we review the notions of mapping class groups, stable compactifications, topological tangential base points, orbifold structures, paths given by standard Dehn twists, fundamental groups, and most importantly, their points of special automorphism groups and the special loci they form.

In order to give a simple topological description of the *stable compactification* of the ordered moduli space $\mathcal{M}_{g,n}$ introduced by Deligne and Mumford, we need to introduce *pants decompositions* of a topological surface S of type (g, n). It is known that the maximal number of disjoint simple closed loops which can be placed on S is $3g - 3 + n$; such a collection cuts S into $2g - 2 + n$ disjoint *pairs of pants*, i.e., spheres with three holes or punctures. A *pants decomposition* is an equivalence class of such unions of circles modulo the action of the pure mapping class group $\Gamma_{g,n}$ of S.

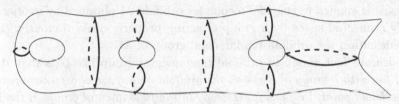

Figure 3.1. A pants decomposition of S of type $(3, 2)$.

These pants decompositions define the underlying topological surfaces of *degenerate stable curves*, which are those curves of type S for which one or more of the loops of some pants decomposition are pinched to a point. Such curves can be equipped with analytic structure, but they do not belong to the moduli space $\mathcal{M}_{g,n}$. Adding all of them to $\mathcal{M}_{g,n}$ forms the *stable (or Deligne–Mumford) compactification* $\overline{\mathcal{M}}_{g,n}$. In particular, the points of *maximal degeneration* in $\overline{\mathcal{M}}_{g,n}$ are those where all loops of some pants decomposition are pinched to points; the analytic structure which can be put on such a degenerate curve is unique (as it is just a union of thrice-punctured Riemann spheres), so that there is exactly one such point in $\overline{\mathcal{M}}_{g,n}$ for each pants decomposition. The *divisor at infinity* $\mathcal{D}_{g,n}^{\infty}$ is the difference $\overline{\mathcal{M}}_{g,n} - \mathcal{M}_{g,n}$.

This procedure gives a natural *stratification* of the compactification $\overline{\mathcal{M}}_{g,n}$. Each stratum is given by specifying (i) a pants decomposition and (ii) which of its loops are pinched to zero. The open stratum is $\mathcal{M}_{g,n}$, where no loops are pinched to zero. Pinching loops cuts S into a union of smaller surfaces $\bigcup S_{g',n'}$, and the corresponding stratum in $\overline{\mathcal{M}}_{g,n}$ is isomorphic to the product of the corresponding $\mathcal{M}_{g',n'}$. If a maximal number of loops, i.e., $3g - 3 + n$ loops forming a pants decomposition, are pinched to zero, we obtain a point of the dimension 0 stratum of $\overline{\mathcal{M}}_{g,n}$, which is the union of the points of *maximal degeneration*.

Roughly — and purely topologically — speaking, *tangential base points* are simply connected regions of $\mathcal{M}_{g,n}$ in the neighborhood of the divisor at infinity $\mathcal{D}_{g,n}^{\infty}$. The most frequently considered tangential base points are in the neighborhood of points of maximal degeneration. One takes a neighborhood of a point x of maximal degeneration in the compactified moduli space and then considers the intersection of this neighborhood with the uncompactified moduli space $\mathcal{M}_{g,n}$. This intersection forms a topological region in $\mathcal{M}_{g,n}$ which can be cut into simply connected pieces; these are the topological tangential base points. It is extremely important to note, however, that the most interesting characteristic of these base points which are "infinitely close to infinity" is that they have a natural algebraic structure for which they can be considered as points in the modular varieties which are "defined over $\mathbb{Q}$" [N1,N2]. Thus the absolute Galois group has a canonical outer action on the algebraic fundamental groups of moduli space based at these tangential base points.

Before proceeding to the examination of the genus zero moduli spaces, we recall some important facts about orbifolds and their fundamental groups. For our purposes, it is enough to consider a complex orbifold M obtained by quotienting a simply connected space by a group G acting properly discontinuously. Then G is by definition the orbifold fundamental group of M.

The delicate fact about an orbifold fundamental group like G is that *it can be identified with a group of loops on the orbifold only if the chosen base point is not an orbifold point, i.e., has no isotropy in the fundamental group.* If the base point is an orbifold point, then there will be nontrivial elements of G which will give trivial paths on M; these are in some sense automorphisms of the special points, i.e., as paths they remain "at the point" but they "do something to the point", so they are not trivial.

3.1. Genus zero, four marked points.

3.1.1. The ordered moduli space $\mathcal{M}_{0,4}$.

THE MAPPING CLASS GROUP $\Gamma_{0,4}$. This group is free on two generators, namely the twists σ_1^2 and σ_2^2 along loops surrounding the first and second, resp. second and third marked points. It has no torsion.

THE STABLE COMPACTIFICATION OF $\mathcal{M}_{0,4}$. We list the stable curves of type $(0,4)$. For this, we consider the possible pants decompositions on the topological sphere $S_{0,4}$ with four marked points. Such pants decompositions consist of a single loop (up to the action of $\Gamma_{0,4}$), so there are only three possibilities depending on whether this loop separates the first and second points from the the third and fourth, or the first and third from the second and fourth, or the first and fourth from the second and third. They are schematically represented by graphs

If S is now a Riemann sphere with marked points $(0, 1, \infty, \lambda)$, let γ denote a geodesic simple closed loop on S separating the four points 0, 1, ∞ and λ into two packets of two. Then λ is paired with one of 0, 1 or ∞. Modifying the analytic structure of S by pinching the γ to a point (i.e., decreasing its length) means that λ eventually becomes identified with the point it is paired with. Thus, the three points of maximal degeneration (and the only degenerate points) correspond to the three degenerate spheres $(0, \lambda, 1, \infty)$ with $\lambda \in \{0, 1, \infty\}$.

The stable compactification $\overline{\mathcal{M}}_{0,4}$ consists of $\mathcal{M}_{0,4}$ together with the divisor at infinity, so here it comes down to simply adding the three points 0, 1 and ∞ to $\mathcal{M}_{0,4} \simeq \mathbb{P}^1 - \{0, 1, \infty\}$, i.e., $\overline{\mathcal{M}}_{0,4}$ is isomorphic to $\mathbb{P}^1$.

TOPOLOGICAL TANGENTIAL BASE POINTS ON $\mathcal{M}_{0,4}$. The neighborhood of each of the three points of maximal degeneration in $\overline{\mathcal{M}}_{0,4}$ is homeomorphic to a disk, and its intersection with $\mathcal{M}_{0,4}$ to a pointed disk (i.e., a disk with a point removed). It is necessary to take two separate pieces of each disk in order to obtain simply connected regions which can serve as base points for fundamental groups. A useful convention for doing so is to consider only the intersection of the *real locus* of each disk with $\mathcal{M}_{0,4}$; one naturally obtains six small segments of the real line on $\mathbb{P}^1\mathbb{C} - \{0, 1, \infty\}$ neighboring 0, 1 and ∞. They are usually denoted $\overrightarrow{01}$, $\overrightarrow{10}$, $\overrightarrow{0\infty}$, $\overrightarrow{\infty 0}$, $\overrightarrow{1\infty}$, $\overrightarrow{\infty 1}$; we write $\hat{B}_{0,4}$ for the set of the six.

ORBIFOLD STRUCTURE. By (2.1.3), the moduli space $\mathcal{M}_{0,4}$ of Riemann spheres with four marked points is isomorphic to $\mathbb{P}^1 - \{0, 1, \infty\}$. The pure mapping class group $\Gamma_{0,4}$ acts not only properly and discontinuously but also freely on the Teichmüller space $\mathcal{T}_{0,4}$, so $\mathcal{M}_{0,4}$ is just an ordinary topological manifold. This holds for all the genus zero ordered moduli spaces.

DEHN TWISTS ON $S_{0,4}$. Consider a topological sphere S with four ordered marked points x_1, x_2, x_3, x_4, which we can consider as lying in a row. Recall that since the Dehn twists σ_1, σ_2 and σ_3 permute these marked points, they correspond to paths on the ordered moduli space $\mathcal{M}_{0,4}$ (starting at some chosen base point), which descend to loops on the unordered space $\mathcal{M}_{0,[4]}$. In Section 3.1.2, we will discuss how to represent these paths on $\mathcal{M}_{0,4}$. In the present section, we restrict ourselves to the loops on $\mathcal{M}_{0,4}$ corresponding to the two generators σ_1^2 and σ_2^2 of the pure mapping class group $\Gamma_{0,4}$, which is free, and as we saw above, acts freely on the upper half-plane.

These elements are Dehn twists along the loops α_1 and α_2 shown in Figure 3.2.

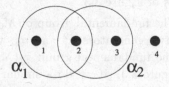

Figure 3.2. The Dehn twists σ_1^2 and σ_2^2, generators of $\Gamma_{0,4}$.

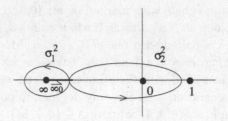

Figure 3.3. The Dehn twists σ_1^2 and σ_2^2 as paths on $\mathcal{M}_{0,4}$.

We show how to parametrize loops on the moduli space

$$\mathcal{M}_{0,4} \simeq \mathbb{P}^1 - \{0,1,\infty\},$$

much as in Proposition 2.3.3, except that we work with the σ_i^2 instead of the σ_i (working with the σ_i themselves when $n = 4$ is more delicate, which is why it is not included in Proposition 2.3.3; see Section 3.1.2 for details.)

Fix $\overrightarrow{\infty 0}$ for our choice of base point; it is represented on $\mathcal{M}_{0,4}$ by a simply-connected region of points of the form $(0,1,\infty,x)$ with $x \in (-\infty,-A)$ for some very large positive number A. The Dehn twist σ_i^2 can be parametrized as a movement of marked points by

$$\left(-\tfrac{1}{2}e^{2\pi it} + \tfrac{1}{2}, \tfrac{1}{2}e^{2\pi it} + \tfrac{1}{2}, \infty, x\right)$$

for $t \in [0,1]$, which by the transformation $e^{-2\pi it}(z + \tfrac{1}{2}e^{2\pi it} - \tfrac{1}{2})$ we bring to the standard form

$$\left(0,1,\infty, e^{-2\pi it}(x + \tfrac{1}{2}e^{2\pi it} - \tfrac{1}{2})\right).$$

Thus, on the moduli space $\mathcal{M}_{0,4}$ parametrized by the fourth component, we have the loop σ_1^2 in Figure 3.3 below, starting at the base point $\overrightarrow{\infty 0}$. Similarly, σ_2^2 can be parametrized by

$$\left(0, 1 - i\frac{(2t-1)^2 - 1}{2t-1}, 1 + i\frac{2t-1}{(2t-1)^2 - 1}, x\right),$$

which can be transformed to

$$\left(0, 1, \infty, \frac{x}{x - 1 - i\frac{2t-1}{(2t-1)^2-1}} \cdot \frac{-i\frac{(2t-1)^2-1}{2t-1} - i\frac{2t-1}{(2t-1)^2-1}}{1 - i\frac{(2t-1)^2-1}{2t-1}}\right),$$

which describes the loop σ_2^2 of Figure 3.3.

FUNDAMENTAL GROUP. The fundamental group of $\mathcal{M}_{0,4}$ is isomorphic to $\Gamma_{0,4}$, which is a free group on two generators σ_1^2 and σ_2^2, which acts freely on the Teichmüller space $\mathcal{T}_{0,4}$. The fundamental group $\pi_1(\mathcal{M}_{0,4}; \overrightarrow{\infty 0})$ is generated by the two loops shown in Figure 3.3. If we do a similar computation, using the base point $\overrightarrow{01}$ and a standard form $(0, x, 1, \infty)$ for representatives of the points of $\mathcal{M}_{0,4}$, we find the more standard identification shown in the following figure.

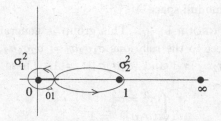

Figure 3.4. The generators of $\pi_1(\mathcal{M}_{0,4}, \overrightarrow{01})$.

POINTS WITH SPECIAL AUTOMORPHISM GROUP. A special feature of Riemann surfaces with ordered marked points is that it can happen that a permutation of the points can be realized as an automorphism of the surface, for instance the rotation of a sphere having n marked points on its equator through an angle of $2\pi/n$. Such points are not orbifold points on the ordered moduli space, but they are preimages of orbifold points on the unordered moduli space, since they have less than $n!$ preimages under the action of S_n. We determine all such points in $\mathcal{M}_{0,4}$. To begin with, we see that the Klein 4-subgroup of S_4 fixes each and every point of $\mathcal{M}_{0,4}$. Indeed, if $(0, 1, \infty, x)$ is a point, then the action of say $(12)(34)$ on it takes it to $(1, 0, x, \infty)$, and then the transformation by the isomorphism $z \mapsto (xz - x)/(z - x)$ brings it back to $(0, 1, \infty, x)$, so it is the same point on moduli space, and this also holds for $(13)(24)$ and $(14)(23)$. Thus, every point of $\mathcal{M}_{0,4}$ has automorphism group at least isomorphic to the Klein 4-group; this is the generic automorphism group. Five points of $\mathcal{M}_{0,4} \simeq \mathbb{P}^1 - \{0, 1, \infty\}$ have special automorphism groups. The first three are those given by representatives $(\tau, 0, 1, \infty)$ with $\tau \in \{1/2, 2, -1\}$, which apart from being fixed by the Klein 4-group are also fixed by the permutations (23), (34) and (24) respectively, forming three different dihedral groups of order 8. These automorphism groups can be identified with the automorphism group of the octahedron, by identifying the octahedron with the sphere with four marked points around the equator and a north and a south pole; the automorphism group is generated by a rotation of order 4 around the north-south axis and a north-south flip of order 2. The two remaining points with special automorphism group are the points $(\tau, 0, 1, \infty)$ with $\tau = exp(\pm 2\pi i/6)$, which are each fixed by the permutation group $\langle (123) \rangle$ as well as the Klein 4-group, forming a group of order 12 isomorphic to the alternating group A_4, realized as the automorphism group of the tetrahedron, by identifying the tetrahedron with the sphere with three marked points around the equator and one at the north pole.

SPECIAL LOCI. Special loci are orbifold points, and there are none on $\mathcal{M}_{0,4}$, which has no orbifold structure. As remarked above, the points of special automorphism group given above determine where the special loci will lie on the unordered moduli space $\mathcal{M}_{0,[4]}$.

3.1.2. The unordered moduli space $\mathcal{M}_{0,[4]}$.

THE MAPPING CLASS GROUP $\Gamma_{0,[4]}$. This group is generated by Dehn twists σ_i for $i = 1, 2, 3$, subject to the relations $\sigma_1\sigma_2\sigma_1 = \sigma_2\sigma_1\sigma_2$, $\sigma_2\sigma_3\sigma_2 = \sigma_3\sigma_2\sigma_3$, $\sigma_1\sigma_3 = \sigma_3\sigma_1$, $\sigma_3\sigma_2\sigma_1^2\sigma_2\sigma_3 = 1$ and $(\sigma_1\sigma_2\sigma_3)^4 = 1$. These relations imply the following:

$$
\begin{cases}
\sigma_1^2 = \sigma_3^2 \\
(\sigma_1\sigma_2)^3 = 1 \\
(\sigma_1\sigma_2\sigma_1)^2 = 1.
\end{cases}
$$

To prove the second relation, we just use the braid relations to show that

$$(\sigma_1\sigma_2)^3(\sigma_3\sigma_2\sigma_1^2\sigma_2\sigma_3) = (\sigma_1\sigma_2\sigma_3)^4 = 1,$$

so $(\sigma_1\sigma_2)^3 = 1$. Then, by the braid relations, we have $(\sigma_1\sigma_2)^3 = \sigma_1^2\sigma_2\sigma_1^2\sigma_2$, so $\sigma_1^2 = \sigma_2^{-1}\sigma_1^{-2}\sigma_2^{-1}$. But the relation $\sigma_3\sigma_2\sigma_1^2\sigma_2\sigma_3 = 1$ shows that $\sigma_3^2 = \sigma_2^{-1}\sigma_1^{-2}\sigma_2^{-1}$, so $\sigma_1^2 = \sigma_3^2$, which gives the first relation. For the third, using only $\sigma_1\sigma_2\sigma_1 = \sigma_2\sigma_1\sigma_2$, we see that

$$(\sigma_1\sigma_2\sigma_1)^2 = (\sigma_1\sigma_2)^3 = 1.$$

The torsion elements in $\Gamma_{0,[4]}$ are as follows. There is only one conjugacy class of elements of order 4, namely the class of $\sigma_1\sigma_2\sigma_3$. There is only one conjugacy class of elements of order 3, namely that of $\sigma_1\sigma_2$. There are two conjugacy classes of order 2, namely that of $(\sigma_1\sigma_2\sigma_3)^2$ and that of $\sigma_1\sigma_2\sigma_1$. (Note that this contradicts the statement of the Corollary on p. 508 of [HM]; their corollary is however valid for $n \geq 5$.)

THE STABLE COMPACTIFICATION OF $\mathcal{M}_{0,[4]}$. Degenerate stable curves with four unordered marked points must correspond to the unique trivalent tree with four unnumbered tails, so there is only one such point. Indeed, this corresponds to the fact that under the morphism $\mathcal{M}_{0,4} \simeq \mathbb{P}^1 - \{0, 1, \infty\} \to \mathcal{M}_{0,[4]}$ by quotienting by the action of S_4, the three points $0, 1, \infty$ all pass to a single point on $\mathcal{M}_{0,[4]}$.

TOPOLOGICAL TANGENTIAL BASE POINTS ON $\mathcal{M}_{0,[4]}$. The neighborhood of the missing point is a disk, so dividing it into two simply connected regions, there are two tangential base points at the maximally degenerate point. Note that only one of these is the image of all six tangential base points on $\mathcal{M}_{0,4}$.

ORBIFOLD STRUCTURE. Let S be a sphere with four marked points x_1, x_2, x_3 and x_4. By definition, the unordered moduli space $\mathcal{M}_{0,[4]}$ is the quotient of the Teichmüller space $\mathcal{T}_{0,4}$ by the action of the full mapping class group $\Gamma_{0,[4]}$, which we recall (Figure 2.1) is generated by Dehn twists σ_i for $i = 1, 2, 3$ along loops passing through the marked points x_i and x_{i+1} respectively. As in the previous section, we have an isomorphism

$$\Gamma_{0,[4]} \xrightarrow{\sim} \pi_1(\mathcal{M}_{0,[4]}, \overrightarrow{\infty 0}).$$

However, there is a fundamental difference between the ordered moduli space $\mathcal{M}_{0,4}$ and the unordered space $\mathcal{M}_{0,[4]} = \mathcal{M}_{0,4}/S_4$, due to the orbifold structure of

$\mathcal{M}_{0,[4]}$ which arises from the fact that $\Gamma_{0,[4]}$ does not act freely on $\mathcal{T}_{0,4}$. Equivalently, since $\Gamma_{0,4}$ acts freely on $\mathcal{T}_{0,4}$ and we have an exact sequence

$$1 \to \Gamma_{0,4} \hookrightarrow \Gamma_{0,[4]} \to S_4 \to 1,$$

the orbifold structure arises because S_4 does not act freely on $\mathcal{M}_{0,4}$.

We saw in the discussion of orbifold fundamental groups given just before Section 3.1.1 that such a group can be identified with a group of loops if the base point is not an orbifold point, i.e., has trivial isotropy. However, in the case of $\mathcal{M}_{0,[4]}$, every point has a nontrivial isotropy subgroup in $\Gamma_{0,[4]}$. Indeed, letting $\sigma_{41} = \sigma_3\sigma_2\sigma_1\sigma_2^{-1}\sigma_3^{-1}$, $\sigma'_{24} = \sigma_3^{-1}\sigma_2\sigma_3$ and $\sigma_{13} = \sigma_2\sigma_1\sigma_2^{-1}$, the three elements

$$a = \sigma_1\sigma_3^{-1}, \quad b = \sigma_{13}\sigma'^{-1}_{24} \quad \text{and} \quad c = \sigma_{41}\sigma_2^{-1}$$

all fix every point of $\mathcal{T}_{0,4}$. We show that these elements are all of order 2, and commute in $\Gamma_{0,[4]}$. For the first one, we use the fact that $\sigma_1^2 = \sigma_3^2$ in $\Gamma_{0,[4]}$, which we saw in the paragraph on the mapping class group above. Then, it follows directly from writing their expressions that the other two elements are of order two, since in fact $b = \sigma_2 a\sigma_2$ and $c = \sigma_3\sigma_2 a\sigma_2^{-1}\sigma_3^{-1}$. Now we show that $\langle a, b, c \rangle$ is a Klein 4-subgroup of $\Gamma_{0,[4]}$. We first show that $ab = c$. In fact, using only the braid relations, we check that

$$ab = \sigma_1\sigma_3^{-1}\sigma_2\sigma_1\sigma_3^{-1}\sigma_2^{-1} = \sigma_1^2\sigma_3^{-2}\sigma_3\sigma_2\sigma_1\sigma_3^{-1}\sigma_2^{-1}\sigma_3^{-1} = \sigma_1^2\sigma_3^{-2}c,$$

and this is equal to c since $\sigma_1^2 = \sigma_3^2$. Thus, we have $ab = c$, so $abab = 1$ since c is of order 2, so $ab = b^{-1}a^{-1} = ba$ since a and b are of order 2, so a and b commute. This shows that $\langle a, b, c \rangle$ is a Klein 4-subgroup.

Now we can give a complete description of the orbifold structure of $\mathcal{M}_{0,[4]}$. Every point of $\mathcal{M}_{0,[4]}$ is an "orbifold point" in the sense that it has nontrivial isotropy in $\Gamma_{0,[4]}$. The special (non-generic) orbifold points are the images of the points with special automorphism group given in $\mathcal{M}_{0,4}$. The three points $x = 1/2, 2, -1$ on $\mathcal{M}_{0,4}$ with dihedral automorphism group of order 8 pass to a single point on $\mathcal{M}_{0,[4]}$, and the two points $j, \bar{j}$ on $\mathcal{M}_{0,4}$ with special automorphism group of order 12 (isomorphic to A_4) pass to a single point on $\mathcal{M}_{0,[4]}$. Topologically, $\mathcal{M}_{0,[4]}$ is a sphere with one missing point at infinity; its orbifold structure is given by these two special points and the generic isotropy group $\mathbb{Z}/2\mathbb{Z} \times \mathbb{Z}/2\mathbb{Z}$ at each point.

This space is not dissimilar to the moduli space of elliptic curves, given as the quotient of the Teichmüller space $\mathcal{T}_{1,1}$, which is again the Poincaré upper half-plane, by the action of $\mathrm{SL}_2(\mathbb{Z})$; it looks like a sphere with one missing point and two "special" points, and there is a nontrivial isotropy group at every point, only it is just $\mathbb{Z}/2\mathbb{Z}$ instead of $\mathbb{Z}/2\mathbb{Z} \times \mathbb{Z}/2\mathbb{Z}$. Indeed, if $\mathcal{M}'_{1,1}$ denotes the reduced orbifold of elliptic curves, we have an orbifold isomorphism $\mathcal{M}'_{1,1} \simeq \mathcal{M}_{0,4}/S_3$, both of these orbifolds having π_1 isomorphic to $\mathrm{PSL}_2(\mathbb{Z})$.

DEHN TWISTS ON $S_{0,[4]}$. By the discussion of the orbifold fundamental group given just before Section 3.1.1, we see that since there is a nontrivial isotropy group at each point of $\mathcal{M}_{0,[4]}$, if we attempt to identify the generating elements $\sigma_1, \sigma_2, \sigma_3$ of $\Gamma_{0,[4]}$ with loops on the moduli space $\mathcal{M}_{0,[4]}$ (or paths on $\mathcal{M}_{0,4}$) based at an arbitrary non-special point, we will not be able to distinguish σ_1 from σ_3. Let us show this explicitly, working on $\mathcal{M}_{0,4}$ (parametrized by the single variable x), with the base point $\overrightarrow{\infty 0}$.

We first parametrize the twists σ_1, σ_2 and σ_3, shown in the following figure, directly on the sphere, as in Proposition 2.3.3.

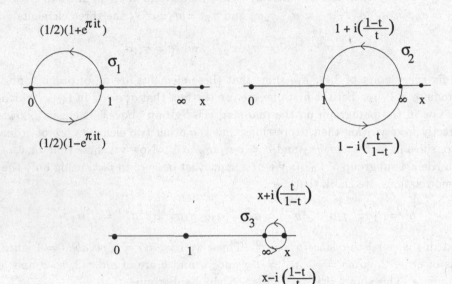

Figure 3.5. The Dehn twists σ_1, σ_2, σ_3.

We obtain

$$\begin{cases} \sigma_1 = -\tfrac{1}{2}e^{\pi it} + \tfrac{1}{2}, \tfrac{1}{2}e^{\pi it} + \tfrac{1}{2}, \infty, x \\[2mm] \sigma_2 = \left(0, 1 - i\,\dfrac{t}{1-t}, 1 + i\,\dfrac{1-t}{t}, x\right) \\[2mm] \sigma_3 = \left(0, 1, x - i\,\dfrac{1-t}{t}, x + i\,\dfrac{t}{1-t}\right). \end{cases}$$

Bringing these parametrizations back to the standard form $(0, 1, \infty, f(t))$, we find

$$\begin{cases} \sigma_1 = \left(0, 1, \infty, e^{-\pi it}\left(x + \tfrac{1}{2}e^{\pi it} - \tfrac{1}{2}\right)\right) \\[3mm] \sigma_2 = \left(0, 1, \infty, \dfrac{x}{tx - t - (1-t)i} \cdot \dfrac{-i(2t^2 - 2t + 1)}{1 - t - it}\right) \\[3mm] \sigma_3 = \left(0, 1, \infty, \dfrac{\big((1-t)x + it\big)\big(t - tx + (1-t)i\big)}{it^2 + i(1-t)^2}\right). \end{cases}$$

These three paths are shown in Figure 3.6, where it can be seen that σ_1 and σ_3 give rise to homotopic paths.

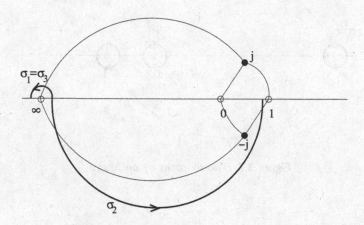

Figure 3.6. The paths σ_1, σ_2 and σ_3 on $\mathcal{M}_{0,4}$.

FUNDAMENTAL GROUP. Figure 3.6 shows the three Dehn twists σ_1, σ_2 and σ_3 as paths on $\mathcal{M}_{0,4}$, starting from the tangential base point $\overrightarrow{\infty 0}$. They are shown in bold lines; σ_1 is the tiny half-circle from the base point $\overrightarrow{\infty 0}$ to the base point $\overrightarrow{\infty 1}$ in the northern hemisphere, σ_3 is indistinguishable from it (as we saw, $\sigma_1\sigma_3^{-1}$ is "invisible" as a path), and σ_2 is the large path from $\overrightarrow{\infty 0}$ to $\overrightarrow{10}$.

The thin lines show the real axis, the line segment from 0 to j, and the curves which are the six images of this line segment under the group of automorphisms of $\mathcal{M}_{0,4} = \mathbb{P}^1 - \{0, 1, \infty\}$, generated by $z \mapsto 1 - z$ and $z \mapsto 1/(1 - z)$. These curves divide $\mathcal{M}_{0,4}$ into six regions, each of which is a fundamental domain for the action of S_4 on $\mathcal{M}_{0,4}$.

In order to represent σ_1 and σ_2 as loops on $\mathcal{M}_{0,[4]}$, we first show, in Figures 3.7 and 3.8, the six paths σ_1 and σ_2 starting from the six tangential base points, obtained as the images of the σ_1 and σ_2 in Figure 3.6 under the automorphism group of $\mathcal{M}_{0,4}$.

Now, to see the loops on $\mathcal{M}_{0,[4]}$, we select one fundamental domain, and use only the trace of the paths σ_1 and σ_2 lying inside it. We find what is shown in Figure 3.9, where σ_1 is represented with dashed curves. Now, we make $\mathcal{M}_{0,[4]}$ by identifying the two edges of this domain coming out of the point j, and folding the real segment $(0, 1)$ in half, pinching at $1/2$ and gluing it together. This reveals the orbifold structure of $\mathcal{M}_{0,[4]}$ as a sphere with one hold and two "pinched" orbifold points coming from j and $1/2$, and the paths σ_1 and σ_2 pass to the loops shown in Figure 3.10.

Finally, we note how much more natural it is to consider the finite order loops $\sigma_1\sigma_2$ and $\sigma_1\sigma_2\sigma_1$ as generators of the fundamental group of $\mathcal{M}_{0,[4]}$; indeed, these

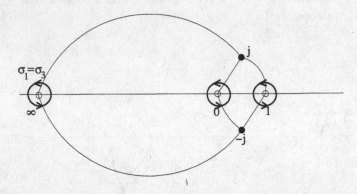

Figure 3.7. The six paths σ_1 on $\mathcal{M}_{0,4}$.

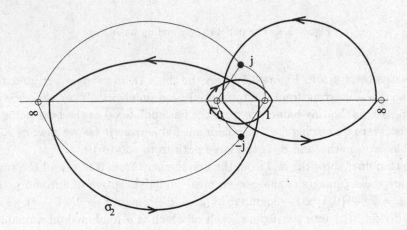

Figure 3.8. The six paths σ_2 on $\mathcal{M}_{0,4}$.

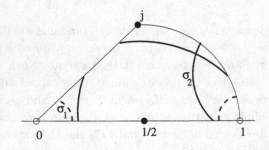

Figure 3.9. The paths σ_1 and σ_2 in a fundamental domain.

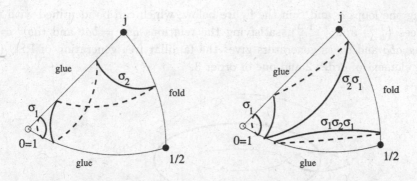

Figure 3.10 (left). The loops σ_1 and σ_2 on $\mathcal{M}_{0,[4]}$. **Figure 3.11** (right). The loops σ_1, $\sigma_1\sigma_2\sigma_1$ and $\sigma_2\sigma_1$ on $\mathcal{M}_{0,[4]}$.

loops, shown in Figure 3.11, are just the finite-order loops surrounding the two special orbifold points.

SPECIAL LOCI. The whole of $\mathcal{M}_{0,[4]}$ is a special locus in the sense that every point is an orbifold point of group $\mathbb{Z}/2\mathbb{Z} \times \mathbb{Z}/2\mathbb{Z}$, this group being identified with the Klein 4-subgroup of S_4. Apart from these generic automorphism groups, as we saw, the points on $\mathcal{M}_{0,[4]}$ having special automorphism group are the two points which are the images of the points $\{1/2, 2, -1\}$ and $\{j, \bar{j}\}$ in $\mathcal{M}_{0,4}$, and these two points have associated isotropy group D_4 and A_4 respectively.

3.2. Genus one, one marked point.

As a topological space, the moduli space $\mathcal{M}_{1,1}$ of one-pointed tori (i.e., elliptic curves) is homeomorphic to $\mathbb{C}$. This is because it is realizable as the quotient of the Poincaré upper half-plane $\mathcal{H}$ by the action of $\mathrm{SL}_2(\mathbb{Z})$ given by

$$z \mapsto \frac{az+b}{cz+d} \quad \text{for} \quad \begin{pmatrix} a & b \\ c & d \end{pmatrix} \in \mathrm{SL}_2(\mathbb{Z}).$$

A fundamental domain for this action has the familiar shape shown in Figure 3.12 (with $j = \exp(2\pi i/6)$). As usual, the domain in this figure is rolled up with the two vertical edges identified and the arc of the boundary circle from j^2 to i identified with the arc from i to j. This makes the fundamental domain appear rather like a sock, with a toe at j and a pointed heel at i.

MAPPING CLASS GROUP. The mapping class group $\Gamma_{1,1}$ is isomorphic to the quotient

$$\mathrm{Diff}(S_{1,1})/\mathrm{Diff}^0(S_{1,1}),$$

which in turn is isomorphic to $\mathrm{PSL}_2(\mathbb{Z})$. It is generated by the two Dehn twists a and

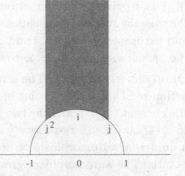

Figure 3.12. The moduli space $\mathcal{M}_{1,1}$.

b along the loops α and β in the figure below, which can be identified with the matrices $\left(\begin{smallmatrix} 1 & 1 \\ 0 & 1 \end{smallmatrix}\right)$ and $\left(\begin{smallmatrix} 1 & 0 \\ -1 & 1 \end{smallmatrix}\right)$, satisfying the relations $aba = bab$ and $(ab)^3 = 1$. Taking aba and ba as generators gives the familiar free generation of $\mathrm{PSL}_2(\mathbb{Z})$ by an element of order 2 and one of order 3.

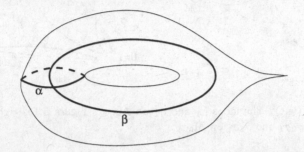

Figure 3.13. Generators of the mapping class group $\Gamma_{1,1}$.

However, there is an interesting subtlety involving the identification of $\mathcal{M}_{1,1}$ with the quotient of the Teichmüller space $\mathcal{T}_{1,1}$ by the mapping class group. Namely, the Teichmüller space $\mathcal{T}_{1,1}$ is the Poincaré upper half-plane $\mathcal{H}$, and we saw above that $\mathcal{M}_{1,1}$ is the quotient of $\mathcal{T}_{1,1}$ by $\mathrm{SL}_2(\mathbb{Z})$. Since $\left(\begin{smallmatrix} -1 & 0 \\ 0 & -1 \end{smallmatrix}\right)$ acts trivially on $\mathcal{T}_{1,1}$, the quotients $\mathcal{T}_{1,1}/\mathrm{SL}_2(\mathbb{Z})$ and $\mathcal{T}_{1,1}/\mathrm{PSL}_2(\mathbb{Z})$ are identical as topological spaces, but they are not identical as orbifolds. The fundamental group of $\mathcal{M}_{1,1}$ is actually isomorphic to $\mathrm{SL}_2(\mathbb{Z})$, and the homomorphism

$$\pi_1(\mathcal{M}_{g,n}) \to \Gamma_{g,n}$$

which is usually an isomorphism has a kernel in this exceptional case.

THE STABLE COMPACTIFICATION OF $\overline{\mathcal{M}}_{1,1}$. It is obtained by adding the missing point to $\mathcal{M}_{1,1}$, so we have $\overline{\mathcal{M}}_{1,1} \simeq \mathbb{P}^1\mathbb{C}$ as topological spaces.

TOPOLOGICAL TANGENTIAL BASE POINTS ON $\mathcal{M}_{1,1}$. The intersection of the neighborhood of the maximally degenerate point ∞ on $\overline{\mathcal{M}}_{1,1}$ with the space $\mathcal{M}_{1,1}$ is a pointed disk, so as before there must be two tangential base points. We take the two half-segments $i\lambda$ and $1/2 + i\lambda$ where $\lambda \in (\Lambda, \infty)$ for some $\Lambda \gg 0$, and we denote them by $\overrightarrow{\infty i}$ and $\overrightarrow{\infty j}$ respectively in analogy with the notation $\overrightarrow{01}$ (i.e., small segments from ∞ towards i and from ∞ towards i).

ORBIFOLD STRUCTURE. The space $\mathcal{M}_{1,1}$ is the orbifold $\mathcal{T}_{1,1}/\mathrm{SL}_2(\mathbb{Z})$. Under the action of $\mathrm{SL}_2(\mathbb{Z})$, every point of $\mathcal{T}_{1,1}$ is fixed by one element of $\mathrm{SL}_2(\mathbb{Z})$, namely the element $\left(\begin{smallmatrix} -1 & 0 \\ 0 & 1 \end{smallmatrix}\right)$, and the two points i and j are fixed by subgroups isomorphic to $\mathbb{Z}/4\mathbb{Z}$ and $\mathbb{Z}/6\mathbb{Z}$ respectively. So the orbifold $\mathcal{M}_{1,1}$ considered in this way has a nontrivial automorphism of order 2 at every point and two additional special points. The automorphism group of each point is equal to the automorphism group of the isomorphism class of elliptic curves associated to that point.

Since $\left(\begin{smallmatrix} -1 & 0 \\ 0 & -1 \end{smallmatrix}\right)$ fixes each point of $\mathcal{H}$, there is a natural surjection

$$\mathcal{M}_{1,1} \simeq \mathcal{T}_{1,1}/\mathrm{SL}_2(\mathbb{Z}) \to \mathcal{T}_{1,1}/\mathrm{PSL}_2(\mathbb{Z}) \simeq \mathcal{M}_{0,[4]} \qquad (3.2.1)$$

which consists in forgetting the automorphism of order 2 at each point.

DEHN TWISTS ON $S_{1,1}$. Consider the tangential base point $\overrightarrow{i\infty}$ on $\mathcal{M}_{1,1}$; it is the preimage under (3.2.1) of the tangential base point on $\mathcal{M}_{0,[4]}$ which is itself the image of the six standard tangential base points on $\mathbb{P}^1 - \{0, 1, \infty\} = \mathcal{M}_{0,4}$. We will determine the paths on $\mathcal{M}_{1,1}$, based at $\overrightarrow{i\infty}$, corresponding to the Dehn twists a and b along the loops α and β of Figure 3.13, and also the paths corresponding to aba and ba. To do this, since (3.2.1) is a one-to-one map, it suffices to study the paths corresponding to σ_1, σ_2, $\sigma_1\sigma_2\sigma_1$ and $\sigma_2\sigma_1$ on $\mathcal{M}_{0,[4]}$. The first two of these are shown in Figure 3.10 and the second two in Figure 3.11.

THE FUNDAMENTAL GROUP OF $\mathcal{M}_{1,1}$. Now consider $a = \left(\begin{smallmatrix} 1 & 1 \\ 0 & 1 \end{smallmatrix}\right)$ and $\left(\begin{smallmatrix} 1 & 0 \\ -1 & 1 \end{smallmatrix}\right)$ as generators of $\mathrm{SL}_2(\mathbb{Z})$ rather than of $\mathrm{PSL}_2(\mathbb{Z})$. The paths on $\mathcal{M}_{1,1}$ corresponding to the generators $aba = \left(\begin{smallmatrix} 0 & 1 \\ -1 & 0 \end{smallmatrix}\right)$ and $ba = \left(\begin{smallmatrix} 1 & 1 \\ -1 & 0 \end{smallmatrix}\right)$ of the fundamental group of $\mathcal{M}_{1,1}$ look the same as figure 3.11 since the only difference is the invisible orbifold structure at every point (see Figure 3.14).

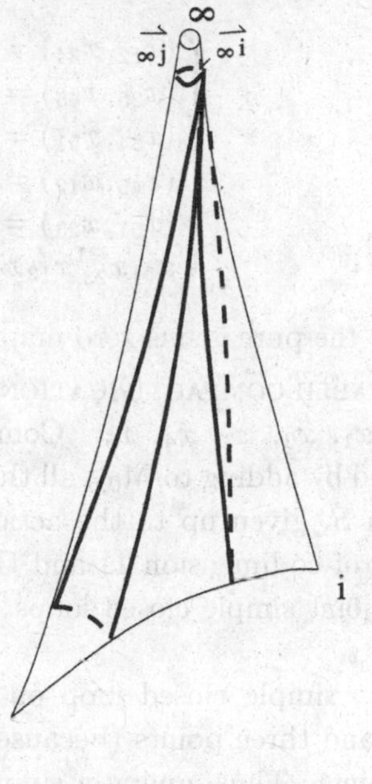

Note that running along the loop around j and then the loop around i gives a loop homotopic to the small ring around ∞ at the top, also that the loop around j is of order 6 and the loop around i is of order 4.

SPECIAL LOCI. The whole of $\mathcal{M}_{1,1}$ is the special orbifold locus of $\{\pm 1\} \in \mathrm{SL}_2(\mathbb{Z})$. On top of this, there are two isolated points which are the special loci of the finite-order elements aba and ba respectively.

Figure 3.14. The fundamental group $\pi_1^{\mathrm{orbifold}}(\mathcal{M}_{1,1}; \overrightarrow{i\infty})$.

3.3. Genus zero, five marked points.

3.3.1. The ordered moduli space $\mathcal{M}_{0,5}$.
THE MAPPING CLASS GROUP $\Gamma_{0,5}$. As we saw in proposition 2.3.5, the full mapping class group $\Gamma_{0,[5]}$ is the quotient of the Artin braid group B_5 on 5 strands, generated by $\sigma_1, \sigma_2, \sigma_3, \sigma_4$, by the two relations $(\sigma_4\sigma_3\sigma_2\sigma_1)^5 = 1$ and $\sigma_4\sigma_3\sigma_2\sigma_1^2\sigma_2\sigma_3\sigma_4 = 1$. There are two nice presentations for its pure subgroup $\Gamma_{0,5}$, one as a semi-direct product of a free group of rank 3 and a free group of rank two, and another which does not clearly reveal the semi-direct product structure, but is more symmetric in terms of the 5 strands. The semi-direct product presentation is given by $F_3 \rtimes F_2$ where the normal F_3 factor is generated by $x_1 = x_{12}$, $x_2 = x_{23}$, $x_3 = x_{24}$, $x_4 = x_{13}$

and $x_5 = x_{34}$, where the x_{ij} are the generators defined earlier. Then $\Gamma_{0,5} = \langle x_4, x_5 \rangle \rtimes \langle x_1, x_2, x_3 \rangle \simeq F_3 \rtimes F_2$, with the following relations:

$$
\begin{cases}
x_{13}^{-1} x_{12} x_{13} = x_{23} x_{12} x_{23}^{-1} \\
x_{13}^{-1} x_{23} x_{13} = x_{23} x_{12} x_{23} x_{12}^{-1} x_{23}^{-1} \\
x_{13}^{-1} x_{24} x_{13} = x_{23} x_{12} x_{23}^{-1} x_{12}^{-1} x_{24} x_{12} x_{23} x_{12}^{-1} x_{23}^{-1} \\
x_{34}^{-1} x_{12} x_{34} = x_{12} \\
x_{34}^{-1} x_{23} x_{34} = x_{23} x_{24} x_{23} x_{24}^{-1} x_{23}^{-1} \\
x_{34}^{-1} x_{24} x_{34} = x_{23} x_{24} x_{23}^{-1}.
\end{cases}
$$

For two elements a and b of a group, let $(a, b) = aba^{-1}b^{-1}$ denote their commutator. Rewriting the above presentation in the five generators x_{12}, x_{23}, x_{34}, x_{45} and x_{51} and using the relations $x_{45} = x_{12} x_{13} x_{23}$ and $x_{51} = x_{23} x_{24} x_{34}$ gives:

$$
\begin{cases}
(x_{12}, x_{34}) = 1 \\
(x_{23}, x_{45}) = 1 \\
(x_{34}, x_{51}) = 1 \\
(x_{45}, x_{12}) = 1 \\
(x_{51}, x_{23}) = 1 \\
x_{51} x_{23}^{-1} x_{12} x_{34}^{-1} x_{23} x_{45}^{-1} x_{34} x_{51}^{-1} x_{45} x_{12}^{-1} = 1.
\end{cases}
$$

Like all the pure genus zero mapping class groups, $\Gamma_{0,5}$ is torsion free.

THE STABLE COMPACTIFICATION OF $\mathcal{M}_{0,5}$. Let S denote a sphere with 5 marked points x_1, x_2, x_3, x_4, x_5. Combinatorially, the divisor at infinity of $\mathcal{M}_{0,5}$ is obtained by adding to $\mathcal{M}_{0,5}$ all the points obtained by pinching one simple closed loop on S, given up to the action of $\Gamma_{0,5}$, to a point (this gives the strata at infinity of codimension 1) and the points obtained by simultaneously pinching two disjoint simple closed loops (given up to $\Gamma_{0,5}$), i.e., a pants decomposition, to points.

Every simple closed loop on S divides the set of marked points into two points and three points (because a loop surrounding 0 or 1 point is homotopic to a point). Thus, giving a simple closed loop on S up to the action of $\Gamma_{0,5}$ is equivalent to giving the two points x_i and x_j it separates from the others. This means that there are $\binom{5}{2} = 10$ such classes of loops on S. Each loop corresponds to a stratum at infinity of complex codimension 1, so complex dimension 1, of $\mathcal{M}_{0,5}$. Pinching a loop surrounding points x_i and x_j to a point is like making x_i and x_j approach each other until they coalesce, so what we are doing is adding to the moduli space of isomorphism classes of 5-tuples of distinct points $(x_1, x_2, x_3, x_4, x_5)$ the 10 new sets of points with $x_i = x_j$ for $1 \le i < j \le 5$.

We examine each of these 10 sets. A single loop cuts $S_{0,5}$ into one sphere with three marked points and a boundary component, and another with two marked points and a boundary component. When the loop is pinched to zero, we obtain a sphere with four marked points and a sphere with three marked points, so the corresponding stratum at infinity is isomorphic to $\mathcal{M}_{0,4} \times \mathcal{M}_{0,3}$.

We show this isomorphism explicitly. Consider for example a pants decomposition whose degenerate loop encloses the points x_1 and x_5. The corresponding stratum of $\overline{\mathcal{M}}_{0,5}$ consists of isomorphism classes of spheres with five marked points $(x_1, x_2, x_3, x_4, x_1)$, and these isomorphism classes are enumerated by the representatives of the form $(\lambda, 0, 1, \infty, \lambda)$, where $\lambda \notin \{0, 1, \infty\}$ (otherwise one has a maximally degenerate point). The set of partially degenerate spheres of type $x_1 = x_5$ is thus isomorphic to $\mathbb{P}^1 - \{0, 1, \infty\}$, i.e., to $\mathcal{M}_{0,4}$, so that the divisor at infinity of $\overline{\mathcal{M}}_{0,5}$ actually contains ten copies of $\mathcal{M}_{0,4}$. This completely describes the codimension one part of the divisor at infinity of $\mathcal{M}_{0,5}$.

Consider the codimension 2 part, corresponding to maximally degenerate points, i.e., pants decompositions. A pants decomposition on S consists of two disjoint loops, and each loop must contain two marked points, so a pants decomposition is uniquely determined by specifying two disjoint pairs of marked points among the five. This makes fifteen pants decompositions, corresponding to fifteen points of maximal degeneration. The pants decomposition given by two loops one surrounding points x_i and x_j and the other points x_k and x_l corresponds to the degenerate sphere with marked points $x_1, \ldots, x_5$ such that $x_i = x_j$ and $x_k = x_l$. These fifteen points must also be added to $\mathcal{M}_{0,5}$ to obtain the stable compactification. They can be visualized as points where the stratum $x_i = x_j$ crosses $x_k = x_l$, with $\{i, j\} \cap \{k, l\} = \varnothing$, so that in fact the stable compactification is obtained by adding ten copies of $\mathbb{P}^1$ to $\mathcal{M}_{0,5}$. Note that by (2.1.3), we have

$$\mathcal{M}_{0,5} \simeq (\mathbb{P}^1 - \{0, 1, \infty\})^2 - \langle x = y \rangle$$
$$= (\mathbb{P}^1\mathbb{C})^2 - \langle x = 0, x = 1, x = \infty, y = 0, y = 1, y = \infty, x = y \rangle,$$

so that $\mathcal{M}_{0,5}$ is naturally obtained from $(\mathbb{P}^1\mathbb{C})^2$ by removing seven lines, and the stable compactification is obtained by adding ten lines to the result (it is, in fact, the blowup of $\mathbb{P}^1 \times \mathbb{P}^1$ at the three points where the diagonal meets $y = 0$, $y = 1$, $y = \infty$).

TOPOLOGICAL TANGENTIAL BASE POINTS ON $\overline{\mathcal{M}}_{0,5}$. We consider only tangential base points in the neighborhood of the fifteen maximally degenerate points of $\overline{\mathcal{M}}_{0,5}$. A detailed reference for the sketch given here is chapters 1 and 2 of [PS]. As for $\mathcal{M}_{0,4}$, we will consider the real locus of the neighborhood of each maximally degenerate point. This real locus falls naturally into four simply connected regions, in the following way. Consider the point of maximal degeneration given by the pants decomposition enclosing points x_1 and x_2 together and points x_4 and x_5 together, so corresponding to the isomorphism class of the sphere with five marked points $(x_2, x_2, x_3, x_4, x_4)$. The representative in standard form is given by $(0, 0, 1, \infty, \infty)$. Points of $\mathcal{M}_{0,5}$ in the neighborhood of this point correspond to isomorphism classes of spheres with five marked points $(\lambda, 0, 1, \infty, \mu)$ where λ is very near 0 and μ is very near ∞ (we are looking in $\mathcal{M}_{0,5}$ so we don't consider the partially degenerate points where $\lambda = 0$ or $\mu = \infty$). Now, the real locus of this neighborhood consists of the points corresponding to spheres with λ and μ real, and these spheres can be classed into four regions according to whether $\lambda < 0$ or $\lambda > 0$, $\mu < \infty$ or $\mu > \infty$ (this second condition is obviously to be interpreted as meaning that μ is negative and $|\mu|$ is large). Clearly these four small regions are disjoint and simply connected. There are four such regions for each of the fifteen maximally degenerate points, so we obtain a set $\hat{B}_{0,5}$ of sixty tangential base points on $\mathcal{M}_{0,5}$, each having an automorphism of order 2.

ORBIFOLD STRUCTURE. As for all the ordered genus zero moduli spaces, the pure mapping class group $\Gamma_{0,5}$ acts freely on the Teichmüller space $\mathcal{T}_{0,5}$, so that $\mathcal{M}_{0,5}$ has no special orbifold points but is simply a topological space, with topological fundamental group $\Gamma_{0,5}$.

DEHN TWISTS ON $S_{0,5}$. Let x be a base point on $\mathcal{M}_{0,5}$. One can parametrize these twists on $S_{0,5}$ and then as loops on the moduli space, just as in Figures 3.2-3.4. We do only x_{12}, starting from the tangential base point given topologically by the set of points $(0, \varepsilon, 1, 1 + \varepsilon, \infty)$ for real small positive values of ε. The Dehn twist x_{12} is parametrized on $S_{0,5}$, as after Figure 3.2, by

$$\left(-\frac{1}{2}\varepsilon e^{2\pi it} + \frac{1}{2}\varepsilon, \frac{1}{2}\varepsilon e^{2\pi it} + \frac{1}{2}\varepsilon, 1, 1 + \varepsilon, \infty \right),$$

which returns to standard form as

$$\left(0, \frac{\varepsilon e^{2\pi it}}{1 + \frac{1}{2}\varepsilon e^{2\pi it} - \frac{1}{2}\varepsilon}, 1, \frac{1 + \frac{1}{2}\varepsilon + \frac{1}{2}\varepsilon e^{2\pi it}}{1 + \frac{1}{2}\varepsilon e^{2\pi it} - \frac{1}{2}\varepsilon}, \infty \right).$$

It is easily seen that λ, starting at ε, describes a small counterclockwise circle around 0, whereas μ describes a small path which is homotopic to the identity on $\mathbb{P}^1 - \{0, 1, \infty\}$.

The fundamental group of $\mathcal{M}_{0,5}$. Another way of visualizing the loop x_{12} above is that it circles around the (missing) codimension 1 stratum $\lambda = 0$ in $\mathcal{M}_{0,5}$. It is actually more revealing to describe the loops corresponding to the generators $x_{i,i+1}$ (with $i \in \mathbb{Z}/5\mathbb{Z}$) this way. Each $x_{i,i+1}$ is a twist along a loop on $S_{0,5}$ surrounding the points x_i and x_{i+1}. Assume that x is the same tangential base point as above. Consider the region on $\mathcal{M}_{0,5}$ described by points $(x_1, x_2, x_3, x_4, x_5)$ with x_i real and $x_1 < x_2 < x_3 < x_4 < x_5$ (in the obvious sense of considering the real axis as a circle with $+\infty = -\infty$). This is a real, simply connected region on $\mathcal{M}_{0,5}$ containing the tangential base point x. It is a pentagonal region in two real dimensions, with five edges bounded by the real loci of the infinite strata $x_i = x_{i+1}$ for $i \in \mathbb{Z}/5\mathbb{Z}$; the five vertices are the maximally degenerate points where the infinite strata intersect. It is easy to describe the homotopy class of the loop on $\mathcal{M}_{0,5}$ corresponding to the Dehn twist $x_{i,i+1} \in \Gamma_{0,5} \simeq \pi_1(\mathcal{M}_{0,5}; x)$; it is given by composing a path γ from x nearly to the stratum $x_i = x_{i+1}$ (which is present only in the compactification of $\mathcal{M}_{0,5}$, of course, but missing from the space itself) lying in the pentagon, with a small loop around the (missing) stratum, and then with γ^{-1}. This is of course independent of the choice of γ since the pentagon is simply connected.

POINTS WITH SPECIAL AUTOMORPHISM GROUP. As in the case of $\mathcal{M}_{0,[4]}$, although there are no special orbifold points on $\mathcal{M}_{0,5}$, nevertheless determining the points of special automorphism group, i.e., those which are fixed by some permutation subgroup of S_5, will give the key to finding the orbifold points on the unordered moduli space $\mathcal{M}_{0,[5]}$.

A permutation τ acts on a point of $\mathcal{M}_{0,5}$, simply by permuting the marked points via

$$\tau\big(x_1, x_2, x_3, x_4, x_5\big) = \big(x_{\tau(1)}, x_{\tau(2)}, x_{\tau(3)}, x_{\tau(4)}, x_{\tau(5)}\big).$$

If the starting point is given in the form of a standard representation (with three components fixed at 0, 1 and ∞, then bringing the result of the permutation back to this form, we obtain a rational expression for the action of τ. We determine this rational expression for $\tau = (12)$ and $\rho_2 = (12345)$:

$$\tau\big(\lambda, 0, 1, \infty, \mu\big) = \Big(0, \lambda, 1, \infty, \mu\Big) \sim \Big(\frac{-\lambda}{1-\lambda}, 0, 1, \infty, \frac{\mu-\lambda}{1-\lambda}\Big),$$

$$\rho\big(\lambda, 0, 1, \infty, \mu\big) = \Big(0, 1, \infty, \mu, \lambda\Big) \sim \Big(\frac{1}{\mu}, 0, 1, \infty, \frac{\lambda-1}{\lambda-\mu}\Big).$$

As in the case of $\mathcal{M}_{0,4}$, the points of special automorphism group are the points which have nontrivial isotropy subgroup under the action of S_5. There are two kinds of such points, isolated ones and those lying on a stratum of real dimension 1. To compute them, one computes the fixed points of each permutation in S_5. For example, the fixed points of ρ are given by (λ, μ) with

$$\lambda = \frac{1}{\mu} \quad \text{and} \quad \mu = \frac{\lambda-1}{\lambda-\mu},$$

so μ is a root of $\mu^3 - 2\mu + 1$. One root is $\mu = 1$, but this is excluded in $\mathcal{M}_{0,5}$, so the two solutions are

$$(\lambda, \mu) = \Big(\frac{1 \pm \sqrt{5}}{2}, \frac{-1 \pm \sqrt{5}}{2}\Big).$$

On the next page we give a table containing all points of special automorphism group on $\mathcal{M}_{0,5}$. Obviously, all powers of a given permutation have the same fixed points, so that it is only necessary to consider the cyclic subgroups of S_5. Each product of two transpositions fixes a stratum of complex codimension 1, whereas each 3-cycle, each 4-cycle and each 5-cycle fixes a conjugate pair of points. Table 1 gives the complete list of fifteen special loci, but we only compute the conjugate pairs for one representative of each type of cycle.

Permutations having other cycle types than those which appear in this table have no fixed points on $\mathcal{M}_{0,5}$, corresponding to the fact that no Riemann surfaces of type $(0, 5)$ have corresponding automorphisms.

(λ, μ)	fixed by
$(-\mu^2 + 2\mu, \mu)$	$\langle(12)(34)\rangle$
$(\lambda, -\lambda^2 + 2\lambda)$	$\langle(25)(34)\rangle$
$(\frac{\mu^2}{2\mu-1}, \mu)$	$\langle(14)(23)\rangle$
$(\lambda, \frac{\lambda^2}{2\lambda-1})$	$\langle(23)(45)\rangle$
(μ^2, μ)	$\langle(13)(24)\rangle$
(λ, λ^2)	$\langle(24)(35)\rangle$
$(\mu - 1, \mu)$	$\langle(13)(25)\rangle$
$(\lambda, \lambda - 1)$	$\langle(12)(35)\rangle$
$(\lambda, \frac{\lambda}{1-\lambda})$	$\langle(14)(35)\rangle$
$(\frac{\mu}{1-\mu}, \mu)$	$\langle(13)(45)\rangle$
$(\lambda, \frac{1}{2-\lambda})$	$\langle(12)(45)\rangle$
$(\frac{1}{2-\mu}, \mu)$	$\langle(14)(25)\rangle$
$(\lambda, \frac{\lambda}{\lambda-1})$	$\langle(15)(34)\rangle$
$(\lambda, 1/\lambda)$	$\langle(15)(24)\rangle$
$(\lambda, 1 - \lambda)$	$\langle(15)(23)\rangle$
$\left(\frac{-1+\sqrt{-3}}{2}, \frac{-1-\sqrt{-3}}{2}\right), \left(\frac{-1-\sqrt{-3}}{2}, \frac{-1+\sqrt{-3}}{2}\right)$	$\langle(135)\rangle$
$\left(\frac{1+\sqrt{5}}{2}, \frac{-1+\sqrt{5}}{2}\right), \left(\frac{1-\sqrt{5}}{2}, \frac{-1-\sqrt{5}}{2}\right)$	$\langle(12345)\rangle$
$(i, -i), (-i, i)$	$\langle(1234)\rangle$

We study the fifteen special loci of dimension 1. They are all identical, since they are images of each other under the automorphism group S_5 of $\mathcal{M}_{0,5}$, so it suffices to study only one, say the locus fixed by $(15)(24)$, given by points in $\mathcal{M}_{0,5}$ of the form $(0, \lambda, 1, 1/\lambda, \infty)$.

For such a point to lie in $\mathcal{M}_{0,5}$, the five components must be distinct, which means that we must have $\lambda \notin \{0, 1, \infty, -1\}$. Thus the special locus of $(15)(24)$ is a copy of $\mathbb{P}^1 - 4$ points. We know that it has exactly fifteen images under S_5, so it must have a global stabilizer of order 8, of which one element of order 2 is $(15)(24)$, which actually fixes the locus pointwise; this stabilizer is the dihedral group of order 8 generated by (15) and (1254). The permutation (15) acts by $\lambda \mapsto -\lambda$ and (1254) acts by $\lambda \mapsto -1/\lambda$. The corresponding special locus in the unordered moduli space $\mathcal{M}_{0,[5]}$, which is really a special orbifold locus as opposed to the locus described here which is merely a locus of points having isotropy group in S_5, is isomorphic to the quotient of $\mathbb{P}^1 - \{0, 1, \infty, -1\}$ by the action of this dihedral group (see Section 3.3.2).

3.3.2. The unordered moduli space $\mathcal{M}_{0,[5]}$. THE MAPPING CLASS GROUP $\Gamma_{0,[5]}$.
The group $\Gamma_{0,[5]}$ is generated by $\sigma_1, \sigma_2, \sigma_3, \sigma_4$ with the usual braid relations $\sigma_i \sigma_{i+1} \sigma_i = \sigma_{i+1} \sigma_i \sigma_{i+1}$ and $\sigma_i \sigma_j = \sigma_j \sigma_i$ if $|i - j| \geq 2$, as well as the center relation $(\sigma_1 \sigma_2 \sigma_3 \sigma_4)^5 = 1$ and the sphere relation $\sigma_4 \sigma_3 \sigma_2 \sigma_1^2 \sigma_2 \sigma_3 \sigma_4 = 1$.

There is only one conjugacy class each of elements of order 2, 3, 4 and 5 in $\Gamma_{0,[5]}$. Generators of these conjugacy classes are given by $(\sigma_1\sigma_2\sigma_3)^2$, $\sigma_1^2\sigma_2\sigma_3$, $\sigma_1\sigma_2\sigma_3$ and $\sigma_1\sigma_2\sigma_3\sigma_4$ respectively.

THE STABLE COMPACTIFICATION OF $\mathcal{M}_{0,[5]}$. We saw that the divisor at infinity of the ordered moduli space $\mathcal{M}_{0,5}$ consisted of ten crossing copies of $\overline{\mathcal{M}}_{0,4}$, corresponding to the ten ways of pinching a loop surrounding two marked points on S, i.e., the ten ways of choosing two points among five. Thus, the natural action of S_5 on the compactification $\overline{\mathcal{M}}_{0,5}$ permutes these ten strata, and so the compactification of $\mathcal{M}_{0,[5]}$ is obtained by adding a single stratum at infinity to $\mathcal{M}_{0,[5]}$.

This stratum looks like the quotient of $\overline{\mathcal{M}}_{0,4}$ by the stabilizer of each stratum of $\overline{\mathcal{M}}_{0,5}$ in S_5. Since there are ten strata, the stabilizer of each one is of order 12. The stratum corresponding to the pair of points i and j is fixed *pointwise* by the transposition (ij), and stabilized globally by the copy of S_3 inside S_5 given by all permutations fixing i and j. Thus the single stratum at infinity of $\overline{\mathcal{M}}_{0,[5]}$ is isomorphic to the quotient of $\overline{\mathcal{M}}_{0,4}$ by a group $S_3 \times \{\pm 1\}$ with S_3 acting as usual and $\{\pm 1\}$ acting trivially. The fifteen points of maximal degeneration on $\overline{\mathcal{M}}_{0,5}$, three of which lie on each of the ten codimension 1 strata, are all mapped to a single point of maximal degeneration in $\overline{\mathcal{M}}_{0,[5]}$.

ORBIFOLD STRUCTURE AND SPECIAL LOCI. The determination of the orbifold structure of $\mathcal{M}_{0,[5]}$ was prepared by the determination of the points of special automorphism group in $\mathcal{M}_{0,5}$, for the orbifold points of $\mathcal{M}_{0,[5]}$ are exactly the images of these.

We saw that there were fifteen one-dimensional loci of points of special automorphism group in $\mathcal{M}_{0,5}$, each of which is the set of fixed points of one of the fifteen products of two transpositions in S_5. They are permuted by the action of S_5. Each one is globally stabilized by a D_4 subgroup of S_5 and pointwise fixed by an element of order 2. To determine the look of the corresponding one-dimensional special locus in $\mathcal{M}_{0,[5]}$, it suffices to describe the quotient of one of these loci by its automorphism group D_4. We consider the locus $(0, \lambda, 1, 1/\lambda, \infty)$, which is equal to $\mathbb{P}^1 - \{0, 1, \infty, -1\}$, is pointwise fixed by $\{1, (15)(24)\}$ and globally stabilized by $\{1, (15), (24), (15)(24), (1254), (1452), (12)(54), (14)(25)\}$, where (15) acts by $\lambda \mapsto -\lambda$ and (1254) acts by $\lambda \mapsto -1/\lambda$. The orbifold quotient can be obtained in two steps. First we quotient by the pointwise action of the subgroup $\{1, (15)(24)\}$, obtaining a space looking identical to $\mathbb{P}^1 - \{0, 1, \infty, -1\}$, but with a $\mathbb{Z}/2\mathbb{Z}$ attached to each point. Then, we quotient this space by the degree 4 map

$$\frac{1}{2}\left(z^2 + \frac{1}{z^2}\right).$$

The result is a $\mathbb{P}^1$ with 2 missing points, 1 and ∞, and one special orbifold point -1, which has only the two preimages i and $-i$. This orbifold is equipped with a $\mathbb{Z}/2\mathbb{Z}$ at each point and a $\mathbb{Z}/4\mathbb{Z}$ at the special orbifold point. This point corresponds to the points of special automorphism group $\mathbb{Z}/4\mathbb{Z}$ in $\mathcal{M}_{0,5}$ given by $(0, i, 1, -i, \infty)$ and $(0, -i, 1, i, \infty)$ (automorphism group generated by (1254)).

Now consider the isolated orbifold points of $\mathcal{M}_{0,[5]}$. We saw that they come from 3-cycles, 4-cycles and 5-cycles in S_5, each of which has one pair of fixed points. Consider the pairs of fixed points of 4-cycles. Obviously all the pairs become identified in $\mathcal{M}_{0,[5]}$. But furthermore, we just saw that the pair corresponding to one particular 4-cycle, (1254), which is given in $\mathcal{M}_{0,5}$ by $(0, i, 1, -i, \infty)$ and $(0, -i, 1, i, \infty)$ becomes a single, orbifold point on the one-dimensional special locus in $\mathcal{M}_{0,[5]}$.

Similarly, the pairs of fixed points of 3-cycles become identified in $\mathcal{M}_{0,[5]}$, so that we only need consider the image of the pair of fixed points of one 3-cycle, say the one given in Section 3.3.1, fixed by (135), namely

$$\left(0, \frac{-1 + \sqrt{-3}}{2}, 1, \frac{-1 - \sqrt{-3}}{2}, \infty\right) \quad \text{and} \quad \left(0, \frac{-1 - \sqrt{-3}}{2}, 1, \frac{-1 + \sqrt{-3}}{2}, \infty\right).$$

These two points become identified in $\mathcal{M}_{0,[5]}$. Furthermore, as they lie on the locus $(0, \lambda, 1, 1/\lambda, \infty)$ in $\mathcal{M}_{0,5}$, their image lies in the one-dimensional special locus in $\mathcal{M}_{0,[5]}$.

The same holds for the pairs of fixed points of the 5-cycles. We consider the pair given in Section 3.3.1, fixed by (12345), namely

$$\left(0, \frac{1 + \sqrt{5}}{2}, 1, \frac{-1 + \sqrt{5}}{2}, \infty\right) \quad \text{and} \quad \left(0, \frac{1 - \sqrt{5}}{2}, 1, \frac{-1 - \sqrt{5}}{2}, \infty\right).$$

These two points pass to a single point in $\mathcal{M}_{0,[5]}$, which also lies on the special locus.

3.4. Genus one, two marked points. The points of the moduli space $\mathcal{M}_{1,[2]}$ are tori with two marked points determined up to translation. Because they are determined only up to translation, we can take a unique representative of each point of $\mathcal{M}_{1,[2]}$ given by an elliptic curve E equipped with two points P and Q such that $P + Q = 0$. Every such curve has an involution (written $z \mapsto -z$ on the fundamental parallelogram, or $(x, y) \mapsto (x, -y)$ in terms of the points), which exchanges P and Q. Thus we see that the ordered moduli space $\mathcal{M}_{1,2}$ is pointwise the same as $\mathcal{M}_{1,[2]}$, but $\mathcal{M}_{1,[2]}$ is equipped with an additional orbifold structure coming from the presence of an involution fixing every point.

THE MAPPING CLASS GROUPS $\Gamma_{1,2}$ AND $\Gamma_{1,[2]}$. The pure mapping class group $\Gamma_{1,2}$ is generated by the Dehn twists σ_1, σ_2, σ_3 along the loops shown in Figure 3.15.

These twists satisfy the usual braid relations between the σ_i, as well as $(\sigma_1\sigma_2\sigma_3)^4 = 1$. Thus $\Gamma_{1,2}$ is isomorphic to the quotient of the Artin braid group B_4 modulo its center. The full mapping class group $\Gamma_{1,[2]}$ is generated by the pure mapping class group together with a single diffeomorphism class exchanging the two marked points; it is convenient to take this generator to be the rotation c around the vertical axis in Figure 3.15. This rotation has order 2

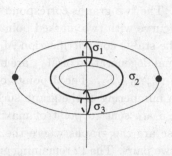

Figure 3.15. Generators of $\Gamma_{1,2}$.

and it is obviously central since it commutes with all the σ_i. Furthermore we have $\Gamma_{1,2} \cap \langle c \rangle = \{1\}$ inside $\Gamma_{1,[2]}$ since no element of the pure mapping class group can permute the marked points. So we find that

$$\Gamma_{1,[2]} \simeq \Gamma_{1,2} \times \langle c \rangle \simeq (B_4/Z) \times \mathbb{Z}/2\mathbb{Z}.$$

The ordered moduli space $\mathcal{M}_{1,2}$ can be identified with the quotient of $\mathcal{M}_{0,5}$ by the subgroup $S_4 \subset S_5$ permuting the first four marked points. Indeed, each point of the quotient $\mathcal{M}_{0,5}/S_4$ is given by an unordered quadruple of distinct points and one additional point; the quadruple determines an elliptic curve (by giving its unordered ramification points), and the fifth point can be identified with the x-coordinate of the point $P - Q$ (which is independent of the choice of P and Q up to translation). In other words, a point of $\mathcal{M}_{0,5}$ represented by $(0, 1, \infty, \lambda, \mu)$ to the (isomorphism class of the) elliptic curve given by the Weierstrass equation $y^2 = x(x - 1)(x - \lambda)$, equipped with any pair of marked points P and Q with $\mu = x(P - Q)$. Obviously, changing the order of the four first points gives rise to an isomorphic elliptic curve; permuting them and returning to standard form only has the effect of replacing λ by one of the six values $1/\lambda$, $1 - \lambda$, $(\lambda - 1)/\lambda$, $\lambda/(\lambda - 1), 1/(1 - \lambda)$, giving an isomorphic elliptic curve.

The maps $\mathcal{M}_{0,5} \to \mathcal{M}_{0,5}/S_4 \simeq \mathcal{M}_{1,2} \to \mathcal{M}_{0,5}/S_5 = \mathcal{M}_{0,[5]}$ correspond to the sequence of homomorphisms of fundamental groups

$$\Gamma_{0,5} \to \Gamma_{1,2} \to \Gamma_{0,[5]}.$$

The unordered moduli space $\mathcal{M}_{1,[2]}$ is equal to the quotient of $M_{1,2}$ by the action of the group $\langle c \rangle$, which fixes each point (acting as the involution of the corresponding elliptic curve).

THE STABLE COMPACTIFICATION OF $\mathcal{M}_{1,2}$. There are two pants decompositions of the topological surface $S_{1,2}$ (shown in Figure 3.15), one obtained by pinching two non-separating loops to points and the other obtained by pinching a separating loop and a non-separating loop. These correspond to the two possible genus 1 graphs with two tails, given by $\propto$ and $\circ\!\!\prec$. We call these two pants decompositions P_1 and P_2. The two graphs correspond to two different types of degeneration of an elliptic curve with two marked points.

The group S_5 acts on the stable compactification $\mathcal{M}_{0,5}$, as does its subgroup S_4, and the points of maximal degeneration of $\mathcal{M}_{0,5}$ are mapped to points of maximal degeneration of $\mathcal{M}_{1,2} \simeq \mathcal{M}_{0,5}/S_4$. The 3 points of maximal degeneration of $\mathcal{M}_{0,5}$ corresponding to a numbering of the graph such that the distinguished (central) tail is numbered 5 are sent to point of maximal degeneration corresponding to P_1, since those are the graphs where the four ramification points numbered 1 to 4 collide in two pairs. The 12 remaining graphs (points of maximal degeneration on $\mathcal{M}_{0,5}$) are all mapped to P_2.

The partially degenerate points are those for which one of the two loops in one of the two pants decompositions is pinched to zero. For the two loops in P_2, this gives two divisors of complex dimension 1. However, in the case of P_1, pinching one or the other of the two loops gives the same divisor, since an automorphism of the point brings one loop to the other.

The loop of P_2 enclosing the two marked points cuts $S_{1,2}$ into an $S_{1,1}$ and an $S_{0,3}$, so the corresponding divisor is a copy of $\mathcal{M}_{1,1}$. The other loop of P_2 cuts $S_{1,2}$ into an $S_{0,4}$, so the corresponding divisor is a copy of $\mathcal{M}_{0,4}$. The loops of P_1 each cut $S_{1,2}$ into an $S_{0,4}$, so there is another copy of $\mathcal{M}_{0,4}$ in the divisor at infinity of $\mathcal{M}_{1,2}$. However, this second copy is self-intersecting, as the point of maximal degeneration lies on it twice.

The stable compactification $\overline{\mathcal{M}}_{1,2}$ is obtained from the stable compactification $\overline{\mathcal{M}}_{0,5}$ by quotienting by the action of S_4. The divisor at infinity is exactly the image of the divisor at infinity of $\overline{\mathcal{M}}_{0,5}$.

TOPOLOGICAL TANGENTIAL BASE POINTS ON $\overline{\mathcal{M}}_{1,2}$. In order to compute the tangential base points of $\mathcal{M}_{1,2}$, it is necessary to understand the topological shape of the neighborhoods of the two points of maximal degeneration, as the images of the neighborhoods of the corresponding points in $\mathcal{M}_{0,5}$. We begin with P_2.

Let P be a point of maximal degeneration of $\overline{\mathcal{M}}_{0,5}$ lying over P_2; say P joins the pair of points 1 and 2 and the pair 4 and 5. Then P is fixed by the dihedral subgroup in S_5 generated by (12), (45) and (1425). To see what subgroup of S_4 fixes P, we take the intersection of this dihedral subgroup with S_4 and it consists only of the identity and the transposition (12). This shows that the neighborhood of P_2 in $\overline{\mathcal{M}}_{1,2}$ is isomorphic to the image of a neighborhood of P in $\overline{\mathcal{M}}_{0,5}$ quotiented by the action of the transposition. Now, the neighborhood of P in $\overline{\mathcal{M}}_{0,5}$ is the product of two disks (pointed disks if we exclude P itself); each disk contains two real segments representing the tangential base points. Retracting the pointed disks to their boundaries, their product is a torus, with four points representing the four tangential base points. The action of the transposition is like $z \mapsto z^2$ on the first disk (or circle) and the identity on the second, so it has an effect on the torus as though one cut through it to make a cylinder, rolled the cylinder over on itself twice and reglued to obtain a new torus. The four points have now been identified two by two to give two points coming from the tangential base points of $\mathcal{M}_{0,5}$, but it is necessary to have four tangential base points on the torus to cut it into simply connected regions.

To summarize, the neighborhood of P_2 in $\overline{M}_{1,2}$ is still a product of two pointed disks, and therefore P_2 is flanked by four tangential base points. Two of them are images of the 48 tangential base points flanking the 12 points of maximal degeneration of $\overline{M}_{0,5}$ lying over P_2, which divide into two orbits of 24 points each under the action of S_4. The other two tangential base points do not come from $\mathcal{M}_{0,5}$. This situation is analogous to the one for $\mathcal{M}_{1,1}$, where one tangential base point comes from the six tangential base points on $\mathcal{M}_{0,4}$ and another one is needed to cut the neighborhood of infinity into simply connected regions.

Now consider the neighborhood of P_1. It is more complicated, because if Q denotes one of the three points of maximal degeneration of $\overline{M}_{0,5}$ lying over P_1, say the one corresponding to the pants decomposition identifying the points 1 and 2 and the points 3 and 4, then Q is fixed by the dihedral group of order 8 generated by (12), (34) and (1324) in S_5, and in fact this subgroup also lies in S_4. Thus, the neighborhood of P_1 is isomorphic to the neighborhood of Q quotiented by this D_4. The elements of the D_4 act on the product of two pointed disks neighboring Q by $z \mapsto z^2$ on each of the disks and by exchanging the two disks. Retracting onto the torus and performing the $z \mapsto z^2$ first sends the torus again to a torus, but where all four tangential base points have been identified. Further quotienting this torus by the order 2 element exchanging the two circles gives a Möbius band. The retraction is quite misleading here since the Möbius band is not orientable, but the neighborhood of P_2 is really the Möbius band times a small segment, which is orientable. The fundamental group of this thickened band is the same as the one of the band itself, namely $\mathbb{Z}$, so it has to be cut into two simply connected pieces. One of these pieces corresponds to the 12 tangential base points flanking the 3 maximally degenerate points of $\overline{M}_{0,5}$ lying over P_1, which form a single orbit under the action of S_4.

SPECIAL LOCI. We saw above that the whole of the moduli space $\mathcal{M}_{1,[2]}$ is a special locus for the canonical involution c. Consider the moduli space $\mathcal{M}_{1,2} = \mathcal{T}_{1,2}/\Gamma_{1,2}$. There is only one special locus of dimension 1 in $\mathcal{M}_{1,2}$, corresponding to the unique conjugacy class of elements of order 2 in $\Gamma_{1,2}$, which is the conjugacy class of $(\sigma_3\sigma_2\sigma_1)^2$.

THEOREM 3.4.1. *The 1-dimensional special locus of $\mathcal{M}_{1,2}$ maps bijectively to the quotient M of $\mathbb{P}^1 - \{0, 1, \infty\}$ by the mapping identifying z with $1/z$, which is given by $\frac{1}{2}(z + \frac{1}{z})$.*

PROOF. Let $\lambda \in \mathbb{P}^1 - \{0, 1, \infty\}$, let E be the elliptic curve of equation $y^2 = x(x - 1)(x - \lambda)$, and let $P \neq Q$ be two distinct points on E. Let μ be the x-coordinate of the point $P - Q$. Note that μ determines the unordered pair of points P, Q exactly up to translation. Indeed, if $P' = P + T$ and $Q' = Q + T$ then $P' - Q' = P - Q$ so $\mu = x(P' - Q') = x(P - Q)$. Conversely, if $x(P - Q) = x(P' - Q')$, then either

$P' - Q' = P - Q$, in which case we have $P' - P = Q' - Q$; setting $T = P' - P$, we have $P' = T + P$ and $Q' = T + Q$, or else $P' - Q' = Q - P$, in which case $P' - Q = Q' - P$; setting $T = P' - Q$, we have $P' = T + Q$ and $Q' = T + P$. Note that $\mu \neq \infty$ ensures that $P \neq Q$.

The moduli space $\mathcal{M}_{1,2}$ can be parametrized by couples (λ, μ) with $\lambda \in \mathbb{P}^1 - \{0, 1, \infty\}$ and $\mu \in \mathbb{C}$, modulo an equivalence relation for which (λ, μ) and (λ', μ') are equivalent if and only the equations $y^2 = x(x-1)(x-\lambda)$ and $y^2 = x(x-1)(x-\lambda')$ describe isomorphic curves, and μ corresponds to μ' under the isomorphism. This is explicitly given by the following equivalence relation grouping the pairs $(\lambda, \mu) \in (\mathbb{P}^1 - \{0, 1, \infty\} \times \mathbb{C})$ into the following groups of six:

$$(\lambda, \mu) \sim (1 - \lambda, 1 - \mu) \sim \left(\frac{1}{1-\lambda}, \frac{1-\mu}{1-\lambda}\right) \sim \left(\frac{1}{\lambda}, \frac{\mu}{\lambda}\right)$$

$$\sim \left(1 - \frac{1}{\lambda}, 1 - \frac{\mu}{\lambda}\right) \sim \left(\frac{\lambda}{\lambda-1}, \left(\frac{\lambda}{\lambda-1}\right)\left(1 - \frac{\mu}{\lambda}\right)\right).$$

The special points in $\mathcal{M}_{1,2}$ are given by the pairs $(\lambda, 0)$, $(\lambda, 1)$ and (λ, λ). They group into sextuplets of the form

$$(\lambda, 0) \sim (1 - \lambda, 1) \sim \left(\frac{1}{1-\lambda}, \frac{1}{1-\lambda}\right) \sim \left(\frac{1}{\lambda}, 0\right) \sim \left(1 - \frac{1}{\lambda}, 1\right) \sim \left(\frac{\lambda}{\lambda-1}, \frac{\lambda}{\lambda-1}\right).$$

There is an obvious isomorphism of this subset of sextuples to M given by associating to a sextuplet the unordered pair of values $\{\lambda, 1/\lambda\}$ obtained by considering only the pair of elements in the sextuplet whose μ-component is equal to 0. $\qquad \Box$

REMARK. It will be shown in Section 4.1 that when the special locus corresponding to a finite-order element φ of a mapping class group is normal in the moduli space, then the fundamental group of the special locus is isomorphic to the normalizer of φ. In the present case, we can take φ to be the order 2 element $(\sigma_3\sigma_2\sigma_1)^2$ of $\Gamma_{1,2}$, and its normalizer is given by

$$\mathrm{Norm}_{\Gamma_{1,2}}(\varphi) = \langle \sigma_3\sigma_2\sigma_3^{-1}, \sigma_3\sigma_1 \rangle.$$

This group contains φ as a central element of order 2, and modulo φ, it is isomorphic to $\pi_1(M) = \langle \tau_1^2, \tau_2 \rangle$ where τ_1 and τ_2 are standard generators of B_3/Z, the Artin braid group on three strands modulo its center.

In order to conclude the determination of the special loci in $\mathcal{M}_{1,2}$, it suffices to list the elements of finite order in $\Gamma_{1,2}$ (since we have already considered the canonical involution in $\Gamma_{1,[2]}$). But apart from the order 2 element $(\sigma_3\sigma_2\sigma_1)^2$ already studied, the only such element (up to conjugacy) is the order 4 element $\sigma_3\sigma_2\sigma_1$. The corresponding special point on $\mathcal{M}_{1,2}$ is the point on the special locus of $(\sigma_3\sigma_2\sigma_1)^2$ described above with λ-value equal to -1; this point corresponds to the elliptic curve of equation $y^2 = x^3 - x$ which has automorphism group $\mathbb{Z}/4\mathbb{Z}$ generated by $(x, y) \mapsto (-x, iy)$, equipped with marked points (∞, ∞) (the origin) and $(0, 0)$. The fundamental parallelogram of this elliptic curve in the upper half-plane $\mathcal{H}$ is the square $\langle 1, i \rangle$, and the marked points are $z = 0$ and $z = (1 + i)/2$, both fixed under the rotation of the square.

3.5. Genus zero, n marked points. In this section, we limit ourselves to investigating the points of special automorphism group and the special loci in the genus zero moduli spaces for arbitrary n. If S is a sphere with n marked points, then a finite-order element of the mapping class group $\Gamma([S])$ is the class of a diffeomorphism which is simply a rotation around an axis:

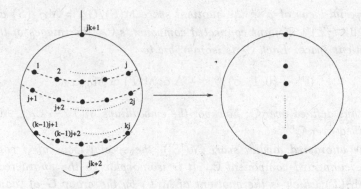

Figure 3.16. A finite-order diffeomorphism in genus zero.

The north and south poles may or may not be marked points, but they are always the only branch points for φ. The permutation associated to a rotation φ is always of the form $c_1 \cdots c_k$, where the c_i are disjoint cycles of length j such that

$$\begin{cases} jk = n & \text{if the north and south poles are not marked} \\ jk = n - 1 & \text{if one of the two poles is marked} \\ jk = n - 2 & \text{if both poles are marked points.} \end{cases}$$

In the following theorem, we compute the points of special automorphism associated to a permutation $[\varphi]$ which is a product of k disjoint cycles of length j only in the case $jk = n - 2$ (i.e., when the two fixed points of φ are marked points), since in the other two cases, the special locus in $\mathcal{M}_{0,n}$ is just the image of the one we compute here in $\mathcal{M}_{0,n+1}$ or $\mathcal{M}_{0,n+2}$, under the morphism given by erasing the extra marked points.

THEOREM 3.5.1. *Let S be a sphere with n marked numbered points, and let φ be the rotation shown in Figure 3.16, with $n = jk + 2$ (i.e. the fixed points of φ are marked points of S), so that*

$$[\varphi] = (1 \cdots j)(j + 1 \cdots 2j) \cdots (j(k-1) + 1 \cdots jk).$$

Let $G_\varphi \subset S_n$ be the subgroup generated by the disjoint cycles $c_1, \ldots, c_k$ of $[\varphi]$. Let T be the orbifold quotient S/φ, which has k marked points with ramification index 1 and 2 marked points with ramification index j.

(i) *The set of fixed points of $[\varphi]$ in the ordered moduli space $\mathcal{M}(S)$ consists of $|(\mathbb{Z}/j\mathbb{Z})^*|$ disjoint connected components given by*

$$\mathcal{C}_\zeta = \Big(1, \zeta, \ldots, \zeta^{j-1}, a_1, a_1\zeta, \ldots, a_1\zeta^{j-1}, \cdots, a_{k-1}, a_{k-1}\zeta, \cdots, a_{k-1}\zeta^{j-1}, 0, \infty\Big),$$
$$(3.5.1)$$

where ζ runs through the primitive j-th roots of unity. Each component is isomorphic to a copy of $(\mathbb{P}^1 - \{0, 1, \zeta, \ldots, \zeta^{j-1}, \infty\})^{k-1}$ minus the $j(k-1)$ lines $a_i = a_r\zeta^s$ for $r \neq i$, $0 \leq s \leq j-1$, and is thus defined over $\mathbb{Q}^{\mathrm{ab}}$.

(ii) *The special locus of φ in the quotient space $\mathcal{M}(S)/G_\varphi = \mathcal{M}_{G_\varphi}(S)$ also consists of $|(\mathbb{Z}/j\mathbb{Z})^*|$ disjoint connected components $\overline{\mathcal{C}}_\zeta$, the images of the $\mathcal{C}_\zeta$ in the quotient space. Each $\overline{\mathcal{C}}_\zeta$ is isomorphic to*

$$(\mathbb{P}^1 - \{0, 1, \infty\})^{k-1} - \Delta \simeq \mathcal{M}(T) \simeq \mathcal{M}_{0,k+2}, \qquad (4.2.2)$$

and is thus defined over $\mathbb{Q}$; however the embeddings $\mathcal{M}(T) \to \overline{\mathcal{C}}_\zeta \subset \mathcal{M}(S)/G_\varphi$ are defined over $\mathbb{Q}^{\mathrm{ab}}$.

(iii) *In the unordered moduli space $\mathcal{M}([S])$, the special locus of φ consists of a single connected component $\mathcal{C}$. It is isomorphic to the unordered moduli space $\mathcal{M}_G(T)$ which is the quotient of $\mathcal{M}(T)$ by the group G of "admissible" permutations, i.e., permutations of marked points of the same ramification index, and the space $\mathcal{M}_G(T)$ and the embedding $\mathcal{M}_G(T) \to \mathcal{C} \subset \mathcal{M}([S])$ are defined over $\mathbb{Q}$.*

PROOF. (i) Let a point on $\mathcal{M}(S)$ be given by its unique representative as a sphere with n marked points $(x_1, \ldots, x_n)$ such that $x_1 = 1$, $x_{n-1} = 0$ and $x_n = \infty$. It is immediate that the components $\mathcal{C}_\zeta$ of (3.5.1) are all disjoint on $\mathcal{M}(S)$, since the points $0, 1, \infty$ are marked.

To see that each point of each component is fixed by the action of the permutation $[\varphi]$, it suffices to notice that the permutation is realized by multiplication by ζ^{-1}. To show that these points are the only points fixed by $[\varphi]$ is a straightforward computation. Finally, a point of $\mathcal{C}_\zeta$ uniquely determines a point $(a_1, \ldots, a_{k-1}) \in (\mathbb{P}^1)^{k-1}$, but conversely, a tuple $(a_1, \ldots, a_{k-1})$ gives rise to a point of $\mathcal{M}(S)$ via (3.5.1) if and only if the components of (3.5.1) are all distinct, i.e., if and the a_i fulfill all the inequality conditions of (i).

(ii) The fact that the $|(\mathbb{Z}/j\mathbb{Z})^*|$ disjoint components remain disjoint in the quotient space $\mathcal{M}(S)/G_\varphi$ follows immediately from the simple remark that each component $\mathcal{C}_\zeta$ is stabilized (not pointwise fixed) by the whole of the group G_φ, so that the components $\overline{\mathcal{C}}_\zeta = \mathcal{C}_\zeta/G_\varphi$ remain disjoint. The action of G_φ on $\mathcal{C}_\zeta$ can be translated into an action on to a copy of $(\mathbb{P}^1 - \{0, 1, \zeta, \ldots, \zeta^{j-1}, \infty\})^{k-1}$ minus the $j(k-2)$ lines $a_i = a_r\zeta^s$ for $r \neq i$, $0 \leq s \leq j-1$; for $i \geq 1$, the $(i+1)$-th generator (i.e., the $(i+1)$-st disjoint cycle of $[\varphi]$) acts via $a_i \mapsto a_i\zeta$ for $1 \leq i \leq k-1$. The quotient is thus isomorphic to $(\mathbb{P}^1 - \{0, 1, \infty\})^{k-1}$, parametrized by $(b_1, \ldots, b_{k-1})$ with $b_i = a_i^j$ for each i, minus the diagonals $b_i = b_j$; this space is isomorphic to

$\mathcal{M}_{0,k+2} \simeq \mathcal{M}(T)$. However, each of the $|(\mathbb{Z}/j\mathbb{Z})^*|$ embeddings $f_\zeta : \mathcal{M}_{0,k+2} \to \overline{\mathcal{C}}_\zeta$ are defined over $\mathbb{Q}^{ab}$ since ζ must be specified.

(iii) Finally, all these components (as well as all those corresponding to other rotations having the same cycle type as φ) become identified in the unordered moduli space $\mathcal{M}([S])$, since for any $i \in (\mathbb{Z}/j\mathbb{Z})^*$, the locus

$$\mathcal{C}_{\zeta^i} = $$
$$\left(1, \zeta^i, \ldots, \zeta^{i(j-1)}, a_1, a_1\zeta^i, \ldots, a_1\zeta^{i(j-1)}, \cdots, a_{k-1}, a_{k-1}\zeta^i, \cdots, a_{k-1}\zeta^{i(j-1)}, 0, \infty\right)$$

differs from

$$\mathcal{C}_\zeta = \left(1, \zeta, \ldots, \zeta^{j-1}, a_1, a_1\zeta, \ldots, a_1\zeta^{j-1}, \cdots, a_{k-1}, a_{k-1}\zeta, \cdots, a_{k-1}\zeta^{j-1}, 0, \infty\right)$$

only by a permutation.

The special locus $\mathcal{C}$ in the unordered moduli space $\mathcal{M}([S])$ is isomorphic to $\mathcal{C}_\zeta$ modulo its stabilizer in $S_n = S_{jk+2}$. Therefore, to determine it, we need to determine its stabilizer.

We begin by determining the order of the stabilizer, which is the quotient of $|S_n|$ by the number of different loci corresponding to groups $\langle \alpha \rangle$ where $\alpha \in S_n$ is a permutation of the same cycle type as $[\varphi]$, i.e., k cycles of length j. So we enumerate the groups $\langle \alpha \rangle$, where α is a product of k j-cycles.

To begin with, there are $\binom{n}{2}$ ways of choosing two indices fixed by such a permutation α. Having fixed such a choice, we count the ways of dividing the permutations of the remaining jk components into groups $\langle \alpha \rangle$. There are jk possible orderings of the remaining indices; packaging them into j-cycles gives a redundant version of the set of possible permutations α of the right form. But each of the k j-cycles can be written in j different ways giving the same permutation (by cyclically permuting them, i.e., (12345) is the same permutation as (23451)), and furthermore any permutation of the k disjoint cycles amongst each other again gives the same permutation. Therefore there are $jk/(k!j^k)$ possible permutations α of the right form, fixing two given indices. We actually want to count the groups $\langle \alpha \rangle$, so we must divide by the number of powers of α which are also products of k j-cycles; this is $\phi(j) = |(\mathbb{Z}/j\mathbb{Z})^*|$. Finally, we have found $\binom{n}{2} \cdot jk/(k!j^k\phi(j))$ different groups $\langle \alpha \rangle$. Since the locus of fixed points of each such cyclic group has $\phi(j)$ connected components, and $n = jk + 2$, we find

$$\frac{(jk+2)!}{2 \cdot k! \cdot j^k}$$

connected components in the ordered moduli space $\mathcal{M}(S)$ for all the different α. The order of the stabilizer of one component is thus equal to $2j^kk!$, and the locus corresponding to the disjoint cycle-type of α is isomorphic to the quotient of $\mathcal{C}_\zeta$ of (3.5.1) by its stabilizer.

We can now easily compute the stabilizer of $\mathcal{C}_\zeta$, one of the loci associated to

$$[\varphi] = (1 \cdots j)(j+1 \cdots 2j) \cdots ((k-1)j \cdots jk).$$

It is generated by three natural subgroups; the first, of order $k!$, corresponding to permuting the k disjoint cycles of $[\varphi]$, the second, of order j^k (which is actually just G_φ), is generated by the j cycles themselves, and the third, of order 2, is generated by the permutation

$$\tau = \begin{cases} (1, jk)(2, jk-1)\cdots((jk/2),(jk+2)/2)\cdot(jk+1, jk+2) & \text{if } jk \text{ is even,} \\ (1, jk)(2, jk-1)\cdots((jk-1)/2,(jk+1)/2)\cdot(jk+1, jk+2) & \text{if } jk \text{ is odd.} \end{cases}$$

This last permutation is easily seen to stabilize $\mathcal{C}_\zeta$ since applying it to the point corresponding to $(a_1, \ldots, a_{k-1})$ gives

$$\left(a_{k-1}\zeta^{j-1}, \ldots, a_{k-1}, a_{k-2}\zeta^{j-1}, \ldots, a_{k-2}, \ldots, a_1\zeta^{j-1}, \ldots, a_1, \zeta^{j-1}, \ldots, 1, \infty, 0\right),$$

and then applying the mapping $z \mapsto a_{k-1}\zeta^{j-1}/z$ gives

$$\left(1, \zeta, \ldots, \zeta^{j-1}, c_1, \ldots, c_1\zeta^{j-1}, \ldots, c_{k-2}, \ldots, c_{k-2}\zeta^{j-1}, c_{k-1}, \ldots, c_{k-1}\zeta^{j-1}\right)$$

with $c_m = a_{k-1}/a_{k-m-1}$ (writing $a_0 = 1$), which also lies on $\mathcal{C}_\zeta$.

It remains only to compute the quotient of $\mathcal{C}_\zeta$ by this stabilizer, by determining the action of the generators of the stabilizer on the a_i. We already saw that

$$\mathcal{C}_\zeta/G_\varphi = \overline{\mathcal{C}}_\zeta \simeq (\mathbb{P}^1 - \{0, 1, \infty\})^{k-1} - \Delta,$$

parametrized by $b_1, \ldots, b_{k-1}$. To quotient by the rest of the stabilizer, we identify $(\mathbb{P}^1 - \{0, 1, \infty\})^{k-1} - \Delta$ with $\mathcal{M}(T) = \mathcal{M}_{0,k+2}$, by identifying the point $(b_1, \ldots, b_{k-1})$ with the orbifold sphere of topological type T, having marked points $(1, b_1, \ldots, b_{k-1}, 0, \infty)$ where 0 and ∞ have associated ramification index j and $1, b_1, \ldots, b_{k-1}$ have associated ramification index 1. Then the permutations of the k disjoint cycles of $[\varphi]$ act like permutations of $b_0 = 1, b_1, \ldots, b_{k-1}$, i.e., we quotient by all possible permutations of the marked points of T of ramification index equal to 1, and the permutation τ acts like

$$(b_0, b_1, \ldots, b_{k-1}, 0, \infty) \xrightarrow{\tau} (b_{k-1}, b_{k-2}, \ldots, b_0, \infty, 0)$$
$$= (1, b_{k-1}/b_{k-2}, \ldots, b_{k-1}/b_0, 0, \infty),$$

the last two tuples representing the same point in the moduli space $\mathcal{M}(T)$ via the transformation $z \to b_{k-1}/z$. Finally, then, we find that the special locus $\mathcal{C} \subset \mathcal{M}([S])$ is isomorphic to $\mathcal{M}(T)/S_k \times (\mathbb{Z}/2\mathbb{Z})$, which is naturally the unordered moduli space $\mathcal{M}([T])$ of the orbifold T, in which marked points of different ramification orders cannot be permuted. This moduli space is defined over $\mathbb{Q}$, as is the isomorphism with $\mathcal{C}$. $\square$

COROLLARY. *The isomorphisms of the (irreducible components of the) special loci with moduli spaces in parts* (ii) *and* (iii) *of Theorem 3.5.1 hold for any finite-order element* $\varphi \in \Gamma([S])$.

PROOF. Any finite-order φ is the conjugate of a φ having a permutation of the type $[\varphi]$ of Theorem 3.5.1, which was fixed only for convenience. Obviously, for any other product of k j-cycles, one needs to modify the expression (3.5.1); everything else in the proof passes to the general case. $\qquad\square$

4. Special Loci and Special Homomorphisms

When dealing with special loci in moduli space, the most natural question to ask is *What varieties can appear?* The brief answer is that *the special loci look very much like moduli spaces themselves.* In fact, the special locus of a finite-order element φ in moduli space of a topological surface S of type (g,n) is closely related to the moduli space of the quotient $T = S/\varphi$. Throughout this section, *we consider only the case where all branch points of the cover $S \to T$ are marked points of T, images of marked points on S*, and we investigate the nature of the special loci in the moduli space of S.

Recall from Section 2.1 that a *good* orbifold $\mathcal{M}$ is the quotient of a simply connected topological space $\mathcal{T}$ by a discrete group Γ acting properly discontinuously but not necessarily freely, and that all the moduli spaces of Riemann surfaces are such orbifolds. The *locus of special orbifold points* on the orbifold $\mathcal{M}$ is the set of points in $\mathcal{M}$ which are the images of points in $\mathcal{T}$ having nontrivial isotropy in Γ. A *special locus* of points in $\mathcal{M}$ is any subset of the locus of special orbifold points.

DEFINITION/NOTATION. Let S be a topological surface of type (g,n), and let $G \subset S_n$ be a subgroup. Let $\mathcal{M}(S)$ denote the ordered moduli space $\mathcal{M}_{g,n}$ (whose points are isomorphism classes of Riemann surfaces of topological type S), and let $\mathcal{M}_G(S)$ denote the quotient of the ordered moduli space $\mathcal{M}(S)$ by the permutation group G. Let $\Gamma(S)$ denote the pure mapping class group $\Gamma_{g,n}$ (isomorphic to $\mathrm{Diff}^+(S)/\mathrm{Diff}^0(S)$, see Section 2.1) and let $\Gamma([S])$ denote the full mapping class group $\Gamma_{g,[n]}$. We also write $\Gamma_G(S)$ for the preimage of G under the canonical surjection $\Gamma([S]) \to S_n$.

If φ is a finite-order element of the full mapping class group $\Gamma([S])$, we let $\mathcal{M}_G(S,\varphi)$ denote the image in $\mathcal{M}_G(S)$ of the set of points in the Teichmüller space $\mathcal{T}(S) = \mathcal{T}_{g,n}$ which are fixed by φ under the canonical action of $\Gamma([S])$ on $\mathcal{T}(S)$ (see Section 2.1). This locus is called the *special locus* of φ in $\mathcal{M}_G(S)$ if and only if $\varphi \in \Gamma_G(S)$, ensuring that the points of the locus are special orbifold points of φ. Otherwise, as we saw in Section 3, we simply call the points of $\mathcal{M}_G(S,\varphi)$ "points with special automorphism group". Note that the special locus $\mathcal{M}_G(S,\varphi)$ depends only on $[\langle\varphi\rangle]$, the conjugacy class in $\Gamma_G(S)$ of the group generated by φ.

Every subgroup of finite order of $\Gamma([S])$ can be realized as an automorphism of some Riemann surface X of topological type S. This result, long known as the *Nielsen realization problem* (and easy for cyclic groups $\langle\varphi\rangle$), was finally proved by S. Kerckhoff ([K1], [K2]). It implies that the special locus of φ consists of isomorphism classes of Riemann surfaces admitting automorphisms of topological type φ. The relation of the special locus $\mathcal{M}_G(S, \varphi)$ to the moduli space of the quotient $T = S/\varphi$ arises from the natural map from the special locus to the moduli space of T given by associating to each point X of the special locus the quotient Riemann surface X/φ. This morphism makes the special locus into a finite cover of the moduli space of T, in which the orbifold structure coming from the presence of the automorphism φ at each point of the special locus may or may not be "trivial" in a certain sense (see Section 4.1). When the map from each irreducible component of the special locus is one-to-one and the orbifold structure is trivial, there is a bijection between each irreducible component and the moduli space of T. The goal of this section is to justify these statements (Section 4.1), to determine these conditions explicitly in terms of fundamental groups in the case where $G = G_\varphi$ is the group generated by the disjoint cycles of φ and $\mathcal{M}(T)$ is the ordered moduli space of T (Section 4.2), to show that in genus zero, the conditions are fulfilled and therefore the bijection exists for all φ (Section 4.3), and to give examples (Sections 4.4 and 4.5).

We consistently use the word *special* to refer to the finite-order diffeomorphisms or automorphisms of Riemann surfaces, as in the terms *curve with special automorphisms* and *special locus associated to a finite-order diffeomorphism*. Therefore, we use the term *special homomorphism between mapping class groups* to denote the homomorphisms between mapping class groups we construct in Section 4.2, which exist whenever there is a bijection between each component of the special locus and the moduli space of T.

4.1. Special loci in moduli space. In Theorem 3.5.1(ii), we saw that each component of the special locus of a finite-order element $\varphi \in \Gamma([S])$ in the genus zero moduli space $\mathcal{M}_\varphi(S)$ is itself isomorphic to the ordered moduli space of $T = S/\varphi$. In this section, we give the general picture underlying this fact. This general picture is weaker than the genus zero case, in that in arbitrary genus, a special locus of $\varphi \in \Gamma([S])$ is *related* to the moduli space $\mathcal{M}(T)$, but not necessarily isomorphic to it.

We begin by citing some important results of González-Díez and Harvey ([GH]; see also [H, Theorem 2 and Corollary 3]), applied to the case where the finite group $G \subset \Gamma([S])$ they consider is the cyclic group $\langle \varphi \rangle$.

THEOREM 4.1.1 (see [GH], [HM], [H]). *Let S be a topological surface of type (g, n), and let φ be a finite-order element of $\Gamma([S])$. Assume that the quotient $T = S/\varphi$ is of genus g' with n' marked points, including all the branch points.*

(i) *Denote by $\mathcal{T}(S)_\varphi$ the subset of points of the Teichmüller space $\mathcal{T}(S) = \mathcal{T}_{g,n}$ fixed by φ, and by $\mathcal{T}(S, \varphi)$ the set of points fixed by any conjugate of φ. Then $\mathcal{T}(S)_\varphi$ is isomorphic to $\mathcal{T}_{g',n'} = \mathcal{T}(T)$ and $\mathcal{T}(S, \varphi)$ is the union of copies $\gamma(\mathcal{T}(T))$ for representatives γ of the $\mathrm{Norm}_{\Gamma([S])}(\varphi)$-cosets in $\Gamma([S])$.*

(ii) *The set of elements of $\Gamma([S])$ globally preserving the subset $\mathcal{T}(S, \varphi)$ in $\mathcal{T}(S)$ is exactly the subgroup $\mathrm{Norm}_{\Gamma([S])}(\varphi)$.*

(iii) *For every $G \subset S_n$ containing the permutation $[\varphi]$ associated to φ, the quotient $\widetilde{\mathcal{M}}_G(S, \varphi) = \mathcal{T}(S, \varphi)/\mathrm{Norm}_{\Gamma_G(S)}(\varphi)$ is isomorphic to the normalisation of the special locus $\mathcal{M}_G(S, \varphi) \subset \mathcal{M}_G(S)$.*

Let G_φ be the subgroup of S_n generated by the disjoint cycles of the permutation $[\varphi]$. Write $\mathcal{M}_\varphi(S)$ for the quotient space $\mathcal{M}_{G_\varphi}(S) = \mathcal{M}(S)/G_\varphi$, and $\mathcal{M}_\varphi(S, \varphi)$ for the whole of the special locus of φ in $\mathcal{M}_\varphi(S)$. We also write $\Gamma_\varphi(S)$ for the group $\Gamma_{G_\varphi}(S)$, the preimage in $\Gamma([S])$ of $\langle[\varphi]\rangle$ under the surjection $\Gamma([S]) \to S_n$.

There is a criterion (see [GH, Section 2]) to determine when the special locus $\mathcal{M}_\varphi(S, \varphi)$ is normal.

(N) $\widetilde{\mathcal{M}}_\varphi(S, \varphi)$ *and* $\mathcal{M}_\varphi(S, \varphi)$ *are distinct if and only if there is a Riemann surface X of topological type S whose automorphism group G contains φ and another element φ' of $\Gamma_\varphi(S)$, such that φ' is a conjugate of φ inside $\Gamma_\varphi(S)$ but not inside G.*

We now introduce a proposition whose proof is the goal of Section 4.2 (see Proposition 4.2.7).

PROPOSITION 4.1.2. *There is a canonical injective homomorphism*

$$\mathrm{Norm}_{\Gamma_\varphi(S)}(\varphi)/\langle\varphi\rangle \hookrightarrow \Gamma(T), \qquad (4.1.1)$$

whose image is of finite index.

This leads to the main result of this section.

PROPOSITION 4.1.3. *The morphism*

$$\mathcal{M}_\varphi(S, \varphi) \to \mathcal{M}(T) \qquad (4.1.2)$$

defined by associating to a point of $\mathcal{M}_\varphi(S, \varphi)$ (which is the isomorphism class of a Riemann surface X admitting φ as an automorphism) the quotient $X/\varphi \in \mathcal{M}(T)$ is a covering map of finite degree.

PROOF. Let X be a point where $\mathcal{M}_\varphi(S, \varphi)$ is not normal, and suppose that $X \mapsto Y \in \mathcal{M}(T)$ under (4.1.2). We lift the morphism (4.1.2) to $\widetilde{\mathcal{M}}_\varphi(S, \varphi)$ by sending all the points lying above X to Y, and we show that the morphism

$$\widetilde{\mathcal{M}}_\varphi(S, \varphi) \to \mathcal{M}(T) \qquad (4.1.3)$$

is a covering map of finite degree. It is enough to show that each irreducible component of $M_\varphi(S, \varphi)$ gives a finite covering. For this, we use Theorem 4.1.1 and Proposition 4.1.2. Firstly, since $\langle\varphi\rangle$ fixes every point of $\mathcal{T}(S)_\varphi$, the action of $\mathrm{Norm}_{\Gamma_\varphi(S)}(\varphi)$ factors through the quotient group $\mathrm{Norm}_{\Gamma_\varphi(S)}(\varphi)/\langle\varphi\rangle$, and there is a canonical one-to-one correspondence

$$\mathcal{T}(S)_\varphi/\mathrm{Norm}_{\Gamma_\varphi(S)}(\varphi) \leftrightarrow \mathcal{T}(S)_\varphi/(\mathrm{Norm}_{\Gamma_\varphi(S)}(\varphi)/\langle\varphi\rangle) \qquad (4.1.4)$$

(the difference between the two spaces is hidden in the orbifold structure due to the action of $\langle \varphi \rangle$ fixing each point).

Now, using Theorem 4.1.1, (4.1.4) and Proposition 4.1.2, we have a natural sequence of morphisms

$$\mathfrak{T}(S)_\varphi \to \mathfrak{T}(S)_\varphi/\mathrm{Norm}_{\Gamma_\varphi(S)}(\varphi) \simeq \widetilde{\mathcal{M}}(S,\varphi) \hookleftarrow \mathfrak{T}(S,\varphi)/(\mathrm{Norm}_{\Gamma_\varphi(S)}(\varphi)/\langle\varphi\rangle)$$

$$\simeq \mathfrak{T}(T)/(\mathrm{Norm}_{\Gamma_\varphi(S)}(\varphi)/\langle\varphi\rangle) \to \mathfrak{T}(T)/\Gamma(T) \simeq \mathcal{M}(T). \quad (4.1.5)$$

This shows that each irreducible component of $\mathcal{M}_\varphi(S,\varphi)$ is a cover of $\mathcal{M}(T)$, and the finiteness of its degree follows from the fact that $\mathrm{Norm}_{\Gamma_\varphi(S)}(\varphi)/\langle\varphi\rangle$ is of finite index in $\Gamma(T)$ (Proposition 4.1.2). □

The goal of the next section, apart from proving Proposition 4.1.2, is to give two conditions on the fundamental groups which are equivalent to two conditions on the covering map (4.1.3), which taken together mean that (4.1.3) is "as close as possible to an isomorphism", given the difference in orbifold structure due to the group $\langle\varphi\rangle$.

4.2. Special homomorphisms between mapping class groups. In this section, we study the *special homomorphisms* between mapping class groups coming from the special morphisms

$$\mathcal{M}_\varphi(S,\varphi) \to \mathcal{M}(T),$$

or rather, from their liftings

$$\widetilde{\mathcal{M}}_\varphi(S,\varphi) \to \mathcal{M}(T)$$

introduced in the previous section. By Theorem 4.1.1, the orbifold fundamental group of each irreducible component of $\widetilde{\mathcal{M}}_\varphi(S,\varphi)$ is $\mathrm{Norm}_{\Gamma_\varphi(S)}(\varphi)$, so the morphism (4.1.3) gives rise to a homomorphism

$$\mathrm{Norm}_{\Gamma_\varphi(S)}(\varphi) \to \Gamma(T).$$

The goal of this section is to study this morphism (using the identification of the mapping class groups with outer automorphism groups of fundamental groups), in order to prove that not only does it factor through $\mathrm{Norm}_{\Gamma_\varphi(S)}(\varphi)/\langle\varphi\rangle$, but that the induced morphism

$$\mathrm{Norm}_{\Gamma_\varphi(S)}(\varphi)/\langle\varphi\rangle \to \Gamma(T)$$

is an injective homomorphism with image of finite index. This is the final result of this section (Proposition 4.2.7) given as Proposition 4.1.2 above. In order to prove it, we characterize the normalizer of φ in $\Gamma_\varphi(S)$ as the set of inertia-preserving outer automorphisms of $\pi_1(S)$ which extend to $\pi_1(T)$ (see Lemma 4.2.3), and work in that context.

Let S be a topological surface of type (g,n), and write $\Gamma([S]) = \Gamma_{g,[n]}$ for the mapping class group of S permuting points and $\Gamma(S) = \Gamma_{g,n}$ for the pure mapping class group. Let φ be a finite-order element of $\Gamma([S])$, and set $T = S/\varphi$;

we assume that all branch points of this cover (and their preimages) are marked points. Let g' denote the genus of T and n' the number of marked points; the fundamental group of T is given by generators and relations as

$$\pi_1(T) = \Big\langle a_1, b_1, \ldots, a_{g'}, b_{g'}, c_1, \ldots, c_{n'} \Big| \prod_{i=1}^{g'}(a_i, b_i)c_1 \cdots c_{n'} = 1 \Big\rangle. \qquad (4.2.1)$$

Recall that the group of inertia-preserving automorphisms of $\pi_1(T)$, denoted

$$\mathrm{Aut}^*([\pi_1(T)]),$$

is defined to be the group of automorphisms

$$\big\{\psi \in \mathrm{Aut}\big(\pi_1(T)\big) \mid \exists \sigma \in S_n \text{ such that } \psi(c_i) \sim c_{\sigma(i)} \text{ for } 1 \leq i \leq n'\big\},$$

where $\sim$ means "is conjugate to". We saw (see Theorem 2.2.1) that

$$\Gamma([T]) = \mathrm{Out}^*([\pi_1(T)]), \qquad (4.2.2)$$

the quotient of $\mathrm{Aut}^*([\pi_1(T)])$ by $\mathrm{Inn}(\pi_1(T))$. As in Section 2.2, we have a natural surjective homomorphism

$$\mathrm{Aut}^*([\pi_1(T)]) \to S_n$$
$$\psi \mapsto \sigma, \qquad (4.2.3)$$

which passes to $\mathrm{Out}^*([\pi_1(T)])$.

Because $T = S/\varphi$, we have the exact sequence

$$1 \to \pi_1(S) \to \pi_1(T) \to \langle\varphi\rangle \to 1. \qquad (4.2.4)$$

The quotient $\langle\varphi\rangle$ is an outer automorphism group of $\pi_1(S)$, which is naturally identified with the subgroup $\langle\varphi\rangle$ of $\Gamma([S]) = \mathrm{Out}^*([\pi_1(S)])$, see the following lemma.

LEMMA 4.2.1. *The homomorphism*

$$\pi_1(T) \hookrightarrow \mathrm{Aut}([\pi_1(S)])$$
$$t \mapsto \mathrm{inn}(t)|_{\pi_1(S)} \qquad (4.2.5)$$

is injective, and its image lies in $\mathrm{Aut}^*([\pi_1(S)])$.

PROOF. Let $\tilde{\varphi}$ be a lifting of φ to $\pi_1(T)$ in (4.2.4). If r is the order of φ, then $\tilde{\varphi}^r \in \pi_1(S)$, so every element of $\pi_1(T)$ can be written $s\tilde{\varphi}^m$ with $s \in \pi_1(S)$ and $0 \leq m \leq r - 1$. Suppose that $s\tilde{\varphi}^m$ maps to the identity in (4.2.5). If $m = 0$, then $s = 1$ because the restriction of (4.2.5) to $\pi_1(S)$ is injective, since $\pi_1(S)$ is center-free. If $1 \leq m \leq r - 1$, then the image of $s\tilde{\varphi}^m$ under the induced homomorphism

$$\pi_1(T) \to \mathrm{Aut}([\pi_1(S)])/\mathrm{Inn}(\pi_1(S))$$

is equal to φ^m, which is nontrivial since $1 \leq m \leq r - 1$, so $s\tilde{\varphi}^m$ cannot lie in the kernel of (4.2.5). To see that the image of (4.2.5) lies in $\mathrm{Aut}^*([\pi_1(S)])$, note that $\pi_1(T)$ is generated by $\pi_1(S)$ and $\tilde{\varphi}$, and the image of $\pi_1(S)$ certainly lies in $\mathrm{Aut}^*([\pi_1(S)])$. But so does the image of $\tilde{\varphi}$, since its reduction modulo $\pi_1(S)$ (identified with $\mathrm{Inn}(\pi_1(S))$) is equal to φ, which is in $\Gamma([S]) = \mathrm{Out}^*([\pi_1(S)])$. $\square$

LEMMA 4.2.2. *Let φ be an element of finite order in $\Gamma([S])$ and $T = S/\varphi$ as usual, so $\pi_1(S) \subset \pi_1(T)$ as in (4.2.4). If an automorphism of $\pi_1(S)$ extends to an automorphism of $\pi_1(T)$, then it extends uniquely.*

PROOF. Let r denote the order of φ, and let $\tilde{\psi}$ be an automorphism of $\pi_1(S)$ extending to $\pi_1(T)$. Suppose that Ψ and Φ are two extensions of $\tilde{\psi}$ to automorphisms of $\pi_1(T)$, and set $\chi = \Psi \circ \Phi^{-1}$, so that χ is an automorphism of $\pi_1(T)$ which restricts to the trivial automorphism on the subgroup $\pi_1(S)$. Let $\tilde{\varphi}$ be a lifting of φ to $\pi_1(T)$, so $\pi_1(T)$ is generated by $\pi_1(S)$ and $\tilde{\varphi}$. Then $\chi(\tilde{\varphi}) = s\tilde{\varphi}^m$ for some $s \in \pi_1(S)$ and some m with $0 \le m \le r - 1$. For all $s' \in \pi_1(S)$, we have $\tilde{\varphi}^{-1}s'\tilde{\varphi} \in \pi_1(S)$ since $\pi_1(S)$ is normal in $\pi_1(T)$, but by assumption, χ fixes elements of $\pi_1(S)$, so

$$\tilde{\varphi}^{-1}s'\tilde{\varphi} = \chi(\tilde{\varphi}^{-1}s'\tilde{\varphi}) = \chi(\tilde{\varphi}^{-1})s'\chi(\tilde{\varphi}) = \tilde{\varphi}^{-m}s^{-1}s's\tilde{\varphi}^m.$$

Thus, $s\tilde{\varphi}^{m-1}$ commutes with s' for all $s' \in \pi_1(S)$, so $s\tilde{\varphi}^{m-1}$ lies in the kernel of the homomorphism (4.2.5). Thus by Lemma 4.2.1, $s\tilde{\varphi}^{m-1} = 1$, so $s = 1$ and $m - 1 = 0$. This means that $\chi(\tilde{\varphi}) = \tilde{\varphi}$, i.e., χ is the identity on all of $\pi_1(T)$. $\square$

LEMMA 4.2.3. *Let φ be a finite-order element of $\Gamma([S])$. Then the group $\mathrm{Norm}_{\Gamma([S])}(\varphi)$ is exactly the subgroup of elements ψ of $\Gamma([S]) = \mathrm{Out}^*([\pi_1(s)])$ whose liftings to automorphisms $\tilde{\psi} \in \mathrm{Aut}^*([\pi_1(S)])$ extend to automorphisms of $\pi_1(T)$.*

PROOF. Let r be the order of φ in $\Gamma([S])$, and let $\tilde{\varphi}$ be a lifting of φ to $\pi_1(T)$ in (4.2.4); we consider $\tilde{\varphi}$ in $\mathrm{Aut}^*([\pi_1(S)])$ via the inclusion

$$\pi_1(T) \xrightarrow{\sim} \mathrm{Inn}(\pi_1(T)) \subset \mathrm{Aut}^*([\pi_1(S)])$$

of Lemma 4.2.1. Let $\psi \in \mathrm{Norm}_{\Gamma([S])}(\varphi)$, so that $\psi\varphi\psi^{-1} = \varphi^m$ for some m. Let $\tilde{\psi}$ be an arbitrary lifting of ψ to $\mathrm{Aut}^*([\pi_1(S)])$; then there exists $s \in \pi_1(S)$ such that

$$\tilde{\psi}\tilde{\varphi}\tilde{\psi}^{-1} = s\tilde{\varphi}^m, \tag{4.2.6}$$

where s is identified with $\mathrm{inn}(s)$ in $\mathrm{Aut}^*([\pi_1(S)])$.

The action of the automorphism $\tilde{\psi}$ on $\pi_1(S)$ is recovered by conjugating the copy of $\pi_1(S)$ inside $\mathrm{Aut}^*([\pi_1(S)])$, identified with $\mathrm{Inn}(\pi_1(S))$, by $\tilde{\psi}$ inside $\mathrm{Aut}^*([\pi_1(S)])$. Indeed, for every $s, s' \in \pi_1(S)$, we have

$$(\tilde{\psi}\,\mathrm{inn}(s)\tilde{\psi}^{-1})(s') = (\tilde{\psi}\,\mathrm{inn}(s))(\tilde{\psi}^{-1}(s')) = \tilde{\psi}(s\tilde{\psi}^{-1}(s')s^{-1})$$
$$= \tilde{\psi}(s)s'\tilde{\psi}(s)^{-1} = \mathrm{inn}(\tilde{\psi}(s))(s').$$

Furthermore, (4.2.6) shows that the automorphism $\tilde{\psi}$ of $\pi_1(S)$ (given by conjugation by $\tilde{\psi}$ inside $\mathrm{Aut}^*([\pi_1(S)])$) preserves $\pi_1(T)$ since $\pi_1(T)$ is generated by $\pi_1(S)$ and $\tilde{\varphi}$, so that it gives an automorphism of $\pi_1(T)$. This proves one direction of the lemma.

Now, we take $\psi \in \Gamma([S]) = \text{Out}^*([\pi_1(S)])$ such that any lifting $\tilde{\psi}$ of ψ to an automorphism of $\pi_1(S)$ extends to an automorphism of $\pi_1(T)$ (note that if one lifting extends then they all do since they differ by elements of $\text{Inn}(\pi_1(S))$). Then since $\pi_1(T)$ is generated by $\pi_1(S)$ and a lifting $\tilde{\varphi}$ of φ to $\pi_1(T)$, we must have $\tilde{\psi}(\tilde{\varphi}) = \tilde{\psi}\tilde{\varphi}\tilde{\psi}^{-1} = s\tilde{\varphi}^m$ for some m and some $s \in \pi_1(S)$. Thus $\psi\varphi\psi^{-1} = \varphi^m$ in $\Gamma([S])$. Since conjugation by $\tilde{\psi}$ is an automorphism, we find that m must be relatively prime to r, so ψ normalizes φ in $\Gamma([S])$. $\square$

DEFINITION. Let $\text{Aut}^*([S/T])$ denote the subgroup of $\text{Aut}^*([\pi_1(T)])$ consisting of elements which preserve the subgroup $\pi_1(S) \subset \pi_1(T)$. This subgroup is of finite index in $\text{Aut}^*(\pi_1(T))$, since $\pi_1(S)$ is of finite index in the finitely generated group $\pi_1(T)$. Restriction from $\pi_1(T)$ to $\pi_1(S)$ gives a homomorphism

$$\text{Aut}^*([S/T]) \to \text{Aut}([\pi_1(S)]).$$

Indeed, we easily see that this restriction actually gives a homomorphism

$$\text{Aut}^*([S/T]) \to \text{Aut}^*([\pi_1(S)]), \tag{4.2.7}$$

since because S is a topological cover of T, any loop surrounding a marked point on S must be a conjugate of a power of a loop surrounding the corresponding marked point of T, so the restriction of an element of $\text{Aut}^*([S/T])$ must then send it to a conjugate of another such loop.

LEMMA 4.2.4. *The homomorphism*

$$\text{Aut}^*([S/T]) \to \text{Aut}^*([\pi_1(S)])$$

of (4.2.7) *is injective.*

PROOF. If $\psi \in \text{Aut}^*([S/T])$ lies in the kernel, it is an element of $\text{Aut}^*([\pi_1(T)])$ that acts like the identity on the subgroup $\pi_1(S)$. Lemma 4.2.2 then shows that ψ is the identity on $\pi_1(T)$. $\square$

LEMMA 4.2.5. $\text{Aut}^*([S/T])$ *can be identified with the subgroup of* $\text{Aut}^*([\pi_1(S)])$ *of automorphisms of $\pi_1(S)$ which extend to $\pi_1(T)$.*

PROOF. The previous lemma shows that we can consider $\text{Aut}^*([S/T])$ as a subgroup of $\text{Aut}^*([\pi_1(S)])$, and clearly every element of this subgroup is an automorphism of $\pi_1(S)$ which extends to $\pi_1(T)$. Therefore, we only need to show the converse, i.e., that if $\psi \in \text{Aut}^*([\pi_1(S)])$ is an automorphism of $\pi_1(S)$ which extends to an automorphism of $\pi_1(T)$, then it lies in $\text{Aut}^*([S/T])$. This means that we must show that if an element $\psi \in \text{Aut}^*([\pi_1(S)])$ extends to $\pi_1(T)$ at all, then it extends to an inertia-preserving automorphism of $\pi_1(T)$, i.e., it gives an element of $\text{Aut}^*([\pi_1(T)])$.

By Lemma 4.2.2, if ψ extends to an automorphism of $\pi_1(T)$, then it does so uniquely. Let Ψ be the unique extension of ψ to $\pi_1(T)$; we need to show that it lies in $\mathrm{Aut}^*([\pi_1(T)])$.

Recall that we assume that all branch points of the cover $S \to T$ are marked points on T, images of marked points of S. Let the n marked points of S be labeled $(x_1, \ldots, x_n)$ and the n' marked points of T be labeled $(y_1, \ldots, y_{n'})$. Let $c_1, \ldots, c_{n'}$ be loops around $y_1, \ldots, y_{n'}$ forming part of a generating system of $\pi_1(T)$ as in (4.2.1), and let $e_1, \ldots, e_n$ be analogous loops around $x_1, \ldots, x_n$ on S. We need to show that for every $k \in \{1, \ldots, n'\}$, $\Psi(c_k)$ is a conjugate of c_l for some $l \in \{1, \ldots, n'\}$. Choose k, and choose an $i \in \{1, \ldots, n\}$ such that the marked point $x_i \in S$ lies over y_k. The restriction of Ψ to the subgroup $\pi_1(S)$ is equal to $\psi \in \mathrm{Aut}^*([\pi_1(S)])$, so we have $\Psi(e_i) = ae_ja^{-1}$ for some $j \in \{1, \ldots, n\}$. Let y_l be the image in T of the marked point $x_j \in S$. Then the loop $e_i \in \pi_1(S)$, seen as an element of $\pi_1(T)$ via $\pi_1(S) \subset \pi_1(T)$, is a conjugate of a power of c_k, say $e_i = bc_k^r b^{-1}$, where r is the ramification index of y_k. Similarly, the loop e_j is a conjugate of a power of c_l, say $e_j = dc_l^s d^{-1}$ where s is the ramification index of y_l. Thus

$$\Psi(e_i) = ae_ja^{-1}$$

can be written as

$$\Psi(bc_k^r b^{-1}) = \Psi(b)\Psi(c_k)^r \Psi(b)^{-1} = adc_l^s d^{-1} a^{-1},$$

so

$$\Psi(c_k)^r = \Psi(b)^{-1}adc_l^s d^{-1} a^{-1}\Psi(b).$$

Now, $\pi_1(T)$ is a free group on finitely many generators, with no torsion, and this means that if x and y are such that $x^r = y^s$ in this group, then there exists an element z such that $x = z^a$ and $y = z^b$ with $ra = sb$. Taking $x = \Psi(c_k)$ and $y = \Psi(b)^{-1}adc_ld^{-1}a^{-1}\Psi(b)$, this shows that there exists $z \in \pi_1(T)$ and integers a, b with $ra = sb$, such that

$$\Psi(c_k) = z^a \quad \text{and} \quad \Psi(b)^{-1}adc_ld^{-1}a^{-1}\Psi(b) = z^b.$$

The right hand element is a simple loop around the point y_l of T, and as (the conjugate of) a fundamental generator of $\pi_1(T)$, it cannot be a nontrivial power of any z, so $b = 1$. Thus

$$\Psi(c_k) \sim c_l^a.$$

Let Φ be the inverse of Ψ, and set $w = \Phi(c_l)$. Then $\Phi(c_l^a) = w^a \sim c_k$, which means that c_k is the a-th power of a conjugate of w, so since again c_k is a fundamental generator of $\pi_1(T)$, we must have $a = 1$ and $\Psi(c_k) \sim c_l$. This shows that $\Psi \in \mathrm{Aut}^*([\pi_1(T)])$. $\square$

LEMMA 4.2.6. *Let* $T = S/\varphi$ *for a finite-order element* $\varphi \in \Gamma([S])$ *as usual. Set*

$$\begin{cases} A = \mathrm{Aut}^*([S/T])/\mathrm{Inn}(\pi_1(S)) \subset \Gamma([S]) \\ B = \mathrm{Aut}^*([S/T])/\mathrm{Inn}(\pi_1(T)) \subset \Gamma([T]). \end{cases}$$

Then $\varphi \in A$ and $\langle\varphi\rangle$ is the kernel of the natural surjection

$$g : A \to B. \qquad (4.2.8)$$

REMARK. By Lemmas 4.2.3 and 4.2.5, $A = \mathrm{Norm}_{\Gamma([S])}(\varphi)$.

PROOF. Let $\tilde{\varphi}$ be a lifting of φ to $\pi_1(T)$. Then by Lemmas 4.2.1 and 4.2.4, we have inclusions

$$\pi_1(T) \subset \mathrm{Aut}^*([S/T]) \subset \mathrm{Aut}^*([\pi_1(S)]),$$

and $\tilde{\varphi} \in \pi_1(T)$, so $\varphi \in \mathrm{Aut}^*([S/T])/\mathrm{Inn}(\pi_1(S)) = A \subset \Gamma([S])$. Now, let ψ be in the kernel of g, and let $\tilde{\psi}$ be a lifting of ψ to $\mathrm{Aut}^*([S/T])$. Then since ψ is in the kernel of g, we must have $\tilde{\psi} = \mathrm{inn}(t)$ for some $t \in \pi_1(T)$, and we can write $t = s\tilde{\varphi}^m$ for some $0 \le m \le r-1$ and some $s \in \pi_1(S)$, i.e.

$$\mathrm{inn}(t) = \tilde{\psi} = \mathrm{inn}(s\tilde{\varphi}^m) \quad \mathrm{in} \quad \mathrm{Inn}(\pi_1(T)) \subset \mathrm{Aut}^*([S/T]) \subset \mathrm{Aut}^*([\pi_1(S)]).$$
$$(4.2.9)$$

Therefore,

$$\psi = \varphi^m \quad \mathrm{in} \quad A = \mathrm{Aut}^*([S/T])/\mathrm{Inn}(\pi_1(S)) \subset \mathrm{Out}^*([\pi_1(S)]) = \Gamma([S]).$$

This shows that the kernel of g lies inside the group $\langle\varphi\rangle \subset A$. To show that it is equal to $\langle\varphi\rangle$, it suffices to check that φ itself lies in the kernel. But that is of course true, since $\tilde{\varphi} \in \pi_1(T) \subset \mathrm{Aut}^*([S/T])$, so we have

$$\mathrm{Aut}^*([S/T]) \to A = \mathrm{Aut}^*([S/T])/\mathrm{Inn}(\pi_1(S)) \to B = \mathrm{Aut}^*([S/T])/\mathrm{Inn}(\pi_1(T))$$

$$\tilde{\varphi} \quad \mapsto \quad \varphi \quad \mapsto \quad 1.$$

This proves the lemma. □

We introduce the following notation for the group B of Lemma 4.2.6. Set

$$\Gamma_{[S/T]} = \mathrm{Aut}^*([S/T])/\mathrm{Inn}(\pi_1(T)) \subset \mathrm{Out}^*([\pi_1(T)]) = \Gamma([T]), \qquad (4.2.10)$$

and

$$\Gamma_{S/T} = \mathrm{Aut}^*(S/T)/\mathrm{Inn}(\pi_1(T)) = \Gamma_{[S/T]} \cap \Gamma(T) \subset \Gamma(T). \qquad (4.2.11)$$

PROPOSITION 4.2.7. *Let G_φ be the subgroup of S_n generated by the disjoint cycles of the permutation $[\varphi]$ associated to φ, and let $\Gamma_\varphi(S)$ be the subgroup of $\Gamma([S])$ which is the preimage of G_φ under the surjection $\Gamma([S]) \to S_n$. We have the isomorphisms*

$$\mathrm{Norm}_{\Gamma([S])}(\varphi)/\langle\varphi\rangle \xrightarrow{\sim} \Gamma_{[S/T]} \subset \mathrm{Out}^*([\pi_1(T)]) = \Gamma([T]), \qquad (4.2.12)$$

$$\mathrm{Norm}_{\Gamma_\varphi(S)}(\varphi)/\langle\varphi\rangle \xrightarrow{\sim} \Gamma_{S/T} \subset \mathrm{Out}^*(\pi_1(T)) = \Gamma(T). \qquad (4.2.13)$$

Furthermore, the subgroups $\Gamma_{[S/T]}$ and $\Gamma_{S/T}$ are of finite index in $\Gamma([T])$ and $\Gamma(T)$, respectively.

PROOF. By Lemmas 4.2.3 and 4.2.5, the subgroup $A \subset \Gamma([S])$ is equal to $\mathrm{Norm}_{\Gamma([S])}(\varphi)$, and by Lemma 4.2.6, since $\Gamma_{[S/T]}$ is the subgroup B, we obtain an injective homomorphism

$$\overline{g} : A/\langle\varphi\rangle \hookrightarrow B$$

which is exactly (4.2.12). For the second statement, we first observe that the restriction of the injection $\overline{g}$ to the subgroup $\mathrm{Norm}_{\Gamma_\varphi(S)}(\varphi)$ produces an image which lies inside $\Gamma_{[S/T]}$, but also inside $\Gamma(T)$. Indeed, let $\psi \in \mathrm{Norm}_{\Gamma_\varphi(S)}(\varphi)$ and let $\tilde{\psi}$ be a lifting of ψ to $\mathrm{Aut}^*([\pi_1(S)])$. Then the permutation associated to ψ (by its action on the loops around marked points of S) lies in G_φ, which means that $\tilde{\psi}$ permutes only loops in $\pi_1(S)$ around points of S which map to a single point in T. Thus the extension of $\tilde{\psi}$ to an automorphism of $\pi_1(T)$ lies in $\mathrm{Aut}^*(\pi_1(T))$, so the image of $\tilde{\psi}$ in $\Gamma([T])$ lies in $\Gamma(T)$.

This shows that the image of $\mathrm{Norm}_{\Gamma_\varphi(S)}(\varphi)$ under (4.2.13) lies inside $\Gamma_{S/T}$. But if ϕ is an element of $\Gamma_{S/T} = \mathrm{Aut}^*(S/T)/\mathrm{Inn}(\pi_1(T))$ and Φ is a lifting of ϕ to $\mathrm{Aut}^*(S/T)$, then Φ extends to $\pi_1(T)$, so the image of Φ in $\mathrm{Aut}^*(S/T)/\mathrm{Inn}(\pi_1(S))$ lies in $\mathrm{Norm}_{\Gamma([S])}(\varphi)$. Since its associated permutation as an inertia-preserving automorphism of T is trivial, its associated permutation as an inertia-preserving automorphism of $\pi_1(S)$ must lie in G_φ, so it in fact lies in $\mathrm{Norm}_{\Gamma_\varphi(S)}(\varphi)$, which proves (4.2.13).

For the last statement, recall that $\mathrm{Aut}^*([S/T])$ has finite index in $\mathrm{Aut}^*([\pi_1(T)])$ since $\pi_1(S)$ is of finite index in $\pi_1(T)$ (see the definition of $\mathrm{Aut}^*([S/T])$), and $\mathrm{Aut}^*(S/T)$ is of finite index in $\mathrm{Aut}^*(\pi_1(T))$. Thus $\mathrm{Aut}^*([S/T])/\mathrm{Inn}(\pi_1(T)) = \Gamma_{[S/T]}$ is of finite index in

$$\mathrm{Aut}^*([\pi_1(T)])/\mathrm{Inn}(\pi_1(T)) = \Gamma([T])$$

and $\mathrm{Aut}^*(S/T)/\mathrm{Inn}(\pi_1(T))$ is of finite index in $\mathrm{Aut}^*(\pi_1(T))/\mathrm{Inn}(\pi_1(T)) = \Gamma(T)$. This concludes the proof. $\qquad\square$

DEFINITION. (1) We say that a finite-order element $\varphi \in \Gamma([S])$ satisfies the *surjectivity condition* if $\Gamma_{S/T} = \Gamma(T)$, i.e., every element of $\mathrm{Aut}^*(\pi_1(T))$ preserves the subgroup $\pi_1(S)$, in other words the homomorphism (4.2.13)

$$\mathrm{Norm}_{\Gamma_\varphi(S)}(\varphi)/\langle\varphi\rangle \to \Gamma(T)$$

is surjective. In geometric terms, this means that the covering map

$$\text{irreducible component of } \widetilde{\mathcal{M}}_\varphi(S,\varphi) \to \mathcal{M}(T)$$

of (4.1.3) is one-to-one, consisting only in forgetting the orbifold structure of $\widetilde{\mathcal{M}}_\varphi(S,\varphi)$ due to the action of φ.

(2) We say that φ satisfies the *splitting condition* if the surjection

$$\mathrm{Norm}_{\Gamma_\varphi(S)}(\varphi) \to \mathrm{Norm}_{\Gamma_\varphi(S)}(\varphi)/\langle\varphi\rangle \simeq \Gamma_{S/T}$$

splits; in other words, if we have a semi-direct product

$$\mathrm{Norm}_{\Gamma_\varphi(S)}(\varphi) \simeq \langle\varphi\rangle \rtimes \Gamma_{S/T}. \qquad (4.2.14)$$

Whenever we have $T = S/\varphi$, where $\varphi \in \Gamma([S])$ satisfies both the surjectivity and the splitting conditions, we have

$$\langle \varphi \rangle \rtimes \Gamma(T) \simeq \operatorname{Norm}_{\Gamma_\varphi(S)}(\varphi) \subset \Gamma([S]);$$

for each choice of splitting, we can thus define (non-canonical) *special homomorphisms*

$$\Gamma(T) \to \Gamma([S]). \tag{4.2.15}$$

In geometric terms, this means that $\widetilde{\mathcal{M}}_\varphi(S, \varphi)$ is as close to $\mathcal{M}(T)$ as possible, in the following sense. The morphism (4.1.3) is given by the "forgetting the action of $\langle \varphi \rangle$" map

$$\mathcal{T}(T)/\operatorname{Norm}_{\Gamma_\varphi(S)}(\varphi) \to \mathcal{T}(T)/\left(\operatorname{Norm}_{\Gamma_\varphi(S)}(\varphi)/\langle \varphi \rangle\right)$$

of (4.1.5). A point of the left-hand space can be represented by a pair (t, h), where t is the image of a point in $\mathcal{T}(T)$ and $h \in \operatorname{Norm}_{\Gamma_\varphi(S)}(\varphi)$, whereas a point of the right-hand space can be represented as $(t, \bar{h})$ where $\bar{h} \in \operatorname{Norm}_{\Gamma_\varphi(S)}(\varphi)/\langle \varphi \rangle$. With the splitting condition, the action of $\langle \varphi \rangle$ can be defined on $\mathcal{M}(T)$, which is the right-hand space, by $\varphi(t, \bar{h}) = (t, \varphi \cdot \bar{h})$, where $\varphi \cdot \bar{h} \in \langle \varphi \rangle \rtimes \operatorname{Norm}_{\Gamma_\varphi(S)}(\varphi)/\langle \varphi \rangle \simeq \operatorname{Norm}_{\Gamma_\varphi(S)}(\varphi)$.

Taken together, the splitting and surjectivity conditions mean that (4.1.3) is as close as possible to an isomorphism with each irreducible component, given the difference of orbifold structures. In Section 4.3 we will show that when S is of genus zero, every finite-order element $\varphi \in \Gamma([S])$ satisfies both the splitting and the surjectivity conditions.

4.3. The case of genus zero. In the case of the genus zero moduli spaces, the results of the previous section become simpler and more explicit; not only are the splitting condition always satisfied (Theorem 4.3.1) as well as the surjectivity condition (Theorem 4.3.2), but the semi-direct product (4.2.14) is always a direct product

$$\operatorname{Norm}_{\Gamma_\varphi(S)}(\varphi) \simeq \langle \varphi \rangle \times \Gamma(T),$$

and an algorithm allows us to compute the splitting morphism $\Gamma(T) \to \Gamma([S])$ explicitly (proof of Theorem 4.3.2), so that we actually obtain a method for computing normalizers of finite-order elements in genus zero mapping class groups.

To begin with, we recall from Section 3.5 that if S is a sphere with n marked points, then a finite-order element of the mapping class group $\Gamma([S])$ is the class of a rotation around an axis (see Figure 4.1 on the next page).

In general, the north and south poles (fixed points of the rotation) may or may not be marked points, but in this article we restrict ourselves to the case where all branch points of T are marked, so we only consider the case where they are both marked. Thus, the permutation associated to φ is always of the form $c_1 \cdots c_k$, where the c_i are disjoint cycles of length j such that $jk = n - 2$.

The *splitting condition* introduced in the previous section has the following strong form in genus zero.

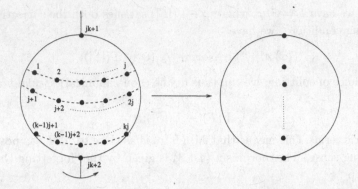

Figure 4.1. A finite-order diffeomorphism in genus zero.

THEOREM 4.3.1. *Let S be a sphere with n marked points, and φ a finite-order diffeomorphism of S whose fixed points are marked. Let $T = S/\varphi$, let $[\varphi]$ denote the permutation associated to φ, and let $G_\varphi \subset S_n$ be the group generated by the disjoint cycles of $[\varphi]$. Let $H_\varphi \subset G_\varphi$ be generated by any choice of all but one of the disjoint cycles of $[\varphi]$ ($H_\varphi = \{\mathrm{id}\}$ if $[\varphi]$ is a single cycle). Then we have the direct product isomorphism*

$$\mathrm{Norm}_{\Gamma_\varphi(S)}(\varphi) = \mathrm{Norm}_{\Gamma_{H_\varphi}(S)}(\varphi) \times \langle \varphi \rangle.$$

PROOF. Take a subgroup $H_\varphi \subset G_\varphi$ as in the statement of the theorem, and consider the two groups

$$\mathrm{Norm}_{\Gamma_{H_\varphi}(S)}(\varphi) \subset \mathrm{Norm}_{\Gamma_\varphi(S)}(\varphi).$$

We will show that

- the first group does not contain any nontrivial power of φ;
- the second group is generated by the first group and the element φ;
- the first group is normal in the second.

These three facts taken together imply that we have a semi-direct product

$$\mathrm{Norm}_{\Gamma_\varphi(S)}(\varphi) = \mathrm{Norm}_{\Gamma_{H_\varphi}(S)}(\varphi) \rtimes \langle \varphi \rangle. \tag{4.3.1}$$

But $\langle \varphi \rangle$ is also normal inside $\mathrm{Norm}_{\Gamma_\varphi(S)}(\varphi)$, naturally! Now, let a lie in the normal subgroup $\mathrm{Norm}_{\Gamma_{H_\varphi}(S)}(\varphi)$; then $\varphi^{-1}a\varphi = b \in \mathrm{Norm}_{\Gamma_{H_\varphi}(S)}(\varphi)$. But we also have $a\varphi a^{-1} = \varphi^m$, since a normalises φ. Thus

$$\varphi^{-1}a\varphi a^{-1} = ba^{-1} = \varphi^{-1}\varphi^m,$$

so this element lies in the intersection of the two groups in (4.3.1), so it is equal to 1, and we have $a = b$ and $m = -1$. Thus a commutes with φ, so all of $\mathrm{Norm}_{\Gamma_{H_\varphi}(S)}(\varphi)$ commutes with φ; in fact this normaliser is a centraliser, and the semi-direct product (4.3.1) is a direct product

$$\mathrm{Norm}_{\Gamma_\varphi(S)}(\varphi) = \mathrm{Norm}_{\Gamma_{H_\varphi}(S)}(\varphi) \times \langle \varphi \rangle \tag{4.3.2}$$

as in the statement of the theorem. Note that this implies that in fact

$$\text{Norm}_{\Gamma_\varphi(S)}(\varphi) = \text{Cent}_{\Gamma_\varphi(S)}(\varphi). \tag{4.3.3}$$

It remains only to prove the three points above. To see that $\text{Norm}_{\Gamma_{H_\varphi}(S)}(\varphi)$ does not contain any power of φ, it suffices to consider the permutations associated to elements of this group; by definition they all lie in H_φ. Now, write $[\varphi] = c_1 \cdots c_k$ as a product of k disjoint cycles c_i each of length j, and assume that H_φ is generated by $c_1, \ldots, c_{k-1}$. Let $\{m_1, \ldots, m_{jk}\} \subset \{1, \ldots, n\}$ be the set of numbers not fixed by the permutation $[\varphi] \in S_n$, i.e., those occurring in the cycles $c_1, \ldots, c_k$. Then none of the m_i are left fixed by any nontrivial power of $[\varphi]$, since the lengths of all the cycles c_i are equal; on the other hand every element of H_φ fixes the numbers m_i occurring in the cycle c_k. This shows that no nontrivial power of $[\varphi]$ lies in H_φ.

For the second point, let $g \in \text{Norm}_{\Gamma_\varphi(S)}(\varphi)$. If its associated permutation $[g]$ lies in H_φ, then by definition $g \in \text{Norm}_{\Gamma_{H_\varphi}(S)}(\varphi)$. If $[g]$ is not in H_φ, then since it is in G_φ, there exists an integer m such that $[g][\varphi]^m \in H_\varphi$, and since $g\varphi^m$ normalises φ, we see that $g\varphi^m \in \text{Norm}_{\Gamma_{H_\varphi}(S)}(\varphi)$. This proves the second point. For the third point, it suffices to check that conjugation by φ preserves the subgroup $\text{Norm}_{\Gamma_{H_\varphi}(S)}(\varphi)$. Let $h \in \text{Norm}_{\Gamma_{H_\varphi}(S)}(\varphi)$; we need to show that $g = \varphi h \varphi^{-1} \in \text{Norm}_{\Gamma_{H_\varphi}(S)}(\varphi)$. But g certainly normalises φ, and its permutation is equal to the permutation of h since $[\varphi]$ commutes with $[h]$ (the group G_φ is abelian). This completes the proof of Theorem 4.3.1. $\qquad\square$

Next, we show that the surjectivity condition always holds in genus zero.

THEOREM 4.3.2. *Let S be a topological sphere with n marked points, and φ a finite-order diffeomorphism of S whose fixed points are marked. Let $T = S/\varphi$. Then the homomorphism*

$$\text{Norm}_{\Gamma_\varphi(S)}(\varphi)/\langle\varphi\rangle \hookrightarrow \Gamma(T),$$

which is injective by Proposition 4.2.7, is also surjective.

PROOF. We present a rather charming method for giving an explicit lift of each of the generators of $\Gamma(T)$ to $\text{Norm}_{\Gamma_\varphi(S)}(\varphi) \subset \Gamma([S])$. These lifts do not necessarily correspond to a splitting homomorphism, i.e., the lifts do not necessarily satisfy the same relations as the generators of $\Gamma(T)$, but we are only proving surjectivity here, so it is enough to show that each generator has a lift.

We saw (Figure 4.1) that φ must be a rotation of permutation $c_1 \cdots c_k$, where each c_i is a cycle of length j, and $jk = n - 2$. Then T has $k + 2$ marked points, so $\Gamma(T)$ is generated by all but one of the x_{ij} such that $1 \leq i < j \leq k + 1$. It is even more convenient to use the following system of generators. Set $y_{1j} = \sigma_{j-1} \cdots \sigma_2 \sigma_1^2 \sigma_2 \cdots \sigma_{j-1}$; then via the relation

$$y_{1j} = x_{1j} x_{2j} \cdots x_{j-1,j},$$

we see that all but one of the elements y_{1j} for $1 \leq j \leq k+1$, together with the x_{ij} for $2 \leq i < j \leq k+1$, also form a system of generators for $\Gamma(T)$.

Now, S is a sphere with ordered numbered points, so $[\varphi]$ is a fixed permutation in S_n. Since we are working up to conjugation, we may assume that

$$[\varphi] = (1, k+1, \cdots, (j-1)k+1)(2, k+2, \cdots, (j-1)k+2) \cdots (k, 2k, \cdots jk),$$

which is more convenient than the numbering of Figure 4.1 for giving a visual representation of the braids. We number the marked points on T by taking the image of the fixed points of φ to be 1 and $k+2$, and the image of the marked point numbered i on S to be numbered $i+1$ on T. Then the generator x_{ij} of $\Gamma(T)$ for $2 \leq i < j \leq k+1$ is represented by a braid, and by a motion of the points, as in Figure 4.2.

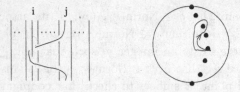

Figure 4.2. The generator x_{ij} of $\Gamma(T)$.

Each of these generators is lifted to a motion of points on S simply by j copies of x_{ij}, as in Figure 4.3.

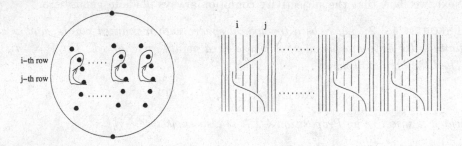

Figure 4.3. Lifting the generator x_{ij} to $\Gamma([S])$.

The element $y_{1j} \in \Gamma(T)$ is shown by the motion of points in the right-hand part of Figure 4.4, and lifts to that of the left-hand side.

Having separately lifted each of the y_{1j} and the x_{ij}, we now make a final remark. In themselves, these liftings do not generate a copy of $\Gamma(T)$ inside $\Gamma([S])$. However, we know that it is enough to take all but one of these generators to generate $\Gamma(T)$. We claim that if we throw out any one of the y_{1j} with $2 \leq j \leq k+1$, then the lifts of the remaining generators as above do generate a copy of $\Gamma(T)$ inside of $\Gamma([S])$, and the k different copies obtained in this way are distinct. Indeed, it is easy to check directly that the lifting obtained by throwing

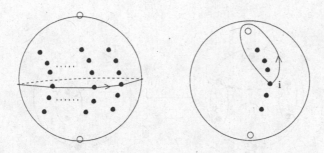

Figure 4.4. Lifting the generator y_{1j} to $\Gamma([S])$.

out y_{1j} is exactly the subgroup $\mathrm{Norm}_{\Gamma_{H_\varphi}(S)}(\varphi)$ where H_φ is generated by all of the cycles c_i except for c_j. This proof thus provides an explicit set of generators for the group $\mathrm{Norm}_{\Gamma_\varphi(S)}(\varphi)$, namely the generators constructed above for any $\mathrm{Norm}_{\Gamma_{H_\varphi}(S)}(\varphi)$, together with φ itself. $\square$

The following corollary is an immediate consequence of Theorems 4.3.1 and 4.3.2.

COROLLARY. *Let S be a sphere with n marked points, and φ a finite-order diffeomorphism of S, and set $T = S/\varphi$. Then*

$$\mathrm{Norm}_{\Gamma_\varphi(S)}(\varphi)/\langle\varphi\rangle \simeq \mathrm{Norm}_{\Gamma_{H_\varphi}(S)}(\varphi) \simeq \Gamma(T),$$

and for each of the k possible choices of $H_\varphi \subset G_\varphi$, we obtain a special homomorphism

$$\Gamma(T) \to \Gamma([S]).$$

4.4. An example in genus zero. S OF TYPE $(0,6)$, T OF TYPE $(0,4)$. The mapping class group $\Gamma([S]) \simeq \Gamma_{0,[6]}$ is generated by σ_1, σ_2, σ_3, σ_4 and σ_5 with the usual braid relations and the relations

$$\sigma_5\sigma_4\sigma_3\sigma_2\sigma_1^2\sigma_2\sigma_3\sigma_4\sigma_5 = (\sigma_1\sigma_2\sigma_3\sigma_4\sigma_5)^6 = 1. \qquad (4.4.1)$$

The element φ we consider in this example is

$$\varphi = \sigma_1\sigma_2\sigma_1\sigma_3\sigma_2\sigma_1\sigma_4\sigma_3\sigma_2\sigma_1.$$

Its associated permutation is $(15)(24)$. Direct computation (manipulation with the braid relations) shows that $\varphi^2 = (\sigma_4\sigma_3\sigma_2\sigma_1)^5$, and this element is equal to 1 in $\Gamma_{0,[6]}$, so φ is of order 2. In fact, φ is the class of the $180°$ rotation Φ around the axis shown in Figure 4.5.

The quotient T of S by Φ has 4 marked points coming from the 6 marked points of S; the north and south poles are fixed points of the rotation Φ, so they have ramification index equal to 2.

We compute the groups $\mathrm{Aut}^*([\pi_1(T)])$, $\mathrm{Aut}^*([S/T])$, $\Gamma_{[S/T]}$, and their pure versions.

Write

$$\pi_1(T) = \langle c_1, c_2, c_3, c_4 \mid c_1c_2c_3c_4 = 1\rangle,$$

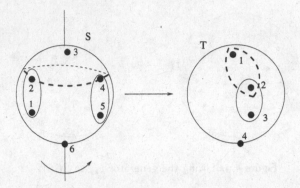

Figure 4.5. The finite-order diffeomorphism Φ lifting φ.

where c_i is a loop around the i-th marked point of T. The group $\mathrm{Aut}^*([\pi_(T)])$ of T is generated by $\mathrm{Inn}(\pi_1(T))$ and by elements τ_1, τ_2 and τ_3, acting via

$$\tau_i(c_j) = \begin{cases} c_{i+1} & \text{if } j = i \\ c_{i+1}^{-1}c_ic_{i+1} & \text{if } j = i+1 \\ c_j & \text{otherwise.} \end{cases} \qquad (4.4.2)$$

Now we realize $\pi_1(S)$ as a subgroup of $\pi_1(T)$, and compute $\mathrm{Aut}^*([S/T])$. Writing

$$\pi_1(S) = \langle x_1, x_2, x_3, x_4, x_5 \mid x_1x_2x_3x_4x_5 = 1 \rangle,$$

then $\pi_1(S)$ can be included in $\pi_1(T)$ via

$$\begin{cases} x_1 = c_1c_2c_3c_2^{-1}c_1^{-1} \\ x_2 = c_1c_2c_1^{-1} \\ x_3 = c_1^2 \\ x_4 = c_2 \\ x_5 = c_3. \end{cases}$$

(To realize the cover $\Phi : S \to T$ geometrically, one can take for example the Riemann surface $\mathbb{P}^1\mathbb{C}$ with ordered marked points $2, 1, 0, -1, -2, \infty$; the loops x_1, x_2, x_3, x_4 and x_5 go around the first five points in order, based at the base point i. Then Φ corresponds to the quotient of this Riemann surface by $z \mapsto z^2$, and the quotient is the projective line $\mathbb{P}^1\mathbb{C}$ with marked points $4, 1, 0$ and ∞, surrounded by loops c_1, c_2, c_3, c_4 based at -1.)

Now, in order to have a chance of restricting to $\pi_1(S)$, i.e., of belonging to $\mathrm{Aut}^*([S/T])$, an element of $\mathrm{Aut}^*([\pi_1(T)])$ cannot permute marked points of T having different ramification indices under φ. Therefore, the primary candidate for the subgroup $\mathrm{Aut}^*([S/T])$ is the subgroup of all elements of $\mathrm{Aut}^*([\pi_1(T)])$ whose permutations lie in the subgroup $\langle (14), (23) \rangle \subset S_4$. This group is generated by a lifting of each of the two transpositions, together with the whole of the pure subgroup $\mathrm{Aut}^*(\pi_1(T))$. A lift of the transposition (23) is given by τ_2, and a lift of (14) is given by the element $\tau_{14} = \tau_3\tau_2\tau_1\tau_2^{-1}\tau_3^{-1}$, whereas $\mathrm{Aut}^*(\pi_1(T))$ is generated by $\mathrm{Inn}(\pi_1(T))$ and by the six elements τ_1^2, $\tau_2\tau_1^2\tau_2^{-1}$, $\tau_3\tau_2\tau_1^2\tau_2^{-1}\tau_3^{-1}$, τ_2^2, $\tau_3\tau_2^2\tau_3^{-1}$, τ_3^2. Therefore, the preimage of $\langle(14),(23)\rangle$ is generated in $\mathrm{Aut}^*([\pi_1(T)])$ by

$$\tau_1^2, \quad \tau_2\tau_1^2\tau_2^{-1}, \quad \tau_3\tau_2\tau_1\tau_2^{-1}\tau_3^{-1}, \quad \tau_2, \quad \tau_3\tau_2^2\tau_3^{-1}, \quad \tau_3^2.$$

We can drop $\tau_2\tau_1^2\tau_2^{-1}$ from this set of generators, and since by the braid relation, we have $\tau_3\tau_2^2\tau_3^{-1} = \tau_2^{-1}\tau_3^2\tau_2$, we can also drop this element. Furthermore, $\tau_1^2 = \tau_3^2$ in $\mathrm{Aut}^*(\pi_1(T))$, so we are left with three generators

$$\tau_1^2, \quad \tau_2, \quad \tau_3\tau_2\tau_1\tau_2^{-1}\tau_3^{-1}.$$

Explicit computation shows that each of these three generators restricts to $\pi_1(S)$ as an element of $\mathrm{Aut}^*([\pi_1(S)])$. Indeed, we have

$$\tau_1^2 : \begin{cases} x_1 \mapsto x_1 \\ x_2 \mapsto x_4 \\ x_3 \mapsto x_4^{-1}x_3x_4 \\ x_4 \mapsto x_4^{-1}x_3^{-1}x_2x_3x_4 \\ x_5 \mapsto x_5 \\ x_6 \mapsto x_6, \end{cases} \qquad \tau_2 : \begin{cases} x_1 \mapsto x_2 \\ x_2 \mapsto x_2^{-1}x_1x_2 \\ x_3 \mapsto x_3 \\ x_4 \mapsto x_5 \\ x_5 \mapsto x_5^{-1}x_4x_5 \\ x_6 \mapsto x_6, \end{cases}$$

$$\tau_3\tau_2\tau_1\tau_2^{-1}\tau_3^{-1} : \begin{cases} x_1 \mapsto x_1 \\ x_2 \mapsto x_2 \\ x_3 \mapsto x_6 \\ x_4 \mapsto (x_1x_2x_3)x_4(x_1x_2x_3)^{-1} \\ x_5 \mapsto (x_1x_2x_3)x_5(x_1x_2x_3)^{-1} \\ x_5 \mapsto (x_1x_2x_3)x_3(x_1x_2x_3)^{-1}. \end{cases}$$

This shows that $\Gamma_{[S/T]} = \langle \tau_1^2, \tau_2, \tau_3\tau_2\tau_1\tau_2^{-1}\tau_3^{-1} \rangle$. Since furthermore we have $\Gamma(T) = \langle \tau_1^2, \tau_2^2 \rangle$, we see that $\Gamma_{S/T} = \Gamma(T)$, so φ satisfies the surjectivity condition of Section 4.2; this condition was of course guaranteed anyway in genus zero by Theorem 4.3.2.

Let $\sigma_1, \ldots, \sigma_5$ be generators of $\mathrm{Aut}^*([\pi_1(S)])$, acting on the x_j exactly as τ_i acts on c_j in (4.4.2). Then it is easy to see that τ_1^2 acts on $\pi_1(S)$ like $\sigma_2\sigma_3\sigma_2$, and τ_2 acts like $\sigma_1\sigma_4$, so τ_2^2 acts like $\sigma_1^2\sigma_4^2$.

Furthermore, since $\Gamma(T)$ is a free group on the generators τ_1^2 and τ_2^2, we obtain a *special homomorphism*

$$\Gamma(T) \to \Gamma([S])$$
$$\tau_1^2 \mapsto \sigma_2\sigma_3\sigma_2 \tag{4.4.3}$$
$$\tau_2^2 \mapsto \sigma_1^2\sigma_4^2.$$

This is an explicit version of the *splitting condition* (2) of Section 4.2, guaranteed in genus zero by Theorem 4.3.1.

We now explain the meaning of these computations in geometric terms, i.e., by determining the points of special automorphism corresponding to φ in $\mathcal{M}(S)$, and the special locus of φ in the quotient moduli space $\mathcal{M}_\varphi(S)$, and relating the image of the splitting map (4.4.3) to the normalizer $\mathrm{Norm}_{\Gamma_\varphi(S)}(\varphi)$.

The permutation $[\varphi] = (15)(24)$, and to find the set of points of special automorphism corresponding to φ, it is enough to parametrize the set of points fixed by $(15)(24)$. We use the unique representative of the form $(1, a, 0, b, c, \infty)$ for each point of $\mathcal{M}(S) = \mathcal{M}_{0,6}$. This is odd-looking but ensures that 0 and ∞ correspond to the third and sixth marked points, i.e., the north and south poles of S which are the fixed points under φ (see Figure 4.5). Applying $(15)(24)$ gives $(c, b, 0, a, 1, \infty)$ which is equivalent to $(1, b/c, 0, a/c, 1/c, \infty)$ via $z \mapsto z/c$. The point $(1, a, 0, b, c, \infty)$ is a point of special automorphism corresponding to φ if and only if it is fixed under the action of $(15)(24)$, i.e., if $c = -1$ and $a = -b$, so the set of special points is given by all points of the form

$$(1, a, 0, -a, -1, \infty). \tag{4.4.4}$$

This special locus has no orbifold structure since $\mathcal{M}(S)$ does not, and it is normal. It is easily seen to be a copy of $\mathbb{P}^1$ with four points removed, namely those corresponding to the values $a = 0, \pm 1, \infty$.

We now consider the special locus $\mathcal{M}_\varphi(S, \varphi)$ inside the quotient moduli space $\mathcal{M}_\varphi(S)$; it is the image of the locus (4.4.4) in the quotient space $\mathcal{M}(S)/G_\varphi$ where $G_\varphi = \langle (15), (24) \rangle$. The permutation $(15)(24)$ fixes each point, whereas (15) and (24) both act via $a \mapsto -a$; thus the points $a = 0$ and $a = \infty$ are fixed under the action of G_φ and the points $a = 1$ and $a = -1$ are exchanged. There are no other fixed points under G_φ. Thus, the special locus we are interested in, $\mathcal{M}_\varphi(S, \varphi)$, is a $\mathbb{P}^1$ with three points removed and a trivial orbifold structure with group $\mathbb{Z}/2\mathbb{Z}$ at each point. This is the geometric structure which is reflected by the fundamental groups since

$$\pi_1(\mathcal{M}_\varphi(S, \varphi)) = \mathrm{Norm}_{\Gamma_\varphi(S)}(\varphi)$$
$$= \langle \sigma_2 \sigma_3 \sigma_2, \sigma_1^2 \sigma_4^2, \varphi \rangle \simeq \langle \varphi \rangle \times \langle \sigma_2 \sigma_3 \sigma_2, \sigma_1^2 \sigma_4^2 \rangle \simeq \mathbb{Z}/2\mathbb{Z} \times F_2.$$

The $F_2 \simeq \Gamma(T)$ factor in this direct product is the image of the splitting homomorphism (4.4.3).

4.5. An example in genus one. S OF TYPE $(1, 2)$, T OF TYPE $(1, 1)$.

Here we quotient a torus S with 2 marked points by the diffeomorphism φ which is the 180° rotation around the axis passing vertically through the hole. The quotient T is a torus with one marked point and no branch points.

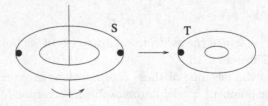

Figure 4.6. A diffeomorphism of type $(1, 2)$.

The fundamental group $\pi_1(T)$ has the well-known presentation

$$\langle a, b, c \mid (a, b)c = 1 \rangle,$$

where (a, b) denotes the commutator $bab^{-1}a^{-1}$. The fundamental group $\pi_1(S)$ has the presentation

$$\langle a', b', c_1, c_2 \mid (a', b')c_1c_2 = 1 \rangle.$$

This group can be realized as a subgroup of $\pi_1(T)$ via the embedding

$$\pi_1(S) \hookrightarrow \pi_1(T)$$
$$a' \mapsto a$$
$$b' \mapsto b^2$$
$$c_1 \mapsto c$$
$$c_2 \mapsto bcb^{-1}.$$

The group $\mathrm{Aut}^*(\pi_1(T))$ is generated by $\mathrm{Inn}(\pi_1(T))$ and by twists τ_1 and τ_2 along the loops α and β in Figure 3.13. In the mapping class group $\Gamma(T)$, these twists satisfy the relations $\tilde{\tau}_1\tilde{\tau}_2\tilde{\tau}_1 = \tilde{\tau}_2\tilde{\tau}_1\tilde{\tau}_2$ and $(\tilde{\tau}_1\tilde{\tau}_2)^3 = 1$, and in fact $\Gamma(T)$ is isomorphic to $\mathrm{PSL}_2(\mathbb{Z})$.

Explicitly, the twists τ_1 and τ_2 act on $\pi_1(T)$ via

$$\tau_1(a) = a, \quad \tau_1(b) = ba^{-1}, \quad \tau_1(c) = c$$

and

$$\tau_2(a) = ab, \quad \tau_2(b) = b, \quad \tau_2(c) = c.$$

Since T has only one marked point, we have $\mathrm{Aut}^*([\pi_1(T)]) = \mathrm{Aut}^*(\pi_1(T))$ and $\mathrm{Aut}^*([S/T]) = \mathrm{Aut}^*(S/T)$. We show that the subgroup $\mathrm{Aut}^*(S/T)$ is given by $\langle \tau_1, \tau_2^2 \rangle$.

We first show that τ_1 lies in $\mathrm{Aut}^*(S/T)$, by computing its restriction to $\pi_1(S)$; we find that

$$\begin{cases} \tau_1(a') = \tau_1(a) = a = a' \\ \tau_1(b') = \tau_1(b^2) = ba^{-1}ba^{-1} = (a')^{-1}c_1b'(a')^{-1} \\ \tau_1(c_1) = \tau_1(c) = c = c_1 \\ \tau_1(c_2) = \tau_1(bcb^{-1}) = ba^{-1}cab^{-1} = b'(a')^{-1}(b')^{-1}c_1^{-1}a', \end{cases}$$

so τ_1 preserves the subgroup $\pi_1(S)$. We next show that τ_2^2 lies in $\mathrm{Aut}^*(S/T)$ by computing its restriction:

$$\begin{cases} \tau_2^2(a') = \tau_2^2(a) = ab^2 = a'b' \\ \tau_2^2(b') = \tau_2^2(b^2) = b^2 = b' \\ \tau_2^2(c_1) = \tau_2^2(c) = c = c_1 \\ \tau_2^2(c_2) = \tau_2^2(bcb^{-1}) = bcb^{-1} = c_2, \end{cases}$$

so again τ_2^2 preserves $\pi_1(S)$, so it lies in $\mathrm{Aut}^*(S/T)$. Now, the subgroup $\langle \tau_1, \tau_2^2 \rangle$ is of index 2 in $\Gamma(T) = \langle \tau_1, \tau_2 \rangle$, so if the subgroup $\mathrm{Aut}^*(S/T)$ was strictly greater than $\langle \tau_1, \tau_2^2 \rangle$, it would be all of $\Gamma(T)$. But this is not the case, since for instance $\tau_2 \notin \mathrm{Aut}^*(S/T)$; indeed, $\tau_2(a) = ab$ and $ab \notin \pi_1(S)$ (otherwise $b \in \pi_1(S)$ and $\pi_1(S) = \pi_1(T)$ which is absurd). Therefore we have shown that

$$\mathrm{Aut}^*(S/T) = \langle \tau_1, \tau_2^2 \rangle.$$

Thus, the surjectivity condition does not hold for this φ.

For the purposes of studying the surjectivity and splitting conditions, we should consider not the mapping class group $\mathrm{Diff}^+([S])/\mathrm{Diff}^0(S)$ but the orbifold fundamental group of the moduli space $\mathcal{M}([S])$. These two groups coincide in general, but in this particular case they differ, as we saw in Section 3.2. The fundamental group $\pi_1(\mathcal{M}([S])$ is generated by τ_1 and τ_2 subject to the braid relation and to the relation $(\tau_1\tau_2)^6 = 1$; it is isomorphic to $\mathrm{SL}_2(\mathbb{Z})$. The above arguments go through and the surjectivity condition naturally still does not hold, however the splitting condition holds. Indeed, τ_1 lifts to the element $\sigma_1\sigma_3$ of $\Gamma([S])$ and τ_2^2 lifts to σ_2. The only relation satisfied by τ_1 and τ_2^2 is $(\tau_1\tau_2^2\tau_1\tau_2^2) = (\tau_1\tau_2)^3 = 1$. So we need to check that $(\sigma_1\sigma_3\sigma_2\sigma_1\sigma_3\sigma_2)^2 = 1$ in $\Gamma([S])$, but this is of course true since $\sigma_1\sigma_3\sigma_2$ is of order four in $\Gamma([S])$ (conjugate by σ_1 to $\sigma_3\sigma_2\sigma_1$).

We give the geometric description of the situation. As we saw in Section 3.4, each point of the moduli space $\mathcal{M}(S)$ of tori with two marked points can be represented uniquely by such a parallelogram, marked with two marked points whose sum is equal to the origin. In other words, a point of $\mathcal{M}(S)$ is determined by a pair (τ, z) of parameters, with τ in the fundamental domain of the action of $\mathrm{SL}_2(\mathbb{Z})$ on $\mathcal{H}$ (see Figure 3.12) defining the elliptic curve, and z a complex number $\neq 0$ in the fundamental parallelogram determined by 1 and τ; the marked points are taken to be z and $-z$ (both considered modulo the period lattice, i.e., in the fundamental parallelogram). The action of φ on the parallelogram is given by $z \mapsto z + \tau/2$. Thus the quotient of the torus by this map identifies the lower and upper half of the parallelogram as in Figure 4.7.

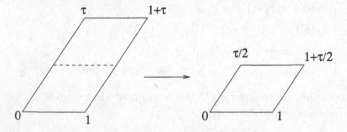

Figure 4.7. Action of φ on a parallelogram.

The points of the moduli space $\mathcal{M}(S)$ which are preserved by the diffeomorphism φ, given by $z \mapsto z + \tau/2$, are those such that the two marked points are exchanged by this diffeomorphism, namely $z + \tau/2 = -z$ modulo the lattice $\langle 1, \tau \rangle$. This shows that the special locus of φ is not the whole moduli space, so it is the special locus of Section 3.4, which as we saw there is isomorphic to the quotient of $\mathbb{P}^1 - \{0, 1, \infty\}$ by $\frac{1}{2}(z^2 + \frac{1}{z^2})$. This corresponds to the isomorphism

$$\pi_1(\mathcal{M}(S, \varphi)) = \langle \tau_1, \tau_2^2 \rangle.$$

References

[B] J. Birman, *Braids, Links and Mapping Class Groups*, Princeton Univ. Press, 1974.

[D] V. G. Drinfel'd, "On quasitriangular quasi-Hopf algebras and a group closely connected with $\mathrm{Gal}(\overline{\mathbb{Q}}/\mathbb{Q})$", *Leningrad Math. J.* **2** (1991), No. 4, 829-860.

[GH] G. González-Díez and W. J. Harvey, "Moduli of Riemann surfaces with symmetry", pp. 75–93 in *Discrete Groups and Geometry*, edited by W. J. Harvey, London Math. Soc. Lecture Note Series **173**, Cambridge Univ. Press, 1992.

[H] W. Harvey, "On branch loci in Teichmüller space", *Trans. Amer. Math. Soc.* **153** (1971), 387–399.

[HM] W. J. Harvey and C. MacLachlan, "On mapping class groups and Teichmüller spaces", *Proc. London Math. Soc.* (3) **30** (1975), 496–512.

[HN] A. Haefliger and Quach Ngoc Du, "Une présentation du groupe fondamental d'une orbifold", pp. 98–107 in *Structure transverse des feuilletages*, Astérisque **116**, Soc. math. de France, Paris, 1984.

[K1] S. Kerckhoff, "The Nielsen realization problem", *Bull. Amer. Math. Soc.* (*N.S.*) **2**:3 (1980), 452–454.

[K2] S. Kerckhoff, "The Nielsen realization problem", *Ann. of Math.* (2) **117**:2 (1983), 235–265.

[M] D. Mumford, *Curves and their Jacobians*, Univ. of Michigan Press, Ann Arbor, 1975.

[Mac] C. Maclachlan, "Modular groups and fibre spaces over Teichmuller spaces", pp. 297–313 in *Discontinuous Groups and Riemann Surfaces*, edited by L. Greenberg, Ann. of Math. Studies **79**, Princeton Univ. Press, 1974.

[N1] H. Nakamura, "Galois rigidity of pure sphere braid groups and profinite calculus" (Appendix), *J. Math. Sci. Univ. Tokyo* **1** (1994), 71–136.

[N2] H. Nakamura, "Galois representations in the profinite Teichmüller modular groups", in *Geometric Galois Actions 1*, edited by L. Schneps and P. Lochak, London Math. Soc. Lecture Note Series **242**, Cambridge Univ. Press, 1997.

[PS] *Triangulations, Courbes Arithmétiques et Théorie des Champs*, edited by L. Schneps, Panoramas et Synthèses **7**, Société math. de France, Paris, 1999.

LEILA SCHNEPS
ÉQUIPE ANALYSE ALGÉBRIQUE
INSTITUT DE MATHÉMATIQUES DE JUSSIEU
UNIVERSITÉ DE PARIS 6
175 RUE DU CHEVALERET
75013 PARIS
FRANCE
leila@math.jussieu.fr

Galois Groups and Fundamental Groups
MSRI Publications
Volume 41, 2003

Cellular Decomposition of Compactified Hurwitz Spaces

MICHEL IMBERT

ABSTRACT. We describe a cellular decomposition of compactified Hurwitz spaces, generalizing the cellular decomposition of moduli spaces of punctured Riemann surfaces $\mathcal{M}_{g,n}$.

The main motivation for this work is the integration of cohomology classes on compactified Hurwitz spaces, and is provided by Witten's conjecture on moduli spaces of Riemann surfaces with spin, and by the fact (proved in this paper) that these spaces are closely related to Hurwitz spaces of Galois cyclic coverings. This article also aims to give all details of the Harer–Kontsevich theorem.

1. Introduction

In the last ten years, the geometry of moduli spaces of punctured Riemann surfaces have seen an increasing interest and known some striking progress [20] in connection with physic's theories [26]. This also concerns some generalizations of moduli spaces of punctured surfaces, like moduli spaces of stable maps, or moduli spaces of Riemann surfaces with spin [27] [17]. In this paper, we firstly show that the topological framework which have allowed M. Kontsevich to compute in a combinatorial way some Chern classes on $\mathcal{M}_{g,n}$ (a key point of his proof of Witten's conjecture [26]) extends to the setting of Hurwitz spaces, and secondly we sketch an analogy between Hurwitz spaces of cyclic coverings and moduli spaces of Riemann surfaces with spin. In a forthcoming work, our results will be used to study the cohomology of Hurwitz spaces for cyclic coverings.

Hurwitz spaces basically consist in equivalence classes of ramified coverings between two compact Riemann surfaces, where $(p_1 : S_1 \to T_1)$ is equivalent to $(p_2 : S_2 \to T_2)$ if there exists two biholomorphisms $f : S_1 \to S_2$ and $h : T_1 \to T_2$ such that $p_2 \circ f = h \circ p_1$. We often denote by g the genus of the total space, and by g' the genus of the base. They are related by the Riemann–Hurwitz formula: $2g - 2 = d(2g' - 2) + B$, where d is the degree of the covering, and B, the total

---

Mathematics Subject Classification: 05C10, 14H30, 14E22, 30F30, 30C62, 32G15.

branching number. We only consider cases $d > 1$ and $B > 0$. Let us denote by $\mathcal{H}(g, g', d)$ the set of such equivalence classes with fixed g, g' and d, and call it a Hurwitz space. There are two ways to study it.

The first one is to work with this equivalence relation. This is the approach of W. J. Harvey [11] [15], and A. Kuribayashi [22] in the setting of Teichmüller theory.

The second one is configurative (only branch points are allowed to move on the base). This is the approach of W. Fulton [9] and M. Fried [8].

In both cases, it is necessary to fix more accurate invariants in equivalence classes of coverings, such as number of branch points, multiplicities, monodromy group. It is in fact the notion of ramification data (a way to encode what actually happens in a neighborhood of a ramification point) which allows us a combinatorial study of Hurwitz spaces, using coverings of fat-graphs.

We show that the graphical description of punctured Riemann surfaces with fat-graphs and lengths of edges, extend to holomorphic maps between compact Riemann surfaces. Punctures come as ramification points and branch points. Coverings of fat-graphs are étale on the underlying graphs, the ramification can be read on faces. We give details on the combinatorial model of Kontsevich's compactification of moduli spaces, and compactify Hurwitz spaces in a similar way.

Let $\mathcal{H}(g, g', G, K, R)$ be the subset of $\mathcal{H}(g, g', d)$ where we have fixed a monodromy of type (G, K, R), for a fixed finite group G, some ramification data R of degree b, and a subgroup K of index d such that $\bigcap_{t \in G} t^{-1} K t = \{e_G\}$.

Then, our main result is a generalization of the Harer–Kontsevich theorem (theorem in appendix B of [20]):

THEOREM 1.1.   *There exist a combinatorial Hurwitz space $\overline{\mathcal{H}}_{g,g'}^{\mathrm{comb}}(G, K, R)$, a suitable compactification $\overline{\mathcal{H}}'(g, g', G, K, R)$ of $\mathcal{H}(g, g', K, G, R)$ and an homeomorphism*

$$\overline{\mathcal{H}}_{g,g'}^{\mathrm{comb}}(G, K, R) \to \overline{\mathcal{H}}'(g, g', G, K, R) \times \mathbb{P}(\mathbb{R}_{>0}^b)$$

leading to a cellular decomposition of $\overline{\mathcal{H}}'(g, g', G, K, R) \times \mathbb{P}(\mathbb{R}_{>0}^b)$, compatible with its orbifold structure.

Spaces $\mathcal{H}(g, g', G, K, R)$ arise as finite quotients of Hurwitz spaces $\mathcal{H}_h(G, R)$, made of genus h compact Riemann surfaces with an holomorphic action of G of ramification data R, defined up to G-equivariant biholomorphism. A space $\mathcal{H}_h(G, R)$ generally splits up into many connected components, each one being the quotient of a Teichmüller space.

All these Hurwitz spaces are closely related to moduli spaces. Firstly, forgetting group actions, we have a map from $\mathcal{H}_h(G, R)$ into $\mathcal{M}_h$. See [11] [15] for this viewpoint. Secondly, $\mathcal{H}_h(G, R)$ projects onto moduli spaces of punctured Riemann surfaces (map a Riemann surface with group action onto the quotient Riemann surface with its set of branch points as punctures).

Since Harer–Kontsévich's theorem is useful in explicit calculations of cohomological invariants of moduli spaces [3] [20], we hope that Theorem 1.1 can be useful in this direction, for example integrating cohomology classes on Hurwitz spaces. Actually, this is already the case for pullback of classes defined on the moduli space of punctured Riemann surfaces.

The first motivation is in fact a conjecture of E. Witten [27] which deals with intersection numbers defined on moduli spaces of punctured Riemann surfaces with spin, closely related to Hurwitz spaces of Galois coverings with cyclic groups.

Section 2 is devoted to Hurwitz spaces. It contains a complete treatment of Hurwitz spaces for non Galois coverings. In Section 3, we show how to adapt Strebel's theorem to coverings and to stable Riemann surfaces. Section 4 is first a review on fat-graphs; then we give the graphical construction of holomorphic maps between compact Riemann surfaces to obtain the main theorem in the non-compactified setting (Theorem 4.8). In Section 5, we prove the continuity and describe the cellular decomposition of decorated Hurwitz spaces. The cellular decomposition is compatible with the orbifold structure. The continuity involves a non trivial construction of quasiconformal homeomorphisms. As consequence of the cellular decomposition, we obtain a combinatorial characterization of their connected components. In Section 6 we show how to extend the setting to suitable compactifications, and finish the proof of Theorem 1.1. We conclude in the last section by the analogy between Hurwitz spaces of Galois cyclic coverings and moduli spaces of spin curves.

Acknowledgments. The author expresses his gratitude to Professor J. Bertin.

2. Hurwitz Spaces

In this section, we recall how to define Hurwitz spaces and their orbifold structure, coming from Teichmüller theory. For the sake of clearness, we first consider Hurwitz spaces of Galois coverings. The main and most natural discrete invariant is the ramification data associated to a group and to any holomorphic action of it on a compact Riemann surface. Used by A. Kuribayashi [22], W. J. Harvey [15] in the context of Teichmüller spaces, the ramification data is also useful in the context of arithmetic geometry after M. Fried [8], see also [5]. We conclude by a study of Hurwitz spaces for non Galois coverings, they arise as finite quotients of Galois Hurwitz spaces.

DEFINITION 2.1. Let G be a finite group. Then an abstract *ramification data* of G is a formal sum $R = \sum_{i=1}^{t} r_i C_i$ where $r_i \in \mathbb{N}^* \ \forall i \in \{1, \ldots, t\}$, and $(C_i)_{i \in \{1, \ldots, t\}}$ is a set of non-trivial and pairwise distinct conjugacy classes of G. The degree of R is $\sum_{i=1}^{t} r_i$.

Let C be a compact Riemann surface, and $\varphi : G \hookrightarrow \text{Aut}(C)$ an effective biholomorphic action of a finite group (finiteness is automatic if $g(C) \geq 2$ by the Hurwitz theorem). We only consider actions with some fixed points.

Denote by p the branched cover $C \to C/\varphi(G)$. Let x be a fixed point of some element of G, and $y = p(x)$. There exist local coordinates (U, z) and (U', z') such that $z(x) = 0$, $z'(y) = 0$, and $z' \circ p = z^{e_x}$. The integer e_x is called the multiplicity of x. The stabilizer G_x of x is a cyclic group of order e_x. We consider the privileged generator τ_x of G_x defined by $\tau_x(z) = \exp\left(\frac{2i\pi}{e_x}\right) z$. This generator is independent of z since it is the unique element of G_x which acts on the holomorphic tangent space of C at x in multiplying by $\exp\left(\frac{2i\pi}{e_x}\right)$.

If $y = g \cdot x$, then $\tau_y = g\tau_x g^{-1}$. We say that the conjugacy class $C(\tau_x)$ of τ_x is the *color* of x (and of its orbit). Two distinct orbits can possess the same color. Let $\{O_i, b_i\}_{i \in \{1,\dots,t\}}$ be the colored orbits of fixed points with multiplicities, and $C(\tau_i)$ be the colors.

Then to the action φ is associated the ramification data:

$$R(\varphi) = \sum_{i=1}^{t} b_i C(\tau_i).$$

Note that a G-equivariant biholomorphism $h : (C, \varphi) \to (C', \varphi')$ preserves the colors, but can permute two orbits of the same color.

DEFINITION 2.2. Let G be a fixed finite group with a fixed ramification data R. Then a (G, R)–Hurwitz space $\mathcal{H}_g(G, R)$ consists in equivalence classes of couples (C, φ) made of a compact Riemann surface C of genus g together with an action $\varphi : G \hookrightarrow \text{Aut}(C)$ such that $R(\varphi) = R$. Two couples are equivalent if they differ by a G-equivariant biholomorphism.

Looking at the action of a stabilizer G_x on the complexified real tangent space $T_x \otimes \mathbb{C}$ shows that orientation preserving equivariant diffeomorphisms preserve the ramification data.

An important tool for studying Hurwitz spaces is Riemann's existence theorem [8] [9] [10], which furnishes a condition of existence for coverings. We state here a Galois and modular version of this theorem. In the following, it is not necessary to specify the base point for fundamental groups.

THEOREM 2.3. *Fix* $R = b_1 C_1 + \cdots + b_t C_t$ *a degree* b *ramification data. There is a bijective correspondence between* $\mathcal{H}_g(G, R)$ *and the set of classes* $[S - Y, \psi]$ *, where* S *is a compact Riemann surface of genus* g', Y *a finite subset of* S *shared into* t *colors,* $\psi : \pi_1(S - Y) \to G$ *an epimorphism with* $\psi(\gamma) \in C_i$ *if* γ *is a canonical loop around some point of color* i; $[S - Y, \psi] = [S' - Y', \psi']$ *if and only if there exist a biholomorphism* $h : S \to S'$ *with* $h(Y) = Y'$ *(preserving the set of colors) and* $\theta \in \text{Int}(G)$ *such that* $\theta \circ \psi = \psi' \circ h_*$.

For convenience, we recall the correspondence:

The map from $\mathcal{H}_g(G, R)$ associates to $[C, \varphi]$ the quotient surface $S = C/\varphi(G)$, and Y is the set of branch points. The genus $g(S) = g'$ is given by Riemann–Hurwitz. The epimorphism ψ is built as follows: given $\gamma \in \pi_1(S - Y, Q)$, we consider its lift α on C from a given point of the fiber over Q. Then we associate to γ the unique element of G which sends $\alpha(0)$ on $\alpha(1)$.

Conversely, given $[S - Y, \psi]$, we build C as the universal covering $\tilde{S}$ of S quotiented by $\ker(\psi)$. Then $\theta \in G$ acts on $[x] \in C$ by $\theta \cdot [x] = [\psi^{-1}(\theta) \cdot x]$.

If $[S, Y, \psi]$ corresponds to $[C, \varphi]$, then the centralizer of $\varphi(G)$ in $\mathrm{Aut}(C)$ quotiented by the center $Z(G)$ is isomorphic to the subgroup of $\mathrm{Aut}(S - Y)$ made of the elements h satisfying $\theta \circ \psi = \psi \circ h_*$ for $\theta \in \mathrm{Int}(G)$.

EXAMPLE 1. We consider the case of cyclic coverings of compact Riemann surfaces of genus g'. Put $G = \mathbb{Z}/n$; we fix the ramification data $R = \sum_{i=1}^{t} b_i[m_i]$ of degree b, where $[m_i] \in \mathbb{Z}/n \setminus \{[0]\}$, and $[m_i] \neq [m_j]$ if $i \neq j$. If e_i is the order of m_i, then the Riemann–Hurwitz formula determines the genus g of such coverings: $2g - 2 = n(2g' - 2) + \sum_{i=1}^{t} n b_i(1 - 1/e_i)$.

If $[C, \varphi] \in \mathcal{H}_g(\mathbb{Z}/n, R)$, and $\sigma = \varphi([1])$, then we set $S = C/\langle \sigma \rangle$. Let $Q = (Q_1, \ldots, Q_b)$ be the set of branch points of $\pi : C \to S$, $\{P_{i,j}\}_j$ the preimages of Q_i, and $d_i = \#\{P_{i,j}\}$, such that $e_i = n/d_i$. The stabilizers $G_{i,j}$ of all preimage points of Q_i are equal to the unique subgroup of order e_i generated by σ^{d_i}.

If $\sigma^{d_i k_i}$ (with $(e_i, k_i) = 1$) is the generator of $G_{i,j}$ which acts on the tangent space $T_{P_{i,j}}(S)$ by $\exp(2i\pi/e_i)$, then we have $R(\varphi) = \sum_i b_i'[d_i k_i]$, and $R(\varphi) = R$ means $m_i = d_i k_i$ and $b_i = b_i'$.

Riemann's existence theorem implies that $\sum_{i=1}^{t} b_i[d_i k_i] \equiv 0(n)$ (in particular $b \geq 2$). Since $[C, \varphi]$ is given by an epimorphism $\pi_1(S \setminus Q) \to \mathbb{Z}/n$, the condition is implied by the surface relation satisfied by $\pi_1(S \setminus Q)$: the image of the product of commutators made of homological loops is trivial.

Using the Theorem 2.3, it is easy to see that the natural projection $\mathcal{H}_g(G, R) \to \mathcal{M}_{g',(b_1,\ldots,b_t)}$ which sends $[C, \varphi]$ to $[C/\varphi(G), Y]$, where Y is the branch locus, is in general a ramified covering. The fiber over $[S, Y]$ is in 1-1 correspondence with the set of epimorphisms $\pi_1(S - Y) \to G$ with ramification data R, up to conjugacy, and up to automorphisms of $[S, Y]$ which do not act trivially on the fundamental group. Generically, there is no such automorphisms, and exceptional Riemann surfaces with such automorphisms make the locus of branching. It may happen that every $[S, Y]$ possess such automorphisms (think to the hyperelliptic involution), then the branch locus is made of exceptional Riemann surfaces with more automorphisms.

In the previous example, if the surjectivity is ensured by the ramification data, then images of homological loops are free, and the degree of $\mathcal{H}_g(\mathbb{Z}/n, R) \to \mathcal{M}_{g',(b_1,\ldots,b_t)}$ is $n^{2g'}$.

We recall how to define the topology of $\mathcal{H}_g(G, R)$ coming from the Teichmüller theory. It turns out that these spaces possess in general many connected components, each one being a quotient of a Teichmüller space.

REMARK. There is a nice description of this by W. J. Harvey [15], in the context of Fuchsian groups. Instead of Teichmüller spaces of Fuchsian groups, we use

Teichmüller spaces of marked Riemann surfaces. This is because, in our context, Riemann surfaces arise as explicit complex structures on smooth surfaces.

DEFINITION 2.4. Let (C, φ) be a compact Riemann surface with a biholomorphic action of G. Then the Hurwitz space $\mathcal{H}_G(C, \varphi)$ of topological type (C, φ) is the subset of $\mathcal{H}_{g(C)}(G, R(\varphi))$ made of couples (T, ρ) such that there is a G-equivariant diffeomorphism between T and C.

We recall that in each Hurwitz space $\mathcal{H}_g(G, R)$, the number $h_g(G, R)$ of distinct topological types (sometimes called the Hurwitz number) is finite. Take $[p_i : C_i \to (S, Y)]_{i=\{1,2\}}$ two elements of $\mathcal{H}_g(G, R)$, and let $\psi_i : \pi_1(S - Y) \to G$ the corresponding epimorphisms (Theorem 2.3). Then the existence of a G-equivariant diffeomorphism $f : C_1 \to C_2$ such that $p_2 \circ f = p_1$ amounts to the existence of an element k in $\mathrm{Diff}^+(S, Y)$ (the subgroup of $\mathrm{Diff}^+(S)$ which preserves Y and its partition in colors), which satisfies $\psi_2 \circ k_* = \psi_1$, up to conjugation. This relies on the theorem on lift of mappings.

Using now Nielsen's theorem (Theorem 1 of [16]), every element of $\mathrm{Diff}^+(S, Y)$ comes from an automorphism of the fundamental group of $S - Y$ which preserves the conjugacy class of loops around points of the same color.

If we denote the quotient of this subgroup by the inner automorphisms by $\mathrm{Out}(\pi_1(S - Y), Y)$, and by $\mathrm{Epi}_R(\pi_1(S - Y), G)$ the set of epimorphims from $\pi_1(S - Y)$ onto G with image of loops around points of Y fixed by the ramification data R, then

$$h_g(G, R) = \#\ \mathrm{Out}(\pi_1(S - Y), Y)\backslash \mathrm{Epi}_R(\pi_1(S - Y), G)/G.$$

To see an element of $\mathrm{Epi}_R(\pi_1(S-Y), G)$ as a collection of elements of G satisfying certain relations (see [8]) allows the computation of $h_g(G, R)$ in some particular case. As an example, the reader can verify that it is one for G abelian and genus zero or one for quotient surfaces. See [6] for examples with two topological types.

Now, Hurwitz spaces with fixed topological type arise as quotients of some Teichmüller spaces.

DEFINITION 2.5. Let G be a fixed finite group, and (C, φ) a reference couple made of a compact Riemann surface C and of a biholomorphic action φ of G on C. The Teichmüller space $\mathcal{T}_G(C, \varphi)$ of C relative to the action φ, consists in classes of 3-tuple (T, ρ, f), where T is a compact Riemann surface, $\rho : G \hookrightarrow \mathrm{Aut}(T)$ is a monomorphism, and $f : C \to T$ is a G-equivariant quasiconformal homeomorphism. Then $(T_i, \rho_i, f_i)_{i=1,2}$ are equivalent if and only if $f_2 \circ f_1^{-1}$ is homotopic to a G-equivariant biholomorphism $h : T_1 \to T_2$.

To illustrate this definition, note that $[T, \rho, f] = [T, \rho, f \circ \varphi(g^{-1})]$ if $g \in Z(G)$ (the center of G). The following lemma justifies the use of quasiconformal homeomorphisms in the previous definition and gives a metric on $\mathcal{T}_G(C, \varphi)$. See [22], Proposition 3.12, for a proof. The author limits itself to cyclic groups, but the proof remains valid for an arbitrary group.

LEMMA 2.6.   *Let $[T_i, \rho_i, f_i]_{i=1,2}$ be two points of $\mathcal{T}_G(C, \varphi)$. Then there exist G-equivariant quasiconformal homeomorphisms homotopic to $f_2 \circ f_1^{-1}$. Among them, there is a single one f_0, for which the infimum of the maximal dilatations is reached, almost everywhere on the surface. Moreover its dilatation K_{f_0} is a constant.*

The distance is then defined by $d([T_1, \rho_1, f_1], [T_2, \rho_2, f_2]) = \ln(K_{f_0})$.

The *relative modular group* of a couple $(C; \varphi)$, denoted by $Mod_G(C, \varphi)$, is the quotient

$$\frac{Z_G(C, \varphi)}{Z_G(C, \varphi) \cap \mathrm{Diff}_+^0(C)},$$

where $Z_G(C, \varphi)$ is itself the quotient of the centralizer of $\varphi(G)$ in $\mathrm{Diff}_+(C)$ by $\varphi(Z(G))$. An element f of $Mod_G(C, \varphi)$ acts on $\mathcal{T}_G(C, \varphi)$ by $f \cdot [T, \rho, h] = [T, \rho, h \circ f^{-1}]$. As said above:

LEMMA 2.7.   *The quotient of $\mathcal{T}_G(C, \varphi)$ by the action of $Mod_G(C, \varphi)$ is in bijection with the Hurwitz space $\mathcal{H}_G(C, \varphi)$. The stabilizer of some fixed point $[T, \rho, h]$ under the action of the relative mapping class group is in bijection with the commutant of $\rho(G)$ in $\mathrm{Aut}(T)$ quotiented by $Z(G)$.*

PROOF.   Let $[C_i, \varphi_i, f_i]$ map to $[C_i, \varphi_i]$ for $i = 1, 2$ and assume that $f : C_1 \to C_2$ is a G-equivariant biholomorphism. Then $m = f_1^{-1} \circ f^{-1} \circ f_2$ is an element of the relative modular group $Mod_G(C, \varphi)$, and $[C_1, \varphi_1, f_1] = [C_2, \varphi_2, f_2 \circ m^{-1}]$ in $\mathcal{T}_G(C, \varphi)$.   □

The Hurwitz space $\mathcal{H}_G(C, \varphi)$ is endowed with the quotient topology. Furthermore, the action of $Mod_G(C, \varphi)$ on Teichmüller space is a discontinuous one (see [16]); hence $\mathcal{H}_G(C, \varphi)$ acquires the structure of an orbifold. On $\mathcal{H}_g(G, R)$ we put the natural topology so that its subsets $\mathcal{H}_G(C, \varphi)$ become its connected components.

THEOREM 2.8.   *Let C be a compact Riemann surface equipped with a biholomorphic action $\varphi : G \hookrightarrow \mathrm{Aut}(C)$, and P be the set of branch points of the quotient surface. Then the Teichmüller spaces $\mathcal{T}_G(C, \varphi)$ and $\mathcal{T}(C/\varphi(G), P)$ are analytically equivalent.*

It is a classical theorem, see Corollary 3 of [15], Proposition 8 of [16].

Note however that Teichmüller spaces considered here are slightly different from the ones in [15] and [16], since we consider group actions up equivariance instead of global conjugacy. However, using the lemma in § 2.3 of [7], it is easy to adapt their proof. For, this lemma of C. J. Earle ensures that an orientation preserving diffeomorphism homotopic to the identity map which normalizes the action of G, commutes with the action of G.

In [11], the authors describe the normalization of the locus in the moduli space $\mathcal{M}_h$ of points with automorphism group G. They also work with fixed topological type. This normalization is the quotient of a Teichmüller space by a relative modular group. Since their relative modular group is a normalizer rather than a centralizer, the Hurwitz spaces $\mathcal{H}_G(C, \varphi)$ defined here are finite Galois coverings of these normalizations (with Galois group $\text{Out}(G)$).

In this work, we restrict ourself to the topological setting on Teichmüller spaces. However, it will be worthwhile to study whether the combinatorial study of Teichmüller spaces (see §5) extends to the analytic setting.

We now turn to the general case, and define a Hurwitz space for non Galois coverings with fixed monodromy action. Again, we need a version of Riemann's existence theorem [8] [9] [10].

THEOREM 2.9. *There is a 1-1 correspondence between $\mathcal{H}(g, g', d)$ and the space of following couples $[S - Y, \psi]$: S is a compact Riemann surface of genus g', Y is a finite subset of S, $\psi : \pi_1(S - Y, q) \to S_d$ is a group homomorphism with transitive image, defined up to base point change; and $[S' - Y', \psi'] = [S - Y, \psi]$ if and only if there exist some biholomorphism $h : S \to S'$ sending Y on Y', and $\theta \in \text{Int}(S_d)$, such that $\theta \circ \psi = \psi' \circ h_*$.*

As usually, the image G of ψ is called the *monodromy group* of the corresponding covering. Let $[p : T \to S] \in \mathcal{H}(g, g', d)$ and $\psi : \pi_1(S - Y, q) \to G \hookrightarrow S_d$ the corresponding homomorphism. Another important data is the conjugacy class of the stabilizer K under the monodromy action of any point in the fiber over the base point q. This is a subgroup of G, of index d, which satisfies: $\bigcap_{t \in G} t^{-1} K t = \{e_G\}$.

Note that the natural action of G on the coset G/K of right classes (G acts by $(h, Kg) \mapsto Kgh^{-1}$), and the identification between right classes and the set $\{1, \ldots, d\}$, furnish the embedding $G \hookrightarrow S_d$: faithfulness of the action is implied by the property of K.

DEFINITION 2.10. We fix G and K a subgroup of index d such that $\bigcap gKg^{-1} = \{e_G\}$. Let $i : G \hookrightarrow S_d$ the embedding given by the action of G on G/K.

- Let $[p : T \to S]$ in $\mathcal{H}(g, g', d)$ with the corresponding $\psi : \pi_1(S - Y, q) \to S_d$. Then the branched covering p is said to have a monodromy of type (G, K) if there exists an isomorphism α between $\text{Im}(\psi)$ and G such that $i \circ \alpha$ is the identity on $\text{Im}(\psi)$ (the corresponding permutation representations are isomorphic).
- $\mathcal{H}(g, g', G, K)$ is the subset of $\mathcal{H}(g, g', d)$ made of elements with monodromy of type (G, K).

For an element of $\mathcal{H}(g, g', G, K)$, the stabilizer K_i of any point x_i in the fiber over the base point q is then conjugated to K via α. We set $\tilde{\psi} = \alpha \circ \psi$.

We cannot define the ramification data in this setting, due to the fact that the equivalence relation involved is the conjugation in the symmetric group, not in G. The solution comes from the operation of Galois closure [10].

Let us consider the embeddings $p_*^i : \pi_1(T - X, x_i) \to \pi_1(S - Y, q)$, and $H_i = p_*^i(\pi_1(T - X, x_i))$. Then these are conjugated subgroups of $\pi_1(S - Y, q)$, $K_i = \psi(H_i)$ and $\ker(\psi) = \cap_i H_i$.

Now $\tilde{\psi} = \alpha \circ \psi$ furnishes, via the theorem 2.3, a Riemann surface C with a well-defined holomorphic action φ of G on C such that $S = C/\varphi(G)$, with a ramification data induced by $\tilde{\psi}$. Clearly, the restriction $\tilde{\psi} : H_i \to K_i$ gives T isomorphic to $C/\varphi(K_i)$.

The Galois covering $C \to S$ is called a Galois closure of $p : T \to S$, this is a minimal Galois covering factorizing by p. We note that if $[\psi_i, \alpha_i]_{i \in \{1,2\}}$ represents two elements of $\mathcal{H}(g, g', G, K)$ in the same class, then $\tilde{\psi}_2 = \alpha_2 \circ \theta \circ \alpha_1^{-1} \circ \tilde{\psi}_1$ for θ in $\mathrm{Int}(S_d)$, and $\alpha_2 \circ \theta \circ \alpha_1^{-1}$ is an element of $\mathrm{Aut}(G, K)$, the subgroup of $\mathrm{Aut}(G)$ whose elements preserve the conjugacy class of K.

Thus a Galois closure $[C, \varphi]$ with ramification data R is defined only modulo the action of $\mathrm{Out}(G, K)$ (since $\mathrm{Int}(G)$ acts trivially).

Let $\mathrm{Out}(G, K, R)$ be the subgroup whose elements also preserve the ramification data R.

PROPOSITION 2.11. *Let (G, K) be as in the previous definition, R a ramification data of G, and h an integer. There is a bijection*

$$\mathcal{H}(g, g', G, K) \longleftrightarrow \coprod_{R/\mathrm{Out}(G,K)} \frac{\mathcal{H}_h(G, R)}{\mathrm{Out}(G, K, R)}$$

PROOF. There is a well-defined map from $\mathcal{H}_h(G, R)$ to $\mathcal{H}(g, g', d)$ sending $[C, \varphi]$ to the branched covering $[C/\varphi(K) \to C/\varphi(G)]$.

If $\psi : \pi_1(S - Y, q) \to G$ is the corresponding epimorphism, then using the action of G on G/K, we also have an homomorphism $\tilde{\psi} : \pi_1(S - Y, q) \to S_d$ with monodromy group G and isotropy group K. Thus we have in fact a map $\mathcal{H}_h(G, R) \to \mathcal{H}(g, g', G, K)$. Note that since the monodromy is determined, we can calculate the genus g, since g' is known from h and R.

This map factorize by the quotient of $\mathcal{H}_h(G, R)$ by $\mathrm{Out}(G, K, R)$. If θ is an element of $\mathrm{Out}(G, K, R)$, then we can extend it to $\bar{\theta} \in \mathrm{Int}(S_d)$. Assuming $\theta(K) = K$, $\bar{\theta}$ is the conjugation by the bijection $Kg \mapsto K\theta(g)$.

We now prove that the map is injective. Assume that $[C_i, \varphi_i] \mapsto [C_i/\varphi_i(K) \to C_i/\varphi_i(G)]$ such that

$$[p_1 : C_1/\varphi_1(K) \to C_1/\varphi_1(G)] = [p_2 : C_2/\varphi_2(K) \to C_2/\varphi_2(G)],$$

i.e., there exist $f : C_1/\varphi_1(K) \to C_2/\varphi_2(K)$ and $h : C_1/\varphi_1(G) \to C_2/\varphi_2(G)$ such that $p_2 \circ f = h \circ p_1$. This means that $\theta \circ \psi_1 = \psi_2 \circ h_*$ for some $\theta \in \mathrm{Int}(S_d)$. Thus we can lift f onto $k : (C_1, \varphi_1) \to (C_2, \varphi_2)$ such that $k \circ (\varphi_1 \circ \theta'(g)) = \varphi_2(g) \circ k$ for $\theta' \in \mathrm{Out}(G, K)$ induced by θ. It remains to show that $\theta'(R) = R$. But k is an equivariant map between $[C_2, \varphi_2]$ and $[C_1, \varphi_1 \circ \theta']$ so that $R(\varphi_2) = R(\varphi_1 \circ \theta')$. Since $R(\varphi_1) = R(\varphi_2)$, we are done.

Finally, the discussion preceding the proposition shows that every element of $\mathcal{H}(g, g', G, K)$ comes exactly from one element of $\mathcal{H}_h(G, R)/\text{Out}(G, K, R)$, provided that R is taken modulo $\text{Out}(G, K)$. □

The image of $\mathcal{H}_h(G, R)/\text{Out}(G, K, R)$ into $\mathcal{H}(g, g', G, K)$ will be denoted by $\mathcal{H}(g, g', G, K, R)$; this is the space involved in our main theorem. We will say that an element of $\mathcal{H}(g, g', G, K, R)$ possess a monodromy of type (G, K, R), but will keep in mind that R is defined modulo $\text{Out}(G, K)$.

EXAMPLE 2. We look at the case of degree d simple coverings, for which the monodromy around each branch point is given by a single transposition of S_d. Thus the preimage of a branch point consists in $d - 2$ points of index one, and one point of index two. The ramification data is $R = b\tau$ where b is the number of branch points and τ is the conjugacy class of transpositions. In this case (see [9] for more details) the monodromy group is S_d, and $K \cong S_{d-1}$, so that $\text{Out}(S_d, K)$ is the trivial group. Also, the spaces $\mathcal{H}_h(S_d, R)$ are connected [9].

3. Strebel Differentials

We recall the main ingredient of cellular decomposition of moduli spaces of pointed Riemann surfaces, i.e., Strebel's theorem [25], and show how to adapt it to the study of Hurwitz spaces.

Let C be a compact Riemann surface of genus g with a set of n punctures $X = \{x_1, \ldots, x_n\}$. Let us consider the set $St(C, X)$ of holomorphic quadratic differentials q on $C \backslash X$ with the following properties:

(1) Its critical graph covers a set of measure zero (q is a Jenkins–Strebel differential).
(2) Each puncture x_i is a double pole with a real negative coefficient $-p_i^2$.
(3) Its characteristic ring domains are n disks punctured at the x_i, described by the closed trajectories.

The elements of $St(C, X)$ are often called *Strebel differentials*.

Recall that the critical graph Γ_q is the union of non closed trajectories, and that the perimeter of a closed trajectory is its length measured with the metric induced by the quadratic differential.

The second condition means that the perimeter of a closed trajectory around x_i is p_i. The third condition says that the critical graph cuts the surface into n punctured disks described by the closed trajectories. In this case, Γ_q consists in segments joining zeroes of the differential. Moreover, all its vertices are at least trivalent. Indeed, vertices are zeroes of q, and this excludes monovalent vertices (corresponding to first order poles) and bivalent vertices (corresponding to some regular points).

THEOREM 3.1. [25] *Let C be a compact Riemann surface of genus g with a set $X = (x_1, \ldots, x_n)$ of n punctures such that $2g - 2 + n > 0$ and $n > 0$. Then* *any n-tuple $(p_1, \ldots, p_n) \in \mathbb{R}^n_{>0}$ of perimeters determines a single element w of $St(C, X)$ such that the perimeter around x_i is p_i.*

The following proposition shows that Strebel differentials behave well under lifting through ramified coverings.

PROPOSITION 3.2. *Let $[p : C \to S]$ be some class of ramified coverings between two compact Riemann surfaces. Let Y be the set of branch points, b its cardinal, and X be the set of ramification points. Denote by e_x the multiplicity of $x \in X$. Then each b-tuple $(p_1, \ldots, p_b)$ of strictly positive real numbers determines a unique couple of Strebel differentials $w_C \in St(C, X)$ and $w_S \in St(S, Y)$, such that $p^*(w_S) = w_C$, and the perimeter of a closed trajectory of w_C around $x \in X$ which maps onto $y_j \in Y$ is $e_x p_j$.*

PROOF. We first apply Strebel's theorem to the surface S endowed with the b-tuple $(p_j)_{j \in \{1, \ldots, b\}}$. We obtain a unique element w_S of $St(S, Y)$ such that $w_S = -(p_j)^2(dz_j/z_j)^2$ around $y_j \in Y$. Then consider the pullback $p^*(w_s) = \Phi(\hat{z})(d\hat{z})^2$. For local coordinates such that $p(\hat{z}) = z$, it is defined by $p^*(w_s) = \Phi(z)(dz)^2$. If precisely $z = \hat{z}^{e_x}$, then $(dz/z)^2 = e_x^2(d\hat{z}/\hat{z})^2$, so that the only poles of order two of $p^*(w_s)$ are the ramification points. Moreover perimeters of $p^*(w_s)$ are the $e_x p_j$. The critical graph of $p^*(w_S)$ is the pullback of Γ_{w_S}, and its characteristic ring domains are necessarily disks punctured at the ramification points, so that $p^*(w_S) \in St(C, X)$. Uniqueness is obtained by a new application of Strebel's theorem. □

If we take $[C, \varphi] \in \mathcal{H}_g(G, R)$, then the same arguments yield a Strebel differential q relative to C and to fixed points, invariant under $\varphi(G)$. Moreover, the action of $\varphi(G)$ on q induce a free isometric action on its critical graph.

We also need a more precise version of Strebel's theorem for the case of Riemann surfaces with nodes. Let T be a compact Riemann surface with $n \geq 1$ punctures $P = (P_1, \ldots, P_n)$, and m ties $Q = (Q_1, \ldots, Q_m)$ (some other distinguished points, distinct from the punctures). Then we consider the set $St'(T, P, Q)$ of quadratic differential, holomorphic on $T \backslash (P \cup Q)$, with the properties 1,2 and 3 stated out above (replace x_i by P_i), but we also demand that ties are vertices or middles of edges of the critical graph. Hence, they are possibly monovalent or bivalent vertices of the critical graph.

THEOREM 3.3. *Let T be a compact Riemann surface of genus g, with a set of $n \geq 1$ punctures $P = (P_1, \ldots, P_n)$, and a set of m ties $Q = (Q_1, \ldots, Q_m)$, such that $2g - 2 + n + m > 0$ and $n \geq 2$ if $g = 0$. Then each b-tuple $(p_1, \ldots, p_n) \in \mathbb{R}^n_{>0}$ determines a unique element of $St'(T, P, Q)$ with perimeters equal to p_i around P_i.*

PROOF. Assume that $m = 2k$. Using the Riemann's existence theorem (theorem 2.3), we take a double covering $p : R \to T$ ramified only at the ties. The surface R is endowed with the unique complex structure such that the projection becomes holomorphic. Put $(S_i^1, S_i^2) = p^{-1}(P_i)$ and $T_j = p^{-1}(Q_j)$. We have $g(R) = 2g - 1 + k$ by Riemann–Hurwitz, and the stability condition asserts that $2g(R) - 2 + 2n > 0$. We apply Strebel's theorem to R: there exists a unique element w of $\mathcal{St}(R, (S_i^1, S_i^2)_i)$ with perimeters equal to $(p_1, p_1, p_2, p_2, \ldots, p_n, p_n)$.

Let α be the order two deck transformation of the covering. Then $\alpha(S_i^1) = S_i^2$, and uniqueness in Strebel's theorem tells us that $\alpha^*(w) = w$. Furthermore, α induces an isometry of the critical graph Γ_w; and since $\alpha(T_j) = T_j$, these points belong to Γ_w. Thus, they are vertices of even valency (possibly bivalent) of Γ_w.

Since $\alpha^*(w) = w$, we get a quadratic differential q on T such that $p^*(q) = w$. The punctures P_i are its poles of order two, the perimeters are given by the p_i, and clearly $q \in \mathcal{St}'(T, P, Q)$. If the valency of the vertex T_j is $2v$ then the valency of Q_j is v, possibly equal to one. Uniqueness of w gives uniqueness of q.

If $m = 2q + 1$, we take a double covering ramified only in $Q_1, \ldots, Q_{2k}$ to bring us back to the previous case.   □

4. Graphical Construction of Coverings

DEFINITION 4.1. [18] [3] A *fat-graph* Γ is given by a finite set $A(\Gamma)$ (of oriented edges) and by two permutations σ_0 and σ_1 of this set, where σ_1 is an involution without fixed points.

All fat-graphs will be connected: we assume that the group generated by σ_0 and σ_1 (the so-called cartographic group) acts transitively on the set of oriented edges.

The geometric edges are the orbits of $\sigma_1(\Gamma)$, and the vertices are those of $\sigma_0(\Gamma)$. We denote by $A_g(\Gamma)$ and $V(\Gamma)$ these sets, and by $a(\Gamma)$ and $v(\Gamma)$ their cardinals. Note that Γ is a graph. If $a \in A(\Gamma)$, we denote by $a(0)$ its origin and by $a(1)$ its end. We take the convention that $a(1)$ is the σ_0-orbit of a and $a(0)$ is the σ_0-orbit of $\sigma_1(a)$. For convenience, we often put $\bar{a} = \sigma_1(a)$.

We define $\sigma_2(\Gamma) = \sigma_1(\Gamma)\sigma_0(\Gamma)^{-1}$, such that $\sigma_0\sigma_1\sigma_2 = 1$. The orbits of $\sigma_2(\Gamma)$ are called the *faces* of Γ. We denote the set of faces by $F(\Gamma)$ and its cardinal by $f(\Gamma)$. The length of a cycle of $\sigma_0(\Gamma)$ (resp. $\sigma_2(\Gamma)$) is the valency of the corresponding vertex (resp. face).

Every face is an oriented loop, for, if $b = \sigma_2(a)$ then $b(0) = a(1)$.

As an example, a graph Γ embedded in a compact oriented surface is a topological realization of a fat-graph. The permutation σ_0 of Γ is given by the projection of the neighborhoods of the vertices on the tangent planes at these points.

The genus $g(\Gamma)$ of the fat-graph Γ is then defined by the Euler formula:

$$2 - 2g(\Gamma) = v(\Gamma) - a(\Gamma) + f(\Gamma).$$

A morphism $f : \Gamma \to \Gamma'$ between two fat-graphs is a map $f : A(\Gamma) \to A(\Gamma')$ which satisfies to $f \circ \sigma_i = \sigma_i' \circ f$ for $\in \{0, 1, 2\}$. If f is bijective, then f is an isomorphism. In this definition of an isomorphism, we assume that faces can be exchanged. If we specify some colors on the set of faces, then we ask that isomorphisms respect these colors. The (full) *automorphism group* $\mathrm{Aut}(\Gamma)$ is the centralizer of σ_1 and σ_2 in the group of all permutations of $A(\Gamma)$.

Figure 1 shows a fat-graph of genus one described by its cartographic group $\sigma_0(\Delta) = (abdc)(\bar{a}\bar{b}\bar{c}\bar{d})$, $\sigma_1(\Delta) = (a\bar{a})(b\bar{b})(c\bar{c})(d\bar{d})$, $\sigma_2(\Delta) = (a\bar{c}\bar{b}\bar{a}d b)(c\bar{d})$.

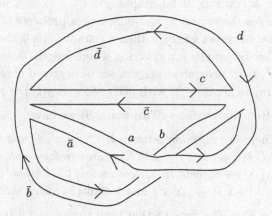

Figure 1. A fat-graph Δ of genus one.

Let Γ be a fat-graph. A metric m on Γ is given by a map $m : A_g(\Gamma) \to \mathbb{P}(\mathbb{R}_{>0}^{a(\Gamma)})$; we say that $m(e)$ is the length of the geometric edge e. We note (Γ, m) a fat-graph endowed with a metric, and we call it a *Riemannian fat-graph*. The perimeter $p(F)$ of a face $F = (a_1, \ldots, a_k)$ of (Γ, m) is defined by $p(F) = \sum_{i=1}^{k} m(\{a_i, \bar{a}_i\})$.

An isomorphism between two Riemannian fat-graphs is called an isometry if it preserves lengths of edges. We denote by $\mathrm{Aut}(\Gamma, m)$ the group of isometric automorphisms of (Γ, m).

Given Γ, we can realize its faces as oriented polygons in the plane, and fill in by punctured disks. Then gluing them with σ_1 give an orientable compact surface $F(\Gamma)$ minus $f(\Gamma)$ points (one for each face), together with an embedding $i : |\Gamma| \hookrightarrow F(\Gamma)$ such that $i(|\Gamma|)$ is a retract by deformation of $F(\Gamma)$. We will see that some metric on Γ determines some complex structure on $F(\Gamma)$.

From the induced isomorphism between first homotopy groups $\pi_1(|\Gamma|)$ and $\pi_1(F(\Gamma))$ we deduce that $g(\Gamma) = g(F(\Gamma))$. For, on one hand, $\pi_1(F(\Gamma))$ is a free group of rank $2g(F(\Gamma)) + f(\Gamma) - 1$, and on the other hand, $\pi_1(|\Gamma|)$ is isomorphic to the fundamental group of Γ which is free of rank $a(\Gamma) - v(\Gamma) + 1 = 2g(\Gamma) + f(\Gamma) - 1$.

Faces of a fat-graph Γ provide some particular elements of the fundamental group $\pi_1(\Gamma, p)$ of Γ. Let $(a_1, a_2, \ldots, a_k)$ a face of Γ, with $a_i = \sigma_2^{i-1}(a_1)$. Join $a_1(0)$ to the base vertex p by an oriented path α, and consider the homotopy class of the oriented loop $\gamma = \alpha a_1 \cdots a_k \bar{\alpha}$. Note that another choice of α leads to a conjugate loop. And if the face is given by $(a_i, \ldots, a_{i-1})$ with $i > 1$, we get the oriented loop $\beta a_i \cdots a_{i-1}\bar{\beta}$, homotopic to $(\beta a_i \cdots a_l)(a_1 \cdots a_{i-1}a_i \cdots a_l)(\bar{a}_l \cdots \bar{a}_i\bar{\beta})$, which is a conjugate loop of γ.

Therefore to each face is associated a well defined conjugacy class of elements of $\pi_1(\Gamma, p)$, which will be called a *loop-face*.

Fat-graphs whose vertices are at least trivalent are called *smooth* fat-graphs.

DEFINITION 4.2.   A covering of fat-graphs $p : \Gamma \to \Delta$ is a morphism of fat-graphs which satisfies the existence and uniqueness property of lifting oriented paths. The cardinal of vertex's fiber is then constant: this is the degree. A covering of Riemannian fat-graphs is a covering where lengths edges are preserved. Define $\mathcal{H}^{\mathrm{comb}}(g, g', d)$ to be equivalence classes of degree d coverings of smooth Riemannian fat-graphs $[p : \Gamma \to \Delta]$ with $g(\Gamma) = g$ and $g(\Gamma') = g'$.

As in the case of Riemann surfaces (Theorem 2.9) we have:

THEOREM 4.3.   *There is a bijection between* $\mathcal{H}^{\mathrm{comb}}(g, g', d)$ *and the set of classes* $[\Delta, l, \psi]$ *where* (Δ, l) *is a genus* g' *smooth Riemannian fat-graph, and* ψ *is a group homomorphism* $\pi_1(\Delta) \to S_d$ *with transitive image;* $[\Delta, l, \psi] = [\Delta', l', \psi']$ *if there exists an isometry* $h : (\Delta, l) \to (\Delta', l')$ *and* $\theta \in \mathrm{Int}(S_d)$ *such that* $\theta \circ \psi = \psi' \circ h_*$.

It will become clear after Theorem 4.8 that both theorems 2.9 and 4.3 are equivalent, using the fact that Riemannian fat-graphs are deformation retract of punctured Riemann surfaces. In fact, it is not difficult to give a direct proof of 4.3. Furthermore, we have the notions of monodromy groups and monodromy actions. Thus we define $\mathcal{H}_{g,g'}^{\mathrm{comb}}(G, K)$ in perfect analogy with $\mathcal{H}(g, g', G, K)$.

We now focus on the case of Galois coverings. Let Γ be a fat-graph, and G be some finite group. We denote by $\varphi : G \hookrightarrow \mathrm{Aut}(\Gamma)$ an action of G on Γ. We call it *quasifree* if $\varphi(G)$ acts freely on the underlying graph (on geometric edges and vertices), but with non trivial isotropy group for each face of Γ (these actions will provide some actions of G on surfaces with fixed points). Then the quotient fat-graph is well defined. In fact, the quotient graph is well defined by freeness of $\varphi(G)$ on edges. And since edges incident on a vertex of the quotient graph are in bijection with edges incident on any vertex of its fiber, we put the induced cyclic ordering. The projection $p : \Gamma \to \Gamma/\varphi(G)$ is a Galois covering of fat-graphs. Two couples (Γ, φ) and (Γ', φ') are equivalent if they differ by a G-equivariant isomorphism.

Inspired by the setting of Riemann surfaces (see the previous section), we look at stabilizers G_F of faces F. Let $\sigma_{2,F}$ be the cycle defining F, of order v_F. Then $g \in G_F$ means that $\varphi(g)\sigma_{2,F}\varphi(g^{-1}) = \sigma_{2,F}$. Hence $\varphi(g) = \sigma_{2,F}^i$ on F, for $i \in \mathbb{Z}/v(F)$, and we have an embedding $\varphi_F : G_F \hookrightarrow \langle \sigma_{2,F} \rangle \cong \mathbb{Z}/v(F)$. Set e_F the order of G_F. Then $\varphi_F^{-1}(\sigma_{2,F}^{v_F/e_F})$ is a privileged generator of G_F, which we denote by τ_F. As before, we say that F and its orbit are colored by $C(\tau_F)$, the conjugacy class of τ_F.

Let $\{O_i, b_i\}_{i \in \{1,\ldots,t\}}$ be the set of colored orbits of faces with multiplicities, and $C(\tau_i)$ be the conjugacy classes of privileged generators.

Then the ramification data $R(\varphi)$ associated to φ is given by

$$R(\varphi) = \sum_{i=1}^{t} b_i C(\tau_i).$$

If $f : (\Gamma, \varphi) \to (\Gamma', \varphi')$ is a G-equivariant isomorphism, then $R(\varphi) = R(\varphi')$. The degree of R is the number of face's orbits, i.e., the number of faces of the quotient fat-graph $\Gamma/\varphi(G)$. Furthermore, note that if $R = \sum_{i=1}^{t} b_i C_i$, and if $e_i = \mathrm{Ord}(C_i)$ is the order of any element in the class C_i, then we simply have $f(\Gamma) = \sum_{i=1}^{t} b_i |G|/e_i$.

Writing the Euler–Poincaré formulas, we recover now the Riemann–Hurwitz formula $2g(\Gamma) - 2 = |G|(2g(\Gamma/\varphi(G)) - 2) + \sum_{i=1}^{b} |G|(1 - 1/e_i)$.

DEFINITION 4.4. Let G be a finite group, and R a ramification data of G. Then $\mathcal{H}_g^{\mathrm{comb}}(G, R)$ is the set of equivalence classes $[\Gamma, l, \varphi]$ where (Γ, l) is a smooth Riemannian fat-graph of genus g, and $\varphi : G \hookrightarrow \mathrm{Aut}(\Gamma, l)$ is a quasifree action of G such that $R(\varphi) = R$. Two elements are equivalent if they differ by a G-equivariant isometry.

We state now a combinatorial version of Riemann's existence theorem for Galois coverings (Theorem 2.3).

THEOREM 4.5. *The space $\mathcal{H}_g^{\mathrm{comb}}(G, R)$ is in 1-1 correspondence with a set made of classes $[\Delta, l, \psi : \pi_1(\Delta) \to G]$, where (Δ, l) is a smooth Riemannian fat-graph of genus g', ψ is an epimorphism, defined up to conjugacy, sending loop-faces on prescribed images (by the ramification data), and such that $[\Delta, l, \psi] = [\Delta', l', \psi']$ if there exists an isometry $h : (\Delta, l) \to (\Delta', l')$ such that $\psi = \psi' \circ h_*$.*

If $\mathrm{Aut}(\Delta, \psi)$ is the subgroup of $\mathrm{Aut}(\Delta)$ made of elements h satisfying $\psi = \psi \circ h_*$, and if $\mathrm{Aut}_G(\Gamma, \varphi)$ is the centralizer of $\varphi(G)$ in $\mathrm{Aut}(\Gamma)$, then we have $\mathrm{Aut}(\Delta, \psi) \cong \mathrm{Aut}_G(\Gamma, \varphi)/Z(G)$.

Again, after Theorem 4.8, it will become clear that Theorems 2.3 and 4.5 are equivalent.

We also have an analog of Proposition 2.11, which allows us to define the set $\mathcal{H}_{g,g'}^{\mathrm{comb}}(G, K, R)$.

EXAMPLE 3. We give a covering of the genus one fat-graph Δ described previously. Given (Δ, ψ) we could build (Γ, φ) such that $\Delta = \Gamma/\varphi(G)$ as the universal covering of Δ quotiented by $\ker(\psi)$. Instead, we use the notion of the Cayley graph, which gives a more tractable method to build (Γ, φ) from (Δ, ψ).

First we choose a presentation of $\pi_1(\Delta, O)$. Take the edge $(b, \bar{b})$ as a maximal tree. Then $\pi_1(\Delta, O) = \langle \delta_a, \delta_c, \delta_d \rangle$ where $\delta_a = a\bar{b}$, $\delta_c = c\bar{b}$, $\delta_d = d\bar{b}$. Take $\gamma_3 = (c\bar{d})$ and $\gamma_4 = (a\bar{c}\bar{b}a\bar{d}\bar{b})$ for loop-faces. An epimorphism ψ from $\pi_1(\Delta, O)$ to G must satisfy $\psi(\gamma_3) = \psi(\gamma_4)^{-1}$. Choose $\psi(\gamma_3) = [1]$ and $\psi(\gamma_4) = [2]$ as ramification data.

This implies $\psi(\delta_c \delta_d^{-1}) = [1]$ and $\psi(\delta_a \delta_c^{-1} \delta_a^{-1} \delta_d) = [2]$. We choose $\psi(\delta_a) = [1]$, $\psi(\delta_c) = [2]$, $\psi(\delta_d) = [1]$. Note that the corresponding element $\Gamma(\psi)$ of

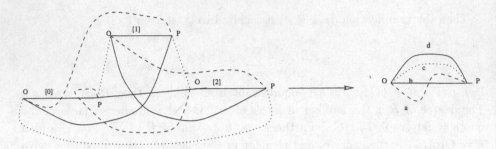

Figure 2. The covering $\Gamma(\Psi) \to \Delta$.

$\mathcal{H}_g^{\mathrm{comb}}(\mathbb{Z}/3, R)$ possess two faces of stabilizers $\mathbb{Z}/3$. Thus $g = 3$ by Riemann–Hurwitz.

To build $\Gamma(\psi)$ (figure 2), we take three copies of the maximal tree, indexed by $[0], [1], [2]$. Join the vertex (O, i) to the vertex (P, j) if $[i] + \psi(\delta_a) = [j]$ to build edges projecting on $(a, \bar{a})$. Idem for c and d. Cyclic orderings around the vertices are giving by these around O and P. The action of G on vertices of $\Gamma(\psi)$ is $g \cdot (O, i) = (O, g \times i)$ (likewise for P).

We describe now the main combinatorial operation on fat-graphs: Whitehead collapses. We generalize them in an obvious way to fat-graphs with a quasifree action of a group G.

DEFINITION 4.6. Let (Γ, φ) a fat-graph with a quasifree action of the (possibly trivial) group G, and $e = (a, \bar{a})$ a geometric edge with $a(0) \neq a(1)$. We denote $O(e)$ the orbit of e under $\varphi(G)$. We assume that $O(e)$ do not contain any loop of Γ. The operation which consists in retracting $O(e)$ and gluing $O(a(0))$ with $O(a(1))$ is called an *equivariant Whitehead collapse* along $O(e)$. We note $W_{O(e)}(\Gamma)$ the new graph acquired in this way.

We group together in a lemma the first properties of this operation.

LEMMA 4.7. (i) *The graph $W_{O(e)}(\Gamma)$ is a fat-graph.*

(ii) *The number of faces and the genus are invariant under equivariant Whitehead collapses.*

(iii) *The quasifree action φ of G on Γ restricts to a quasifree action of G on $W_{O(e)}(\Gamma)$. Moreover the ramification data is preserved.*

PROOF. For the first point, we describe the cartographic group of $W_{O(e)}(\Gamma)$.

Let $(a_1^j, \ldots, a_k^j)$ and $(b_1^j, \ldots, b_l^j)$ for $j \in \{1, \ldots, |G|\}$ be the G-orbit of $a(0)$ and $a(1)$, such that $\{a_1^j\}_j$ (resp. $\{b_1^j\}_j$) is the orbit of $\bar{a}$ (resp. a). Then retracting the orbit $O(e)$ give the new vertices $s_j = (a_2^j, \ldots, a_k^j, b_2^j, \ldots, b_l^j)$. The graph $W_{O(e)}(\Gamma)$ is the fat-graph with same σ_1 private of $O(e) = \{(a_1^j, b_1^j)\}_j$, and same σ_0 except for the orbits of $a(0)$ and $a(1)$ replaced by the new orbit $\{s_j\}_j$.

As we can retract neither a loop, nor an orbit which contains some faces, we have $f(W_{O(e)}(\Gamma)) = f(\Gamma)$. Since $a(W_{O(e)}(\Gamma)) = a(\Gamma) - |G|$ and $s(W_{O(e)}(\Gamma)) = s(\Gamma) - |G|$, we have $g(W_{O(e)}(\Gamma)) = g(\Gamma)$ by Euler–Poincaré.

The action of G on $W_{O(e)}(\Gamma)$ is defined by restriction of φ on $A(\Gamma)\backslash(O(a) \cup O(\bar{a}))$.

Let $\sigma_{2,k}$ be the cycle of $\sigma_2(\Gamma)$ defining a face F_k, of order v_k. The privileged generator τ_k of the stabilizer is defined such that

$$\sigma_{2,k} = (a_1 a_2 \cdots a_{v_k - 1} \tau_k(a_1) \tau_k(a_2) \cdots \tau_k^{(v_k/e_k)-1}(a_1) \cdots \tau_k^{(v_k/e_k)-1}(a_{v_k - 1})).$$

Retracting the orbit of a_i do not disturb the action of τ_k on $\sigma_{2,k}$, and thus the ramification data is preserved. □

If $[\Gamma, m, \varphi] \in \mathcal{H}_g^{\mathrm{comb}}(G, R)$, then to each equivariant Whitehead collapses along $O(e)$, we associate $[W_{O(e)}(\Gamma), m, \varphi] \in \mathcal{H}_g^{\mathrm{comb}}(G, R)$, setting $m(\varphi(t)(e)) = 0$ for all $t \in G$.

Recall that $\mathcal{M}_{g',(b_1,\ldots,b_t)}^{\mathrm{comb}}$ consist in isometry classes of smooth Riemannian fat-graphs $[\Gamma, m]$ of genus g', with b faces shared in t colors, where isometries respect these colors. We state our main theorem (without compactifications):

THEOREM 4.8.   *Let G be a finite group, $R = b_1 C_1 + \cdots + b_t C_t$ a degree b ramification data, and K a subgroup of G such that $\bigcap_{t \in G} tKt^{-1} = \{e_G\}$.*

   Then we have the following commutative diagram

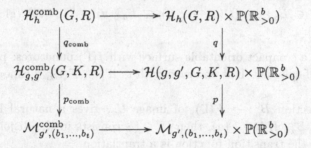

Horizontal maps are homeomorphisms, and vertical ones are ramified coverings.

The vertical maps have been already defined: $q([C, \varphi]) = [C/\varphi(K) \to C/\varphi(G)]$, and $p \circ q([C, \varphi]) = [C/\varphi(G), Y]$. The maps from the right to the left are given by Theorem 3.1 and Proposition 3.2. Next theorem gives the inverse maps. At the level of moduli spaces of punctured Riemann surfaces, the construction is known since [13]. More details are given in [20] and [23]. The continuity is discussed in the next section.

THEOREM 4.9.   *Let $[\Gamma, l, \varphi] \in \mathcal{H}_h^{\mathrm{comb}}(G, R)$. There exist a compact Riemann surface $\overline{\|\Gamma\|}$ endowed with $\hat{\varphi} : G \hookrightarrow \mathrm{Aut}(\overline{\|\Gamma\|})$ and $w_\Gamma \in \mathcal{S}t(\overline{\|\Gamma\|}, X)$, where $X = \{x_1, \ldots, x_{f(\Gamma)}\}$ is the set of fixed points of $\hat{\varphi}(G)$, such that:*

(1) *The critical graph of w_Γ is the image of a topological realization of Γ by an embedding $\iota : |\Gamma| \hookrightarrow \|\Gamma\|$ such that $\overline{\|\Gamma\|} - \iota(|\Gamma|)$ is a disjoint union of $f(\Gamma)$ disks, punctured at the x_i. Moreover $l_{w_\Gamma}(\iota(|a|)) = l(a)$, where l_{w_Γ} is the metric induced by w_Γ.*

(2) $\varphi(g)^*(w_\Gamma) = w_\Gamma \ \forall g \in G$, the action of $\hat{\varphi}(G)$ on $\iota(|\Gamma|)$ is the realization of the action of $\varphi(G)$ on Γ.

(3) If $[S, \psi]$ is another element of $\mathcal{H}_h(G, R)$ with preceding properties for a Strebel differential v, then there exists a single G-equivariant biholomorphism $h : S \to \overline{\overline{\|\Gamma\|}}$ such that $h^*(w_\Gamma) = v$.

PROOF. We first describe $\overline{\overline{\|\Gamma\|}}$ in terms of complex local coordinates (see [23] for a discussion along the same lines). If $a \in A(\Gamma)$, we denote by $F(a)$ the unique face in which it lies, and p_a the perimeter of this face. We associate to each $a \in A(\Gamma)$ the following closed strip embedded in the complex plane $\mathbb{C}$: $B_a = [0, p_a] \times [0, \infty[$. Moreover we establish a bijection $|F(a)| \to [0, p_a] \times \{0\}$ such that $|a|$ maps to $[0, l(a)] \times \{0\}$, $|\sigma_2(a)|$ maps to $[l(a), l(a) + l(\sigma_2(a))]$, and so on...

Define $F(\Gamma) = \left(\coprod_{a \in A(\Gamma)} B_a \right) / \sim$, where $\sim$ stands for the following identifications:

(1) $(0, t) \sim (p_a, t) \ \forall t \in [0, \infty[$ (Form half-cylinders).
(2) Let $b = \sigma_2^i(a) \in F(a)$ and $\alpha = \sum_{j=0}^{i-1} l(\sigma_2^j(a))$.
   Then $(x, t) \sim (x - \alpha, t) \ \forall x \in [\alpha, p_a]$, and $(x, t) \sim (p_a - \alpha + x, t) \ \forall x \in [0, \alpha]$, $\forall t \in [0, \infty[$.
(3) $\{(x, 0) \in B_a\} \sim \{(l(a) - x, 0) \in B_{\bar{a}}\} \ \forall x \in [0, l(a)]$ (glue the cylinders with σ_1).

So $F(\Gamma)$ is a compact orientable surface with $f(\Gamma)$ punctures: points at infinity of each half-cylinder. It is endowed with $\iota : |\Gamma| \hookrightarrow F(\Gamma)$, a canonical embedding of Γ.

The injection $B_a^0 \hookrightarrow F(\Gamma)$, of image U_a, gives a natural local coordinate $u_a : U_a \to B_a^0$. If $b \in F(a)$, then $U_a \cap U_b$ consists in the disjoint union of two strips, and the transition function is a translation.

Let $V_a \subset F(\Gamma)$ be the image of the infinite strip $]0, l(a)[\times]-\infty, +\infty[$. We define the local coordinate by $v_a = u_a$ on $V_a \cap U_a$, and $v_a = p(F(\bar{a})) - u_{\sigma_2(\bar{a})}$ on $V_a \cap U_{\sigma_2(\bar{a})}$. These coordinates are clearly holomorphically compatible with (U_a, u_a). Moreover, in these coordinates, we have the flat quadratic differential $(du_a)^2 = (dv_a)^2$, denoted by w_Γ.

Let now $s \in V(\Gamma)$ be a k-valent vertex with $k \geq 3$, and $(a_1, \ldots, a_k)$ the oriented edges pointing towards it such that $\sigma_0(a_i) = a_{i+1}$ for $i \in \{1, \ldots, k-1\}$ and $\sigma_0(a_k) = a_1$. Note that $\sigma_2(a_i) = \bar{a}_{i-1}$. Let

$$P_{a_i} =]-l(a_i), 0] \times]-\infty, +\infty[\ \cup \ ]0, l(a_{i-1})[\times]0, +\infty[$$

and let T_{a_i} its image in $F(\Gamma)$. We define $t_{a_i} : T_{a_i} \to P_{a_i}$ by $t_{a_i} = u_{a_i} - l(a_i)$ on the image of $]-l(a_i), l(a_{i-1})[\times]0, +\infty[$ and $t_{a_i} = v_{a_i} - l(a_i)$ on the image of $]-l(a_i), 0[\times]-\infty, +\infty[$.

Let $E_s = \bigcup_{j=1}^k T_{a_j}$, then the coordinate around s is $\xi_s : E_s \to \mathbb{C}$ defined by $\xi_s = \exp(2i\pi j/k) t_{a_j}^{2/k}$ on T_{a_j}.

Next we study the quadratic differential w_Γ. Since $\xi_s = ct^{2/k}$ where c is a constant, we deduce that $(dt)^2 = (k/2c)^2 \xi_s^{k-2} (d\xi_s)^2$, that is to say k-valent vertices are zeroes of order $k-2$ of w_Γ. Furthermore its closed trajectories are images in $\|\Gamma\|$ of the segments $[(0,t); (p_a, t)]$ for $t > 0$, and $\iota(|\Gamma|)$ is its critical graph. By construction, lengths of edges measured with the metric induced by w_Γ are these specified by the metric l. Finally, the change of coordinates from $]0, p_a[\times]0, \infty[$ onto a disk centered on the origin: $u_a \mapsto \zeta = \exp(2i\pi u_a/p_a)$ yields the following form of w_Γ near x_a: $(du_a)^2 = -\left(\frac{p_a}{2\pi}\right)^2 (d\zeta/\zeta)^2$. To sum up, we have $w_\Gamma \in \mathcal{S}t(\overline{\|\Gamma\|}, (x_1, \ldots, x_{f(\Gamma)}))$.

If $f : (\Gamma, l) \to (\Gamma_0, l_0)$ is an isometry, and as a consequence, $l_0 = cl$ with $c \in \mathbb{R}_{>0}$, then the homothety $h_c : B_a^0 \to B_{f(a)}^0$ gives a biholomorphism $\hat{f} : \overline{\|\Gamma\|} \to \overline{\|\Gamma_0\|}$. Thus to every isometric automorphism $\varphi(g)$, we associate a biholomorphic automorphism $\hat{\varphi}(g)$. The action on the closed trajectories of level $t > 0$ is defined to be the same than the action on the graph. The action on the punctures is specified by that on the corresponding faces. By construction, $\varphi(g)^*(w_\Gamma) = w_\Gamma \ \forall g \in G$.

It remains to prove the uniqueness property. Drop group actions to simplify. Let $k : |\Gamma| \hookrightarrow S$ be the embedding of Γ such that $k(|\Gamma|)$ is the critical graph of v. There is a single isotopy class of homeomorphisms $f : \|\Gamma\| \to S$ satisfying $f \circ \iota = k$ (existence is trivial, and uniqueness comes from the Alexander's lemma, which ensures that two orientation preserving homeomorphisms between topological disks which agree on the boundary of the first one are isotopic). Then we find only one biholomorphism in this isotopy class. Indeed, let y_j be the puncture of S corresponding to some face F_j of Γ. Let $\|F_j\|$ (resp. D_j) be the pointed disk by x_j (resp. y_j). Take $\xi_j : \|F_j\| \to D(0,1)$ a local coordinate as above, i.e., such that $(du_a)^2 = -p_j^2 (d\xi_j/\xi_j)^2$. By hypothesis, there exist a local coordinate $z_j : D_j \to D(0,1)$ such that $v = -p_j^2 (dz_j/z_j)^2$. Then define f restricted to $\|F_j\|$ by $f = z_j^{-1} \circ \xi_j$. This is a biholomorphism. The local coordinates ξ_j and z_j extend to the boundary with $\xi_j \circ \iota(|a|) = z_j \circ k(|a|)$, so that f is globally defined with $f \circ \iota = k$. $\qquad \square$

For non galoisian coverings, we have the following version of the theorem: to each $[p : (\Gamma, l) \to (\Delta, m)] \in \mathcal{H}_{g,g'}^{\mathrm{comb}}(G, K, R)$ is associated $[\hat{p} : \overline{\|\Gamma\|} \to \overline{\|\Delta\|}] \in \mathcal{H}_{g,g'}(G, K, R)$ such that $w_\Gamma = \hat{p}^*(w_\Delta)$.

PROOF OF THEOREM 4.8. The geometrical realization map

$$[\Gamma, l, \varphi] \mapsto \{[\overline{\|\Gamma\|}, \hat{\varphi}], (p_1, \ldots, p_b)\}$$

(where p_i is the perimeter in the orbit i) is well-defined, since a G-equivariant isometry yield a G-equivariant biholomorphism. The fact that this is the inverse map is ensured by uniqueness in Strebel theorem, and uniqueness in the geometrical realization (property 3 of the last theorem). Again, uniqueness in

Strebel theorem gives the commutativity of the diagram. It remains to prove the continuity; this is done in Theorem 5.2.

5. Cellular Decomposition

We now describe the cellular decomposition of decorated Hurwitz spaces. This leads to a characterization of its connected components. Also, we show that the computation of their orbifold characteristic relies on the computation of the degree between Hurwitz spaces and moduli spaces.

DEFINITION 5.1. Let $(\Gamma_0, l_0, \varphi_0) \in \mathcal{H}_g^{\text{comb}}(G, R)$ and $(\|\overline{\Gamma_0}\|, \hat{\varphi}_0)$ its geometrical realization. The combinatorial relative Teichmüller space $\mathcal{T}_G^{\text{comb}}(\Gamma_0, l_0, \varphi_0)$ is made of quadruplets (Γ, l, φ, f), where $(\Gamma, l, \varphi) \in \mathcal{H}_g^{\text{comb}}(G, R)$, together with a G-equivariant quasiconformal homeomorphism $f : \|\overline{\Gamma_0}\| \to \|\Gamma\|$. Two quadruplets $(\Gamma_i, l_i, \varphi_i, f_i)_{i=1,2}$ are equivalent if and only if there exists a G-equivariant isometry $h : (\Gamma_1, l_1, \varphi_1) \to (\Gamma_2, l_2, \varphi_2)$ such that $f_2 \circ f_1^{-1} = \hat{h}$ up to homotopy.

The relative modular group acts obviously on this Teichmüller space. Denote by $\mathcal{H}_G^{\text{comb}}(\Gamma_0, l_0, \varphi_0)$ the quotient space.

The bijection $\mathcal{H}_g^{\text{comb}}(G, R) \to \mathcal{H}_g(G, R) \times \mathbb{P}(\mathbb{R}_{>0}^b)$ comes in fact from equivariant bijections $\mathcal{T}_G^{\text{comb}}(\Gamma_0, l_0, \varphi_0) \to \mathcal{T}_G(\|\overline{\Gamma_0}\|, \hat{\varphi}_0) \times \mathbb{P}(\mathbb{R}_{>0}^b)$ with respect to the actions of the relative modular group, $((\Delta, m, \psi, f)$ maps to $(\|\overline{\Delta}\|, \hat{\psi}, f))$, and from the induced bijections $\mathcal{H}_G^{\text{comb}}(\Gamma_0, l_0, \varphi_0) \to \mathcal{H}_G(\Gamma_0, l_0, \varphi_0) \times \mathbb{P}(\mathbb{R}_{>0}^b)$.

Continuity. The following result gives control on variations of complex structures on $\|\Gamma\|_l$ with the variations of the metric l.

THEOREM 5.2. *If* $f : (\Gamma, l, \varphi) \to (\Gamma_0, l_0, \varphi_0)$ *is composed of equivariant isomorphisms and equivariant Whitehead collapses, then there exists a G-equivariant K-quasiconformal homeomorphism* $\hat{f} : \|\Gamma\| \to \|\overline{\Gamma_0}\|$ *such that* $K(\hat{f}) \to 1$ *when* $|l - l_0| \to 0$.

PROOF. We can restrict ourself to the case where either f is an isomorphism which alters only one edge's length, or f is a Whitehead collapse on one edge, keeping constant other lengths. For, in the general case, we decompose f, and use the fact that $g \circ h$ is $K(g)K(h)$-quasiconformal if g (resp. h) is $K(g)$- (resp. $K(h)$-) quasiconformal.

We use complex structures defined in the proof of Theorem 4.9, but with the change of coordinates

$$z_a = \frac{p(F(a))}{2\pi} \exp(2i\pi u_a / p(F(a))).$$

We will always use the power function like this: when the argument is fixed, it consists in multiplying it with the exponent.

First Case: Let $e = (a, \bar{a})$ be the edge such that $l(f(e)) = l(e) + \varepsilon$ and $F(a) = (a_1, a_2, \ldots, a_k = a)$ be the face which contains a. Set $l_k = l(a)$, $p = p(F(a))$, $R = p/(2\pi)$, $R_\varepsilon = (p+\varepsilon)/(2\pi)$. We first assume that $k \geq 2$.

Then we define $\hat{f}_F : D(O, R) \to D(O, R_\varepsilon)$ by

$$\frac{R_\varepsilon}{R} z^{\frac{p}{p+\varepsilon}} |z|^{\frac{\varepsilon}{p+\varepsilon}} \text{ for } 0 \leq \arg(z) \leq 2\pi(p - l_k)/p,$$

$$\frac{R_\varepsilon}{R} \frac{z^{\frac{p}{p+\varepsilon}(1+\varepsilon/l_k)}}{|z|^{\frac{\varepsilon}{p+\varepsilon}(p/l_k-1)}} \times \exp\left(2i\pi \frac{\varepsilon}{p+\varepsilon}\left(1 - \frac{p}{l_k}\right) \right) \text{ for } 2\pi(p - l_k)/p \leq \arg(z) \leq 2\pi.$$

This is well defined for $\arg(z) = 2\pi$ and $\arg(z) = 2\pi(p - l_k)/p$.

We do the same construction for the face $F(\bar{a})$, and take identity for other faces.

In the case $k = 1$, take $\hat{f}_F(z) = (p + \varepsilon)z/p$, and note that $F(\bar{a})$ can not be monovalent.

We have a similar construction in the case where a and $\bar{a}$ belong to the same face.

We have to check that the $\hat{f}_F$ glue to defined $\hat{f}$ between $\|\Gamma\|$ and $\|\Gamma_0\|$.

Firstly, the homeomorphisms $\hat{f}_F$ extend on boundaries of disks, and vertices maps onto vertices.

Without loss of generality, we consider the case where a and $\bar{a}$ do not belong to the same face. If we set $\theta = \arg(z)$ and $\alpha = \arg(\hat{f}_F(z))$, we have

$$\alpha = \theta \frac{p}{p+\varepsilon} \text{ or } \alpha = \frac{\theta p(1 + \varepsilon/l_k) + c}{p+\varepsilon}.$$

with c some constant. Since $\theta = 2\pi x/p$, and $\alpha = 2\pi y/(p + \varepsilon)$, we deduce that, with canonical coordinates, $\hat{f}_F$ is the identity or an affine map ($x \mapsto y = x(1 + \varepsilon/l_k) + c$) on the segment $[0, p]$. As values at the vertices are fixed, affine maps induced by $\hat{f}_F$ and $\hat{f}_{F'}$ equal on edges.

We have to calculate the dilatations $K(\hat{f}(z))$. We can restrict the computation to angular sectors where $\hat{f}$ is differentiable, and use following lemma, whose proof is direct.

LEMMA 5.3. *Let T the open set of the complex plane defined by $0 < \alpha < \arg(z) < \beta < 2\pi$ and $0 \leq |z| < 1$. Let $g : T \to T'$ be some diffeomorphism of the form $g(z) = cz^a \bar{z}^b$, where a, b, c are real numbers such that $|a| > |b|$. Then g is K-quasiconformal with $K = (|a|+|b|)/(|a|-|b|)$.*

In the case where a and $\bar{a}$ do not belong to the same face, we obtain for example

$$K(\hat{f}(z)) = \frac{p+\varepsilon}{p} \text{ for } 0 < \arg(z) < 2\pi(p - l_k)/p,$$

$$K(\hat{f}(z)) = \frac{l_k}{l_k + \varepsilon} \frac{p+\varepsilon}{p} \text{ for } 2\pi(p - l_k)/p < \arg(z) < 2\pi.$$

Second case: Let (Γ, l) be a smooth Riemannian fat-graph and $(W_e(\Gamma), l)$ the smooth Riemannian fat-graph obtained by collapsing the edge $e = (a, \bar{a})$. Let

p be the perimeter of the face $F(a) = (a_1, \ldots, a_k = a)$. We set $a_i' = f(a_i)$ for $i \in \{1, \ldots, k-1\}$. We assume $l(a_i) = l(a_i')$, and denote it by l_i; we also set $l_k = l(a_k)$.

For brevity, we will give details only for $k \geq 3$ and $\bar{a} \notin F(a)$ (the procedure is totally similar for other cases).

Again, we build $\hat{f}$ on each face, but firstly in disks of radii $\eta_F p(F)/2\pi$ with $0 < \eta_F < 1$. Then we should define $\hat{f}$ on the complementary of these disks, a ribbon which contains the geometric realization of the fat-graph.

In the disk corresponding to the face $F(a)$, we define $\hat{f}_{F(a)}$ by

$$\frac{p - l_k}{p} \frac{z^{\frac{p}{p-l_k}}}{|z|^{\frac{l_k}{p-l_k}}} \quad \text{for } 0 \leq \arg(z) \leq 2\pi \frac{p - l_k - l_{k-1}}{p},$$

$$\frac{p - l_k}{p} \frac{z^{\frac{p l_{k-1}}{(p-l_k)(l_{k-1}+l_k)}}}{|z|^{\frac{l_k(l_k+l_{k-1}-p)}{(p-l_k)(l_{k-1}+l_k)}}} \times e^{2i\pi l_k \frac{p-l_k-l_{k-1}}{(p-l_k)(l_{k-1}+l_k)}} \quad \text{for } 2\pi \frac{p-l_k-l_{k-1}}{p} \leq \arg(z) \leq 2\pi.$$

This defines a homeomorphism onto its image, a disk of radius $\eta_{F(a)}(p - l_k)/2\pi$. We make the same construction for $F(\bar{a})$ and take the identity for other faces.

Using Lemma 5.3, the desired properties for $K(\hat{f}(z))$ are immediate.

The transformation along closed trajectories with canonical coordinates is

$$x' = \begin{cases} x & \text{for } 0 \leq x \leq p - l_k - l_{k-1}, \\ \dfrac{x l_{k-1}}{l_{k-1} + l_k} + \dfrac{l_k(p - l_k - l_{k-1})}{l_{k-1} + l_k} & \text{for } p - l_k - l_{k-1} \leq x \leq p. \end{cases} \tag{$*$}$$

We define a closed neighborhood R of $(a, \bar{a})$ which contains an open neighborhood of vertices $a(0)$ et $a(1)$, and included in the ribbon defined upon.

Set $b_0 = \sigma_2^{-1}(\bar{a})$ and $b_1 = \sigma_2(\bar{a})$, of lengths m_0 and m_1. Recall that $a_{k-1} = \sigma_2^{-1}(a)$ and $a_1 = \sigma_2(a)$ of lengths l_1 et l_{k-1}. Set $p = p(F(a))$ and $p' = p(F(\bar{a}))$. For shortness we assume that $a(0)$ et $a(1)$ are trivalent. Then $\bar{a}_{k-1} = \sigma_2(\bar{b}_1)$ and $\bar{b}_1$ belong to a common face F_3 of perimeter p_3, and $\bar{a}_1$ and $\bar{b}_0$ belong to a common face F_4 of perimeter p_4. Let $\mu \in]0, \min(m_0, m_1, l_1, l_{k-1})[$ be a real number.

The closed set R is made of four glued quadrangles $(R_i)_{i \in \{1,\ldots,4\}}$ (see Figure 3). The size of R is parametrized by $l_k, \mu, \eta, \eta', \eta_3, \eta_4$.

We send R on the corresponding set of $\|W_e(\Gamma)\|$ in a compatible way with the definition of $\hat{f}$ on the boundary; see formula $(*)$.

We choose as image of the edge e a segment of vertical trajectory, and whose length is $2c(l_k)$, where $c(l_k)$ goes to zero as l_k goes to zero.

The image of the $(R_i)_{i \in \{1,\ldots,4\}}$ are the polygons $(P_i)_{i \in \{1,\ldots,4\}}$ of figure 3.

To estimate the dilatation K_i of $\hat{f}$ on R_i, we can use the notion of modulus of $\Lambda(\Gamma)$ for a family of piecewise smooth paths describing an open disk of the plane, defined by L. V. Ahlfors in chapter I, §D of [2].

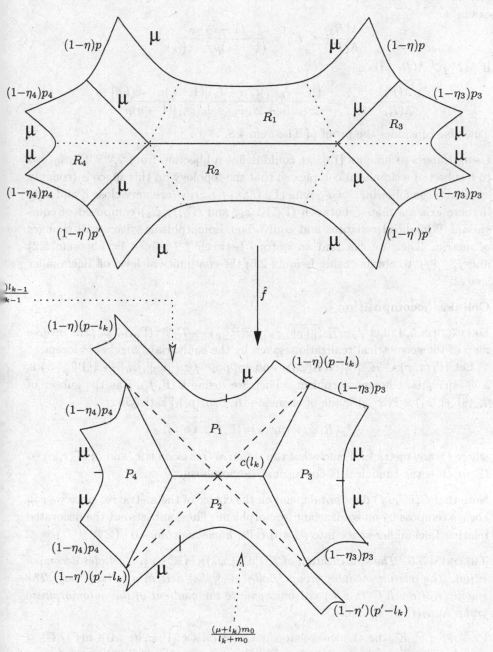

Figure 3. Homeomorphism in a neighborhood of the collapsed edge.

For R_1 and R_3 we choose paths between the sides of lengths $2\mu + l_k$ and 2μ, respectively. We put P_i between rectangles to estimate $\Lambda(P_i)$. We calculate for example

$$\frac{\Lambda(R_3)}{\Lambda(P_3)} \leq K_3 = \frac{(1 - \eta_3)p_3}{(1 - \eta_3)p_3 - c(l_k)}.$$

If $\Lambda(P_1) \geq \Lambda(R_1)$, then

$$\frac{\Lambda(P_1)}{\Lambda(R_1)} \leq K_1 = \frac{(l_k + l_{k-1})(2\mu + l_k)[(1 - \eta)p + c(l_k)]}{[\mu(l_k + 2l_{k-1}) + l_k l_{k-1}](1 - \eta)p}. \qquad \square$$

This also concludes the proof of Theorem 4.8. $\qquad \square$

Using Harer's techniques [13], we could define a bijection from $\mathcal{T}_G^{\mathrm{comb}}(\Gamma_0, l_0, \varphi_0)$ to a subset of a simplicial complex so that the topology on this space is (roughly speaking) the following: two points $(\Gamma_i, l_i, \varphi_i, f_i)_{i=\{1,2\}}$ are very close if and only if there exists a map f between $(\Gamma_1, l_1, \varphi_1)$ and $(\Gamma_2, l_2, \varphi_2)$ composed of equivariant Whitehead collapses and equivariant isomorphisms with small changes of metric. Then we can find an isotopy between $\hat{f}$ (defined in Theorem 5.2) and $f_1^{-1} \circ f_2$, to obtain (using Lemma 2.6) the continuity at level of Teichmüller spaces.

Cellular decomposition

DEFINITION 5.4. Let $J_\varphi : \mathcal{T}_G(\overline{\|\Gamma_0\|}, \hat{\varphi}_0) \times \mathbb{P}(\mathbb{R}_{>0}^b) \to \mathcal{T}_G^{\mathrm{comb}}(\Gamma_0, l_0, \varphi_0)$ the inverse map of the geometrical realization, given by the equivariant Strebel theorem.

Let $(\Gamma, m, \varphi) \in \mathcal{H}_{g(\Gamma_0)}^{\mathrm{comb}}(G, R(\varphi_0))$ and suppose $f : (\overline{\|\Gamma_0\|}, \hat{\varphi}_0) \to (\overline{\|\Gamma\|}_m, \hat{\varphi})$ is a G-equivariant homeomorphism. Then we define $C(\Gamma, f, \varphi)$ as the subset of $\mathcal{T}_G(\overline{\|\Gamma_0\|}, \hat{\varphi}_0) \times \mathbb{P}(\mathbb{R}_{>0}^b)$ made of elements $[R, \psi, g, (p_i)_i]$ satisfying

$$J_\varphi(R, \psi, g, (p_i)_i) = (\Gamma, l, \varphi, \hat{i} \circ f),$$

where l is any metric on Γ such that the action of G is isometric, and $i : (\Gamma, l, \varphi) \to (\Gamma, m, \varphi)$ is the canonical G-equivariant isomorphism.

Note that $C(\Gamma, \varphi, f)$ does not depend on the choice of the metric m, since we can always compose by an equivariant isomorphism. These subsets cut the decorated relative Teichmüller space into (open) cells, homeomorphic to $\mathbb{P}(\mathbb{R}_{>0}^{a(\Gamma)/|G|})$.

THEOREM 5.5. *The stratification of* $\mathcal{T}_G(\overline{\|\Gamma_0\|}, \hat{\varphi}_0) \times \mathbb{P}(\mathbb{R}_{>0}^b)$ *is a cellular decomposition. The relative modular group* $\mathrm{Mod}_G(\overline{\|\Gamma_0\|}, \hat{\varphi}_0)$ *acts in a cellular way. The stabilizer of a cell* $C(\Gamma, \varphi, h)$ *is isomorphic to the quotient of the automorphism group* $\mathrm{Aut}_G(\Gamma, \varphi)$ *by* $Z(G)$.

PROOF. Let K_m be the m-skeleton made of cells $C(\Gamma, \varphi, h)$ with $a(\Gamma)/|G| \leq m + 1$, and take such a cell with $a(\Gamma)/|G| = m + 1$. We build the attachment map of this cell. Let $l = (l_1, \ldots, l_m)$ a metric on Γ, where l_i is the constant length in the orbit of a geometrical edge e_i, and $\sum_i l_i = 1$. We now allow $l_i = 0$, except for orbits those quotient edge is a loop.

If $l_i > 0 \; \forall i \in \{1, \ldots, m\}$, then we map l to the corresponding element of $C(\Gamma, \varphi, h)$.

If $l_i = 0$ for some i, we map l to the corresponding element of $C(W_{O(e_i)}(\Gamma), \varphi_i, \hat{w} \circ h)$, where φ_i is the induced action of G, and $w : (\Gamma, l, \varphi) \to (W_{O(e_i)}(\Gamma), l, \varphi_i)$ is the equivariant Whitehead collapse.

This defines a continuous map $\prod_{i=1}^{n}]0, 1[\times \prod_{i=n+1}^{m} [0, 1[\to K_m$, which maps $\prod_{i=1}^{m}]0, 1[$ homeomorphically onto $C(\Gamma, \varphi, h)$, and its complementary in K_{m-1}.

The cellular action of the relative modular group is $[\theta, C(\Gamma, \varphi, h)] \mapsto C(\Gamma, \varphi, h \circ \theta^{-1})$.

We consider the center $e(\Gamma, \varphi, h)$ of the cell. This is the point defined by the unitary metric: $l(e) = 1 \; \forall e \in A_g(\Gamma)$. For this special metric, all the automorphisms are isometric. The center consists in a compact Riemann surface R marked by $f : \overline{\|\Gamma_0\|} \to R$, equipped with an action of G and a Strebel differential q_R stable under this action. The critical graph of q_R is an embedding of Γ which realizes the unitary metric. Thus we have $\mathrm{Aut}_G(\Gamma, \varphi) \cong \mathrm{Aut}_G(R, q_R) = \mathrm{Aut}(e(\Gamma, \varphi, h))$.

But if $\theta \in \mathcal{Mod}_G(\overline{\|\Gamma_0\|}, \hat{\varphi}_0)$, then $\theta \cdot e(\Gamma, \varphi, h) = e(\Gamma, \varphi, h \circ \theta^{-1})$, so that $\mathrm{Aut}(e(\Gamma, \varphi, h))/Z(G) \cong \mathrm{Stab}(C(\Gamma, \varphi, h))$, from which we deduce the result. $\square$

On the decorated Hurwitz space $\mathcal{H}_G(\overline{\|\Gamma_0\|}, \hat{\varphi}_0) \times \mathbb{P}(\mathbb{R}_{>0}^b)$, we obtain the induced cellular decomposition.

The cellular decomposition is compatible with the orbifold structure in the sense of Thurston: the finite groups associated to each point of the orbifold are constant along each cell. We also emphasize that the notion of cellular decomposition used here coincide with the classical one only for compact spaces.

Cells $C(\Gamma, \varphi)$ are parametrized by metrics on the quotient fat-graph. Thus they are homeomorphic to $C(\Gamma/\varphi(G))$. Moreover, $C(\Gamma', \varphi')$ is in the boundary of $C(\Gamma, \varphi)$ if and only if (Γ', φ') is obtained from (Γ, φ) by equivariant Whitehead collapses. Then $C(\Gamma'/\varphi'(G))$ is in the boundary of $C(\Gamma/\varphi(G))$. Thus, in a way, the projection $\mathcal{H}_g(G, R) \to \mathcal{M}_{g', (b_1, \ldots, b_t)}$ is a cellular one; the same fact holds for the natural morphisms between Hurwitz spaces obtained by restriction to a subgroup, or by quotient by a normal subgroup.

Top-dimensional cells of moduli spaces and Hurwitz spaces are yielded by trivalent fat-graphs (maximal number of edges).

Generalizing the notion of flip, we show that equivariant flips characterize connected components of $\mathcal{H}_g(G, R)$, described in § 2. Two top-dimensional cells $C(\Gamma, \varphi)$ and $C(\Gamma', \varphi')$ share a codimension one cell $C(\Gamma_0, \varphi_0)$ if and only if there exists $e \in A_g(\Gamma)$ and $e' \in A_g(\Gamma')$ such that $[W_{O(e)}(\Gamma)] = [\Gamma_0, \varphi_0] = [W_{O(e')}(\Gamma')]$. We say that (Γ, φ) differs from (Γ', φ') by an *equivariant flip*.

The flips are sometimes referred as elementary moves [24], or as Whitehead moves, the result of a Whitehead collapse and a Whitehead inflation.

PROPOSITION 5.6. *Two points of $\mathcal{H}_g(G, R)$ lie in the same connected component if and only if they belong to the adherence of top-dimensional cells indexed by* *trivalent fat-graphs, endowed with a quasifree action of G of ramification data R, which differ one from the other by a finite sequence of equivariant flips.*

PROOF. The condition is sufficient since equivariant Whitehead collapses give equivariant homeomorphisms which preserve the topological type.

The condition is necessary: let $C(\Gamma_1, \varphi_1)$ and $C(\Gamma_2, \varphi_2)$ be two top dimensional cells of $\mathcal{H}_G(S_0, \varphi_0)$, a connected component of $\mathcal{H}_g(G, R)$. Let a_i be an adherent point of $C(\Gamma_i, \varphi_i)$. Join a_i to a point b_i of $C(\Gamma_i, \varphi_i)$ if necessary. Then join b_1 to b_2 by a path intersecting only cells of codimension zero and one. Then (Γ_1, φ_1) differs from (Γ_2, φ_2) by a finite sequence of equivariant flips.   □

To illustrate the fact that the combinatorial description of Hurwitz spaces encodes their orbifold structure, we link their orbifold Euler characteristic, to those of moduli spaces.

W. Thurston has extended the notion of Euler characteristic to orbifolds $\mathcal{O}$ which possess some cellular decomposition $(C_i)_i$ compatible with its orbifold structure. Then

$$\chi_{\text{orb}}(\mathcal{O}) = \sum_i \frac{(-1)^{\dim(C_i)}}{|G(C_i)|},$$

where $G(C_i)$ is the finite group associated to each cell.

Here, using Theorem 5.5, we have

$$\chi_{\text{orb}}(\mathcal{H}_g(G, R) \times \mathbb{P}(\mathbb{R}_{>0}^b)) = \sum_{[\Gamma_j, \varphi_j]} \frac{(-1)^{a(\Delta_j)-1} \times |Z(G)|}{|\text{Aut}_G(\Gamma_j, \varphi_j)|}$$

where $\Delta_j = \Gamma_j/\varphi_j$. Then, since $\chi_{\text{orb}}(\mathbb{P}(\mathbb{R}_{>0}^b)) = (-1)^{b-1}$, using the Euler formula, we deduce:

$$\chi_{\text{orb}}(\mathcal{H}_g(G, R)) = \sum_{[\Gamma_j, \varphi_j]} \frac{(-1)^{s(\Delta_j)} \times |Z(G)|}{|\text{Aut}_G(\Gamma_j, \varphi_j)|}.$$

We resum on the isomorphism classes $[\Gamma_j/\varphi_j = \Delta_j]$ indexing the cellular decomposition of $\mathcal{M}_{g',(b_1,\ldots,b_t)}$, and then we use the relation between automorphisms groups (Theorem 4.5) to obtain

$$\chi_{\text{orb}}(\mathcal{H}_g(G, R)) = \sum_{[\Delta]} \frac{(-1)^{s(\Delta)}}{|\text{Aut}(\Delta)|} \sum_{\psi} \frac{|\text{Aut}(\Delta)|}{|\text{Aut}(\Delta, \psi)|}$$

where the ψ describe the set of conjugacy classes of epimorphisms from $\pi_1(\Delta)$ onto G with images of loop-faces fixed by the ramification data. But

$$\sum_{\psi} \frac{|\text{Aut}(\Delta)|}{|\text{Aut}(\Delta, \psi)|}$$

is the class equation for the action of $\text{Aut}(\Delta)$ on this set. Thus this set is independent from Δ. Denote it by $\text{Epi}_{g'}(G, R)$, and let d_{orb} be its cardinal. We have proved:

PROPOSITION 5.7. $\chi_{orb}(\mathcal{H}_g(G,R)) = d_{orb} \times \chi_{orb}(\mathcal{M}_{g',(b_1,\ldots,b_t)})$.

This means that Hurwitz spaces are less singular than moduli spaces. The rational number $\chi_{orb}(\mathcal{M}_{g',(b_1,\ldots,b_t)})$ is calculated in [14] (see also [3] [20]).

The computation of d_{orb} is a non trivial one. The cardinal of $\mathrm{Hom}_{g'}(G,R)$ (replace epimorphisms by homomorphisms) is calculated in [19] (see also [12]), it only depends on the irreducible complex representations of G. For symmetric groups, these cardinals appear as coefficient of a generating series coming from a matrix model [21], closely related to Yang–Mills theory for surfaces [12].

We conclude with the non-galoisian case. We also have a cellular decomposition of $\mathcal{H}(g,g',G,K,R) \times \mathbb{P}(\mathbb{R}^b_{>0})$ into cells $C(p)$, where $[p : \Gamma \to \Delta]$ is an equivalence class of degree d covering with a monodromy of type (G,K,R). These cells are parametrized by edges' lengths of Δ. Performing Whitehead collapses along e on Δ, and thus Whitehead collapses along $p^{-1}(e)$ on Γ, give a cell $C(q)$ incident on $C(p)$, where $q : W_{p^{-1}(e)}(\Gamma) \to W_e(\Delta)$.

Furthermore, we see that $\mathrm{Out}(G,K,R)$ acts cellularly on $\mathcal{H}_h(G,R) \times \mathbb{P}(\mathbb{R}^b_{>0})$ by $\theta \cdot C(\Gamma,\varphi) = C(\Gamma, \varphi \circ \theta^{-1})$. Hence

$$\chi_{orb}(\mathcal{H}(g,g',K,G,R)) = \chi_{orb}(\mathcal{H}_h(G,R))/\mathrm{Out}(G,K,R).$$

REMARK. An important step in Kontsevich results [20] is the explicit computation of cohomology classes on the cellular decomposition of moduli space of punctured Riemann surfaces. The cohomology classes are first Chern classes of line bundles over moduli spaces, whose fibers are the cotangent spaces at the i-th puncture.

It follows from our work, that the pullback of these classes on Hurwitz spaces, can be computed in the same way.

6. Compactification

The convenient tool to describe compactifications of moduli spaces, or of Hurwitz spaces, is again a graphical one: the modular graph (terminology of Y. Manin). To avoid any confusion, we keep the Greek letters for fat-graphs. For a graph E, we denote by $V(E)$ the set of vertices, $v(E)$ the number of vertices, v_s the valency of the vertex s, $a(E)$ the number of edges, and $h^1(E) = a(E) - v(E) + 1$.

Moduli spaces

DEFINITION 6.1. A modular graph E of type (g',b) is a connected graph together with two maps $g : V(E) \to \mathbb{N}$ and $P : V(E) \to \mathcal{P}(\{1,\ldots,b\})$ such that

- $\{1,\ldots,b\} = \coprod_v P(v)$,
- $2g(s) - 2 + v_s + \#P(s) > 0 \ \forall s \in V(E)$,
- $g' = \sum_{s \in V(E)} g(s) + h^1(E)$.

Two such modular graphs (E, g, P) and (H, h, Q) are isomorphic if there exists an isomorphism of graphs $i : E \to H$ with $i \circ g = h \circ i$ and $i \circ P = Q \circ i$.

We have the following well-known description of the Deligne–Mumford–Knudsen compactification of moduli spaces: $\overline{\mathcal{M}}_{g',b} = \coprod_E \mathcal{M}_{g',b}(E)$, where E runs over the modular graphs of type (g', b), and $\mathcal{M}_{g',b}(E)$ consists in all stable Riemann surfaces built as follows:

- Associate to each vertex s a compact Riemann surface R_s of genus $g(s)$, with $\#P(s)$ punctures and v_s ties.
- Associate to each edge an identification of the corresponding ties.

  Such objects are connected compact Riemann surfaces C with punctures and singular points, called nodes, obtained by identification of ties. We denote them by (C, i_C), where i_C is the identification of ties.
- Associate to each isomorphism, an homeomorphism $h : C \to C'$ such that $h \circ i_C = i_{C'} \circ h$, biholomorphic when restricted to each component.

Each element of $\overline{\mathcal{M}}_{g',b}$ comes from pinching some boundary curves of a pants decomposition of a smooth surface [4].

Then the neighborhood of a node looks like $\{(y : U \to D) \times (z : V \to \overline{D})/yz = 0\}$, where U and V are some disks in the surface pointed by the node N, D is the unit disk of the complex plane, $\overline{D}$ is the same disk but with the conjugate complex structure, and $y(N) = z(N) = 0$ [4].

In our context of marked Riemann surfaces, we define the topology of $\overline{\mathcal{M}}_{g',b}$ with quasiconformal deformations.

First we have the following definition of L. Bers [4]. Let S_1 and S_2 be two stable curves of type (g', b), and N_i be the set of nodes of S_i. A *deformation* $f : S_1 \to S_2$ is a surjective map with $f(N_1) \subset N_2$, such that the preimage of a node is either a node or a Jordan curve (non null-homotopic, non homotopic to some puncture), and such that f is an homeomorphism component by component, when restricted to $S_1 - f^{-1}(N_2)$.

Then after the work of W. Abikoff (see § 1.3 and Theorem 1 of [1]), S_1 is close to S_2 if f^{-1} is $(1 + \varepsilon)$-quasiconformal on each component of $S_2 - K$ with K a compact neighborhood of N_2.

Maybe some components of such a stable Riemann surface do not have any puncture. These components cannot be parametrized by means of fat-graphs, and this leads M. Kontsevich [20] to take a quotient of $\overline{\mathcal{M}}_{g',b}$: homeomorphisms $h : (C, i_C) \to (D, i_D)$ restricted to any component without puncture may be just an homeomorphism instead of a biholomorphism. Kontsevich's compactification $\overline{\mathcal{M}}'_{g',b}$ is then the quotient of $\overline{\mathcal{M}}_{g',b}$ by the closure (of the graph) of this new equivalence relation.

At the level of modular graphs, this means that we can retract any edge a joining two vertices s_1 and s_2 with $P(s_i) = \varnothing$ into a new single vertex s, setting $g(s) = g(s_1) + g(s_2)$ if $s_1 \neq s_2$ and $g(s) = g(s_1) + 1$ if a is a loop.

Thus, in $\overline{\mathcal{M}}'_{g',b}$, if a stable Riemann surface possess a component C_s without punctures, we can forget C_s, but keep as data in the modular graph its genus and the number of ties.

If γ is a simple loop which separates a Riemann surface into two stable components C_1 and C_2 such that C_1 is without punctures, then, from a topological viewpoint in $\overline{\mathcal{M}}'_{g',b}$, pinching γ to a point is the same as collapsing C_1 into a point.

We now precise the intricate definition of $\overline{\mathcal{M}}^{\text{comb}}_{g',b}$ given in [20]. The role of ties is played by distinguished vertices.

DEFINITION 6.2. A stable fat-graph of type (g',b) is made of a modular graph (E,g,P) of type (g',b) such that:

- To each vertex s with $P(s) \neq \varnothing$ is associated a Riemannian fat-graph Γ_s, with $\#P(s)$ faces, genus $g(s)$, and v_s distinguished vertices (maybe monovalent or bivalent, the other ones at least trivalent.)
- To each edge joining vertices s_i with $P(s_i) \neq \varnothing$ is associated an identification of the corresponding distinguished vertices.

Two stable fat-graphs $(E,g,P,(\Gamma)_s)$ and $(F,l,Q,(\Delta)_s)$ are equivalent if there exist $h : (E,g,P) \to (F,l,Q)$ or $(F,l,Q) \to (E,g,P)$ composed of isomorphisms and retraction of edges joining vertices s_i with $P(s_i) = \varnothing$, such that Γ_s and $\Delta_{h(s)}$ are isometric if $P(s) \neq \varnothing$.

We now have to precise how smooth fat-graphs of type (g',b) degenerates into stable fat-graphs of same type. We have to distinguish two cases.

Firstly, we can perform Whitehead's collapses on loops which do not bound any face. Let e such a loop incident on a vertex v in a fat-graph Γ. After the retraction, we disconnect the graph on v to produce two distinguished vertices. Precisely, if $e = (a, \bar{a})$, such that $\sigma_2(a) \neq a$ and $\sigma_2(\bar{a}) \neq \bar{a}$, then, as a σ_0-orbit, $v = (a_1, \ldots, a_j, a, b_1, \ldots, b_k, \bar{a})$ with $j,k \geq 1$ (because for example $\sigma_0(a) = \bar{a}$ implies $\sigma_2(\bar{a}) = \bar{a}$, which is forbidden). Then the new distinguished vertices are $(a_1, \ldots, a_j)$ and $(b_1, \ldots, b_k)$. They are possibly monovalent or bivalent.

At the level of modular graphs we create an edge and a new vertex, or a loop, with appropriate data.

Secondly, we can collapse a subgraph Δ of Γ into a vertex which becomes distinguished (provided that the number of faces keeps constant). At the level of modular graphs, we create an edge and a monovalent vertex s with $g(s) = g(\Delta)$ and $P(s) = \varnothing$.

Then, we have Kontsevich's theorem, which furnishes a cellular decomposition of the whole space $\overline{\mathcal{M}}'_{g',b} \times \mathbb{P}(\mathbb{R}^b_{>0})$.

THEOREM 6.3. *There is an homeomorphism* $\overline{\mathcal{M}}^{\text{comb}}_{g',b} \to \overline{\mathcal{M}}'_{g',b} \times \mathbb{P}(\mathbb{R}^b_{>0})$

PROOF. Fix a modular graph (E,g,P). The map from right to left is given by the Theorem 3.3 applied to each component associated to a vertex s with

$P(s) \neq \varnothing$. For the inverse map, we have to say something about monovalent and bivalent vertices.

If s is a bivalent vertex of a Riemannian fat-graph (Γ, l), with edges a and b incident on it, then $\|\Gamma\|$ is biholomorphic to $\|\Gamma'\|$, where Γ' is the Riemannian fat-graph obtained from (Γ, l) in canceling s, edges a and b becoming a new edge of length $l(a) + l(b)$.

Let s be a monovalent vertex of (Γ, l), and a be the unique edge incident on it. If D_s is a small disk centered in s, then $(D_s \cap U_a, u_a^2)$ is a local coordinate at s, holomorphically compatible with the other ones, and s becomes a first order pole of w_Γ.

Continuity can be sketched as follows. We associate to each edge's retraction a deformation whose dilation is controlled in function of edge's length (as in Theorem 5.2). $\square$

Hurwitz spaces. From now on, we fix $R = b_1 C_1 + \cdots + b_t C_t$ a degree b ramification data of a finite group G. We first have to extend the ramified covering $\mathcal{H}_h(G, R) \to \mathcal{M}_{g', \{b_1, \ldots, b_t\}}$ to a suitable compactification $\overline{\mathcal{H}}_h(G, R)$.

Since every stable Riemann surface of $\overline{\mathcal{M}}_{g', \{b_1, \ldots, b_t\}}$ is obtained by retraction of some loops on a smooth Riemann surface of $\mathcal{M}_{g', \{b_1, \ldots, b_t\}}$, we build elements of $\overline{\mathcal{H}}_h(G, R)$ from smooth ones in retracting some orbits of loops. The following result is important since it gives the stability condition for actions of G on stable Riemann surfaces.

PROPOSITION 6.4. *Let* $[C, \varphi] \in \mathcal{H}_h(G, R)$. *Then the stabilizer* $G(L)$ *of a loop* L *is either cyclic or dihedral.*

PROOF. Assume that $G(L)$ is non trivial. Then $G(L)$ stabilizes a small part of cylinder whose fundamental group is generated by L. Using uniformization, we can choose C biholomorphic to a regular ring $\{z \in \mathbb{C} : r < |z| < 1\}$ so that L becomes the circle of radius $\sqrt{r}$. Using uniformization again, the finite order automorphisms of this ring stabilizing L are rotations of finite order, and the symmetry $z \mapsto r/z$. $\square$

Note that rotations do not possess fixed points in the ring, and stabilizes each part of this ring. On the contrary, the symmetry exchanges both parts and possess two fixed points at diametrically opposite points of L. In fact, if $G(L) \cong D_n$ the dihedral group of order $2n$, there is $2n$ fixed points on L. We assimilate the case where $G(L)$ is the cyclic group generated by the symmetry described just upon to the case of dihedral stabilizers (thus we exclude this case when we talk of cyclic stabilizer).

Hence, we cannot retract loops with dihedral stabilizers if we want to keep constant the ramification data. The phenomenon of collision of ramification points is also relevant in the context of algebraic geometry [5].

We can say something more in the case of cyclic stabilizers. Let C_1 and C_2 be the components attached by the node N, obtained by retraction of L. Denote by N_1 (resp. N_2) the corresponding ties, and by ξ_1 (resp. ξ_2) the privileged generators of stabilizer $G(N_1)$ (resp. $G(N_2)$) (here, the privileged generator has the same meaning than in §2). Then $\xi_1 = \xi_2^{-1}$ in G (just recall the description of the neighborhood of a node).

DEFINITION 6.5. • An admissible symmetric subsurface decomposition D of $[C, \varphi] \in \mathcal{H}_h(G, R)$ is a subsurface decomposition of $C - X$ (where X is the set of fixed points), symmetric with respect to the holomorphic action of G, and such that every curve of the decomposition does not intersect the set X. Such a decomposition is called maximal if the quotient decomposition of $C/\varphi(G)$ is a pants decomposition.

• Let (C, i_C) be a compact Riemann surface with nodes. An action $\varphi : G \hookrightarrow$ homeo(C, i_C) is said to be stable if $\varphi(G)$ acts holomorphically on each component, if the node stabilizers are all cyclic, with the preceding property on stabilizers of ties, and if $2g_i - 2 + F_i + L_i > 0$ (L_i is the cardinal of nodes on the i-th component, F_i is the cardinal of smooth fixed points).

• $\overline{\mathcal{H}}_h(G, R)$ is the set of equivalence classes of $[(C, i_C), \varphi]$ where (C, i_C) is a genus g compact Riemann surface with nodes, and φ is a stable action with $R(\varphi) = R$. Two elements define the same class if they differ by a G-equivariant homeomorphism which is biholomorphic on each component.

Then every element $[C, \varphi]$ of $\overline{\mathcal{H}}_h(G, R)$ with nodes come from pinching curves orbits of an admissible symmetric and maximal subsurface decomposition of a smooth $[C_t, \varphi_t]$. The condition $2g_i - 2 + F_i + L_i > 0$ correspond to the fact that curves of the decomposition are not homotopic to a smooth point (fixed or not). The cyclicity of node's stabilizers corresponds to the fact that curves of the decomposition do not contain any fixed point, since curves with dihedral stabilizers contain fixed points.

With this definition, $\overline{\mathcal{H}}_h(G, R) \to \overline{\mathcal{M}}_{g', (b_1, \dots, b_t)}$ is a ramified covering, thus $\overline{\mathcal{H}}_h(G, R)$ is a compact space. The map is well-defined, because $2g_i - 2 + F_i + L_i = |G_i|(2g'_i - 2 + f_i + l_i)$ by Riemann–Hurwitz, where G_i is the stabilizer of C_i, f_i is the number of branch points (resp. ties) on $C_i/\varphi(G_i)$.

Again, elements of $\overline{\mathcal{H}}_h(G, R)$ are well-described in terms of modular graphs (F, g, P), equipped with a G-action. A map P from $V(F)$ orbits to $\mathcal{P}(\{1, \dots, b\})$ gives the spreading of ramification data. We put (forgetting the multiplicities) $R_{P(s)} = \sum_{i \in P(s)} C_i$.

Hypothesis of stability for the group actions on Riemann surfaces with nodes impose conditions on modular graphs. Denote by G_s and G_a the stabilizer of $s \in V(F)$ and $a \in A(F)$, respectively. The groups G_a are cyclic. We have excluded elements θ such that $\theta(a) = \bar{a}$, so that $G_a \cong G_{\bar{a}}$ and $G_a \hookrightarrow G_{a(0)}$, $G_a \hookrightarrow G_{a(1)}$. Moreover, if we consider ties stabilizers, and a distinguished generator (which acts on the tangent plane in multiplying by a fixed root of unity), then we can assign to each oriented edge e this generator g_e, and $g_{\bar{e}} = g_{e^{-1}}$.

Now $\overline{\mathcal{H}}_h(G, R) = \coprod_{F,g,P,\rho} \mathcal{H}_{h,G,R}(F, g, P, \rho)$ where (F, g, P, ρ) runs over isomorphism classes of decorated modular graphs of genus h with an action $\rho : G \hookrightarrow \mathrm{Aut}(F, g, P)$ with preceding properties, and such that $R = \sum_s R_{P(s)}$. The decoration consists in assigning to each oriented edge e a generator g_e of G_e such that $g_{\bar{e}} = g_{e^{-1}}$. The stratum $\mathcal{H}_{h,G,R}(F, g, P, \rho)$ is made of all stable Riemann surfaces with stable actions whose associated decorated modular graph is (F, g, P, ρ).

DEFINITION 6.6. $\overline{\mathcal{H}}'_h(G, R)$ is the quotient of $\overline{\mathcal{H}}_h(G, R)$ by the closure of the following equivalence relation: two elements are identified if they differ by an equivariant homeomorphism which is biholomorphic only when restricted to components with smooth fixed points.

Again, this means that we can retract edges in the modular graph, in fact, orbits of edges joining vertices s with $P(s) = \varnothing$.

We now define the suitable extension $\overline{\mathcal{H}}_h^{\mathrm{comb}}(G, R)$ of $\mathcal{H}_h^{\mathrm{comb}}(G, R)$.

DEFINITION 6.7. An element of $\overline{\mathcal{H}}_h^{\mathrm{comb}}(G, R)$ of topological type (F, g, P, ρ) is made as follows:

- Associate to each vertex s of F with $P(s) \neq \varnothing$, a Riemannian fat-graph Γ_s of genus $g(s)$, with v_s distinguished vertices, endowed with $\varphi_s : G_s \hookrightarrow \mathrm{Aut}(\Gamma_s)$, quasifree, except for distinguished vertices, and such that $R(\varphi_s) = R_{P(s)}$.
- A distinguished vertex $v \in V(\Gamma_s)$ given by an edge a incident on s has stabilizer G_a. Edges of F yield the identification of distinguished vertices. If θ_a (resp. $\theta_{\bar{a}}$) are the privileged generators of G_a (resp. $G_{\bar{a}}$), then we ask that $\theta_a = \theta_{\bar{a}}^{-1}$.
- If $\rho(\theta)(s) = t$, then $\varphi(\theta) : \Gamma_s \to \Gamma_t$ must be an equivariant isometry if $P(s) \neq \varnothing$.

As in the case of moduli spaces, these stable fat-graphs are all obtained by retractions of edges, in fact retractions of orbits which do not contain any face, or by equivariant retraction of subgraphs.

We obtain the analog of Theorem 6.3 in the case of Galois coverings.

THEOREM 6.8. *There is an homeomorphism* $\overline{\mathcal{H}}_h^{\mathrm{comb}}(G, R) \to \overline{\mathcal{H}}'_h(G, R) \times \mathbb{P}(\mathbb{R}_{>0}^b)$

PROOF. Fix a modular graph (F, g, P, ρ). Then we apply Theorem 4.8 component by component, with the same discussion than for theorem 6.3. $\square$

We define $\overline{\mathcal{H}}(g, g', G, K, R)$ as the coverings of stable Riemann surfaces got from elements of $\mathcal{H}(g, g', G, K, R)$ by retraction of loops on the base (resp. orbits of loops on the total space) which do not contain any branch points (resp. ramification points). We could describe its elements by coverings of modular graphs. Then $\overline{\mathcal{H}}'(g, g', G, K, R)$ is the quotient by the closure of the equivalence relation where coverings need to be holomorphic only on components with ramification points.

Similarly, $\overline{\mathcal{H}}_{g,g'}^{\mathrm{comb}}(G, K, R)$ is defined as coverings of stable fat-graphs obtained by retraction on the base either of loops which do not bound any face or of subgraphs which do not contain any face, and by suitable retraction of fibers on the total space.

END OF PROOF OF THEOREM 1.1. The action of $\text{Out}(G, K, R)$ on $\mathcal{H}_h(G, R)$ extends on $\overline{\mathcal{H}}'_h(G, R)$. For, a stable Riemann surface (with action of G) of $\overline{\mathcal{H}}'_h(G, R)$ is obtained from a Riemann surface (with action of G) of $\mathcal{H}_h(G, R)$ by pinching some curves orbits of an admissible maximal and symmetric subsurface decomposition (Definition 6.5).

Thus we have an homeomorphism from the quotient $\overline{\mathcal{H}}'_h(G, R)/\text{Out}(G, K, R)$ to $\overline{\mathcal{H}}'(g, g', G, K, R)$. The map is surjective: an element $(p_0 : T_0 \to S_0)$ comes from a smooth $(p_t : T_t \to S_t)$ got from (C_t, φ_t) in $\mathcal{H}_h(G, R)$ (Proposition 2.11). Then $(p_0 : T_0 \to S_0)$ is the image of (C_0, φ_0). Similarly, the injectivity is deduced from injectivity in the smooth case.

Now the same fact holds for combinatorial spaces, so that theorem 6.8 gives Theorem 1.1. $\square$

7. Moduli Spaces of Curves with Cyclic Group Actions

We now emphasize on Hurwitz spaces $\mathcal{H}_g(\mathbb{Z}/n, R)$, a special case of own interest, due to their striking analogy with moduli space of curves with spin ([17] [27]).

Let G be a fixed finite group, together with a fixed ramification data R. Consider the Hurwitz space $\mathcal{H}_g(G, R)$, assumed to be non empty. Let $[C, \varphi] \in \mathcal{H}_g(G, R)$, and $\pi : [C, \varphi] \to S = C/\varphi(G)$, the associated ramified covering.

Denote by $\mathcal{O}_C$ the sheaf of holomorphic functions on C, and let $\pi_*(\mathcal{O}_C)$ the image direct sheaf. This is a locally free sheaf of G-module, with dimension $|G|$. If $\hat{G}$ denotes the set of the irreducible complex representations of G, $\chi : G \to \text{GL}(V_\chi)$, then $\pi_*(\mathcal{O}_C) = \bigoplus_{\chi \in \hat{G}} L_\chi \otimes V_\chi$, where L_χ is the χ-isotopic subbundle, with rank $\deg(\chi) = \dim(V_\chi)$, and V_χ is the constant bundle with fiber V_χ.

Assume now that G is the cyclic group $\mathbb{Z}/n$. We refer to the example 1 in §2. Put $\xi = \exp(2i\pi/n)$ and $\sigma = \varphi([1])$. Then $\pi_*(\mathcal{O}_C) = \bigoplus_0^{n-1} L_k$, where L_k is the line bundle of holomorphic germs corresponding to the eigenvalue ξ^k. We have $L_0 = \mathcal{O}_S$.

Describing the image of $L_1^{\otimes n} \hookrightarrow L_0$ involves the ramification data $R(\varphi)$. More precisely, we have:

PROPOSITION 7.1. *If* $G = \mathbb{Z}/n$, *and* $R = \sum_1^b [m_i]$ *with* $[m_i] \neq [m_j]$ *for* $i \neq j$, *then* $L_1^{\otimes n} \cong \mathcal{O}_S(-\sum_1^b m_i x_i)$.

PROOF. We study the line bundle L_j in a neighborhood U of a branch point Q. We refer to the notations of example 1 in §2. Let us drop the index i. Set $d = (m, n)$, $e = n/d$, and $k = m/d$. By definition, $\pi_*(\mathcal{O}_C)(U) = \mathcal{O}_C(\pi^{-1}(U))$. Since π is a covering, we choose $\pi^{-1}(U) = \coprod_j V_j$. Let $\{P_1, \ldots, P_{d-1}\}$ be the fiber over Q, and z_j a local coordinate at P_j with $Z_j(P_j) = 0$ and such that $(P_j, z_j) = \sigma^j(P_0, z_0)$. Take also a local coordinate u at Q such that $u(Q) = 0$ and $u = z_j^e$.

Furthermore, we have $\sigma^{kd}(z_0) = \xi^d(z_0)$, or $\sigma^d(z_0) = \xi^{\nu d}z_0$ if ν denotes the inverse of k modulo e.

Extend each z_j on $\coprod V_k$ by $z_j = 0$ on V_k if $k \neq j$. We have to decompose in proper subspaces the space of holomorphic germs defined on $\coprod V_k$.

We use the notation $\langle ik \rangle$ for ik modulo e.

Recall that the function $X_i = z_0^{\langle ik \rangle} + \xi^{-i}z_1^{\langle ik \rangle} + \cdots + \xi^{-(d-1)i}z_{d-1}^{\langle ik \rangle}$ is a frame of L_i and $(X_i)_{i \in \{0,\ldots,n-1\}}$ is a frame of $\pi_*(\mathcal{O}_C)$.

Then for $i = 1$, we have $X_1 = z_0^k + \xi^{-1}z_1^k + \cdots + \xi^{-(d-1)}z_{d-1}^k$, and $X_1^j = z_0^{jk} + \xi^{-j}z_1^{jk} + \cdots + \xi^{-(d-1)j}z_{d-1}^{jk}$ (since $z_j = 0$ on V_k for $k \neq j$, products $z_k z_j$ are zero).

Set $jk = [jk/e]e + \langle jk \rangle$, with $[jk/e]$ the integral part of jk modulo e; then we obtain $X_1^j = X_j \times u^{[jk/e]}.*)$, and in particular $X_1^n = u^m$. $\square$

Moduli spaces of genus g and b-pointed curves with data $(n, (m_i)_{i \in \{1,\ldots,b\}})$ are defined to be equivalence classes of couples made of a Riemann surface S and of a line bundle L on S satisfying $L^{\otimes n} \cong K(-\sum m_i x_i)$, where K is the canonical line bundle. As a corollary, if the quotient surfaces are punctured torus, then the Hurwitz space $\mathcal{H}_g(\mathbb{Z}/n, R)$ is isomorphic to a moduli space of genus one curves with spin.

But the main point in the striking analogy in any genus between both spaces, for example by restriction to a subgroup, or considering the behavior of their compactification.

To emphasize, we conclude by a remark (without proof) on the analogy between the compactification of Hurwitz spaces described here, and compactifications of moduli space of curves with spins defined in [27] and [17].

The boundary of the compactification of $\mathcal{H}_g(\mathbb{Z}/n, R)$ divides into two parts, depending on locally freeness of a n-th root L of $\mathcal{O}(-\sum m_i x_i)$ at a node. The case where L is locally free (resp. non locally free) at a node q corresponds to the case where every node in the preimage of q by a stable covering has trivial (resp. non trivial) stabilizer.

These parts should be called Ramond–Ramond and Neveu–Schwarz, as in reference [27].

References

[1] W. Abikoff, "Degenerating families of Riemann surfaces", Ann. of Math. (2) **105** (1977), 29–44.

[2] L. V. Ahlfors, *Lectures on quasiconformal mappings*, Van Nostrand, Princeton, 1966.

[3] M. Bauer and C. Itzykson, "Triangulations", in *The Grothendieck theory of dessins d'enfants*, edited by L. Schneps, London Math. Soc. Lecture Notes Series **200**, Cambridge Univ. Press, Cambridge, 1995.

[4] L. Bers, "Finite dimensional Teichmüller spaces and generalizations", *Bull. Amer. Math. Soc.* **5** (1981), 131–172.

upon completion). So taking spectra, we also obtain an equivalence between *algebraic branched covers* and *analytic branched covers*.

This formal argument can be summarized informally as follows:

GENERAL PRINCIPLE 2.2.4. *An equivalence of tensor categories of modules induces a corresponding equivalence of categories of algebras, of branched covers, and of Galois branched covers for any given finite Galois group.*

The last point (about Galois covers) holds because an equivalence of categories between covers automatically preserves the Galois group.

In order to obtain Riemann's Existence Theorem, one more step is needed, viz. passage from branched covers of a curve X to étale (or unramified) covers of an open subset of X. For this, recall that an algebraic branched cover is locally a covering space (in the metric topology) precisely where it is étale, by the Inverse Function Theorem. Conversely, an étale cover of a Zariski open subset of X extends to an algebraic branched cover of X (by taking the normalization in the function field of the cover). Such an extension also exists for analytic covers of curves, since it exists *locally* over curves. (Namely, a finite covering space of the punctured disc $0 < |z| < 1$ extends to an analytic branched cover of the disc $|z| < 1$, since the covering map — being bounded and holomorphic — has a removable singularity [Ru, Theorem 10.20].) Thus the above equivalence for branched covers induces an equivalence of categories between finite étale covers of a smooth complex algebraic curve X, and finite analytic covering maps to X^h. That is, the categories (i) and (ii) in Riemann's Existence Theorem are equivalent; and this completes the proof of that theorem.

Apart from Riemann's Existence Theorem, GAGA has a number of other applications, including several proved in [Se3]. Serre showed there that if V is a smooth projective variety over a number field K, and if X is the complex variety obtained from V via an embedding $j : K \hookrightarrow \mathbb{C}$, then the Betti numbers of X are independent of the choice of j [Se3, Cor. to Prop. 12]. Serre also used GAGA to obtain a proof of Chow's Theorem [Ch] that every closed analytic subset of $\mathbb{P}_{\mathbb{C}}^n$ is algebraic [Se3, Prop. 13], as well as several corollaries of that result. In addition, he showed that if X is a projective algebraic variety, then the natural map $H^1(X, \mathrm{GL}_n(\mathbb{C})) \to H^1(X^h, \mathrm{GL}_n(\mathbb{C})^h)$ is bijective [Se3, Prop. 18]. As a consequence, the set of isomorphism classes of rank n algebraic vector bundles over X (in the Zariski topology) is in natural bijection with the set of isomorphism classes of rank n analytic vector bundles over X^h (in the metric topology). In a way, this is surprising, since the corresponding assertion for *covers* is false (because all covering spaces in the Zariski topology are trivial, over an irreducible complex variety).

Having completed the proofs of GAGA and Riemann's Existence Theorem, we return to the proofs of Theorems 2.2.2 and 2.2.3.

PROOF OF THEOREM 2.2.2. First we reduce to the case $X = \mathbb{P}_{\mathbb{C}}^r$ as in Step 2 of the proof of Theorem 2.2.1, using that $H^q(X, \mathcal{F}) = H^q(\mathbb{P}_{\mathbb{C}}^r, j_* \mathcal{F})$ if $j : X \hookrightarrow \mathbb{P}_{\mathbb{C}}^r$, and similarly for X^h.

Second, we verify the result directly for the case $\mathcal{F} = \mathcal{O}$ and $\mathcal{F}^h = \mathcal{H}$, for all $q \geq 0$. The case $q = 0$ is clear, since then both sides are just $\mathbb{C}$, because X is projective (and hence compact). On the other hand if $q > 0$, then $H^q(X, \mathcal{O}) = 0$ by the (algebraic) cohomology of projective space [Hrt2, Chap. III, Theorem 5.1], and $H^q(X^h, \mathcal{H}) = 0$ via Dolbeault's Theorem [GH, p. 45].

Third, we verify the result for the sheaf $\mathcal{O}(n)$ on $X = \mathbb{P}_{\mathbb{C}}^r$. This step uses induction on the dimension r, where the case $r = 0$ is trivial. Assuming the result for $r - 1$, we need to show it for r. This is done by induction on $|n|$; for ease of presentation, assume $n > 0$ (the other case being similar). Let E be a hyperplane in $\mathbb{P}_{\mathbb{C}}^r$; thus $E \approx \mathbb{P}_{\mathbb{C}}^{r-1}$. Tensoring the exact sequence $0 \to \mathcal{O}(-1) \to \mathcal{O} \to \mathcal{O}_E \to 0$ with $\mathcal{O}(n)$, we obtain an associated long exact sequence $(*)$ which includes, in part:

$$H^{q-1}(E, \mathcal{O}_E(n)) \to H^q(X, \mathcal{O}(n-1)) \to H^q(X, \mathcal{O}(n))$$
$$\to H^q(E, \mathcal{O}_E(n)) \to H^{q+1}(X, \mathcal{O}(n-1))$$

Similarly, replacing $\mathcal{O}$ by $\mathcal{H}$, we obtain an analogous long exact sequence $(*)^h$; and there are (commuting) maps ε from each term in $(*)$ to the corresponding term in $(*)^h$. By the inductive hypotheses, the map ε is an isomorphism on each of the outer four terms above. So by the Five Lemma, ε is an isomorphism on $H^q(X, \mathcal{O}(n))$.

Fourth, we handle the general case. By a vanishing theorem of Grothendieck ([Gr1]; see also [Hrt2, Chap. III, Theorem 2.7]), the q-th cohomology vanishes for a Noetherian topological space of dimension n if $q > n$. (Cf. [Hrt2, p. 5] for the definition of *dimension*.) So we can proceed by descending induction on q. Since $\mathcal{F}$ is coherent, it is a quotient of a sheaf $\mathcal{E} = \bigoplus_i \mathcal{O}(n_i)$ [Hrt2, Chap. II, Cor. 5.18], say with kernel $\mathcal{N}$. The associated long exact sequence includes, in part:

$$H^q(X, \mathcal{N}) \to H^q(X, \mathcal{E}) \to H^q(X, \mathcal{F}) \to H^{q+1}(X, \mathcal{N}) \to H^{q+1}(X, \mathcal{E})$$

The (commuting) homomorphisms ε map from these terms to the corresponding terms of the analogous long exact sequence of coherent $\mathcal{H}$-modules on X^h. On the five terms above, the second map ε is an isomorphism by the previous step; and the fourth and fifth maps ε are isomorphisms by the descending inductive hypothesis. So by the Five Lemma, the middle ε map is surjective. This gives the surjectivity part of the result, for an arbitrary coherent sheaf $\mathcal{F}$. In particular, surjectivity holds with $\mathcal{F}$ replaced by $\mathcal{N}$. That is, on the first of the five terms in the exact sequence above, the map ε is surjective. So by the Five Lemma, the middle ε is injective; so it is an isomorphism. $\square$

Concerning Theorem 2.2.3, that result is equivalent to the following assertion:

THEOREM 2.2.5. *Let* $X = \mathbb{P}_{\mathbb{C}}^r$ *or* $(\mathbb{P}_{\mathbb{C}}^r)^h$, *and let* $\mathcal{M}$ *be a coherent sheaf on* X. *Then there is an* n_0 *such that for all* $n \geq n_0$ *and all* $q > 0$, *we have* $H^q(X, \mathcal{M}(n)) = 0$.

In the algebraic setting, Theorem 2.2.5 is due to Serre; cf. [Hrt2, Chap. III, Theorem 5.2]. In the analytic setting, this is Cartan's "Theorem B" ([Se1], exp. XVIII of [Ca2]); cf. also [GH, p. 700].

The proof of Theorem 2.2.3 is easier in the algebraic situation than in the analytic one. In the former case, the proof proceeds by choosing generators of stalks $\mathcal{M}_\xi$; multiplying each by an appropriate monomial to get a global section of some $\mathcal{M}(n)$; and using quasi-compactness to require only finitely many sections overall (also cf. [Hrt2, Chapter II, proof of Theorem 5.17]). But this strategy fails in the analytic case because the local sections are not rational, or even meromorphic; and so one cannot simply clear denominators to get a global section of a twisting of $\mathcal{M}$.

The proof in the analytic case proves Cartan's Theorems A and B (i.e. 2.2.3 and 2.2.5) together, by induction on r. Denoting these assertions in dimension r by (A_r) and (B_r), the proof in ([Se1], exp. XIX of [Ca3]) proceeds by showing that $(A_{r-1}) + (B_{r-1}) \Rightarrow (A_r)$ and that $(A_r) \Rightarrow (B_r)$. Since the results are trivial for $r = 0$, the two theorems then follow; and as a result, GAGA follows as well. Serre's later argument in [Se3] is a variant on this inductive proof that simultaneously proves GAGA and Theorems A and B (i.e. Theorems 2.2.1, 2.2.3, and 2.2.5 above).

Theorems A and B were preceded by a non-projective version of those results, viz. for polydiscs in $\mathbb{C}^r$, and more generally for Stein spaces (exp. XVIII and XIX of [Ca2]; cf. also [GuR, pp. 207, 243]). There too, the two theorems are essentially equivalent. Also, no twisting is needed for Theorem B in the earlier version because the spaces were not projective there.

The proof of Theorem A in this earlier setting uses an "analytic patching" argument, applied to overlapping compact sets K', K'' on a Stein space X. In that situation, one considers metric neighborhoods U', U'' of K', K'' respectively, and one chooses generating sections $f_1', \ldots, f_k' \in \mathcal{M}(U')$ and $f_1'', \ldots, f_k'' \in \mathcal{M}(U'')$ for the given sheaf $\mathcal{M}$ on U', U'' respectively. From this data, one produces generating sections $g_1, \ldots, g_k \in \mathcal{M}(U)$, where U is an open neighborhood of $K = K' \cup K''$. This is done via Cartan's Lemma on matrix factorization, which says (for appropriate choice of K', K'') that every element $A \in GL_n(K' \cap K'')$ can be factored as a product of an element $B \in GL_n(K')$ and an element $C \in GL_n(K'')$. That lemma, which can be viewed as a multiplicative matrix analog of Cousin's Theorem [GuRo, p. 32], had been proved earlier in [Ca1], with this application in mind; and a special case had been shown even earlier in [Bi]. See also [GuRo, Chap. VI, § E]. (Cartan's Lemma is also sometimes called Cartan's "attaching theorem", where *attaching* is used in essentially the same sense as *patching* here.)

Cartan's Lemma can be used to prove these earlier versions of Theorems A and B by taking bases f_i' and f_i'' over U' and U'', and letting A be the transition matrix between them (i.e. $\vec{f'} = A\vec{f''}$, where $\vec{f'}$ and $\vec{f''}$ are the column vectors with entries f_i' and f_i'' respectively). The generators g_i as above can then be defined as the sections that differ from the f'''s by B^{-1} and from the f''''s by C (i.e. $\vec{g} = B^{-1}\vec{f} = C\vec{f''}$). The g_i's are then well-defined wherever either the f'''s or f''''s are — and hence in a neighborhood of $K = K' \cup K''$. This matrix factorization strategy also appears elsewhere, e.g. classically, concerning the Riemann-Hilbert problem, in which one attempts to find a system of linear differential equations whose monodromy representation of the fundamental group is a given representation (this being a differential analog of the inverse Galois problem for covers).

This use of Cartan's Lemma also suggests another way of restating GAGA, in the case where a projective variety X is covered by two open subsets X_1, X_2 that are strictly contained in X. The point is that if one gives coherent analytic (sheaves of) modules over X_1 and over X_2 together with an isomorphism on the overlap, then there is a unique coherent *algebraic* module over X that induces the given data compatibly. Of course by definition of coherent sheaves, there is such an *analytic* module over X (and similarly, we can always reduce to the case of two metric open subsets X_1, X_2); but the assertion is that it is algebraic.

To state this compactly, we introduce some categorical terminology. If $\mathcal{A}, \mathcal{B}, \mathcal{C}$ are categories, with functors $f : \mathcal{A} \to \mathcal{C}$ and $g : \mathcal{B} \to \mathcal{C}$, then the 2-*fibre product* of $\mathcal{A}$ and $\mathcal{B}$ over $\mathcal{C}$ (with respect to f, g) is the category $\mathcal{A} \times_{\mathcal{C}} \mathcal{B}$ in which an *object* is a pair $(A, B) \in \mathcal{A} \times \mathcal{B}$ together with an isomorphism $\iota : f(A) \xrightarrow{\sim} g(B)$ in $\mathcal{C}$; and in which a *morphism* $(A, B; \iota) \to (A', B'; \iota')$ is a pair of morphisms $A \to A'$ and $B \to B'$ that are compatible with the ι's. For any variety [resp. analytic space] X, let $\mathfrak{M}(X)$ denote the category of algebraic [resp. analytic] coherent modules on X. (Similarly, for any ring R, we write $\mathfrak{M}(R)$ for the category of finitely presented R-modules. This is the same as $\mathfrak{M}(\mathrm{Spec}\ R)$.) In this language, GAGA and its generalizations to algebras and covers can be restated thus:

THEOREM 2.2.6. *Let X be a complex projective algebraic variety, with metric open subsets X_1, X_2 such that $X = X_1 \cup X_2$; let X_0 be their intersection. Then the natural base change functor*

$$\mathfrak{M}(X) \to \mathfrak{M}(X_1) \times_{\mathfrak{M}(X_0)} \mathfrak{M}(X_2)$$

is an equivalence of categories. Moreover the same holds if $\mathfrak{M}$ is replaced by the category of finite algebras, or of finite branched covers, or of Galois covers with a given Galois group.

Here X is regarded as an algebraic variety, and the X_i's as analytic spaces (so that the left hand side of the equivalence consists of algebraic modules, and the objects on the right hand side consist of analytic modules). In the case of curves, each X_i is contained in an affine open subset U_i (unless $X_i = X$), so coherent sheaves of modules on X_i can be identified with coherent modules over the ring $\mathcal{H}(X_i)$; thus we may identify the categories $\mathfrak{M}(X_i)$ and $\mathfrak{M}(\mathcal{H}(X_i))$.

The approach in Theorem 2.2.6 will be useful in considering analogs of GAGA in Sections 3 and 4 below.

2.3. Complex patching and constructing covers. Consider a Zariski open subset of the Riemann sphere, say $U = \mathbb{P}^1_{\mathbb{C}} - \{\xi_1, \ldots, \xi_r\}$. By Riemann's Existence Theorem, every finite covering space of U is given by an étale morphism of complex algebraic curves. Equivalently, every finite branched cover of $\mathbb{P}^1_{\mathbb{C}}$, branched only at $S = \{\xi_1, \ldots, \xi_r\}$, is given by a finite dominating morphism from a smooth complex projective curve Y to $\mathbb{P}^1_{\mathbb{C}}$. As discussed in Section 2.1, passage from topological to analytic covers is the easier step, but it requires knowledge of what topological covering spaces exist (essentially via knowledge of the fundamental group, which is understood via loops). Passage from analytic covers to algebraic covers is deeper, and can be achieved using GAGA, as discussed in Section 2.2.

Here we consider how covers can be constructed from this point of view using complex analytic patching, keeping an eye on possible generalizations. In particular, we raise the question of how to use these ideas to understand covers of curves that are not defined over the complex numbers.

We begin by elaborating on the bijection described in Corollary 2.1.2.

Taking $U = \mathbb{P}^1_{\mathbb{C}} - \{\xi_1, \ldots, \xi_r\}$ as above, choose a base point $\xi_0 \in U$. The topological fundamental group $\pi_1(U, \xi_0)$ is then the discrete group

$$\langle x_1, \ldots, x_r \mid x_1 \cdots x_r = 1 \rangle,$$

as discussed in Section 2.1. Up to isomorphism, the fundamental group is independent of the choice of ξ_0, and so the mention of the base point is often suppressed; but fixing a base point allows us to analyze the fundamental group more carefully. Namely, we may choose a "bouquet of loops" at ξ_0 (in M. Fried's terminology [Fr1]), consisting of a set of counterclockwise loops $\sigma_1, \ldots, \sigma_r$ at ξ_0, where σ_j winds once around ξ_j and winds around no other ξ_k; where the support of the σ_j's are disjoint except at ξ_0; where $\sigma_1 \cdots \sigma_r$ is homotopic to the identity; and where the homotopy classes of the σ_j's (viz. the x_j's) generate $\pi_1(U, \xi_0)$. In particular, we can choose σ_j to consist of a path ϕ_j from ξ_0 to a point ξ'_j near ξ_j, followed by a counterclockwise loop λ_j around ξ_j, followed by ϕ_j^{-1}. The term "bouquet" is natural with this choice of loops (e.g. in the case that $\xi_0 = 0$ and $\xi_j = e^{j\pi i/r}$, with $j = 1, \ldots, r$, and where each ϕ_j is a line segment from ξ_0 to $(1 - \varepsilon)\xi_j$ for some small positive value of ε).

Let $f : V \to U$ be a finite Galois covering space, say with Galois group G. Then $\pi_1(V)$ is a subgroup N of finite index in $\pi_1(U)$, and $G = \pi_1(U)/N$. Let $g_1, \ldots, g_r \in G$ be the images of $x_1, \ldots, x_r \in \pi_1(U)$, and let m_j be the order of g_j. Thus $(g_1, \ldots, g_r)$ is the branch cycle description of $V \to U$; i.e. the G-Galois cover $V \to U$ corresponds to the uniform conjugacy class of $(g_1, \ldots, g_r)$ in

Corollary 2.1.2. By Riemann's Existence Theorem 2.1.1, the cover $V \to U$ can be given by polynomial equations and regarded as a finite étale cover. Taking the normalization of $\mathbb{P}^1_\mathbb{C}$ in V, we obtain a smooth projective curve Y containing V as a Zariski open subset, and a G-Galois connected branched covering map $f : Y \to \mathbb{P}^1_\mathbb{C}$ which is branched only over $S = \{\xi_1, \ldots, \xi_r\}$.

In the above notation, with $\sigma_j = \phi_j \lambda_j \phi_j^{-1}$ (and multiplying paths from left to right), we can extend ϕ_j to a path ψ_j from ξ_0 to ξ_j in $\mathbb{P}^1_\mathbb{C}$. The path ψ_j can be lifted to a path $\tilde{\psi}_j$ in Y from a base point $\eta_0 \in Y$ over ξ_0, to a point $\eta_j \in Y$ over ξ_j. The element g_j generates the inertia group A_j of η_j (i.e. the stabilizer of η_j in the group G). If X_j is a simply connected open neighborhood of ξ_j that contains no other ξ_k, then the topological fundamental group of $X_j - \{\xi_j\}$ is isomorphic to $\mathbb{Z}$. So $f^{-1}(X_j)$ is a union of homeomorphic connected components, each of which is Galois and cyclic of order m_j over X_j, branched only at ξ_j. The component Y_j of $f^{-1}(X_j)$ that contains η_j has stabilizer $A_j = \langle g_j \rangle \subset G$, and by Kummer theory it is given by an equation of the form $s_j^{m_j} = t_j$, if t_j is a uniformizer on X_j at ξ_j. Moreover g_j acts by $g_j(s_j) = e^{2\pi i/m_j} s_j$. So $f^{-1}(X_j)$ is a (typically disconnected) G-Galois cover of X_j, consisting of a disjoint union of copies of the m_j-cyclic cover $Y_j \to X_j$, indexed by the left cosets of A_j in G. We say that $f^{-1}(X_j)$ is the G-Galois branched cover of X_j that is *induced* by the A_j-Galois cover $Y_j \to X_j$; and we write $f^{-1}(X_j) = \mathrm{Ind}_{A_j}^G Y_j$. Similarly, if U' is a simply connected open subset of U (and so U' does not contain any branch points ξ_j), then $f^{-1}(U')$ is the trivial G-Galois cover of U', consisting of $|G|$ copies of U' permuted simply transitively by the elements of G; this cover is $\mathrm{Ind}_1^G U'$.

Since the complex affine line is simply connected, the smallest example of the above situation is the case $r = 2$. By a projective linear change of variables, we may assume that the branch points are at $0, \infty$. The fundamental group of $U = \mathbb{P}^1_\mathbb{C} - \{0, \infty\}$ is infinite cyclic, so a finite étale cover is cyclic, say with Galois group C_m; and the cover has branch cycle description $(g, g^{-1}) = (g, g^{m-1})$ for some generator g of the cyclic group C_m. This cover is given over U by the single equation $y^m = x$. So no patching is needed in this case. (If we instead take two branch points $x = c_0, x = c_1$, with $c_0, c_1 \in \mathbb{C}$, then the equation is $y^m = (x - c_0)(x - c_1)^{m-1}$ over $\mathbb{P}^1_\mathbb{C}$ minus the two branch points.)

The next simplest case is that of $r = 3$. This is the first really interesting case, and in fact it is key to understanding cases with $r > 3$. By a projective linear transformation we may assume that the branch locus is $\{0, 1, \infty\}$. We consider this case next in more detail:

EXAMPLE 2.3.1. We give a recipe for constructing Galois covers of $U = \mathbb{P}^1_\mathbb{C} - \{0, 1, \infty\}$ via patching, in terms of the branch cycle description of the given cover.

The topological fundamental group of U is $\langle \alpha, \beta, \gamma \,|\, \alpha\beta\gamma = 1 \rangle$, and this is isomorphic to the free group on two generators, viz. α, β. If we take $z = 1/2$ as the base point for the fundamental group, then these generators can be taken to be counterclockwise loops at $1/2$ around $0, 1$, respectively. The paths ψ_0, ψ_1

as above can be taken to be the real line segments connecting the base point to $0, 1$ respectively, and ψ_∞ can be taken to be the vertical path from $1/2$ to "$1/2 + i\infty$".

Let G be a finite group generated by two elements a, b. Let $c = (ab)^{-1}$, so that $abc = 1$. Consider the connected G-Galois covering space $f : V \to U$ with branch cycle description (a, b, c), and the corresponding branched cover $Y \to \mathbb{P}^1_\mathbb{C}$ branched at S. As above, after choosing a base point $\eta \in Y$ over $1/2 \in \mathbb{P}^1_\mathbb{C}$ and lifting the paths ψ_j, we obtain points $\eta_0, \eta_1, \eta_\infty$ over $0, 1, \infty$, with cyclic stabilizers $A_0 = \langle a \rangle$, $A_1 = \langle b \rangle$, $A_\infty = \langle c \rangle$ respectively. Let ι be the path in $\mathbb{P}^1_\mathbb{C}$ from 0 to 1 corresponding to the real interval $[0, 1]$, and let $\tilde{\iota}$ be the unique path in Y that lifts ι and passes through η. Observe that the initial point of $\tilde{\iota}$ is η_0, and the final point is η_1.

Consider the simply connected neighborhoods $X_0 = \{z \in \mathbb{C} \mid \mathrm{Re}\, z < 2/3\}$ of 0, and $X_1 = \{z \in \mathbb{C} \mid \mathrm{Re}\, z > 1/3\}$ of 1. We have that $X_0 \cup X_1 = \mathbb{C}$, and $U' := X_0 \cap X_1$ is contained in U. Also, $U = U_0 \cup U_1$, where $U_j = X_j - \{j\}$ for $j = 0, 1$. By the above discussion, $f^{-1}(X_0) = \mathrm{Ind}_{A_0}^G Y_0$, where $Y_0 \to X_0$ is a cyclic cover branched only at 0, and given by the equation $y_0^m = x$ (where m is the order of a). Similarly $f^{-1}(X_1) = \mathrm{Ind}_{A_1}^G Y_1$, where the branched cover $Y_1 \to X_1$ is given by $y_1^n = x - 1$ (where n is the order of b). Since the overlap $U' = X_0 \cap X_1$ does not meet the branch locus S, we have that $f^{-1}(U')$ is the trivial G-Galois cover $\mathrm{Ind}_1^G U'$. These induced covers have connected components that are respectively indexed by the left cosets of $A_0, A_1, 1$; and the identity coset corresponds to the component respectively containing η_0, η_1, η. Observe that the identity component of $\mathrm{Ind}_1^G U'$ is contained in the identity components of the other two induced covers, because $\tilde{\iota}$ passes through η_0, η_1, η.

Turning this around, we obtain the desired "patching recipe" for constructing the G-Galois cover of U with given branch cycle description (a, b, c): Over the above open sets U_0 and U_1, take the induced covers $\mathrm{Ind}_{A_0}^G V_0$ and $\mathrm{Ind}_{A_1}^G V_1$, where $V_0 \to U_0$ and $V_1 \to U_1$ are respectively given by $y_0^m = x$ and $y_1^n = x - 1$, and where $A_0 = \langle a \rangle$, $A_1 = \langle b \rangle$. Pick a point η over $1/2$ on the identity components of each of these two induced covers; thus $g(\eta)$ is a well-defined point on each of these induced covers, for any $g \in G$. The induced covers each restrict to the trivial G-Galois cover on the overlap $U' = U_0 \cap U_1$; now paste together the components of these trivial covers by identifying, for each $g \in G$, the component of $\mathrm{Ind}_{A_0}^G V_0$ containing $g(\eta)$ with the component of $\mathrm{Ind}_{A_1}^G V_1$ containing that point. The result is the desired cover $V \to U$. $\qquad\square$

The above example begins with a group G and a branch cycle description (a, b, c), and constructs the cover $V \to U = \mathbb{P}^1_\mathbb{C} - \{0, 1, \infty\}$ with that branch cycle description. In doing so, it gives the cover locally in terms of equations over two topological open discs U_0 and U_1, and instructions for patching on the overlap. Thus it gives the cover analytically (not algebraically, since the U_i's are not Zariski open subsets).

The simplest specific instance of the above example uses the cyclic group $C_3 = \langle g \rangle$ of order 3, and branch cycle description (g, g, g). Over U_0 the cover is given by (one copy of) $y_0^3 = x$; and over U_1 it is given by $y_1^3 = x - 1$. Here, over U_i, the generator g acts by $g(y_i) = \zeta_3 y_i$, where $\zeta_3 = e^{2\pi i/3}$. By GAGA, the cover can be described algebraically, i.e. by polynomials over Zariski open sets. And in this particular example, this can even be done globally over U, by the single equation $z^3 = x(x - 1)$ (where $g(z) = \zeta_3 z$). Here $z = y_0 f_0(x)$ on U_0, where $f_0(x)$ is the holomorphic function on U_0 such that $f_0(0) = -1$ and $f_0^3 = x - 1$; explicitly, $f_0(x) = -1 + \frac{1}{3}x + \frac{1}{9}x^2 + \cdots$ in a neighborhood of $x = 0$. Similarly, $z = y_1 f_1(x)$ on U_1, where $f_1(x)$ is the holomorphic function on U_1 such that $f_1(1) = 1$ and $f_1^3 = x$; here $f_1(x) = 1 + \frac{1}{3}(x - 1) - \frac{1}{9}(x - 1)^2 + \cdots$ in a neighborhood of $x = 1$. (Note that for this very simple cover, the global equation can be written down by inspection. But in general, for non-abelian groups, the global polynomial equations are not at all obvious from the local ones, though by GAGA they must exist.)

Example 2.3.1 requires GAGA in order to pass from the analytic equations (locally, on metric open subsets) to algebraic equations that are valid on a Zariski open dense subset. In addition, it uses ideas of topology — in particular, knowledge of the fundamental group, and the existence of open sets that overlap and together cover the space U. In later sections of this paper, we will discuss the problem of performing analogous constructions over fields other than $\mathbb{C}$, in order to understand covers of algebraic curves over those fields. For that, we will see that often an analog of GAGA exists — and that analog will permit passage from "analytic" covers to algebraic ones. A difficulty that has not yet been overcome, however, is how to find analogs of the notions from topology — both regarding explicit descriptions of fundamental groups and regarding the need for having overlapping open sets (which in non-archimedean contexts do not exist in a nontrivial way). One way around this problem is to consider only certain types of covers, for which GAGA alone suffices (i.e. where the information from topology is not required). The next example illustrates this.

EXAMPLE 2.3.2. Let G be a finite group, with generators $g_1, \ldots, g_r$ (whose product need not be 1). Let $S = \{\xi_1, \ldots, \xi_{2r}\}$ be a set of $2r$ distinct points in $\mathbb{P}^1_{\mathbb{C}}$, and consider the G-Galois covering space $V \to U = \mathbb{P}^1_{\mathbb{C}} - S$ with branch cycle description

$$(g_1, g_1^{-1}, g_2, g_2^{-1}, \ldots, g_r, g_r^{-1}), \tag{$*$}$$

with respect to a bouquet of loops $\sigma_1, \ldots, \sigma_{2r}$ at a base point $\xi_0 \in U$. Let $Y \to \mathbb{P}^1_{\mathbb{C}}$ be the corresponding branched cover. This cover is well defined since the product of the entries of $(*)$ is 1, and it is connected since the entries of $(*)$ generate G. The cover can be obtained by a "cut-and-paste" construction as follows: Choose disjoint simple (i.e. non-self-intersecting) paths $s_1, \ldots, s_r$ in $\mathbb{P}^1_{\mathbb{C}}$, where s_j begins at ξ_{2j-1} and ends at ξ_{2j}. Take $|G|$ distinct copies of $\mathbb{P}^1_{\mathbb{C}}$, indexed by the elements of G. Redefine the topology on the disjoint union of

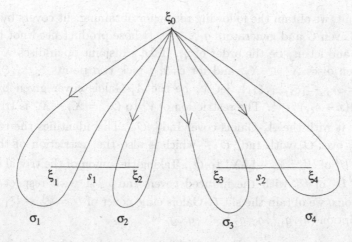

Figure 2.3.3. Base of a slit cover of $\mathbb{P}^1_{\mathbb{C}}$ with slits s_1 from ξ_1 to ξ_2, and s_2 from ξ_3 to ξ_4; and with generators g_1, g_2, corresponding to the loops σ_1, σ_3, respectively. (Here the inverses g_1^{-1}, g_2^{-1} correspond to the loops σ_2, σ_4.)

these copies by identifying the right hand edge of a "slit" along s_j on the g-th copy of $\mathbb{P}^1_{\mathbb{C}}$ to the left hand edge of the "slit" along s_j on the gg_j-th copy of $\mathbb{P}^1_{\mathbb{C}}$ (with the orientation as one proceeds along the slits). The resulting space maps to $\mathbb{P}^1_{\mathbb{C}}$ in the obvious way, and away from S it is the G-Galois covering space of $\mathbb{P}^1_{\mathbb{C}} - S$ with branch cycle description $(*)$. Because of this construction, we will call covers of this type *slit covers* [Ha1, 2.4]. (The corresponding branch cycle descriptions $(*)$ have been referred to as "Harbater-Mumford representatives" [Fr3].)

Now choose disjoint simply connected open subsets $X_j \subset \mathbb{P}^1_{\mathbb{C}}$ for $j = 1, \ldots, r$, such that $\xi_{2j-1}, \xi_{2j} \in X_j$. (If ξ_{2j-1} and ξ_{2j} are sufficiently close for all j, relative to their distances to the other ξ_k's, then the X_j's can be taken to be discs.) In the above cut-and-paste construction, the paths $s_1, \ldots, s_r$ can be chosen so that the support of s_j is contained in X_j, for each j. Each X_j contains a strictly smaller simply connected open set X_j^* (for instance, a smaller disc) which also contains the support of s_j, and whose closure $\bar{X}_j^*$ is contained in X_j. Let $U' = \mathbb{P}^1_{\mathbb{C}} - \bigcup \bar{X}_j^*$. In the cut-and-paste construction of $V \to U$, we have that the topology of the disjoint union of the $|G|$ copies of $\mathbb{P}^1_{\mathbb{C}}$ is unaffected outside of the union of the $\check{X}_j^*$'s; and so the restriction of $V \to U$ to U' is a trivial cover, viz. $\mathrm{Ind}_1^G U'$. Suppose that ξ_j is not the point $x = \infty$ on $\mathbb{P}^1_{\mathbb{C}}$; thus ξ_j corresponds to a point $x = c_j$, with $c_j \in \mathbb{C}$. Let m_j be the order of g_j, and let A_j be the subgroup of G generated by g_j. Then the restriction of $V \to U$ to $U_j = X_j \cap U$ is given by $\mathrm{Ind}_{A_j}^G V_j$, where $V_j \to U_j$ is the A_j-Galois étale cover given by $y_j^{m_j} = (x - c_{2j-1})(x - c_{2j})^{m_j - 1}$ (as in the two branch point case, discussed just before Example 2.3.1).

As a result, we obtain the following recipe for obtaining slit covers by analytic patching: Given G and generators $g_1, \ldots, g_r$ (whose product need not be 1), let $A_j = \langle g_j \rangle$, and let m_j be the order of g_j. Take r disjoint open discs X_j, choose smaller open discs $X_j^* \subset X_j$, and for each j pick two points $\xi_{2j-1}, \xi_{2j} \in X_j^*$. Over $U_j = X_j - \{\xi_{2j-1}, \xi_{2j}\}$, let V_j be the A_j-Galois cover given by $y_j^{m_j} = (x - c_{2j-1})(x - c_{2j})^{m_j - 1}$. The restriction of V_j to $O_j := X_j - \bar{X}_j^*$ is trivial, and we identify it with the A_j-Galois cover $\mathrm{Ind}_1^{A_j} O_j$. This identifies the restriction of $\mathrm{Ind}_{A_j}^G V_j$ over O_j with $\mathrm{Ind}_1^G O_j$ — which is also the restriction of the trivial cover $\mathrm{Ind}_1^G U'$ of $U' = \mathbb{P}_{\mathbb{C}}^1 - \bigcup \bar{X}_j^*$ to O_j. Taking the union of the trivial G-Galois cover $\mathrm{Ind}_1^G U'$ of U' with the induced covers $\mathrm{Ind}_{A_j}^G V_j$, with respect to these identifications, we obtain the slit G-Galois étale cover of $U = \mathbb{P}_{\mathbb{C}}^1 - \{\xi_1, \ldots, \xi_{2r}\}$ with description $(g_1, g_1^{-1}, g_2, g_2^{-1}, \ldots, g_r, g_r^{-1})$. $\square$

The slit covers that occur in Example 2.3.2 can also be understood in terms of degeneration of covers — and this point of view will be useful later on, in more general settings. Consider the G-Galois slit cover $V \to U = \mathbb{P}_{\mathbb{C}}^1 - S$ with branch cycle description $(*)$ as in Example 2.3.2; here $S = \{\xi_1, \ldots, \xi_{2r}\}$ and s_j is a simple path connecting ξ_{2j-1} to ξ_{2j}, with the various s_j's having disjoint support. This cover may be completed to a G-Galois branched cover $Y \to \mathbb{P}_{\mathbb{C}}^1$, with branch locus S, by taking the normalization of $\mathbb{P}_{\mathbb{C}}^1$ in (the function field of) V. Now deform this branched cover by allowing each point ξ_{2j} to move along the path s_j backwards toward ξ_{2j-1}. This yields a one (real) parameter family of irreducible G-Galois slit covers $Y_t \to \mathbb{P}_{\mathbb{C}}^1$, each of which is trivial outside of a union of (shrinking) simply connected open sets containing ξ_{2j-1} and (the moving) ξ_{2j}. In the limit, when ξ_{2j} collides with ξ_{2j-1}, we obtain a finite map $Y_0 \to \mathbb{P}_{\mathbb{C}}^1$ which is unramified away from $S' := \{\xi_1, \xi_3, \ldots, \xi_{2r-1}\}$, such that Y_0 is connected; G acts on Y_0 over $\mathbb{P}_{\mathbb{C}}^1$, and acts simply transitively away from S'; and the map is a trivial cover away from S'. In fact, Y_0 is a union of $|G|$ copies of $\mathbb{P}_{\mathbb{C}}^1$, indexed by the elements of G, such that the g-th copy meets the gg_j-th copy over ξ_{2j-1}. The map $Y_0 \to \mathbb{P}_{\mathbb{C}}^1$ is a *mock cover* [Ha1, §3], i.e. is finite and generically unramified, and such that each irreducible component of Y_0 maps isomorphically onto the base (here, $\mathbb{P}_{\mathbb{C}}^1$). This degeneration procedure can be reversed: starting with a connected G-Galois mock cover which is built in an essentially combinatorial manner in terms of the data $g_1, \ldots, g_r$, one can then deform it near each branch point to obtain an irreducible G-Galois branched cover branched at $2r$ points with branch cycle description $(g_1, g_1^{-1}, g_2, g_2^{-1}, \ldots, g_r, g_r^{-1})$. This is one perspective on the key construction in the next section, on formal patching.

As discussed before Example 2.3.2, slit covers do not require topological input — i.e. knowledge of the explicit structure of topological fundamental groups, or the existence of overlapping open discs containing different branch points — unlike the general three-point cover in Example 2.3.1. Without this topological input, for general covers one obtains only the equivalence of algebraic and analytic covers in Riemann's Existence Theorem — and in particular, we do not

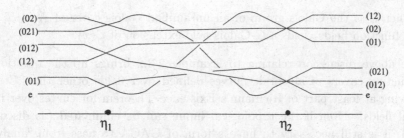

Figure 2.3.4. A mock cover of the line, with Galois group S_3, branched at two points η_1, η_2. The sheets are labeled by the elements of S_3. The cyclic subgroups $\langle(01)\rangle, \langle(012)\rangle$ are the stablizers on the identity sheet over η_1, η_2, respectively.

obtain the corollaries to Riemann's Existence Theorem in Section 2.1. But one can obtain those corollaries as they relate to slit covers, without topological input. Since only half of the entries of the branch cycle description can be specified for a slit cover, such results can be regarded as a "Half Riemann Existence Theorem"; and can be used to motivate analogous results about fundamental groups for curves that are not defined over $\mathbb{C}$, where there are no "loops" or overlapping open discs. (Indeed, the term "half Riemann Existence Theorem" was first coined by F. Pop to refer to such an analogous result [Po2, Main Theorem]; cf. § 4.3 below). In particular, we have the following variant on Corollary 2.1.3:

THEOREM 2.3.5 (ANALYTIC HALF RIEMANN EXISTENCE THEOREM). *Let* $r \geq 1$, *let* $S = \{\xi_1, \ldots, \xi_{2r}\}$ *be a set of $2r$ distinct points in* $\mathbb{P}_{\mathbb{C}}^1$, *and let* $U = \mathbb{P}_{\mathbb{C}}^1 - S$. *Let* $\hat{F}_r$ *be the free profinite group on generators* $x_1, \ldots, x_r$. *Then* $\hat{F}_r$ *is a quotient of the étale fundamental group of* U.

Namely, let G be any finite quotient of $\hat{F}_r$. That is, G is a finite group together with generators $g_1, \ldots, g_r$. Consider the G-Galois slit cover with branch cycle description $(g_1, g_1^{-1}, g_2, g_2^{-1}, \ldots, g_r, g_r^{-1})$. As G and its generators vary, these covers form an inverse subsystem of the full inverse system of covers of U; and the inverse limit of their Galois groups is $\hat{F}_r$.

Here, in order for this inverse system to make sense, one can first fix a bouquet of loops around the points of S; or one can fix a set of disjoint simple paths s_j from ξ_{2j-1} to ξ_{2j} and consider the corresponding set of slit covers. But to give a non-topological proof of this result (which of course is a special case of Corollary 2.1.3), one can instead give compatible local Kummer equations for the slit covers and then use GAGA; or one can use the deformation construction starting from mock covers, as sketched above. These approaches are in fact equivalent, and will be discussed in the next section in a more general setting.

Observe that the above "half Riemann Existence Theorem" is sufficient to prove the inverse Galois problem over $\mathbb{C}(x)$, which appeared above, as Corollary 2.1.4 of (the full) Riemann's Existence Theorem. Namely, for any finite group G, pick a set of r generators of G (for some r), and pick a set S of $2r$ points in

$\mathbb{P}^1_{\mathbb{C}}$. Then G is the Galois group of an unramified Galois cover of $\mathbb{P}^1_{\mathbb{C}} - S$; and taking function fields yields a G-Galois field extension of $\mathbb{C}(x)$.

The above discussion relating to Example 2.3.2 brings up the question of constructing covers of algebraic curves defined over fields other than $\mathbb{C}$, and of proving at least part of Riemann's Existence Theorem for curves over more general fields. Even if the topological input can be eliminated (as discussed above), it is still necessary to have a form of GAGA to pass from "analytic" objects to algebraic ones. The "analytic" objects will be defined over a topology that is finer than the Zariski topology, and with respect to which modules and covers can be constructed locally and patched. It will also be necessary to have a structure sheaf of "analytic" functions on the space under this topology.

One initially tempting approach to this might be to use the étale topology; but unfortunately, this does not really help. One difficulty with this is that a direct analog of GAGA does not hold in the étale topology. Namely, in order to descend a module from the étale topology to the Zariski topology, one needs to satisfy a descent criterion [Gr5, Chap. VIII, § 1]. In the language of Theorem 2.2.6, this says that one needs not just agreement on the overlap $X_1 \times_X X_2$ between the given étale open sets, but also on the "self-overlaps" $X_1 \times_X X_1$ and $X_2 \times_X X_2$, which together satisfy a compatibility condition. (See also [Gr3], in which descent is viewed as a special case of patching, or "recollement".) A second difficulty is that in order to give étale open sets $X_i \to X$, one needs to understand covers of X; and so this introduces an issue of circularity into the strategy for studying and constructing covers.

Two other approaches have proved quite useful, though, for large classes of base fields (though not for all fields). These are the Zariski-Grothendieck notion of formal geometry, and Tate's notion of rigid geometry. Those approaches will be discused in the following sections.

3. Formal Patching

This section and the next describe approaches to carrying over the ideas of Section 2 to algebraic curves that are defined over fields other than $\mathbb{C}$. The present section uses formal schemes rather than complex curves, in order to obtain analogs of complex analytic notions that can be used to obtain results in Galois theory. The idea goes back to Zariski; and his notion of a "formal holomorphic function", which uses formal power series rather than convergent power series, is presented in Section 3.1. Grothendieck's strengthening of this notion is presented in Section 3.2, including his formal analog of Serre's result GAGA (and the proof presented here parallels that of GAGA, presented in Section 2.2). These ideas are used in Section 3.3 to solve the geometric inverse Galois problem over various fields, using ideas motivated by the slit cover construction of Section 2.3. Further applications of these ideas are presented later, in Section 5.

3.1. Zariski's formal holomorphic functions.

In order to generalize analytic notions to varieties over fields other than $\mathbb{C}$, one needs to have "small open neighborhoods", and not just Zariski open sets. One also needs to have a notion of ("analytic") functions on those neighborhoods.

Unfortunately, if there is no metric on the ground field, then one cannot consider discs around the origin in $\mathbb{A}^1_k$, for example, or the rings of power series that converge on those discs. But one can consider the ring of *all* formal power series, regarded as analytic (or holomorphic) functions on the spectrum of the complete local ring at the origin (which we regard as a "very small neighborhood" of that point). And in general, given a variety V and a point $\nu \in V$, we can consider the elements of the complete local ring $\hat{\mathcal{O}}_{V,\nu}$ as holomorphic functions on Spec $\hat{\mathcal{O}}_{V,\nu}$.

While this point of view can be used to study local behaviors of varieties near a point, it does not suffice in order to study more global behaviors locally and then to "patch" (as one would want to do in analogs of GAGA and Riemann's Existence Theorem), because these "neighborhoods" each contain only one closed point. The issue is that a notion of "analytic continuation" of holomorphic functions is necessary for that, so that holomorphic functions near one point can also be regarded as holomorphic functions near neighboring points.

This issue was Zariski's main focus during the period of 1945-1950, and it grew out of ideas that arose from his previous work on resolution of singularities. The question was how to extend a holomorphic function from the complete local ring at a point $\nu \in V$ to points in a neighborhood. As he said later in a preface to his collected works [Za5, pp. xii-xiii], "I sensed the probable existence of such an extension provided the analytic continuation were carried out along an algebraic subvariety W of V." That is, if W is a Zariski closed subset of V, then it should make sense to speak of "holomorphic functions" in a "formal neighborhood" of W in V.

These formal holomorphic functions were defined as follows ([Za4, Part I]; see also [Ar5, p. 3]): Let W be a Zariski closed subset of a variety V. First, suppose that V is affine, say with ring of functions R, so that $W \subset V$ is defined by an ideal $I \subset R$. Consider the ring of rational functions g on V that are regular along W; this is a metric space with respect to the I-adic metric. The space of *strongly holomorphic functions* f *along* W (in V) is defined to be the metric completion of this space (viz. it is the space of equivalence classes of Cauchy sequences of such functions g). This space is also a ring, and can be identified with the inverse limit $\varprojlim R/I^n$.

More generally, whether or not V is affine, one can define a *(formal) holomorphic function along* W to be a function given locally in this manner. That is, it is defined to be an element $\{f_\omega\} \in \prod_{\omega \in W} \hat{\mathcal{O}}_{V,\omega}$ such that there is a Zariski affine open covering $\{V_i\}_{i \in I}$ of V together with a choice of a strongly holomorphic function $\{f_i\}_{i \in I}$ along $W_i := W \cap V_i$ in V_i (for each $i \in I$), such that f_ω is the image of f_i in $\hat{O}_{V,\omega}$ whenever $\omega \in W_i$. These functions also form a ring, denoted $\hat{O}_{V,W}$. Note that $\hat{O}_{V,W}$ is the complete local ring $\hat{O}_{V,\omega}$ if $W = \{\omega\}$. Also, if U is an affine open subset of W, and $U = \tilde{U} \cap W$ for some open subset $\tilde{U} \subset V$, then the ring $\hat{O}_{\tilde{U},U}$ depends only on U, and not on the choice of $\tilde{U}$; so we also denote this ring by $\hat{O}_{V,U}$, and call it the ring of holomorphic functions along U in V.

REMARK 3.1.1. Nowadays, if I is an ideal in a ring R, then the I-adic completion of R is defined to be the inverse limit $\varprojlim R/I^n$. This modern notion of formal completion is equivalent to Zariski's above notion of metric completion via Cauchy sequences, which he first gave in [Za2, § 5]. But Zariski's approach more closely paralleled completions in analysis, and fit in with his view of formal holomorphic functions as being analogs of complex analytic functions. (Prior to his giving this definition, completions of rings were defined only with respect to *maximal* ideals.) In connection with his introduction of this definition, Zariski also introduced the class of rings we now know as *Zariski rings* (and which Zariski had called "semi-local rings"): viz. rings R together with a non-zero ideal I such that every element of $1 + I$ is a unit in R [Za2, Def. 1] . Equivalently [Za2, Theorem 5], these are the I-adic rings such that I is contained in (what we now call) the Jacobson radical of R. Moreover, every I-adically complete ring is a Zariski ring [Za2, Cor. to Thm. 4]; so the ring of strictly holomorphic functions on a closed subset of an affine variety is a Zariski ring.   □

A deep fact proved by Zariski [Za4, § 9, Thm. 10] is that every holomorphic function along a closed subvariety of an affine scheme is strongly holomorphic. So those two rings of functions agree, in the affine case; and the ring of holomorphic functions along $W = \operatorname{Spec} R/I$ in $V = \operatorname{Spec} R$ can be identified with the formal completion $\varprojlim R/I^n$ of R with respect to I.

EXAMPLE 3.1.2. Consider the x-axis $W \approx \mathbb{A}^1_k$ in the x, t-plane $V = \mathbb{A}^2$. Then W is defined by the ideal $I = (t)$, and the ring of holomorphic functions along W in V is $A_1 := k[x][\![t]\!]$. Note that every element of A_1 can be regarded as an element in $\hat{O}_{W,\nu}$ for every point $\nu \in W$; and in this way can be regarded as an analytic continuation of (local) functions along the x-axis. Intuitively, the spectrum S_1 of A_1 can be viewed as a thin tubular neighborhood of W in V, which "pinches down" as $x \to \infty$. For example, observe that the elements x and $x - t$ are non-units in A_1, and so each defines a proper ideal of A_1; and correspondingly, their loci in $S_1 = \operatorname{Spec} A_1$ are non-empty (and meet the x-axis at the origin). On the other hand, $1 - xt$ is a unit in A_1, with inverse $1 + xt + x^2t^2 \cdots$, so its locus in S_1 is empty; and geometrically, its locus in V (which is a hyperbola) approaches the x-axis only as $x \to \infty$, and so misses the ("pinched down") spectrum of A_1. One can similarly consider the ring $A_2 = k[x^{-1}][\![t]\!]$; its spectrum S_2 is a thin neighborhood of $\mathbb{P}^1_k - (x = 0)$ which "pinches down" near $x = t = 0$. (See Figure 3.1.4.)   □

EXAMPLE 3.1.3. Let V' be the complement of the t-axis ($x = 0$) in the x, t-plane $\mathbb{A}^2$, and let $W' \subset V'$ be the locus of $t = 0$. Then the ring of holomorphic functions along W' in V' is $A_0 := k[x, x^{-1}][[t]]$. Geometrically, this is a thin tubular neighborhood of W' in V', which "pinches down" in two places, viz. as x approaches either 0 or ∞. (Again, see Figure 3.1.4.) Observe that Spec A_0 is *not* a Zariski open subset of Spec A_1, where A_1 is as in Example 3.1.2. In particular, A_0 is much larger than the ring $A_1[x^{-1}]$; e.g. $\sum_{n=1}^{\infty} x^{-n} t^n$ is an element of A_0 but not of $A_1[x^{-1}]$. Intuitively, $S_0 := $ Spec A_0 can be viewed as an "analytic open subset" of $S_1 = $ Spec A_1 but not a Zariski open subset — and similarly for S_0 and $S_2 = $ Spec A_2 in Example 3.1.2. Moreover S_0 can be regarded as the "overlap" of S_1 and S_2 in $\mathbb{P}^1_{k[[t]]}$. This will be made more precise below.   □

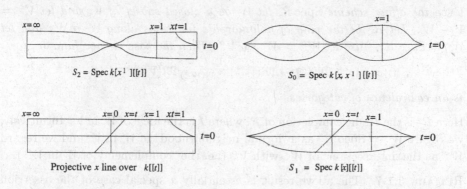

Figure 3.1.4. A covering of $\mathbb{P}^1_{k[[t]]}$ (lower left) by two formal patches, namely $S_1 = $ Spec $k[x][[t]]$ and $S_2 = $ Spec $k[1/x][[t]]$. The "overlap" S_0 is Spec $k[x, 1/x][[t]]$. See Examples 3.1.2 and 3.1.3.

REMARK 3.1.5. Just as the ring $A_0 = k[x, x^{-1}][[t]]$ in Example 3.1.3 is much larger than $A_1[x^{-1}]$, where $A_1 = k[x][[t]]$ as in Example 3.1.2, it is similarly the case that the ring A_1 is much larger than the ring $T := k[[t]][x]$ (e.g. $\sum_{n=1}^{\infty} x^n t^n$ is in A_1 but not in T). The scheme Spec T can be identified with the affine line over the complete local ring $k[[t]]$, and is a Zariski open subset of $\mathbb{P}^1_{k[[t]]}$ (given by $x \neq \infty$). This projective line over $k[[t]]$ can be viewed as a thin but uniformly wide tubular neighborhood of the projective x-line $\mathbb{P}^1_k$, and its affine open subset Spec T can correspondingly be viewed as a uniformly wide thin tubular neighborhood of the x-axis $\mathbb{A}^1_k$ (with no "pinching down" near infinity). As in Example 2, we have here that Spec A_1 is not a Zariski open subset of Spec T, and instead it can be viewed as an "analytic open subset" of Spec T. □

Using these ideas, Zariski proved his Fundamental Theorem on formal holomorphic functions [Za4, §11, p. 50]: If $f : V' \to V$ is a projective morphism of varieties, with V normal and with the function field of V algebraically closed in that of V', and if $W' = f^{-1}(W)$ for some closed subset $W \subset V$, then the natural map $\hat{O}_{V,W} \to \hat{O}_{V',W'}$ is an isomorphism. (See [Ar5, pp. 5–6] for a sketch of the proof.) This result in turn yielded Zariski's Connectedness Theorem [Za4, § 20, Thm. 14] (cf. also [Hrt2, III, Cor. 11.3]), and implied Zariski's Main Theorem (cf. [Hrt2, III, Cor. 11.4]).

The general discussion above suggests that it should be possible to prove an analog of GAGA that would permit patching of modules using formal completions. And indeed, there is the following assertion, which is essentially a result of Ferrand and Raynaud (cf. [FR, Prop. 4.2]). Here the notation is as at the end of Section 2.2 above, and this result can be viewed as analogous to the version of GAGA given by Theorem 2.2.6.

PROPOSITION 3.1.6 (FERRAND–RAYNAUD). *Let R be a Noetherian ring, let V be the affine scheme Spec R, let W be a closed subset of V, and let $V^\circ = V - W$. Let R^* be the ring of holomorphic functions along W in V, and let $W^* = $ Spec R^*. Also let $W^\circ = W^* \times_V V^\circ$. Then the base change functor*

$$\mathfrak{M}(V) \to \mathfrak{M}(W^*) \times_{\mathfrak{M}(W^\circ)} \mathfrak{M}(V^\circ)$$

is an equivalence of categories.

Here R^* is the I-adic completion of R, where I is the ideal of W in V. Intuitively, we regard $W^* = $ Spec R^* as a "formal neighborhood" of W in V, and we regard W° as the "intersection" of W^* with V° (i.e. the "complement" of W in W^*).

REMARK 3.1.7. The above result is essentially a special case of the assertion in [FR, Prop. 4.2]. That result was stated in terms of cartesian diagram of categories, which is equivalent to an assertion concerning 2-fibre products (i.e. the way Proposition 3.1.6 above is stated). The main difference between the above result and [FR, Prop. 4.2] is that the latter result allows W^* more generally to be any scheme for which there is a flat morphism $f : W^* \to V$ such that the pullback $f_W : W^* \times_V W \to W$ is an isomorphism — which is the case in the situation of Proposition 3.1.6 above. Actually, though, [FR, Prop. 4.2] assumes that $f : W^* \to V$ is *faithfully* flat (unlike the situation in Proposition 3.1.6). But this extra faithfulness hypothesis is unnecessary for their proof; and in any event, given a flat morphism $f : W^* \to V$ such that f_W is an isomorphism, one can replace W^* by the disjoint union of W^* and V°, which is then faithfully flat — and applying [FR, Prop. 4.2] to that new W^* gives the desired conclusion for the original W^*.   □

The following result of Artin [Ar4, Theorem 2.6] generalizes Proposition 3.1.6:

PROPOSITION 3.1.8. *In the situation of Proposition 3.1.6, let $\tilde{V}$ be a scheme and let $f : \tilde{V} \to V$ be a morphism of finite type. Let $\tilde{W}^*, \tilde{V}^\circ, \tilde{W}^\circ$ be the pullbacks of W^*, V°, W° with respect to f. Then the base change functor*

$$\mathfrak{M}(\tilde{V}) \to \mathfrak{M}(\tilde{W}^*) \times_{\mathfrak{M}(\tilde{W}^\circ)} \mathfrak{M}(\tilde{V}^\circ)$$

is an equivalence of categories.

Note that $\tilde{V}^\circ = \tilde{V} - \tilde{W}$ in Proposition 3.1.8, where $\tilde{W} = f^{-1}(W)$.

As an example of this result, let V be a smooth n-dimensional affine scheme over a field k, let W be a closed point ω of V, and let $\tilde{V}$ be the blow-up of V at W. So $\tilde{W}$ is a copy of $\mathbb{P}_k^{n-1}$; $W^* = \mathrm{Spec}\, \hat{O}_{V,\omega}$; and $\tilde{W}^*$ is the spectrum of a "uniformly wide tubular neighborhood" of $\tilde{W}$ in $\tilde{V}$. Here $\tilde{W}^*$, which is irreducible, can be viewed as a "twisted version" of $\mathbb{P}_{k[\![s]\!]}^{n-1}$; cf. [Hrt2, p. 29, Figure 3] for the case $n = 2$. According to Proposition 3.1.8, giving a coherent module on $\tilde{V}$ is equivalent to giving such modules on $\tilde{W}^*$ and on the complement of $\tilde{W}$, with agreement on the "overlap" $\tilde{W}^\circ$.

While the two preceding propositions required V to be affine, this hypothesis can be dropped if W is finite:

COROLLARY 3.1.9. *Let V be a Noetherian scheme, and let W be a finite set of closed points in V. Let R^* be the ring of holomorphic functions along W in V, let $W^* = \mathrm{Spec}\, R^*$, let $V^\circ = V - W$, and let $W^\circ = W^* \times_V V^\circ$.*

(a) *Then the base change functor*

$$\mathfrak{M}(V) \to \mathfrak{M}(W^*) \times_{\mathfrak{M}(W^\circ)} \mathfrak{M}(V^\circ)$$

is an equivalence of categories.

(b) *Let $\tilde{V}$ be a scheme and let $f : \tilde{V} \to V$ be a morphism of finite type. Let $\tilde{W}^*, \tilde{V}^\circ, \tilde{W}^\circ$ be the pullbacks of W^*, V°, W° with respect to f. Then the base change functor*

$$\mathfrak{M}(\tilde{V}) \to \mathfrak{M}(\tilde{W}^*) \times_{\mathfrak{M}(\tilde{W}^\circ)} \mathfrak{M}(\tilde{V}^\circ)$$

is an equivalence of categories.

SKETCH OF PROOF. For part (a), we may cover V by finitely many affine open subsets $V_i = \mathrm{Spec}\, R_i$, with R_i Noetherian. Applying Proposition 3.1.6 to each V_i and $W_i := V_i \cap W$, we obtain equivalences over each V_i. These equivalences agree on the overlaps $V_i \cap V_j$ (since each is given by base change), and so together they yield the desired equivalence over V, in part (a). Part (b) is similar, using Proposition 3.1.8. $\qquad\qquad\square$

Unfortunately, while the above results are a kind of GAGA, permitting the patching of modules, they do not directly help to construct covers (via the General Principle 2.2.4); and so they do not directly help prove an analog of Riemann's Existence Theorem. The reason is that these results require that a module be given over a Zariski open subset V° (or $\tilde{V}^\circ$), viz. the complement of the given closed subset W (or $\tilde{W}$). And a normal cover $Z \to V$ is determined by its restriction to a dense open subset V° (viz. it is the normalization of V in the function field of the cover — which is the same as the function field of the restriction). So these results provide a cover $Z \to V$ only in circumstances in which one already has the cover in hand.

Instead, in order to use Zariski's approach to obtain results about covers, we will focus on spaces such as $\mathbb{P}^1_{k[\![t]\!]}$ (and see the discussion in the Remark 3.1.5 above). In that situation, Grothendieck has proved a "formal GAGA", which we discuss next. That result yields a version of Riemann's Existence Theorem for many fields other than $\mathbb{C}$. Combining that approach with the above results of Ferrand–Raynaud and Artin yields even stronger versions of "formal GAGA"; and those formal patching results have been used to prove a number of results concerning covers and fundamental groups over various fields (as will be discussed later).

3.2. Grothendieck's formal schemes. Drawing on Zariski's notion of formal holomorphic functions, Grothendieck introduced the notion of *formal scheme*, and provided a framework for proving a "formal GAGA" that is sufficient for establishing formal analogs of (at least parts of) Riemann's Existence Theorem. In his paper of the same name [Gr2], Grothendieck announced his result GFGA ("géometrie formelle et géométrie algébrique"), and sketched how it leads to results about covers and fundamental groups of curves. The details of this GFGA result appeared later in EGA [Gr4, III, Cor. 5.1.6], and the result in that form has become known as Grothendieck's Existence Theorem. In SGA 1 [Gr5], the details about the results on covers and fundamental groups appeared.

To begin with, fix a Zariski closed subset W of a scheme V. Let $\mathcal{O}_{\mathfrak{V}} = \mathcal{O}_{\mathfrak{V},W}$ be the sheaf of holomorphic functions along W in V. That is, for every Zariski open subset $U \subset W$, let $\mathcal{O}_{\mathfrak{V}}(U)$ be the ring $\hat{\mathcal{O}}_{V,U}$ of holomorphic functions along U in V. Thus $\mathcal{O}_{\mathfrak{V}} = \varprojlim_n \mathcal{O}_V/\mathcal{J}^{n+1}$, where $\mathcal{J}$ is the sheaf of ideals of $\mathcal{O}_V$ defining W in V. The ringed space $\mathfrak{V} := (W, \mathcal{O}_{\mathfrak{V}})$ is defined to be the *formal completion of V along W*.

The simplest example of this takes V to be the affine t-line over a field k, and W to be the point $t = 0$. Here we may identify $\mathcal{O}_{\mathfrak{V}}$ with the ring $k[\![t]\!] = \varprojlim_n O_n$, where $O_n = k[t]/(t^{n+1})$. Here $n = 0$ corresponds to W, and $n > 0$ to infinitesimal thickenings of W. The kernel I_m of $O_m \to O_0$ is the ideal tO_m, and the kernel of $O_m \to O_n$ is $t^{n+1}O_m = I_m^{n+1}$.

As a somewhat more general example, let A be a ring that is complete with respect to an ideal I. Then $W = \operatorname{Spec} A/I$ is a closed subscheme of $V = \operatorname{Spec} A$, consisting of the prime ideals of A that are open in the I-adic topology. The formal completion $\mathfrak{V} = (W, \mathcal{O}_{\mathfrak{V}})$ of V along W consists of the underlying topological space W together with a structure sheaf whose ring of global sections is A. This formal completion is also called the *formal spectrum* of A, denoted $\operatorname{Spf} A$. (For example, if $A = k[x][\![t]\!]$ and $I = (t)$, then the underlying space of $\operatorname{Spf} A$ is the affine x-line over k, and its global sections are $k[x][\![t]\!]$.)

Note that the above definition of formal completion relies on the idea that the geometry of a space is captured by the structure sheaf on it, rather than on the underlying topological space. Indeed, the underlying topological space of $\mathfrak{V}$ is the same as that of W; but the structure sheaf $\mathcal{O}_{\mathfrak{V}}$ incorporates all of the information in the spectra of $\hat{\mathcal{O}}_{V,U}$ — and thus it reflects the local geometry of V near W.

More generally, suppose we are given a topological space X and a sheaf of topological rings $\mathcal{O}_{\mathfrak{X}}$ on X. Suppose also that $\mathcal{O}_{\mathfrak{X}} = \varprojlim_n \mathcal{O}_n$, where $\{\mathcal{O}_n\}_n$ is an inverse system of sheaves of rings on X such that $(X, \mathcal{O}_n)$ is a scheme X_n for each n; and that for $m \geq n$, the homomorphism $\mathcal{O}_m \to \mathcal{O}_n$ is surjective with kernel $\mathcal{I}_m^{n+1}$, where $\mathcal{I}_m = \ker(\mathcal{O}_m \to \mathcal{O}_0)$. Then the ringed space $\mathfrak{X} := (X, \mathcal{O}_{\mathfrak{X}})$ is a *formal scheme*. In particular, in the situation above for $W \subset V$ (taking $\mathcal{O}_n = \mathcal{O}_V/\mathcal{I}^{n+1}$), the formal completion $\mathfrak{V} = (W, \mathcal{O}_{\mathfrak{V}})$ of V along W is a formal scheme.

If W is a closed subset of a scheme V, with formal completion $\mathfrak{V}$, then to every sheaf $\mathcal{F}$ of $\mathcal{O}_V$-modules on V we may canonically associate a sheaf $\hat{\mathcal{F}}$ of $\mathcal{O}_{\mathfrak{V}}$ modules on $\mathfrak{V}$. Namely, for every n let $\mathcal{F}_n = \mathcal{F} \otimes_{\mathcal{O}_V} \mathcal{O}_V/\mathcal{I}^{n+1}$, where $\mathcal{I}$ is the sheaf of ideals defining W. Then let $\hat{\mathcal{F}} = \varprojlim_n \mathcal{F}_n$. Note that $\mathcal{O}_{\mathfrak{V}} = \hat{\mathcal{O}}_V$. Also observe that if $\mathcal{F}$ is a coherent $\mathcal{O}_V$-module, then $\hat{\mathcal{F}}$ is a coherent $\mathcal{O}_{\mathfrak{V}}$-module (i.e. it is locally of the form $\mathcal{O}_{\mathfrak{V}}^m \to \mathcal{O}_{\mathfrak{V}}^n \to \hat{\mathcal{F}} \to 0$).

THEOREM 3.2.1 (GFGA, GROTHENDIECK EXISTENCE THEOREM). *Let A be a Noetherian ring that is complete with respect to a proper ideal I, let V be a proper A-scheme, and let $W \subset V$ be the inverse image of the locus of I. Let $\mathfrak{V} = (W, \mathcal{O}_{\mathfrak{V}})$ be the formal completion of V along W. Then the functor $\mathcal{F} \mapsto \hat{\mathcal{F}}$, from the category of coherent $\mathcal{O}_V$-modules to the category of coherent $\mathcal{O}_{\mathfrak{V}}$-modules, is an equivalence of categories.*

Before turning to the proof of Theorem 3.2.1, we discuss its content and give some examples, beginning with an application:

COROLLARY 3.2.2. [Gr2, Cor. 1 to Thm. 3] *In the situation of Theorem 3.2.1, the natural map from closed subschemes of V to closed formal subschemes of $\mathfrak{V}$ is a bijection.*

Namely, such subschemes [resp. formal subschemes] correspond bijectively to coherent subsheaves of $\mathcal{O}_V$ [resp. of $\mathcal{O}_{\mathfrak{V}}$]. So this is an immediate consequence of the theorem.

This corollary may seem odd, for example in the case where V is a curve over a complete local ring A, and W is thus a curve over the residue field of A — since then, the only reduced closed subsets of W (other than W itself) are finite sets of points. But while distinct closed subschemes of V can have the same intersection with the topological space W, the structure sheaves of their restrictions will be different, and so the induced formal schemes will be different.

Theorem 3.2.1 can be viewed in two ways: as a thickening result (emphasizing the inverse limit point of view), and as a patching result (emphasizing the analogy with the classical GAGA of Section 2.2).

From the point of view of thickening, given $W \subset V$ defined by a sheaf of ideals $\mathfrak{I}$, we have a sequence of subschemes $V_n = \underline{\mathrm{Spec}}\; \mathcal{O}_V/\mathfrak{I}^{n+1}$. Each V_n has the same underlying topological space (viz. that of $\overline{W} = V_0$), but has a different structure sheaf. The formal completion $\mathfrak{V}$ of V along W can be regarded as the direct limit of the schemes V_n. What Theorem 3.2.1 says is that under the hypotheses of that result, to give a coherent sheaf $\mathcal{F}$ on V is equivalent to giving a compatible set of coherent sheaves $\mathcal{F}_n$ on the V_n's (i.e. the restrictions of $\mathcal{F}$ to the V_n's). The hard part (cf. the proof below) is to show the existence of a coherent sheaf $\mathcal{F}$ that restricts to a given compatible set of coherent sheaves $\mathcal{F}_n$. And later, the result will tell us that to give a branched cover of V is equivalent to giving a compatible system of covers of the V_n's.

On the other hand, the point of view of patching is closer to that of Zariski's work on formal holomorphic functions. Given $W \subset V$, we can cover W by affine open subsets U_i. By definition, giving a coherent formal sheaf on W amounts to giving finitely presented modules over the rings $\hat{\mathcal{O}}_{V,U_i}$ that are compatible on the overlaps (i.e. over the rings $\hat{\mathcal{O}}_{V,U_{ij}}$, where $U_{ij} = U_i \cap U_j$). So Theorem 3.2.1 says that to give a coherent sheaf $\mathcal{F}$ on V is equivalent to giving such modules locally (i.e. the pullbacks of $\mathcal{F}$ to the "formal neighborhoods" Spec $\hat{\mathcal{O}}_{V,U_i}$ with agreements on the "formal overlaps" Spec $\hat{\mathcal{O}}_{V,U_{ij}}$). The same principle will be applied later to covers.

EXAMPLE 3.2.3. Let k be a field, let $A = k[\![t]\!]$, and let $V = \mathbb{P}^1_A$, the projective x-line over A. So W is the projective x-line over k. Let $\mathfrak{V}$ be the formal completion of V at W. Theorem 3.2.1 says that giving a coherent $\mathcal{O}_V$-module is equivalent to giving a coherent $\mathcal{O}_{\mathfrak{V}}$-module.

From the perspective of thickening, to give a coherent $\mathcal{O}_{\mathfrak{V}}$-module $\mathcal{F}$ amounts to giving an inverse system of coherent modules $\mathcal{F}_n$ over the V_n's, where V_n is the projective x-line over $k[t]/(t^{n+1})$. Each finite-level thickening $\mathcal{F}_n$ gives more and more information about the given module, and in the limit, the theorem says that the full $\mathcal{O}_V$-module $\mathcal{F}$ is determined.

For the patching perspective, cover W by two open sets U_1 (where $x \neq \infty$) and U_2 (where $x \neq 0$), each isomorphic to the affine k-line. The corresponding rings of holomorphic functions are $k[x][\![t]\!]$ and $k[x^{-1}][\![t]\!]$, while the ring of holomorphic functions along the overlap $U_0 : (x \neq 0, \infty)$ is $k[x, x^{-1}][\![t]\!]$. As in Examples 3.1.2 and 3.1.3, the spectra S_1, S_2 of the first two of these rings can be viewed as tubular neighborhoods of the two affine lines, pinching down near $x = \infty$ and near $x = 0$ respectively. The spectrum S_0 of the third ring (the "formal overlap") can be viewed as a tubular neighborhood that pinches down near both 0 and ∞. (See Figure 3.1.4.) These spectra can be viewed as "analytic open subsets" of V, which cover V (in the sense that the disjoint union $S_1 \cup S_2$ is faithfully flat over

V) — and the theorem says that giving coherent modules over S_1 and S_2, which agree over S_0, is equivalent to giving a coherent module over V. $\qquad\qquad\square$

From the above patching perspective, Theorem 3.2.1 can be rephrased as follows, in a form that is useful in the case of relative dimension 1. In order to be able to apply it to Galois theory (in Section 3.3 below), we state it as well for algebras and covers.

THEOREM 3.2.4. *In the situation of Theorem 3.2.1, suppose that U_1, U_2 are affine open subsets of W such that $U_1 \cup U_2 = W$, with intersection U_0. For $i = 0, 1, 2$, let S_i be the spectrum of the ring of holomorphic functions along U_i in V. Then the base change functor*

$$\mathfrak{M}(V) \to \mathfrak{M}(S_1) \times_{\mathfrak{M}(S_0)} \mathfrak{M}(S_2)$$

is an equivalence of categories. Moreover the same holds if $\mathfrak{M}$ is replaced by the category of finite algebras, or of finite branched covers, or of Galois covers with a given Galois group.

Compare this with the restatement of the classical GAGA at Theorem 2.2.6, and with the results of Ferrand–Raynaud and Artin (Propositions 3.1.6 and 3.1.8). See also Figure 3.1.4 for an illustration of this result in the situation of the above example. As in Theorem 2.2.6, the above assertions for algebras and covers follow formally from the result for modules, via the General Principle 2.2.4. (Cf. also [Ha2, Proposition 2.8].)

REMARKS 3.2.5. (a) Theorem 3.2.1 does not hold if the properness hypothesis on V is dropped. For example, Corollary 3.2.2 is false in the case that $A = k[\![t]\!]$ and $V = \mathbb{A}_A^1$ (since the subscheme $(1 - xt)$ in V induces the same formal subscheme of $\mathcal{V}$ as the empty set). Similarly, Theorem 3.2.4 does not hold as stated if V is not proper over A (and note that $S_1 \cup S_2$ is not faithfully flat over V in this situation). But a variant of Theorem 3.2.4 does hold if V is affine: namely there is still an equivalence if $\mathfrak{M}(V)$ is replaced by $\mathfrak{M}(S)$, where S is the ring of holomorphic functions along W in V. This is essentially a restatement of Zariski's result that holomorphic functions on an affine open subset of W are strongly holomorphic. It is also analogous to the version of Cartan's Theorem A for Stein spaces [Ca2] (cf. the discussion near the end of Section 2.2 above).

(b) The main content of Theorem 3.2.1 (or Theorem 3.2.4) can also be phrased in affine terms in the case of relative dimension 1. For instance, in the situation of the above example with $A = k[\![t]\!]$ and $V = \mathbb{P}_A^1$, a coherent module $\mathcal{M}$ over V is determined up to twisting by its restriction to $\mathbb{A}_A^1 = \operatorname{Spec} A[x]$. Letting S_0, S_1, S_2 be as in the example, and restricting to the Zariski open subset $\mathbb{A}_A^1$, we obtain an equivalence of categories

$$\mathfrak{M}(R) \to \mathfrak{M}(R_1) \times_{\mathfrak{M}(R_0)} \mathfrak{M}(R_2) \qquad\qquad\qquad (*)$$

where $R = k[\![t]\!][x]$; $R_1 = k[x][\![t]\!]$; $R_2 = k[x^{-1}][\![t]\!][x]$; and $R_0 = k[x, x^{-1}][\![t]\!]$. (Here we adjoin x in the definition of R_2 because of the restriction to $\mathbb{A}^1_A$.) In this situation, one can directly prove a formal version of Cartan's Lemma, viz. that every element of $\mathrm{GL}_n(R_0)$ can be written as the product of an element of $\mathrm{GL}_n(R_1)$ and an element of $\mathrm{GL}_n(R_2)$. This immediately gives the analog of $(*)$ for the corresponding categories of finitely generated *free* modules, by applying this formal Cartan's Lemma to the transition matrix between the bases over R_1 and R_2. (Cf. the discussion in Section 2.2 above, and also [Ha2, Prop. 2.1] for a general result of this form.) Moreover, combining this formal Cartan's Lemma with the fact that every element of R_0 is the sum of an element of R_1 and an element of R_2, one can deduce all of $(*)$, and thus essentially all of Theorem 3.2.1 in this situation. (See [Ha2, Proposition 2.6] for the general result, and see also Remark 1 after the proof of Corollary 2.7 there.)

(c) Using the approach of Remark (b), one can also prove analogous results where Theorem 3.2.1 does not apply. For example, let A and B be subrings of $\mathbb{Q}$, let $D = A \cap B$, and let C be the subring of $\mathbb{Q}$ generated by A and B. (For instance, take $A = \mathbb{Z}[1/2]$ and $B = \mathbb{Z}[1/3]$, so $C = \mathbb{Z}[1/6]$ and $D = \mathbb{Z}$.) Then "Cartan's Lemma" applies to the four rings $A[\![t]\!]$, $B[\![t]\!]$, $C[\![t]\!]$, $D[\![t]\!]$ (as can be proved by constructing the coefficients of the entries of the factorization, inductively). So by [Ha2, Proposition 2.6], giving a finitely generated module over $D[\![t]\!]$ is equivalent to giving such modules over $A[\![t]\!]$ and $B[\![t]\!]$ together with an isomorphism between the modules they induce over $C[\![t]\!]$.

Another example involves the ring of convergent arithmetic power series $\mathbb{Z}\{t\}$, which consists of the formal power series $f(t) \in \mathbb{Z}[\![t]\!]$ such that f converges on the complex disc $|t| < 1$. (Under the analogy between $\mathbb{Z}$ and $k[x]$, the ring $\mathbb{Z}[\![t]\!]$ is analogous to $k[x][\![t]\!]$, and the ring $\mathbb{Z}\{t\}$ is analogous to $k[\![t]\!][x]$.) Then with A, B, C, D as in the previous paragraph, "Cartan's Lemma" applies to $A[\![t]\!]$, $B\{t\}$, $C[\![t]\!]$, $D\{t\}$ [Ha2, Prop. 2.3]. As a consequence, the analog of Theorem 3.2.4 holds for these rings: viz. giving a finitely presented module over $D\{t\}$ is equivalent to giving such modules over $A[\![t]\!]$ and $B\{t\}$ together with an isomorphism between the modules they induce over $C[\![t]\!]$ [Ha5, Theorem 3.6]. □

The formal GAGA (Theorem 3.2.1) above can be proved in a way that is analogous to the proof of the classical GAGA (as presented in Section 2.2). In particular, there are two main ingredients in the proof. The first is:

THEOREM 3.2.6 (GROTHENDIECK). *In the situation of Theorem 3.2.1, if $\mathcal{F}$ is a coherent sheaf on V, then the natural map $\varepsilon : H^q(V, \mathcal{F}) \to H^q(\mathfrak{V}, \hat{\mathcal{F}})$ is an isomorphism for every $q \geq 0$.*

This result was announced in [Gr2, Cor. 1 to Thm. 2] and proved in [Gr4, III, Prop. 5.1.2]. Here the formal H^q's can (equivalently) be defined either via Čech cohomology or by derived functor cohomology. The above theorem is analogous to Theorem 2.2.2, concerning the classical case; and like that result, it is proved by descending induction on q (using that $H^q = 0$ for q sufficiently large). As in Section 2.2, it is the key case $q = 0$ that is used in proving GAGA. That case is known as Zariski's Theorem on Formal Functions [Hrt2, III, Thm. 11.1, Remark 11.1.2]; it generalizes the original version of Zariski's Fundamental Theorem on formal holomorphic functions [Za4, §11, p. 50], which is the case $q = 0$ and $\mathcal{F} = \mathcal{O}_V$, and which was discussed in Section 3.1 above.

The second key ingredient in the proof of Theorem 3.2.1 is analogous to Theorem 2.2.3:

THEOREM 3.2.7. *In the situation of Theorem* 3.2.1 (*with V assumed projective over A*), *let $\mathcal{M}$ be a coherent $\mathcal{O}_V$-module or a coherent $\mathcal{O}_{\mathfrak{V}}$-module. Then for $n \gg 0$ the twisted sheaf $\mathcal{M}(n)$ is generated by finitely many global sections.*

Once one has Theorems 3.2.6 and 3.2.7 above, the projective case of Theorem 3.2.1 follows from them in exactly the same manner that Theorem 2.2.1 (classical GAGA) followed from Theorems 2.2.2 and 2.2.3 there. The proper case can then be deduced from the projective case using Chow's Lemma [Gr4, II, Thm. 5.6.1]; cf. [Gr4, III, 5.3.5] for details.

SKETCH OF PROOF OF THEOREM 3.2.7. In the algebraic case (i.e. for $\mathcal{O}_V$-modules), the assertion is again Serre's result [Hrt2, Chap. II, Theorem 5.17]; cf. Theorem 2.2.3 above in the algebraic case. In the formal case (i.e. for $\mathcal{O}_{\mathfrak{V}}$-modules), the assertion is a formal analog of Cartan's Theorem A (cf. the analytic case of Theorem 2.2.3). The key point in proving this formal analog (as in the analytic version) is to obtain a twist that will work for a given sheaf, even though the sheaf is not algebraic and we cannot simply clear denominators (as in the algebraic proof).

To do this, first recall that a formal sheaf $\mathcal{M}$ corresponds to an inverse system $\{\mathcal{M}_i\}$ of sheaves on the finite thickenings V_i. By the result in the algebraic case (applied to V_i), we have that for each i there is an n such that $\mathcal{M}_i(n)$ is generated by finitely many global sections. But we need to know that there is a single n that works for *all* i, and with compatible finite sets of global sections. The strategy is to pick a finite set of generating sections for $\mathcal{M}_0(n)$ for some n (and these will exist if n is chosen sufficiently large); and then inductively to lift them to sections of the $\mathcal{M}_i(n)$'s, in turn. If this is done, Theorem 3.2.7 follows, since the lifted sections automatically generate, by Nakayama's Lemma.

In order to carry out this inductive lifting, first reduce to the case $V = \mathbb{P}_A^m$ for some m, as in Section 2.2 (viz. embedding the given V in some $\mathbb{P}_A^m$ and extending the module by 0). Now let $\operatorname{gr} A$ be the associated graded ring to A and let $\operatorname{gr} \mathcal{O} = (R/I)\mathcal{O} \oplus (I/I^2)\mathcal{O} \oplus \cdots$ (where $\mathcal{O} = \mathcal{O}_V$). Also write $\operatorname{gr} \mathcal{M} = \mathcal{M}_0 \oplus (I/I^2)\mathcal{M}_1 \oplus \cdots$. Since $\mathcal{M}$ is a coherent $\mathcal{O}_{\mathfrak{V}}$-module, it follows that $\operatorname{gr} \mathcal{M}$ is a coherent $\operatorname{gr} \mathcal{O}$-module on $\mathbb{P}_{\operatorname{gr} \mathcal{O}}^m$. So by the algebraic analog of Cartan's Theorem B (i.e. by Serre's result [Hrt2, III, Theorem 5.2]), there is an integer n_0 such that for all $n \geq n_0$, $H^1(\mathbb{P}_{\operatorname{gr} \mathcal{O}}^m, \operatorname{gr} \mathcal{M}(n)) = 0$. But $\operatorname{gr} \mathcal{M}(n) = \bigoplus_i (I^i/I^{i+1})\mathcal{M}_i(n)$, and so each $H^1(\mathbb{P}^m_{A/I^{i+1}}, (I^i/I^{i+1})\mathcal{M}_i(n)) = 0$. By the long exact sequence associated to the short exact sequence $0 \to (I^i/I^{i+1})\mathcal{M}_i(n) \to \mathcal{M}_i(n) \to \mathcal{M}_{i-1}(n) \to 0$, this H^1 is the obstruction to lifting sections of $\mathcal{M}_{i-1}(n)$ to sections of $\mathcal{M}_i(n)$. So choosing such an n which is also large enough so that $\mathcal{M}_0(n)$ is generated by its global sections, we can carry out the liftings inductively and thereby obtain the formal case of Theorem 3.2.7. (Alternatively, one can proceed as in Grothendieck [Gr4, III, Cor. 5.2.4], to prove this formal analog of Cartan's Theorem A via a formal analog of Cartan's Theorem B [Gr4, III, Prop. 5.2.3].) □

As indicated above, Grothendieck's Existence Theorem is a strong enough form of "formal GAGA" to be useful in proving formal analogs of (at least parts of) the classical Riemann Existence Theorem. (This will be discussed further in Section 3.3.) But for certain purposes, it is useful to have a variant of Theorem 3.2.4 that allows U_1 and S_1 to be more local. Namely, rather than taking U_1 to be an affine open subset of the closed fibre, and S_1 its formal thickening, we would instead like to take U_1 to be the spectrum of the complete local ring in the closed fibre at some point ω, and S_1 its formal thickening (viz. the spectrum of the complete local ring at ω in V). In the relative dimension 1 case, the "overlap" U_0 of U_1 and U_2 is then the spectrum of the fraction field of the complete local ring at ω in the closed fibre, and S_0 is its formal thickening.

More precisely, in the case that V is of relative dimension 1 over A, there is the following formal patching result. First we introduce some notation and terminology. If ω is a closed point of a variety V_0, then $\mathcal{K}_{V_0,\omega}$ denotes the total ring of fractions of the complete local ring $\hat{O}_{V_0,\omega}$ (and thus the fraction field of $\hat{O}_{V_0,\omega}$, if the latter is a domain). Let A be a complete local ring with maximal ideal $\mathfrak{m}$, let V be an A-scheme, and let V_n be the fibre of V over $\mathfrak{m}^{n+1}$ (regarding $V_n \subset V_{n+1}$). Let $\omega \in V_0$, and let ω' denote $\operatorname{Spec} \mathcal{K}_{V_0,\omega}$. Then the ring of holomorphic functions in V at ω' is defined to be $\hat{O}_{V,\omega'} := \varprojlim \mathcal{K}_{V_n,\omega}$. (For example, if $A = k[\![t]\!]$ and V is the affine x-line over A, and if ω is the point $x = t = 0$, then $\omega' = \operatorname{Spec} k((x))$, $\mathcal{K}_{V_n,\omega} = k((x))[t]/(t^{n+1})$, and the ring of holomorphic functions at ω' is $\hat{O}_{V,\omega'} = k((x))[\![t]\!]$.)

THEOREM 3.2.8. *Let V be a proper curve over a complete local ring A, let V_0 be the fibre over the closed point of $\operatorname{Spec} A$, let W be a non-empty finite set of closed points of V_0, and let $U = V_0 - W$. Let W^* be the union of the spectra of the complete local rings $\hat{O}_{V,\omega}$ for $\omega \in W$. Let $U^* = \operatorname{Spec} \hat{O}_{V,U}$, and let $W'^* = \bigcup_{\omega \in W} \operatorname{Spec} \hat{O}_{V,\omega'}$, where $\omega' = \operatorname{Spec} \mathcal{K}_{V_0,\omega}$ as above. Then the base change functor*

$$\mathfrak{M}(V) \to \mathfrak{M}(W^*) \times_{\mathfrak{M}(W'^*)} \mathfrak{M}(U^*)$$

is an equivalence of categories. The same holds for finite algebras and for (Galois) covers.

This result appeared as [Ha6, Theorem 1], in the special case that V is regular, $A = k[\![t_1, \ldots, t_n]\!]$ for some field k and some $n \geq 0$, and where attention is re-

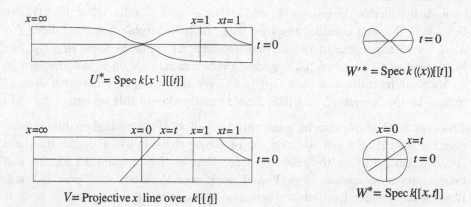

Figure 3.2.9. Example 3.2.10 of Theorem 3.2.8, with $V = \mathbb{P}^1_{k[\![t]\!]}$, $W =$ one point. Here $\mathbb{P}^1_{k[\![t]\!]}$ is covered by the small patch $W^* = \operatorname{Spec} k[\![x, t]\!]$ and the larger patch $U^* = \operatorname{Spec} k[1/x][\![t]\!]$; their overlap is $W'^* = \operatorname{Spec} k(\!(x)\!)[\![t]\!]$ (upper right). Compare Fig. 3.1.4.

stricted to projective modules. The proof involved showing that the appropriate form of Cartan's Lemma is satisfied. In the form above, the result appeared at [Pr1, Theorem 3.4]. There, it was assumed that the complete local ring A is a discrete valuation ring, but that hypothesis was not necessary for the proof there. Namely, the proof there first showed the result for $A/\mathfrak{m}^n$, where $\mathfrak{m}$ is the maximal ideal of A, using Corollary 3.1.9(a) (to the result of Ferrand and Raynaud [FR]); and afterwards used Grothendieck's Existence Theorem (Theorem 3.2.1 above) to pass to A. (This use of [FR] was suggested by L. Moret-Bailly.)

EXAMPLE 3.2.10. Let k be a field, let $A = k[\![t]\!]$, and let $V = \mathbb{P}^1_A$ (the projective x-line over $k[\![t]\!]$), with closed fibre $V_0 = \mathbb{P}^1_k$ over $(t = 0)$. Let W consist of the single point ω where $x = t = 0$. In the notation of Theorem 3.2.8, $W^* = \operatorname{Spec} k[\![x, t]\!]$, which can be viewed as a "small neighborhood" of ω. The formal completion of V along $U := V_0 - W$ is $U^* = \operatorname{Spec} k[1/x][\![t]\!]$, whose "overlap" with W^* is $W'^* = \operatorname{Spec} k(\!(x)\!)[\![t]\!]$. (See Figure 3.2.9.) According to Theorem 3.2.8, giving a coherent module on V is equivalent to giving finite modules over W^* and over U^* together with an isomorphism on their pullbacks ("restrictions") to W'^*. The same holds for covers; and this permits modifying a branched cover of V near ω, e.g. by adding more inertia there; see Remarks 5.1.6(d,e). $\square$

EXAMPLE 3.2.11. Let k, A be as in Example 3.2.10, and let V be an irreducible normal curve over A, with closed fibre V_0. Then V_0 is a k-curve which is connected (by Zariski's Connectedness Theorem [Za4, § 20, Thm. 14], [Hrt2, III, Cor. 11.3]) but not necessarily irreducible; let $V_1, \ldots, V_r$ be its irreducible components. The singular locus of V is a finite subset of V_0, and it includes all the points where irreducible components V_i of V_0 intersect. Let W be a finite subset of V_0 that contains this singular locus, and contains at least one smooth point on each irreducible component V_i of V_0. For $i = 1, \ldots, r$ let $W_i = V_i \cap W$, let $U_i = V_i - W_i$, and consider the ring $\hat{O}_{V,U_i}$ of holomorphic functions along U_i. Also, for each point ω in W, we may consider its complete local ring $\hat{O}_{V,\omega}$ in V. According to Theorem 3.2.8, giving a coherent module on V is equivalent to giving finite modules over each $\hat{O}_{V,U_i}$ and over each $\hat{O}_{V,\omega}$ together with isomorphisms on the "overlaps". See [HS] for a formalization of this set-up. $\square$

Theorem 3.2.8 above can be generalized to allow V to be higher dimensional over the base ring A. In addition, by replacing the result of Ferrand–Raynaud (Proposition 3.1.6) by the related result of Artin (Proposition 3.1.8), one can take a proper morphism $\tilde{V} \to V$ and work over $\tilde{V}$ rather than over V itself. Both of these generalizations are accomplished in the following result:

THEOREM 3.2.12. *Let $(A, \mathfrak{m})$ be a complete local ring, let V be a proper A-scheme, and let $f : \tilde{V} \to V$ be a proper morphism. Let W be a finite set of closed points of V; let $\tilde{W} = f^{-1}(W) \subset \tilde{V}$; let $W^* = \bigcup_{\omega \in W} \mathrm{Spec}\, \hat{O}_{V,\omega}$; and let $\tilde{W}^* = \tilde{V} \times_V W^*$. Let $\tilde{\mathcal{U}}$ [resp. $\tilde{\mathcal{U}}^*$] be the formal completion of $\tilde{V} - \tilde{W}$ [resp. of $\tilde{W}^* - \tilde{W}$] along its fibre over $\mathfrak{m}$. Then the base-change functor*

$$\mathfrak{M}(\tilde{V}) \to \mathfrak{M}(\tilde{W}^*) \times_{\mathfrak{M}(\tilde{\mathcal{U}}^*)} \mathfrak{M}(\tilde{\mathcal{U}})$$

is an equivalence of categories. The same holds for finite algebras and for (Galois) covers.

Note that the scheme $U^* = \mathrm{Spec}\, \hat{O}_{V,U}$ in the statement of Theorem 3.2.8 is replaced in Theorem 3.2.12 by a formal scheme, because the complement of W in the closed fibre of V will no longer be affine, if V is not a curve over its base ring (and so the ring $\hat{O}_{V,U}$ of Theorem 4 would not be defined here). Similarly, the scheme W'^* in Theorem 3.2.8 is also replaced by a formal scheme in Theorem 3.2.12.

PROOF. For $n \geq 0$ let $\tilde{V}_n$ and $\tilde{W}_n^*$ be the pullbacks of $\tilde{V}$ and $\tilde{W}^*$, respectively, over $A_n := A/\mathfrak{m}^{n+1}$. Also, let $\tilde{U}_n = \tilde{V}_n - \tilde{W}$ and $\tilde{U}_n^* = \tilde{W}_n^* - \tilde{W}$; thus the formal schemes $\tilde{\mathcal{U}}, \tilde{\mathcal{U}}^*$ respectively correspond to the inverse systems $\{\tilde{U}_n\}_n, \{\tilde{U}_n^*\}_n$.

For every n, we have by Corollary 3.1.9(b) (to Artin's result, Proposition 3.1.8) that the base change functor

$$\mathfrak{M}(\tilde{V}_n) \to \mathfrak{M}(\tilde{W}_n^*) \times_{\mathfrak{M}(\tilde{U}_n^*)} \mathfrak{M}(\tilde{U}_n)$$

is an equivalence of categories. By definition of coherent modules over a formal scheme, we have that $\mathfrak{M}(\tilde{\mathcal{U}}) = \varprojlim \mathfrak{M}(\tilde{U}_n)$ and $\mathfrak{M}(\tilde{\mathcal{U}}^*) = \varprojlim \mathfrak{M}(\tilde{U}_n^*)$. Moreover, $\tilde{V}$ is proper over A; so Grothendieck's Existence Theorem (Theorem 3.2.1 above) implies that the functor $\mathfrak{M}(\tilde{V}) \to \varprojlim \mathfrak{M}(\tilde{V}_n)$ is an equivalence of categories. So it remains to show that the corresponding assertion holds for $\mathfrak{M}(\tilde{W}^*)$; i.e. that $\mathfrak{M}(\tilde{W}^*) \to \varprojlim \mathfrak{M}(\tilde{W}_n^*)$ is an equivalence of categories.

It suffices to prove this equivalence in the case that W consists of just one point ω; and we now assume that. Let $T = \hat{O}_{V,\omega}$, and let $\mathfrak{m}_\omega$ be the maximal ideal of T, corresponding to the closed point ω. Also, let $\mathfrak{n} = \mathfrak{m}T \subset T$ (where $\mathfrak{m}$ still denotes the maximal ideal of A). Thus $\mathfrak{n} \subset \mathfrak{m}_\omega$, and so T is complete with respect to $\mathfrak{n}$. Also, $\tilde{W}^*$ is proper over the Noetherian $\mathfrak{n}$-adically complete ring T, and $\tilde{W}_n^*$ is the pullback of $\tilde{W}^* \to W^* = \operatorname{Spec} T$ over $T/\mathfrak{n}^{n+1}$. So it follows from Grothendieck's Existence Theorem 3.2.1 that the desired equivalence $\mathfrak{M}(\tilde{W}^*) \to \varprojlim \mathfrak{M}(\tilde{W}_n^*)$ holds. This proves the result in the case of modules.

The analogs for algebras, covers, and Galois covers follow as before using the General Principle 2.2.4. $\qquad\square$

EXAMPLE 3.2.13. Let k, A be as in Examples 3.2.10 and 3.2.11, and let $V = \mathbb{P}_k^n$ for some $n \geq 1$, with homogeneous coordinates $x_0, \ldots, x_n$. Let W consist of the closed point ω of V where $x_1 = \cdots = x_n = t = 0$, and let $f : \tilde{V} \to V$ be the blow-up of V at ω. Let $V_0 = \mathbb{P}_k^n$ be the closed fibre of V over $(t = 0)$. For $i = 1, \ldots, n$, let U_i be the affine open subset of V_0 given by $x_i \neq 0$, and consider the ring $\hat{O}_{V,U_i}$ of holomorphic functions along U_i in V. Also consider the complete local ring $\hat{O}_{V,\omega} = k[\![x_1, \ldots, x_n, t]\!]$ at ω in V, and consider the pullback $\tilde{W}^*$ of $\tilde{V}$ over $\hat{O}_{V,\omega}$ (whose fibre over the closed point ω is a copy of $\mathbb{P}_k^n$). According to Theorem 3.2.12, giving a coherent module over V is equivalent to giving finite modules over the rings $\hat{O}_{V,U_i}$, and a coherent module over $\tilde{W}^*$, together with compatible isomorphisms on the overlaps. (This uses that giving a coherent module on the formal completion of $V - W$ along its closed fibre is equivalent to giving compatible modules over the completions at the U_i's; here we also identify $\tilde{V} - f^{-1}(W)$ with $V - W$.)

In particular, if $n = 1$, then $\tilde{V}$ is an irreducible A-curve whose closed fibre consists of two projective lines meeting at one point (one being the proper transform of the given line V_0, and the other being the exceptional divisor). This one-dimensional case is also within the context of Example 3.2.11, and so Theorem 3.2.8 could instead be used. (See also the end of Example 4.2.4 below.) $\square$

REMARK 3.2.14. The above formal patching results (Theorems 3.2.4, 3.2.8, 3.2.12) look similar, though differing in terms of what types of "patches" are allowed. In each case, we are given a proper scheme V over a complete local ring A, and the assertion says that if a module is given over each of two patches (of a given form), with agreement on the "overlap", then there is a unique coherent module over V that induces them compatibly. Theorem 3.2.4 (a reformulation of Grothendieck's Existence Theorem) is the basic version of formal patching, modeled after the classical result GAGA in complex patching (see Theorem 2.2.6, where two metric open sets are used as patches). In Theorem 3.2.4, the patches correspond to thickenings along Zariski open subsets of the closed fibre of V; see Example 3.2.3 above and see Figure 3.1.4 for an illustration. This basic type of formal patching will be sufficient for the results of Section 3.3 below, on the realization of Galois groups, via "slit covers".

More difficult results about fundamental groups, discussed in Section 5 below, require Theorems 3.2.8 or 3.2.12 instead of Theorem 3.2.4 (e.g. Theorem 5.1.4 and Theorem 5.3.1 use Theorem 3.2.8, while Theorem 5.3.9 uses Theorem 3.2.12). In Theorem 3.2.8 above, one of the patches is allowed to be much smaller than in Theorem 3.2.4, viz. the spectrum of the complete local ring at a point, if the closed fibre is a curve; see Examples 3.2.10 and 3.2.11 above, and see Figure 3.2.9 above for an illustration. Theorem 3.2.12 is still more general, allowing the closed fibre to have higher dimension, and also allowing a more general choice of "small patch" because of the choice of a proper morphism $\tilde{V} \to V$; see Example 3.2.13 above. The advantage of these stronger results is that the overlap of the patches is "smaller" than in the situation of Theorem 3.2.4, and therefore less agreement is required between the given modules. This gives greater applicability to the patching method, in constructing modules or covers with given properties. (Recall that the similar-looking patching results at the end of Section 3.1, which allow the construction of modules by prescribing them along and away from a given closed set, do not directly give results for covers; but they were used, together with Grothendieck's Existence Theorem, in proving Theorems 3.2.8 and 3.2.12 above.)   □

3.3. Formal patching and constructing covers.

The methods of Section 3.2 allow one to construct covers of algebraic curves over various fields other than the complex numbers. The idea is to use the approach of Section 2.3, building "slit covers" using formal patching rather than analytic patching (as was used in Section 2). This will be done by relying on Grothendieck's Existence Theorem, in the form of Theorem 3.2.4. (As will be discussed in Section 5, by using variants of Theorem 3.2.4, in particular Theorems 3.2.8 and 3.2.12, it is possible to make more general constructions as well. See also [Ha6], [St1], [HS1], and [Pr2] for other applications of those stronger patching results, concerning covers with given inertia groups over certain points, or even unramified covers of projective curves.)

The first key result is this:

THEOREM 3.3.1 [Ha4, Theorem 2.3, Corollary 2.4]. *Let R be a normal local domain other than a field, such that R is complete with respect to its maximal ideal. Let K be the fraction field of R, and let G be a finite group. Then G is the Galois group of a Galois field extension L of $K(x)$, which corresponds to a Galois branched cover of $\mathbb{P}^1_K$ with Galois group G. Moreover L can be chosen to be regular, in the sense that K is algebraically closed in L.*

Before discussing the proof, we give several examples:

EXAMPLE 3.3.2. (a) Let $K = \mathbb{Q}_p$, or a finite extension of $\mathbb{Q}_p$, for some prime p. Then every finite group is a Galois group over $\mathbb{P}^1_K$ (i.e. of some Galois branched cover of the K-line), and so is a Galois group over $K(x)$.

(b) Let k be a field, let n be a positive integer, and let $K = k((t_1, \ldots, t_n))$, the fraction field of $k[\![t_1, \ldots, t_n]\!]$. Then every finite group is a Galois group over $\mathbb{P}^1_K$, and so over $K(x)$.

(c) If K is as in Example (b) above, and if $n > 1$, then every finite group is a Galois group over K (and not just over $K(x)$, as above). The reason is that K is separably Hilbertian, by Weissauer's Theorem [FJ, Theorem 14.17]. That is, every separable field extension of $K(x)$ specializes to a separable field extension of K, by setting $x = c$ for an appropriate choice of $c \in K$; such a specialization of a Galois field extension is then automatically Galois. (The condition of being separably Hilbertian is a bit weaker than being Hilbertian, but is sufficient for dealing with Galois extensions. See [FJ, Chapter 11], [Vö, Chapter 1], or [MM, Chapter IV, §1.1] for more about Hilbertian and separably Hilbertian fields.)

This example remains valid more generally, where the coefficient field k is replaced by any Noetherian normal domain A that is complete with respect to a prime ideal. Moreover if A is not a field, then the condition $n > 1$ can even be weakened to $n > 0$. In particular, if K is the fraction field of $\mathbb{Z}[\![t]\!]$ (a field which is much smaller than $\mathbb{Q}((t))$), then every finite group is the Galois group of a regular cover of $\mathbb{P}^1_K$, and is a Galois group over K itself. The proof of this generalization uses formal A-schemes, and parallels the proof of Theorem 1; see [Le].

(d) Let K be the ring of algebraic p-adics (i.e. the algebraic closure of $\mathbb{Q}$ in $\mathbb{Q}_p$), or alternatively the ring of algebraic Laurent series in n-variables over a field k (i.e. the algebraic closure of $k(t_1, \ldots, t_n)$ in $k((t_1, \ldots, t_n))$). Then every finite group is a Galois group over $\mathbb{P}^1_K$. More generally this holds if K is the fraction field of R, a normal henselian local domain other than a field. This follows by using Artin's Algebraization Theorem ([Ar3], a consequence of Artin's Approximation Theorem [Ar2]), in order to pass from formal elements to algebraic ones. See [Ha4, Corollary 2.11] for details. In the case of algebraic power series in $n > 1$ variables, Weissauer's Theorem then implies that every finite group is a Galois group over K, as in Example (c). $\square$

Theorem 3.3.1 also implies that all finite groups are Galois groups over $K(x)$ for various other fields K, as discussed below (after the proof).

Theorem 3.3.1 can be proved by carrying over the slit cover construction of Section 2.3 to the context of formal schemes. Before doing so, it is first necessary to construct cyclic covers that can be patched together (as in Example 2.3.2). Rather than using complex discs as in §2.3, we will use "formal open subsets", i.e. we will take the formal completions of $\mathbb{P}^1_R$ along Zariski open subsets of the closed fibre $\mathbb{P}^1_k$ (where k is the residue field of R). In order to be able to use Grothedieck's Existence Theorem to patch these covers together, we will want the cyclic covers to agree on the "overlaps" of these formal completions — and this will be accomplished by having them be trivial on these overlaps (just as in Example 2.3.2).

In order to apply Grothedieck's Existence Theorem, we will use it in the case of Galois branched covers (rather than for modules), as in Theorem 3.2.4. There, it was stated just for two patches U_1, U_2 and their overlap U_0; but by induction, it holds as well for finitely many patches, provided that compatible isomorphisms are given on overlaps (and cf. the statement of Theorem 3.2.1).

Grothedieck's Existence Theorem will be applied to the following proposition, which yields the cyclic covers $Y \to \mathbb{P}^1$ that will be patched together in order to prove Theorem 3.3.1. The desired triviality on overlaps will be guaranteed by the requirement that the closed fibre $\phi_k : Y_k \to \mathbb{P}^1_k$ of the branched cover $\phi : Y \to \mathbb{P}^1_R$ be a *mock cover*; i.e. that the restriction of ϕ_k to each irreducible component of Y_k be an isomorphism. This condition guarantees that if $U \subset \mathbb{P}^1_k$ is the complement of the branch locus of ϕ_k, then the restriction of ϕ_k to U is trivial; i.e. $\phi_k^{-1}(U)$ just consists of a disjoint union of copies of U.

PROPOSITION 3.3.3 [Ha4, Lemma 2.1]. *Let $(R, \mathfrak{m})$ be a normal complete local domain other than a field, with fraction field K and residue field $k = R/\mathfrak{m}$. Let $S \subset \mathbb{P}^1_k$ be a finite set of closed points, and let $n > 1$. Then there is a cyclic field extension L of $K(x)$ of degree n, such that the normalization of $\mathbb{P}^1_R$ in L is an n-cyclic Galois branched cover $Y \to \mathbb{P}^1_R$ whose closed fibre $Y_k \to P^1_k$ is a mock cover that is unramified over S.*

PROOF. We follow the proof in [Ha4], first observing that we are reduced to the situation that n is a prime power p^r. (Namely, if $n = \prod p_i^{r_i}$, and if $Y_i \to \mathbb{P}^1_R$ are $p_i^{r_i}$-cyclic covers, then we may take Y to be the fibre product of the Y_i's over $\mathbb{P}^1_R$.)

The easiest case is if the field K contains a primitive n-th root of unity ζ_n. Then we may take L to be the field obtained by adjoining an n-th root of $f(x)(f(x) - \alpha)^{n-1}$, where $f(x) \in R[x]$ does not vanish at any point of S, and where $\alpha \in \mathfrak{m} - \{0\}$. (For example, if k is infinite, we may choose $f(x) = x - c$ for some $c \in R$; compare Example 2.3.2.)

Next, suppose that K does not contain a primitive n-th root of unity but that p is not equal to the characteristic of K. Then we can consider $K' = K[\zeta_n]$, and will construct an n-cyclic Kummer extension of $K'(x)$ which descends to a desired extension of $K(x)$. This will be done using constructions in [Slt] to find an element $g(x) \in R[\zeta_n, x]$ such that the extension $y^n - g(x)$ of $R[\zeta_n, x]$ descends to an n-cyclic extension of $R[x]$ whose closed fibre is a mock cover.

Specifically, first suppose that p is odd. Let s be the order of the cyclic group $\mathrm{Gal}(K'/K)$, with generator $\tau : \zeta_n \mapsto \zeta_n^m$. Choose $\alpha \in \mathfrak{m} - \{0\}$ and let $b = f(x)^n - \zeta_n p^2 \alpha$, for some $f(x) \in R[x]$ which does not vanish on S. Let L' be the n-cyclic field extension of $K'(x)$ given by adjoining an n-th root of $M(b) = b^{m^{s-1}} \tau(b)^{m^{s-2}} \cdots \tau^{s-2}(b)^m \tau^{s-1}(b)$. Then $L' = L \otimes_K K'$ for some n-cyclic extension L of $K(x)$, by [Slt, Theorem 2.3]. (Note that the branch locus of the associated cover, which is given by $M(b) = 0$, is invariant under τ. Here the various powers of the factors of $M(b)$ are chosen so that τ will commute with the generator of $\mathrm{Gal}(L'/K'(x))$, given by $y \mapsto \zeta_n y$. These two facts enable the Kummer cover of the K'-line to descend to a cyclic cover of the K-line.)

On the other hand, suppose $p = 2$. If K contains a square root of -1 then $\mathrm{Gal}(K'/K)$ is again cyclic, so the same proof as in the odd case works. Otherwise, if $n = 2$ then take the extension of $K(x)$ given by adjoining a square root of $f(x)^2 - 4\alpha$. If $n = 4$, then adjoin a fourth root of $(f(x)^4 + 4i\alpha)^3(f(x)^4 - 4i\alpha)$ to $K'(x)$; this descends to a 4-cyclic extension of $K(x)$ by [Slt, Theorem 2.4]. If $n = 2^r$ with $r \geq 3$, then $\mathrm{Gal}(K'/K)$ is the product of a cyclic group of order 2 with generator $\kappa : \zeta_n \mapsto \zeta_n^{-1}$, and another of order $s \leq 2^{n-2}$ with generator $\zeta_n \mapsto \zeta_n^m$ for some $m \equiv 1 \pmod 4$. Take $b = f(x)^n + 4\zeta_n\alpha$ and $a = b^{2^{n-1}+1}\kappa(b)^{2^{n-1}-1}$; and (in the notation of the odd case) consider the extension of $K'(x)$ given by adjoining an n-th root of $M(a)$. By [Slt, Theorem 2.7], this descends to an n-cyclic extension of $K(x)$.

Finally, there is the case that p is equal to the characteristic of K. If $n = p$, we can adjoin a root of an Artin–Schreier polynomial $y^p - f(x)^{p-1}y - \alpha$, where $f(x) \in R[x]$ and $\alpha \in \mathfrak{m} - \{0\}$. More generally, with $n = p^r$, we can use Witt vectors, by adjoining the roots of the Witt coordinates of $\mathrm{Fr}(y) - f(x)^{p-1}y - \alpha$, where $f(x)$ and y denote the elements of the truncated Witt ring $W_r(R[x, y_0, \ldots, y_{r-1}])$ with Witt coordinates $(f(x), 0, \ldots, 0)$ and $(y_0, \ldots, y_n)$ respectively, and where Fr denotes Frobenius.

In each of these cases, one checks that the extension L of $K(x)$ has the desired properties. (See [Ha4, Lemma 2.1] for details.) $\qquad\square$

Using this result together with Grothendieck's Existence Theorem (for covers), one easily obtains Theorem 3.3.1:

PROOF OF THEOREM 3.3.1. Let G be a finite group, and let $g_1, \ldots, g_r$ be generators. Let H_i be the cyclic subgroup of G generated by g_i. By Proposition 3.3.3, for each i there is an irreducible normal H_i-Galois cover $Y_i \to \mathbb{P}^1_R$ whose closed fibre is a mock cover of $\mathbb{P}^1_k$; moreover these covers may be chosen inductively so as to have disjoint branch loci B_i (by choosing them so that the branch loci along the closed fibre are disjoint). For $i = 1, \ldots, r$, let $U_i = \mathbb{P}^1_R - \bigcup_{j \neq i} B_j$, let R_i be the ring of holomorphic functions on U_i along its closed fibre (i.e. the $\mathfrak{m}$-adic completion of the ring of functions on U_i), and let $\hat{U}_i = \mathrm{Spec}\, R_i$. Also let $U_0 = \mathbb{P}^1_R - \bigcup_{j=1}^r B_j$ (so that $U_0 = U_i \cap U_j$ for any $i \neq j$), let $H_0 = 1 \subset G$, and let $Y_0 = \mathbb{P}^1_R$. Then the restriction $\hat{Y}_i = Y_i \times_{\mathbb{P}^1_R} \hat{U}_i$ is an irreducible normal H_i-Galois cover, and we may identify the pullback $\hat{Y}_i \times_{\hat{U}_i} \hat{U}_0$ with the trivial cover $\hat{Y}_0 = \mathrm{Ind}_1^{H_i}\hat{U}_0$. Finally, let $\hat{Z}_i = \mathrm{Ind}_{H_i}^G \hat{Y}_i$; this is a (disconnected) G-Galois cover of $\hat{U}_i$, equipped with an isomorphism $\hat{Z}_i \times_{\hat{U}_i} \hat{U}_0 \xrightarrow{\sim} \hat{Z}_0$. By Grothendieck's Existence Theorem for covers (see Theorem 3.2.4), there is a unique G-Galois cover $Z \to \mathbb{P}^1_R$ whose restriction to $\hat{U}_i$ is $\hat{Z}_i$, compatibly. This cover is connected since its closed fibre is (because $H_1, \ldots, H_r$ generate G); it is normal since each $\hat{Z}_i$ is; and so it is irreducible (being connected and normal). The closed fibre of

Z is a mock cover (and so reducible), since the same is true for each $\hat{Z}_i$; and so K is algebraically closed in the function field L of Z. So L is as desired in Theorem 3.3.1.   □

REMARK 3.3.4. A variant approach to Theorem 3.3.1 involves proving a modification of Proposition 3.3.3 — viz. requiring that Y_k contains a k-point that is not in the ramification locus of $Y_k \to \mathbb{P}^1_k$, rather than requiring that $Y_k \to \mathbb{P}^1_k$ is a mock cover. This turns out to be sufficient to obtain Theorem 3.3.1, e.g. by showing that after a birational change of variables on $\mathbb{P}^1$, the cover Y is taken to a cover whose closed fibre is a mock cover (and thereby recapturing the original proposition above). This modified version of the proposition can be proved by first showing that there is *some* n-cyclic extension of $K(x)$, e.g. as in [FJ, Lemma 24.46]; and then adjusting the extension by a "twist" in order to obtain an unramified rational point [HV, Lemma 4.2(a)]. (In general, this twisting method works for abelian covers, and so in particular for cyclic covers.) This modified proposition first appeared in [Li], where it was used to provide a proof of Theorem 3.3.1 using rigid analytic spaces, rather than formal schemes. See Theorem 4.3.1 below for a further discussion of this.   □

As mentioned just after the statement of Theorem 3.3.1 above, that result can be used to deduce that many other fields K have the same inverse Galois property, even without being complete. In particular:

COROLLARY 3.3.5 [Ha3, Corollary 1.5]. *Let k be an algebraically closed field. Then every finite group is a Galois group over $k(x)$; or equivalently, it is the Galois group of some branched cover of the k-line.*

In the case of $k = \mathbb{C}$, this result is classical, and was the subject of Section 2 above, where the proof involved topology and analytic patching. For a more general algebraically closed field, the proof uses Theorem 3.3.1 above and a trick that relies on the fact that every finite extension is given by finitely many polynomials (also used in the remark after Corollary 2.1.5):

PROOF OF COROLLARY 3.3.5. Let $R = k[\![t]\!]$ and $K = k(\!(t)\!)$. Applying Theorem 3.3.1 to R and a given finite group G, we obtain an irreducible G-Galois branched cover $Y \to \mathbb{P}^1_K$ such that K is algebraically closed in its function field. This cover is of finite type, and so it is defined (as a G-Galois cover) over a k-subalgebra A of K of finite type; i.e. there is an irreducible G-Galois branched cover $Y_A \to \mathbb{P}^1_A$ such that $Y_A \times_A K \approx Y$ as G-Galois branched covers of $\mathbb{P}^1_K$. By the Bertini–Noether Theorem [FJ, Prop. 9.29], there is a non-zero element $\alpha \in A$ such that the specialization of Y_A to any k-point of $\operatorname{Spec} A[\alpha^{-1}]$ is (geometrically) irreducible. Any such specialization gives an irreducible G-Galois branched cover of $\mathbb{P}^1_k$.   □

In fact, as F. Pop later observed [Po4], the proof of the corollary relied on k being algebraically closed only to know that every k-variety with a $k(\!(t)\!)$-point has a k-point. So for any field k with this more general property (a field k that is "existentially closed in $k((t))$"), the corollary holds as well. Moreover the resulting Galois extension of $k(x)$ can be chosen to be regular, i.e. with k algebraically closed in the extension, by the geometric irreducibility assertion in the Bertini–Noether Theorem. Pop proved [Po4, Proposition 1.1] that the fields k that are existentially closed in $k((t))$ can be characterized in another way: they are precisely those fields k with the property that every smooth k-curve with a k-rational point has infinitely many k-rational points. He called such fields "large", because they are sufficiently large within their algebraic closures in order to recapture the finite-type argument used in the above corollary. (In particular, if k is large, then any extension field of k, contained in the algebraic closure of k, is also large [Po4, Proposition 1.2].) Thus we obtain the following strengthening of the corollary:

THEOREM 3.3.6 [Po4]. *Let k be a large field, and let G be a finite group.*

(a) *Then G is the Galois group of a Galois field extension L of $k(x)$, and the extension may be chosen to be regular.*

(b) *If k is (separably) Hilbertian, then G is a Galois group over k.*

Here part (b) follows from part (a) as in Example 3.3.2(c).

EXAMPLE 3.3.7. (a) Let K be a complete valuation field. Then K is large by [Po4, Proposition 3.1], the basic idea being that K satisfies an Implicit Function Theorem (and so one may move a K-rational point a bit to obtain other K-rational points). So every finite group is a Galois group over $K(x)$, by Theorem 3.3.6. In particular, this is true for the fraction field K of a complete discrete valuation ring R — as was already shown in Theorem 3.3.1. On the other hand, Theorem 3.3.6 applies to complete valuation fields K that are not of that form.

(b) More generally, a henselian valued field K (i.e. the fraction field of a henselian valuation ring) is large by [Po4, Proposition 3.1]. So again, every finite group is a Galois group over $K(x)$. If the valuation ring is a discrete valuation ring, then this conclusion can also be deduced using the Artin Algebraization Theorem, as in Example 3.3.2(d). But as in Example (a) above, K is large even if it is not discretely valued (in which case the earlier example does not apply).

(c) It is immediate from the definition that a field k will be large if it is PAC (pseudo-algebraically closed); i.e. if every smooth geometrically integral k-variety has a k-point. Fields that are PRC (pseudo-real closed) or PpC (pseudo-p-adically closed) are also large. In particular, the field of all totally real algebraic numbers is large, and so is the field of totally p-adic algebraic numbers (i.e. algebraic numbers α such that $\mathbb{Q}(\alpha)$ splits completely over the prime p). Hence every finite group is a Galois group over $k(x)$, where k is any of the above fields. And if k is Hilbertian (as some PAC fields are), then every finite group is therefore a Galois group over k. See [Po4, Section 3] and [MB1, Thm. 1.3] for details.

(d) Let K be a field that contains a large subfield K'. If K is algebraic over K' then K is automatically large [Po4, Proposition 1.2]; but otherwise K need not be large (e.g. $\mathbb{C}(t)$ is not large). Nevertheless, every finite group is the Galois group of a regular branched cover of $\mathbb{P}^1_K$. The reason is that this property holds for K'; and the function field F of the cover of $\mathbb{P}^1_{K'}$ is linearly disjoint from K over K', because K' is algebraically closed in F (by regularity). In particular, we may use this approach to deduce Theorem 3.3.1 from Theorem 3.3.2, since every normal complete local domain R other than a field must contain a complete discrete valuation ring R_0 — whose fraction field is large. (Namely, if R contains a field k, then take $R_0 = k[\![t]\!]$ for some non-zero element t in the maximal ideal of R; otherwise, R contains $\mathbb{Z}_p$ for some p.) Similarly, we may recover Example 3.3.2(d) in this way (taking the algebraic Laurent series in $k((t_1))$), even though it is not known whether $k((t_1, \ldots, t_n))$ and its subfield of algebraic Laurent series are large. (Note that $k((t_1, \ldots, t_n))$ is not a valuation field for $n > 1$, unlike the case of $n = 1$.) $\qquad\qquad\square$

REMARKS 3.3.8. (a) An arithmetic analog of Example 3.3.7(b) holds for the ring $T = \mathbb{Z}\{t\}$ of power series over $\mathbb{Z}$ convergent on the open unit disc. Namely, replacing Grothendieck's Existence Theorem by its arithmetic analog discussed in Remark 3.2.5(c) above, one obtains an analog of Theorem 3.3.1 above for $\mathbb{Z}\{t\}$ [Ha5, Theorem 3.7]; i.e. that every finite group is a Galois group over the fraction field of $\mathbb{Z}\{t\}$ (whose model over Spec $\mathbb{Z}\{t\}$ has a mock fibre modulo (t)). Moreover, the construction permits one to construct the desired Galois extension L of frac T so that it remains a Galois field extension, with the same Galois group, even after tensoring with the fraction field of $T_r = \mathbb{Z}_{r+}[\![t]\!]$, the ring of power series over $\mathbb{Z}$ convergent on a neighborhood of the closed disc $|t| \leq r$. (Here $0 < r < 1$.) Even more is true: Using an arithmetic analog of Artin's Approximation Theorem (see [Ha5, Theorem 2.5]), it follows that these Galois extensions L_r of T_r can simultaneously be descended to a compatible system of Galois extensions L_r^h of frac T_r^h, where T_r^h is the ring of *algebraic* power series in T_r. Surprisingly, the intersection of the rings T_r^h has fraction field $\mathbb{Q}(t)$ [Ha2, Theorem 3.5] (i.e. every algebraic power series over $\mathbb{Z}$ that converges on the open unit disc is *rational*). So since the Galois extensions L, L_r, L_r^h (for $0 < r < 1$) are all compatible, this suggests that it should be possible to descend the system $\{L_r^h\}$ to a Galois extension L^h of $\mathbb{Q}(t)$. If this could be done, it would follow that every finite group would be a Galois group over $\mathbb{Q}(t)$ and hence over $\mathbb{Q}$ (since $\mathbb{Q}$ is Hilbertian). See [Ha5, Section 4] for a further discussion of this (including examples that demonstrate pitfalls).

(b) The field $\mathbb{Q}^{ab}$ (the maximal abelian extension of $\mathbb{Q}$) is known to be Hilbertian [Vö, Corollary 1.28] (and in fact any abelian extension of a Hilbertian field is Hilbertian [FJ, Theorem 15.6]). It is *conjectured* that $\mathbb{Q}^{ab}$ is large; and if it is, then Theorem 3.3.6(b) above would imply that every finite group is a Galois group over $\mathbb{Q}^{ab}$. Much more is believed: The Shafarevich Conjecture asserts that the absolute Galois group of $\mathbb{Q}^{ab}$ is a *free* profinite group on countably many generators. This conjecture has been posed more generally, to say that if K is a global field, then the absolute Galois group of K^{cycl} (the maximal cyclotomic extension of K) is a free profinite group on countably many generators. (Recall that $\mathbb{Q}^{ab} = \mathbb{Q}^{cycl}$, by the Kronecker–Weber Theorem in number theory.) The Shafarevich Conjecture (along with its generalization to arbitrary number fields) *remains open* — though it too would follow from knowing that $\mathbb{Q}^{ab}$ is large (see Section 5). On the other hand, the generalized Shafarevich Conjecture has been proved in the geometric case, i.e. for function fields of curves [Ha10] [Po1] [Po3]; see Section 5 for a further discussion of this. □

As another example of the above ideas, consider covers of the line over *finite* fields. Not surprisingly (from the terminology), finite fields $\mathbb{F}_q$ are not large. And it *is unknown* whether every finite group G is a Galois group over $k(x)$ for every finite field k. But it is known that every finite group G is a Galois group over $k(x)$ for *almost* every finite field k:

PROPOSITION 3.3.9 (FRIED–VÖLKLEIN, JARDEN, POP). *Let G be a finite group. Then for all but finitely many finite fields k, there is a regular Galois field extension of $k(x)$ with Galois group G.*

PROOF. First consider the case that k ranges just over prime fields $\mathbb{F}_p$. By Example 3.3.2(d) (or by Theorem 3.3.6 and Example 3.3.7(b) above), G is a regular Galois group over the field $\mathbb{Q}((t))^h(x)$, where $\mathbb{Q}((t))^h$ is the field of algebraic Laurent series over $\mathbb{Q}$ (the t-adic henselization of $\mathbb{Q}(t)$). Such a G-Galois field extension is finite, so it descends to a G-Galois field extension of $K(x)$, where K is a finite extension of $\mathbb{Q}(t)$ (in which $\mathbb{Q}$ is algebraically closed, since $K \subset \mathbb{Q}((t))$). This extension of $K(x)$ can be interpreted as the function field of a G-Galois branched cover $Z \to \mathbb{P}^1_V$; here V is a smooth projective curve over $\mathbb{Q}$ with function field K, viz. a finite branched cover of the t-line, say of genus g (see Figure 3.3.10). For all points $\nu \in V$ outside some finite set Σ, the fibre of Z over ν is an irreducible G-Galois cover of $\mathbb{P}^1_{k(\nu)}$, where $k(\nu)$ is the residue field at ν. By taking a normal model $\mathcal{Z} \to \mathcal{V}$ of $Z \to V$ over $\mathbb{Z}$, we may consider the reductions V_p and Z_p for any prime p. For all primes p outside some finite set S, the reduction V_p is a smooth connected curve over $\mathbb{F}_p$ of genus g; the reduction Z_p is an irreducible G-Galois branched cover of $\mathbb{P}^1_{V_p}$; and any specialization of this cover away from the reduction Σ_p of Σ is an irreducible G-Galois cover of the line. According to the Weil bound in the Riemann Hypothesis for curves over finite fields [FJ, Theorem 3.14], the number of k-points on a k-curve of genus g is at least $|k| + 1 - 2g\sqrt{|k|}$. So for all $p \notin S$ with $p > (2g + \deg(\Sigma))^2$, the curve Z_p has an $\mathbb{F}_p$-point that does not lie in the reduction of Σ. The specialization at that point is a regular G-Galois cover of $\mathbb{P}^1_{\mathbb{F}_p}$, corresponding to a regular G-Galois field extension of $\mathbb{F}_p(x)$.

For the general case, observe that if G is a regular Galois group over $\mathbb{F}_p(x)$, then it is also a regular Galois group over $\mathbb{F}_q(x)$ for every power q of p (by base

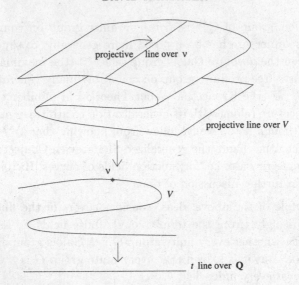

projective line over ν

projective line over V

ν

V

t line over $\mathbf{Q}$

Figure 3.3.10. Base of the Galois cover $Z \to \mathbb{P}^1_V$ in the first case of the proof of Proposition 3.3.9. For most choices of ν in V, the restriction of Z over ν is an irreducible cover of the projective line; and for most primes p, the same is true for its reduction mod p.

change). Now consider the finitely many primes p such that G is not known to be a regular Galois group over $\mathbb{F}_p(x)$. Arguing as above (but using $\mathbb{F}_p((t))$ instead of $\mathbb{Q}((t))$), we obtain a geometrically irreducible G-Galois cover $Y_p \to \mathbb{P}^1_{W_p}$, for some $\mathbb{F}_p$-curve W_p. Again using the Weil bound, there is a constant c_p such that if q is a power of p and $q > c_p$, then W_p has an $\mathbb{F}_q$-point at which Y_p specializes to a regular G-Galois cover of $\mathbb{P}^1_{\mathbb{F}_q}$. So if c is chosen larger than each of the finitely many c_p's (as p ranges over the exceptional set of primes), then G is a regular Galois group over $k(x)$ for every finite field k of order $\geq c$. $\qquad \square$

REMARK 3.3.11. (a) The above result can also be proved via ultraproducts, viz. using that a non-principal ultraproduct of the $\mathbb{F}_q$'s is large (and even PAC); see [FV1, § 2.3, Cor. 2]. In [FV1], just the case of prime fields was shown. But Pop showed that the conclusion holds for general finite fields (as in the statement of Proposition 3.3.9), using ultraproducts.

(b) It is *conjectured* that in fact there are *no* exceptional finite fields in the above result, i.e. that every finite group is a Galois group over each $\mathbb{F}_q(x)$. But at least, it would be desirable to have a better understanding of the possible exceptional set. For this, one could try to make more precise the sets S and Σ in the above proof, and also the bound on the exceptional primes. (The bound in the above proof is certainly not optimal.) $\qquad \square$

REMARK 3.3.12. (a) The class of large fields also goes under several other names in the literature. Following the introduction of this notion by Pop in [Po4] under the name "large", D. Haran and M. Jarden referred to such fields as "ample" [HJ1];

P. Dèbes and B. Deschamps called them fields with "automatique multiplication des points lisses existants" (abbreviated AMPLE) [DD]; J.-L. Colliot-Thélène has referred to such fields as "epais" (thick); L. Moret-Bailly has called them "fertile" [MB2]; and the present author has even suggested that they be called "pop fields", since the presence of a single smooth rational point on a curve over such a field implies that infinitely many rational points will "pop up".

(b) By whatever name, large fields form the natural context to generalize Corollary 3.3.5 above. As noted in Example 3.3.7(d), the class of fields K that *contain* large subfields also has the property that every finite group is a regular Galois group over $K(x)$; and this class is general enough to subsume Theorem 3.3.1, as well as Theorem 3.3.6. On the other hand, this Galois property holds for the fraction field of $\mathbb{Z}[\![t]\!]$, as noted at the end of Example 3.3.2(c); but that field is not known to contain a large subfield. Conjecturally, *every* field K has the regular Galois realization property (see [Ha9, §4.5]; this conjecture has been referred to as the regular inverse Galois problem). But that degree of generality seems *very far from being proved* in the near future.

(c) In addition to yielding regular Galois realizations, large fields have a stronger property: that *every finite split embedding problem is properly solvable* (Theorem 5.1.9 below). *Conjecturally*, all fields have this property (and this conjecture subsumes the one in Remark (b) above). See Section 5 for more about embedding problems, and for other results in Galois theory that go beyond Galois realizations over fields. The results there can be proved using patching theorems from Section 3.2 (including those at the end of §3.2, which are stronger than Grothendieck's Existence Theorem).   □

We conclude this section with a reinterpretation of the above patching construction in terms of thickening and deformation. Namely, as discussed after Theorem 3.2.1 (Grothendieck's Existence Theorem), that earlier result can be interpreted either as a patching result or as a thickening result. Theorem 3.3.1 above, and its Corollary 3.3.5, relied on Grothendieck's Existence Theorem, and were presented above in terms of patching. It is instructive to reinterpret these results in terms of thickening, and to compare these results from that viewpoint with the slit cover construction of complex covers, discussed in Section 2.3.

Specifically, the proof of Theorem 3.3.1 above yields an irreducible normal G-Galois cover $Z \to \mathbb{P}^1_R$ whose closed fibre is a connected mock cover $Z_0 \to \mathbb{P}^1_k$. Viewing $\operatorname{Spec} R$ as a "small neighborhood" of $\operatorname{Spec} k$, we can regard $\mathbb{P}^1_R$ as a "tubular neighborhood" of $\mathbb{P}^1_k$; and the construction of $Z \to \mathbb{P}^1_R$ can be viewed as a thickening (or deformation) of $Z_0 \to \mathbb{P}^1_k$, built in such a way that it becomes irreducible (by making it locally irreducible near each of the branch points). Regarding formal schemes as thickenings of their closed fibres (given by a compatible sequence of schemes over the $R/\mathfrak{m}^i$), this construction be viewed as the result of infinitesimal thickenings (over each $R/\mathfrak{m}^i$) which in the limit give the desired cover of $\mathbb{P}^1_R$.

From this point of view, Corollary 3.3.5 above can be viewed as follows: As before, take $R = k[\![t]\!]$ and as above obtain an irreducible normal G-Galois cover $Z \to \mathbb{P}^1_R$. Since this cover is of finite type, it is defined over a $k[t]$-subalgebra E of R of finite type (i.e. there is a normal irreducible G-Galois cover $Z_E \to \mathbb{P}^1_E$ that induces $Z \to \mathbb{P}^1_R$), such that there is a maximal ideal $\mathfrak{n}$ of E with the property that the fibre of $Z_E \to \mathbb{P}^1_E$ over the corresponding point $\xi_\mathfrak{n}$ is isomorphic to the closed fibre of $Z \to \mathbb{P}^1_R$ (viz. it is the mock cover $Z_0 \to \mathbb{P}^1_k$). The cover $Z_E \to \mathbb{P}^1_E$ can be viewed as a family of covers of $\mathbb{P}^1_k$, parametrized by the variety $V = \operatorname{Spec} E$, and which provides a deformation of $Z_0 \to \mathbb{P}^1_k$. A generically chosen member of this family will be an irreducible cover of $\mathbb{P}^1_k$, and this G-Galois cover is then as desired.

In the case that $k = \mathbb{C}$, we can be even more explicit. There, we are in the easy case of Proposition 3.3.3 above, where the field contains the roots of unity, ramification is cyclic, and cyclic extensions are Kummer. So choosing generators $g_1, \ldots, g_r$ of G of orders $n_1, \ldots, n_r$, and choosing corresponding branch points $x = a_1, \ldots, a_r$ for the mock cover $Z_0 \to \mathbb{P}^1_{\mathbb{C}}$, we may choose $Z \to \mathbb{P}^1_R$ so that it is given locally by the (normalization of the) equation $z_i^{n_i} = (x - a_i)(x - a_i - t)^{n_i - 1}$ in a neighborhood of a point over $x = a_i, t = 0$ (and so the mock cover is given locally by $z_i^{n_i} = (x - a_i)^{n_i}$). By Artin's Algebraization Theorem [Ar3] (cf. Example 3.3.2(d) above), this cover descends to a cover $Z \to \mathbb{P}^1_{R^h}$, where $R^h \subset R = \mathbb{C}[\![t]\!]$ is the ring of algebraic power series. Since that cover is of finite type, it can be defined over a $\mathbb{C}[t]$-subalgebra of R^h of finite type; i.e. the cover further descends to a cover $Y_C \to \mathbb{P}^1_C$, where C is a complex curve together with a morphism $C \to \mathbb{A}^1_{\mathbb{C}} = \operatorname{Spec} \mathbb{C}[t]$, and where the fibre of Y_C over some point $\xi \in C$ over $t = 0$ is the given mock cover $Z_0 \to \mathbb{P}^1_{\mathbb{C}}$. This family $Y_C \to \mathbb{P}^1_C$ can be viewed as a family of covers of $\mathbb{P}^1_{\mathbb{C}}$ deforming the mock cover; and this deformation takes place by allowing the positions of the branch points to move. By the choice of local equations, if we take a typical point on C near ξ, the corresponding cover has $2r$ branch points $x = a_1, a'_1, \ldots, a_r, a'_r$, with branch cycle description

$$(g_1, g_1^{-1}, \ldots, g_r, g_r^{-1}) \tag{$*$}$$

(see Section 2.1 and the beginning of Section 2.3 for a discussion of branch cycle descriptions). So this is a slit cover, in the sense of Example 2.3.2. See also the discussion following that example, concerning the role of the mock cover as a degeneration of the typical member of this family (in which a'_i is allowed to coalesce with a_i).

For more general fields k, we may not be in the easy case of Proposition 3.3.3, and so may have to use more complicated branching configurations. As a result, the deformed covers may have more than $2r$ branch points, and they may come in clusters rather than in pairs. Moreover, while the tamely ramified branch points will move in $\mathbb{P}^1$ as one deforms the cover, wildly ramified branch points can stay at the same location (with just the Artin–Schreier polynomial changing; see the last case in the proof of Proposition 3.3.3).

Still, in the tame case, by following this construction with a further doubling of branch points, it is possible to pair up the points of the resulting branch locus so that the resulting cover has "branch cycle description" of the form $(h_1, h_1^{-1}, \ldots, h_N, h_N^{-1})$, where each h_i is a power of some generator g_j. (Here, since we are not over $\mathbb{C}$, the notion of branch cycle description will be interpreted in the weak sense that the entries of the description are generators of inertia groups at some ramification points over the respective branch points.) This leads to a generalization of the "half Riemann Existence Theorem" (Theorem 2.3.5) from $\mathbb{C}$ to other fields. Such a result (though obtained using the rigid approach rather than the formal approach) was proved by Pop [Po2]; see Section 4.3 below.

The construction in the tame case can be made a bit more general by allowing the r branch points $x = a_i$ of the mock cover to be deformed with respect to independent variables. For example, in the case $k = \mathbb{C}$, we can replace the ring R by $k[[t_1, t_1', \ldots, t_r, t_r']]$ and use the (normalization of the) local equation $z_i^{n_i} = (x - a_i - t_i)(x - a_i - t_i')^{n_i - 1}$ in a neighborhood of a point over $x = a_i$ on the closed fibre $\underline{t} = \underline{t}' = 0$. Using Artin's Algebraization Theorem, we obtain a $2r$-dimensional family of covers that deform the given mock cover, with each of the r mock branch points splitting in two, each moving independently. The resulting family $Z \to \mathbb{P}_V^1$ is essentially a component of a *Hurwitz family* of covers (e.g. see [Fu1] and [Fr1]), which is by definition a total family $Y \to \mathbb{P}_H^1$ of covers of $\mathbb{P}^1$ over the moduli space H for branched covers with a given branch cycle description and variable branch points (the *Hurwitz space*). Here, however, a given cover is permitted to appear more than once in the family (though only finitely often), and part of the boundary of the Hurwitz space is included (in particular, the point of the parameter space V corresponding to the mock cover). That is, there is a finite-to-one morphism $V \to \bar{H}$, where $\bar{H}$ is the compactification of H. From this point of view, the desirability of using branch cycle descriptions of the form (∗) is that one can begin with an easily constructed mock cover, and use it to construct algebraically a component of a Hurwitz space with this branch cycle description. See [Fr3] for more about this point of view.

As mentioned above, still more general formal patching constructions of covers can be performed if one replaces Grothendieck's Existence Theorem by the variations at the end of Section 3.2. In particular, one can begin with a given irreducible cover, and then modify it near one point (e.g. by adding ramification there). Some constructions along these lines will be discussed in Section 5, in connection with the study of fundamental groups.

4. Rigid Patching

This section, like Section 3, discusses an approach to carrying over the ideas of Section 2 from complex curves to more general curves. The approach here is due to Tate, who introduced the notion of rigid analytic spaces. The idea here is to consider power series that converge on metric neighborhoods on curves over a valued field, and to "rigidify" the structure to obtain a notion of "analytic continuation". Tate's original point of view, which is presented in Section 4.1, is rather intuitive. But the details of carrying it out become somewhat complicated, as the reader will see (particularly with regard to the precise method of rigidifying "wobbly spaces"). A simplified approach, due to Grauert, Remmert, and Gerritzen, is discussed later in Section 4.1, including their approach to a rigid analog of GAGA. Section 4.2 then discusses a later reinterpretation of rigid geometry that is due to Raynaud, and which establishes a kind of "dictionary" between the formal and rigid set-ups (and allows rigid GAGA to be deduced from formal GAGA). Applications to the construction of Galois covers of curves are then presented in Section 4.3, including a version of the (geometric) regular inverse Galois problem, and Pop's Half Riemann Existence Theorem. Additional applications of both rigid and formal geometry to Galois theory appear afterwards, in Section 5.

4.1. Tate's rigid analytic spaces.
Another approach to generalizing complex analytic notions to spaces over other fields is provided by Tate's rigid analytic spaces. As in the formal approach discussed in Section 3, the rigid approach allows "small neighborhoods" of points, and permits objects (spaces, maps, sheaves, covers) to be constructed by giving them locally and giving agreement on overlaps (i.e. "patching"). Here the small neighborhoods are metric discs, rather than formal neighborhoods of subvarieties, as in the formal patching approach.

This approach was introduced by Tate in [Ta], a 1962 manuscript which he never submitted for publication. The manuscript was circulated in the 1960's by IHES, with the notation that it consisted of "private notes of J. Tate reproduced with(out) his permission". Later, the paper was published in Inventiones Mathematicae on the initiative of the journal's editors, who said in a footnote that they "believe that it is in the general interests of the mathematical community to make these notes available to everyone".

Tate's approach was motivated by the problem of studying bad reduction of elliptic curves (what we now know as the study of Tate curves; see e.g. [BGR, 9.7]). The idea is to work over a field K that is complete with respect to a non-trivial non-archimedean valuation — e.g. the p-adics, or the Laurent series over a coefficient field k. On spaces defined over such a field K, one can consider discs defined with respect to the metric on K; and one can consider "holomorphic functions" on those discs, viz. functions given by power series that are convergent there. One then wants to work more globally by means of analytic continuation, and to carry over the classical results over $\mathbb{C}$ (e.g. those of Section 2 above) to this context. As a result, one hopes to obtain a GAGA-type result, a version of Riemann Existence Theorem, the realization of all finite groups as Galois groups over $K(x)$, etc.

There are difficulties, however, that are caused by the fact that the topology on K is totally disconnected. For example, on the affine K-line, consider the characteristic function f_D of the open unit disc $|x| < 1$; i.e. $f(x) = 1$ for $|x| < 1$, and $f(x) = 0$ for $|x| \geq 1$. Then this function is continuous, and in a neighborhood of each point $x = x_0$ it is given by a power series. (Namely, on the open disc of radius 1 about $x = x_0$, it is identically 1 or identically 0, depending on whether or not $|x_0|$ is less than 1.) This is quite contrary to the situation over $\mathbb{C}$, where a holomorphic function is "rigid", in the sense that it is determined by its values on any open disc. Thus, if one proceeds in the obvious way, objects will have a strictly local character, and there will be no meaningful "patching".

Tate used two ideas to deal with this problem. The first of these is to consider functions that are locally given on *closed* discs, rather than on open discs, and to require agreement on overlapping boundaries. Note, though, that because the metric is non-archimedean, closed discs are in fact open sets. The second idea is to restrict the set of allowable maps between spaces, by choosing a class of maps that fulfills certain properties and creates a "rigid" situation.

Concerning the first of these ideas, let $K\{x\}$ denote the subring of $K[\![x]\!]$ consisting of power series that converge on the closed unit disc $|x| \leq 1$. Because the metric is non-archimedean, this ring consists precisely of those series $\sum_{i=0}^{\infty} a_i x^i$ for which $a_i \to 0$ as $i \to \infty$. Similarly, the power series in $K[\![x_1, \ldots, x_n]\!]$ that converge on the closed polydisc where each $|x_i| \leq 1$ form the ring $K\{x_1, \ldots, x_n\}$ of series $\sum a_{\underline{i}} x^{\underline{i}}$, where $\underline{i}$ ranges over n-tuples of non-negative integers, and where $a_{\underline{i}} \to 0$ as $\underline{i} \to \infty$. As an example, if $K = k((t))$ for some field k, then $K\{x\} = k[x][\![t]\!][t^{-1}]$. (Verification of this equality is an exercise left to the reader.)

If $0 < r_1 \leq r_2$, then we may also consider the closed annulus $\{x \mid r_1 \leq |x| \leq r_2\}$. Since the metric is non-archimedean, this is an open subset, which we may consider even when $r_1 = r_2$. In particular, in the case $r_1 = r_2 = 1$, we may consider the ring $K\{x, x^{-1}\} = K\{x, y\}/(xy - 1)$ of functions converging on the annulus; this consists of doubly infinite series $\sum_{i=-\infty}^{\infty} a_i x^i$ such that $a_i \to 0$ as $|i| \to \infty$. Similarly, we may consider the ring $K\{x_1, \ldots, x_n, x_1^{-1}, \ldots, x_n^{-1}\} = K\{x_1, \ldots, x_n, y_1, \ldots, y_n\}/(x_i y_i - 1)$ of functions on the "poly-annulus" $|x_i| = 1$ (with $i = 1, \ldots, n$). In the case that $K = k((t))$, we have that $K\{x, x^{-1}\} = k[x, x^{-1}][\![t]\!][t^{-1}]$. (Verification of this is again left to the reader. In this situation, the one-dimensional rings $K\{x\}$ and $K\{x, x^{-1}\}$ are obtained by inverting t in the two-dimensional rings $k[x][\![t]\!]$ and $k[x, x^{-1}][\![t]\!]$; cf. Figure 3.1.4 above.)

In order to consider more general analytic "varieties" over K, Tate considered quotients of the rings $K\{x_1, \ldots, x_n\}$ by ideals. He referred to such quotients by saying that they were of *topologically finite type*; these are also now referred to as *affinoid algebras* [BGR] or as *Tate algebras* [Ra1] (though the latter term is sometimes used only for the ring $K\{x_1, \ldots, x_n\}$ itself [BGR]). Tate showed that a complete K-algebra A is an affinoid algebra if and only if it is a finite extension of some $K\{x_1, \ldots, x_n\}$ [Ta, Theorem 4.4]; and in this case A is Noetherian, every ideal is closed, and the residue field of every maximal ideal is finite over K [Ta, Theorem 4.5]. The association $A \mapsto \operatorname{Max} A$ is a contravariant functor from affinoid algebras to sets, where $\operatorname{Max} A$ is the maximal spectrum of A. (The map $\operatorname{Max} B \to \operatorname{Max} A$ associated to $\phi : A \to B$ is denoted by ϕ°, and is called *rigid*.) Since A/ξ is a finite extension L of K for any $\xi \in \operatorname{Max} A$, we may consider $f(\xi) \in L$ and $|f(\xi)| \in \mathbb{R}$ for any $f \in A$ (and thus regard A as a ring of functions on $\operatorname{Max} A$). By an *affinoid variety*, we then mean a pair $\operatorname{Sp} A := (\operatorname{Max} A, A)$, where A is an affinoid algebra.

Tate defined an *affine subset* $Y \subset \operatorname{Max} A$ to be a subset for which there is an affinoid algebra A_Y that represents the functor $h_Y : B \mapsto \{\phi : A \to B \,|\, \phi^\circ(\operatorname{Max} B) \subset Y\}$; i.e. such that $h_Y(B) = \operatorname{Hom}(A_Y, B)$. (This is called an *affinoid subdomain* in [BGR].) A *special affine subset* $Y \subset \operatorname{Max} A$ is a subset of the form

$$Y = \{\xi \in \operatorname{Max} A : |f_i(\xi)| \leq 1 \,(\forall i), \; |g_j(\xi)| \geq 1 \,(\forall j)\},$$

where $(f_i), (g_j)$ are finite families of elements of A. (These are called *Laurent domains* in [BGR].) Tate showed [Ta, Proposition 7.2] that every special affine subset is affine, viz. that if Y is given by $(f_i), (g_j)$ as above, then $A_Y = A\{f_i; g_j^{-1}\} := A\{x_i; y_j\}/(f_i - x_i, 1 - g_j y_j)$. Moreover if Y is an affine subset of $\operatorname{Max} A$, then the canonical map $\operatorname{Max} A_Y \to Y$ is a bijection [Ta, Proposition 7.3]. In fact, it is a homeomorphism [Ta, Cor. 2 to Prop. 9.1], if we give $\operatorname{Max} A$ the topology in which a fundamental system of neighborhoods of a point ξ_0 is given by sets of the form $U_\varepsilon(g_1, \ldots, g_n) = \{\xi \in \operatorname{Max} A : |g_i(\xi)| < \varepsilon$ for $1 \leq i \leq n\}$, where $\varepsilon > 0$ and where $g_1, \ldots, g_n \in A$ satisfy $g_i(\xi_0) = 0$.

Tate defined Čech cohomology for coverings of affinoid varieties $V = (\operatorname{Max} A, A)$ by finitely many affine subsets, and proved his Acyclicity Theorem [Ta, Theorem 8.2], that $H^i(\mathfrak{V}, \mathcal{O}) = 0$ for $i > 0$; here $\mathcal{O}$ is the presheaf that associates to any affine subset its affinoid algebra, and $\mathfrak{V}$ is a finite covering of V by special affine subsets. (In fact, this holds even with a finite covering of V by affine subsets; see [BGR, §8.2, Theorem 1].) As a consequence, for such a covering $\mathfrak{V}$ of V and any A-module M of finite type, $H^0(\mathfrak{V}, \tilde{M})$ is isomorphic to M, and $H^i(\mathfrak{V}, \tilde{M}) = 0$ for $i > 0$ [Ta, Theorem 8.7]; here $\tilde{M}$ is the presheaf $Y \to M \otimes_A A_Y$ for Y an affine subset of V. These are analogs of the usual facts for the cohomology of affine varieties. Moreover, they imply that $\mathcal{O}$ and $\tilde{M}$ are sheaves. In particular [BGR, §8.2, Corollary 2], if $f, g \in A$ agree on each member U_i of a finite affine covering of V, then they are equal; and if for every i we are given a function f_i on U_i, with agreements on the overlaps, then they may be "patched" — i.e. there is a function $f \in A$ which restricts to each f_i.

As might be expected, if U is an affine open subset of an affinoid variety V, then the map $A_U \to \Gamma(U, \mathcal{O})$ is injective. Unfortunately, it is not surjective, e.g. because of characteristic functions like f_D, mentioned at the beginning of this section. Moreover, the functor $A \mapsto \operatorname{Max} A$ is faithful, but not fully faithful [Ta,

Corollary 2 to Proposition 9.3]; i.e. not every K-ringed space morphism between two affinoid varieties is induced by a homomorphism between the corresponding rings of functions. Because of this phenomenon, if one defines more global analytic K-spaces simply by considering ringed K-spaces that are locally isomorphic to affinoid varieties, then one instead obtains a theory of "wobbly analytic spaces", rather than rigid ones.

In order to "rigidify" these wobbly spaces, Tate introduced the second of the two ideas mentioned earlier — viz. shrinking the class of allowable morphisms between such spaces, in such a way that in the case of affinoid varieties, the allowable morphisms are precisely the rigid ones (i.e. those induced by homomorphisms of the underlying algebras). He did this in a series of steps, which he said followed "fully and faithfully a plan furnished by Grothendieck" [Ta, § 10]. First, he defined [Ta, Definition 10.1] an *h-structure* θ on a wobbly analytic space V to be a choice of a subset $V^\theta(A) \subset \operatorname{Hom}(\operatorname{Max} A, V)$ (of *structural* maps) for every affinoid K-algebra A, such that every point of V is in the image of some open structural immersion, and such that the composition of a rigid map of affinoids with a structural map is structural. An *h-space* is a wobbly analytic space together with an h-structure, and a *morphism* of h-spaces $(V, \theta) \to (V', \theta')$ is a ringed space morphism $V \to V'$ which pulls back structural maps to structural maps. If V, V' are affinoid, then a morphism of h-spaces between them is the same as a rigid morphism between them [Ta, Corollary to Prop. 10.4].

Next, Tate defined a *special covering* of an h-space [Ta, Def. 10.9] to be one that is obtained by taking a finite covering by special affine subsets, then repeating this process on each of those subsets, a finite number of times. An h-space V is then said to be *special* [Ta, Def. 10.12] if it has the property that a ringed space morphism $\operatorname{Max} B \to V$ is structural if and only if its restriction to each member of any special covering of $\operatorname{Max} B$ is structural. An open covering of an h-space V is *admissible* if its pullback by any structural morphism has a refinement that is a special covering. A *semi-rigid* analytic space V over K is a special h-space that has an admissible covering by affine open h-spaces. Finally, a *rigid* analytic space is a semi-rigid space V such that the above admissible covering has the property that the intersection of any two members is semi-rigid [Ta, Definition 10.16].

This rather cumbersome approach to rigidifying "wobbly spaces" was simplified and extended in a number of papers in the 1960's and 1970's, particularly in [GrRe1], [GrRe2], [GG]. From this point of view, the key idea is that analytic continuation on rigid spaces is permitted only with respect to "admissible" coverings by affinoid varieties, and where the only morphisms permitted between affinoid varieties are the rigid ones (i.e. those induced by homomorphisms between the corresponding affinoid algebras). To make sense of "admissibility", the notion of Grothendieck topology was used.

Recall (e.g. from [Ar1] or [Mi]) that a Grothendieck topology is a generalization of a classical topology on a space X, in which one replaces the collection of open sets $U \subset X$ by a collection of (admissible) maps $U \to X$, and in which certain families of such maps $\{V_i \to U\}_{i \in I}$ are declared to be *(admissible) coverings* (of U). This notion was originally introduced in order to provide a framework for the étale topology and for étale cohomology, which for algebraic varieties behaves much like classical singular cohomology in algebraic topology (unlike Zariski Čech cohomology).

In the case of rigid analytic spaces, a less general notion of Grothendieck topology is needed, in which the maps $U \to X$ are just inclusions of (certain) subsets of X, so that one speaks of "admissible subsets" of X [GuRo, §9.1]. According to the definition of a Grothendieck topology, the admissible subsets U and the admissible coverings of the U's satisfy several properties:

- the intersection of two admissible subsets is admissible;
- the singleton $\{U\}$ is an admissible covering of a set U;
- choosing an admissible covering of each member of an admissible covering together gives an admissible covering; and
- the intersection of an admissible covering of U with an admissible subset $V \subset U$ is an admissible covering of V.

Here, though, several additional conditions are imposed [BGR, p. 339]:

- the empty set and X are admissible subsets of X;
- if V is a subset of an admissible $U \subset X$ and if the restriction to V of every member of some admissible covering of U is an admissible subset of X, then V is an admissible subset of X; and
- a family of admissible subsets $\{U_i\}_{i \in I}$ whose union is an admissible subset U, and which admits a refinement that is an admissible covering of U, is itself an admissible covering.

In this framework, a *rigid analytic space* is a locally ringed space $(V, \mathcal{O}_V)$ under a Grothendieck topology as above, with respect to which V has an admissible covering $\{V_i\}_{i \in I}$ where each $(V_i, \mathcal{O}_V|_{V_i})$ is an affinoid variety $\operatorname{Sp} A_i = (\operatorname{Max} A_i, A_i)$. (Here $A_i = \mathcal{O}_V|_{V_i}$.) A *morphism* of rigid analytic spaces $(V, \mathcal{O}_V) \to (W, \mathcal{O}_W)$ is a morphism (f, f^*) as locally ringed spaces. Thus morphisms between affinoid spaces are required to be rigid (i.e. of the form (ϕ°, ϕ), for some algebra homomorphism ϕ), and global morphisms are locally rigid with respect to an admissible covering. Analogously to the classical and formal cases, a *coherent sheaf* $\mathcal{F}$ (of $\mathcal{O}_V$-modules) is an $\mathcal{O}_V$-module that is locally (with respect to an admissible covering) of the form $\mathcal{O}_V^r \to \mathcal{O}_V^s \to \mathcal{F} \to 0$. In the case of an affinoid variety $\operatorname{Sp} A = (\operatorname{Max} A, A)$, coherent sheaves are precisely those of the form $\tilde{M}$, where M is a finite A-module [FP, III, 6.2].

Rigid analogs of key results in the classical and formal situations (cf. Sections 2.2 and 3.2 above) have been proved in this context. A rigid version of Cartan's Lemma on matrix factorization [FP, III, 6.3] asserts that if $V = \operatorname{Sp} A$ is an affinoid variety and $f \in A$, and if we let V_1 [resp. V_2] be the set where $|f| \leq 1$

[resp. $|f| \geq 1$], then every invertible matrix in $GL_n(\mathcal{O}(V_1 \cap V_2))$ that is sufficiently close to the identity can be factored as the product of invertible matrices over $\mathcal{O}(V_1)$ and $\mathcal{O}(V_2)$. There are also rigid analogs of Cartan's Theorems A and B, proved by Kiehl [Ki2]; they assert that a coherent sheaf $\mathcal{F}$ is generated by its global sections, and that $H^i(V, \mathcal{F}) = 0$ for $i > 0$, for "quasi-Stein" rigid analytic spaces V. (These are rigid spaces V that can be written as an increasing union of affinoid open subsets U_i that form an admissible covering of V, and such that $\mathcal{O}(U_{i+1})$ is dense in $\mathcal{O}(U_i)$. Compare Cartan's original version for complex Stein spaces [Ca2] discussed in § 2.2 above.) Kiehl also proved [Ki1] a rigid analog of Zariski's Theorem on Formal Functions [Hrt2, III, Thm. 11.1], which together with Cartan's Theorem B (or Theorem A) was used to obtain GAGA classically. And indeed, there is a rigid analog of GAGA (or in this case, a "GRGA": géométrie rigide et géométrie algébrique) [Köp], asserting the equivalence between coherent rigid sheaves and coherent algebraic sheaves of modules over a projective algebraic K-variety. Thus, to give a coherent sheaf over such a variety, it suffices to give it over the members of an admissible covering (viewing the variety as a rigid analytic space), and giving the patching data on the overlaps.

As in Sections 2 and 3 above, it would be desirable to use these results in order to obtain a version of Riemann's Existence Theorem, which would classify covers. Ideally, this should be precise enough to give an explicit description of the tower of Galois groups of covers of a given space; and that description should be analogous to Corollary 2.1.2, the explicit form of the classical Riemann's Existence Theorem given at the beginning of Section 2.1. Unfortunately, to give such an explicit description, one needs to have a notion of a "topological fundamental group", and one needs to be able to compute that group explicitly. But unlike the complex case, one does not have such a notion, or computation, over more general fields K (in particular, because we cannot speak of "loops"). Thus, in this context, one does not have a full analog of Riemann's Existence Theorem 2.1.1, because one cannot assert an equivalence between finite rigid analytic covering maps and finite topological covering spaces. Still, one can ask for an analog of the first part of Theorem, 2.1.1 viz. an equivalence between finite étale covers of an algebraic curve V over K, and finite analytic covering maps of V (viewed as a rigid analytic space).

Such a result has been obtained (with some restrictions) by Lütkebohmert [Lü2]. As in the proof of the complex version (see Section 2.2), the proof proceeds using GAGA (here, the rigid version discussed above). Namely, as in the complex case, once one has the equivalence of categories that GAGA provides for sheaves of modules, one also obtains an equivalence (as a purely formal consequence) for sheaves of algebras, and hence for branched covers. But as in the complex case, GAGA applies to projective curves, but not to affine curves. So GAGA shows that there is an equivalence between branched (algebraic) covers of a projective K-curve X, and rigid analytic branched covers of the curve. Then to prove the desired portion of Riemann's Existence Theorem, it remains to show (both in the algebraic and rigid analytic settings) that covers of X branched only at a finite set B are equivalent to unramified covers of $V = X - B$ (i.e. that every unramified cover of V extends uniquely to a branched cover of X). In Section 2.2, we saw that this is immediate in the algebraic context, and follows easily from complex analysis in the analytic setting. But in the rigid analytic setting, this extension result for rigid analytic covers is harder, and moreover requires that the characteristic of K is 0.

Specifically, if char $K = 0$, then unramified rigid covers of an affine K-curve $V = X - B$ do extend (uniquely) to rigid branched covers of the projective curve X; and so finite étale covers of V are equivalent to finite unramified rigid analytic covers of V. Moreover this generalizes to higher dimensions, where V is any K-scheme that is locally of finite type over K [Lü2, Theorem 3.1]. But there are counterexamples, even for curves, if char $K = p$. For example, let $K = k((t))$, let V be the affine x-line over K, and consider the rigid unramified covering map $W \to V$ given by $y^p - y = \sum_{i=1}^{\infty} t^{(p+1)^i} x^{p^i}$. Then this map does not extend to a finite (branched) cover of the projective line, and so is not induced by any algebraic cover of V [Lü2, Example 2.10]. On the other hand, if one restricts attention to *tamely* ramified covers, then the desired equivalence between rigid and algebraic unramified covers does hold [Lü2, Theorem 4.1]. (Note that the above wildly ramified example does not contradict rigid GAGA, since that result applies in the projective case, whereas this example is affine.)

Still, we do not have an explicit description of the rigid analytic covers of a given curve (even apart from the difficulty with wildly ramified covers); so this result does not give explicit information about Galois groups and fundamental groups for K-curves (as a full rigid analog of Corollary 2.1.2 would). We return to this issue in Section 4.3, after considering another approach to rigid analytic spaces in Section 4.2.

4.2. Rigid geometry via formal geometry.

Tate's rigid analytic spaces can be reinterpreted in terms of Grothendieck's formal schemes. This reinterpretation was outlined by Raynaud in [Ra1], and worked out in greater detail by Bosch, Lütkebohmert, and Raynaud in [Lü1], [BLü1], [BLü2], [BLüR1], [BLüR2]. (See also [Ra2, § 3]; and Chapters 1 and 2, by M. Garuti [Ga] and Y. Henrio [He], in [BLoR].) As Tate said in [Ta], his approach was motivated by a suggestion of Grothendieck; and according to the introduction to [BLü1], Grothendieck's goal was to associate a generic fibre to a formal scheme of finite type. So this approach may actually be closer to Grothendieck's original intent than the more analytic framework discussed above.

The basic idea of this approach can be seen by revisiting examples from Sections 3.2 and 4.1. In Example 3.2.3, it was seen that $k[x][[t]]$ is the ring of formal functions along the affine x-line in the x, t-plane over a field k, or equivalently that its spectrum is a formal thickening of the affine x-line. The corresponding ring for the affine x^{-1}-line (i.e. the formal thickening of the projective x-line minus $x = 0$) is $k[x^{-1}][\![t]\!]$, and the ring corresponding to the overlap (i.e. the formal thickening of $\mathbb{P}^1 - \{0, \infty\}$) is $k[x, x^{-1}][\![t]\!]$. On the other hand, as seen in Section 4.1, if t is inverted in each of these three rings, one obtains the rings of functions on three affinoids over $K = k((t))$: the disc $|x| \leq 1$; the disc $|x^{-1}| \leq 1$ (i.e. $|x| \geq 1$ together with the point at infinity); and the "annulus" $|x| = 1$. In each of these two contexts (formal and rigid), the first two sets cover the projective line (over $R := k[\![t]\!]$ and $K = \operatorname{frac} R$, respectively), and the third set is their "overlap". The ring of holomorphic functions on an affinoid set over K can (at least in this example) be viewed as the localization, with respect to t, of the ring of formal functions on an affine open subset of the closed fibre on an R-scheme. Correspondingly, an affinoid can be viewed as the generic fibre of the spectrum of the ring of formal functions (in the above example, a curve being the general fibre of a surface). Intuitively, then, a rigid analytic space over K is the general fibre of a (formal) scheme over R. (See Figure 4.2.1.)

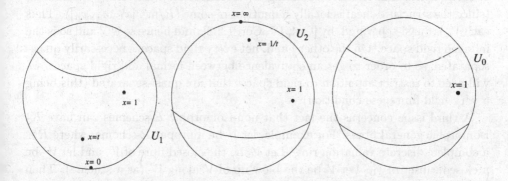

Figure 4.2.1. A rigid covering of $\mathbb{P}^1_K$ (viewed as a sphere, in analogy with the complex case). The patches U_1, U_2 are discs around 0 and ∞, with rings of functions $k[x][\![t]\!][1/t]$ and $k[1/x][\![t]\!][1/t]$ (see §4.1). The overlap U_0 is an annulus containing the point $x = 1$, with ring of functions $k[x, 1/x][\![t]\!][1/t]$. Compare Fig. 3.1.4 and see Example 4.2.3 below.

The actual correspondence between formal schemes and rigid analytic spaces is a bit more complicated, because of several issues. The first concerns which base rings and fields are involved. Formal schemes are defined over complete local rings R, while rigid analytic spaces are defined over complete valuation fields K. The fraction field of a complete discrete valuation ring R is a discrete valuation field K, and every such K arises from such an R. But general valuation fields are not fraction fields of complete local rings, and the fraction fields of general complete local rings are not valuation fields. So in stating the correspondence, we restrict here to the case of a complete discrete valuation ring R, say with maximal ideal $\mathfrak{m}$ (though one can consider, somewhat more generally, a complete height 1 valuation ring R).

Secondly, in order for a formal space to induce a rigid space, it must locally induce affinoid K-algebras, i.e. K-algebras that are of topologically finite type. Correspondingly, we say that an R-algebra A is of *topologically finite type* if it is a quotient of the $\mathfrak{m}$-adic completion of some $R[x_1, \ldots, x_n]$. Observe that this $\mathfrak{m}$-adic completion is a subring of $R[\![x_1, \ldots, x_n]\!]$, and in fact consists precisely of those power series $\sum_{\underline{i} \in \mathbb{N}^n} a_{\underline{i}} x^{\underline{i}}$, where $a_{\underline{i}} \to 0$ as $\underline{i} \to \infty$. It is then easy to verify that $A \otimes_R K$ is an affinoid K-algebra, for any R-algebra A that is of topologically finite type. (This is in contrast to the full rings of power series, where $K[\![x_1, \ldots, x_n]\!]$ is much larger than $R[\![x_1, \ldots, x_n]\!] \otimes_R K$.) A formal R-scheme $\mathcal{V}$ is *locally of topologically finite type* if in a neighborhood of every point, the structure sheaf $\mathcal{O}_{\mathcal{V}}$ is given by an R-algebra that is of topologically finite type. Such a formal scheme is said it be of *topologically finite type* if in addition it is quasi-compact. Thus formal schemes that are of topologically finite type induce quasi-compact rigid spaces.

The condition of a formal R-scheme $\mathcal{V}$ being locally of topologically finite type in turn implies that the corresponding $R/\mathfrak{m}^n$-schemes V_n are locally Noetherian (since the structure sheaf is locally a quotient of some $(R/\mathfrak{m}^n)[x_1, \ldots, x_n]$). Thus each V_n is quasi-separated, by [Gr4, IV, Cor. 1.2.8]; and hence so is $\mathcal{V}$ and so is the induced rigid space. On the other hand, not every rigid space is necessarily quasi-separated; so in order to get an equivalence between formal and rigid spaces, we will need to restrict attention to rigid spaces that are quasi-separated (this being a very mild finiteness condition).

A third issue concerns the fact that non-isomorphic R-schemes can have K-isomorphic general fibres. For example, let V be a proper R-scheme, where R is a complete discrete valuation ring. Let V_0 be the closed fibre of V, and let W be a closed subset of V_0. Let $\tilde{V}$ be the blow up of V along W (as a scheme). Then V and $\tilde{V}$ have the same general fibre. But they are not isomorphic as R-schemes (if the codimension of W in V is at least 2), since $\tilde{V}$ has an exceptional divisor over the blown up points. Hence they do not correspond to isomorphic formal schemes.

In order to deal with this third issue, the strategy is to regard two R-schemes as equivalent if they have a common *admissible* blow-up (i.e. a blow up at a closed subset of the closed fibre). Thus given two R-schemes V, V', to give a morphism from the equivalence class of V to that of V' is to give an admissible blow up $\tilde{V} \to V$ together with a morphism of R-schemes $\tilde{V} \to V'$. Here $V, \tilde{V}, V'$ induce formal R-schemes $\mathcal{V}, \tilde{\mathcal{V}}, \mathcal{V}'$ (given by the direct limit of the fibres $V_n, \tilde{V}_n, V_n'$ over $\mathfrak{m}^n$), and we regard the induced pair $(\tilde{\mathcal{V}} \to \mathcal{V}, \tilde{\mathcal{V}} \to \mathcal{V}')$ as a morphism between the equivalence classes of $\mathcal{V}, \mathcal{V}'$. Equivalently, we are considering morphisms from the class of $\mathcal{V}$ to the class of $\mathcal{V}'$, in the localization of the category of formal R-schemes with respect to the class of admissible formal blow-ups $\tilde{\mathcal{V}} \to \mathcal{V}$. (The *localization* is the category in which those blow-ups are formally inverted. Such a localization automatically exists, according to [Hrt1]; though to be set-theoretically precise, one may wish to work within a larger "universe" [We, Remark 10.3.3].)

Here, for a formal scheme $\mathcal{V}$ induced by a proper R-scheme V, one can correspondingly define admissible blow-ups of $\mathcal{V}$ as the morphisms of formal schemes induced by admissible blow-ups of V. Alternatively, and for a more general formal R-scheme $\mathcal{V}$, admissible blow-ups can be defined directly, despite the fact that the topological space underlying $\mathcal{V}$ is just the closed fibre of the associated R-scheme (if there is one). Namely, the blow-up can be defined algebraically, analogously to the usual definition for schemes. First, observe that if A is a complete R-algebra, then the closed subsets of the closed fibre of Spec A correspond to ideals of A that are *open* in the topology induced by that of R. Now recall [Hrt2, Chap. II, p. 163] that if V is a Noetherian scheme, and $\mathcal{I}$ is a coherent sheaf of ideals on V, then the blow-up of V at $\mathcal{I}$ is Proj $\mathcal{J}$, where $\mathcal{J}$ is the sheaf of graded algebras $\mathcal{J} = \bigoplus_{d \geq 0} \mathcal{I}^d$. So given a formal R-scheme $\mathcal{V}$ and a sheaf $\mathcal{I}$ of open ideals of $\mathcal{O}_{\mathcal{V}}$, define the blow-up of $\mathcal{V}$ along $\mathcal{I}$ to be the formal scheme associated to the direct system of $R/\mathfrak{m}^n$-schemes Proj $\mathcal{J}_n$, where $\mathcal{J}_n = \bigoplus_{d \geq 0} (\mathcal{I}^d \otimes_{\mathcal{O}_{\mathcal{V}}} \mathcal{O}_{\mathcal{V}}/\mathfrak{m}^n)$. We call such a blow-up of the formal scheme $\mathcal{V}$ *admissible*. This agrees with the previous definition, for formal schemes $\mathcal{V}$ induced by R-schemes V.

A fourth issue, which is similar to the third, is that an R-scheme V may have an irreducible component that is contained in the closed fibre V_0. In that case, the general fibre of V "does not see" that component, and so cannot determine V (or the induced formal scheme). So we avoid this case, by requiring that the formal scheme $\mathcal{V}$ have the property that its structure sheaf $\mathcal{O}_{\mathcal{V}}$ has no $\mathfrak{m}$-torsion. We call the formal scheme $\mathcal{V}$ *admissible* if it has this property and is of locally of topologically finite type. (So quasi-compact admissible is the same as $\mathfrak{m}$-torsion-free plus topologically finite type.)

With these restrictions and adjustments, the equivalence between formal and rigid spaces takes place. Consider an admissible formal R-scheme $\mathcal{V}$, whose underlying topological space is a k-scheme V_0 (where $k = R/\mathfrak{m}$). For any affine open subset $U \subset V_0$, let A be the ring of formal functions along U. So A is topologically of finite type, and has no $\mathfrak{m}$-torsion; and $A \otimes_R K$ is an affinoid K-algebra. In the notation of Section 4.1, Sp $A = (\mathrm{Max}\, A, A)$ is an affinoid variety. This construction is compatible with shrinking U, and so from $\mathcal{V}$ we obtain a rigid analytic space, which we denote by $\mathcal{V}^{\mathrm{rig}}$. There is then the following key theorem of Raynaud [Ra1] (see also [BLü1, Theorem 4.1], for details):

THEOREM 4.2.2 (RAYNAUD). *Let R be a complete valuation ring of height 1 with fraction field K. Let* For_R *be the category of quasi-compact admissible formal R-schemes, and let* For'_R *be the localization of* For_R *with respect to admissible formal blow-ups. Let* Rig_K *be the category of quasi-compact quasi-separated rigid analytic K-spaces. Then the functor* rig : $\mathrm{For}_R \to \mathrm{Rig}_K$ *given by $\mathcal{V} \mapsto \mathcal{V}^{\mathrm{rig}}$ induces an equivalence of categories* $\mathrm{For}'_R \to \mathrm{Rig}_K$.

(Alternatively, the conclusion of the theorem could be stated by saying that rig : $\text{For}_R \to \text{Rig}_K$ is a localizing functor with respect to all admissible blow-ups, rather than speaking in terms of For'_R.)

In particular, if V is a proper R-scheme, and if $\mathcal{V}$ is the associated formal scheme, then $\mathcal{V}^{\text{rig}}$ is the rigid analytic space corresponding to the generic fibre V_K of V.

More generally, one can turn the above result around and make it a *definition*, to make sense of rigid analytic spaces over the fraction field K of a Noetherian complete local ring R which is not necessarily a valuation ring (e.g. $k[[x_1, \ldots, x_n]]$, where k is a field and $n > 1$). That is, for such a ring R and fraction field K, one can simply *define* the category Rig_K of rigid analytic K-spaces to be the category For'_R, obtained by localizing the category For_R of formal R-schemes with respect to admissible blow-ups [Ra1], [BLü1], [Ga]. The point is that formal schemes make sense in this context, and thus the notion of rigid spaces can be extended to this situation as well. (Of course, by Raynaud's theorem, the two definitions are equivalent in the case that R is a complete discrete valuation ring.)

The advantage to Raynaud's approach to rigid analytic spaces is it permits them to be studied using Grothendieck's results on formal schemes in EGA [Gr4]. It also permits the use of results in EGA on proper schemes over complete local rings, because of the equivalence of those schemes with formal schemes via by Grothendieck's Existence Theorem ([Gr2], [Gr4, III, Cor. 5.1.6]; see also Section 3.2 above). In particular, Grothendieck's Existence Theorem and Raynaud's theorem above together imply the rigid GAGA result (for projective spaces) discussed in Section 4.1 above. Moreover, Raynaud's approach permits the use of the rigid point of view over more general fields than Tate's original approach did, though with some loss of analytic flavor. Indeed, from this point of view, the rigid and formal contexts are not so different, though there is a difference in terms of intuition. Another difference is that in the formal context one works on a fixed R-model of a space, whereas in the rigid context one works just over K (and thus blow-ups are already included in the geometry).

We conclude this discussion by giving two examples comparing formal and rigid GAGA on the line, beginning with the motivating situation discussed earlier:

EXAMPLE 4.2.3. Let k be a field; $R = k[[t]]$; $K = k((t))$; and $V = \mathbb{P}^1_R$. Let x be a parameter on V, and $y = x^{-1}$. So V is covered by two copies of the affine line over R, the x-line and the y-line, intersecting where $x, y \neq 0$. Letting $\mathcal{V}$ be the formal scheme associated to V, there is the induced rigid analytic space $V^{\text{rig}} := \mathcal{V}^{\text{rig}}$, viz. $\mathbb{P}^1_K$. According to rigid GAGA, giving a coherent sheaf on V^{rig} is equivalent to giving finite modules over (the rings of functions on) the admissible sets U_1 : $|x| \leq 1$ and $U_2 : |y| \leq 1$, with agreement on the overlap $U_0 : |x| = |y| = 1$. Here $U_1 = \text{Sp}\, K\{x\}$, $U_2 = \text{Sp}\, K\{y\}$, and $U_0 = \text{Sp}\, K\{x,y\}/(xy - 1)$. Geometrically (and intuitively), U_1 and U_2 are discs centered around $x = 0, \infty$ respectively (the "south and north poles"), and U_0 is an annulus (a band around the "equator", if $\mathbb{P}^1_K$ is viewed as a "sphere"; see Figure 4.2.1 above).

On the formal level, U_1 is the general fibre of $S_1 = \operatorname{Spec} k[x][\![t]\!]$, the formal thickening of the affine x-line (which pinches down near $x = \infty$). Similarly, U_2 is the general fibre of $S_2 = \operatorname{Spec} k[y][\![t]\!]$, the formal thickening of the affine y-line (which pinches down near $x = 0$). And U_0 is the general fibre of $S_0 = \operatorname{Spec} k[x, x^{-1}][\![t]\!]$, the formal thickening of the line with both 0 and ∞ deleted (and which pinches down near both points — cf. Figure 3.1.4). According to formal GAGA (i.e. Grothendieck's Existence Theorem; cf. Theorems 3.2.1 and 3.2.4), giving a coherent sheaf on V is equivalent to giving finite modules over S_1 and S_2 with agreement on the "overlap" S_0.

In the formal context, even less data is needed in order to construct a coherent sheaf on V — and this permits more general constructions to be performed (e.g. see [Ha6]). Namely, let $\hat{S}_1 = \operatorname{Spec} k[\![x, t]\!]$, the complete local neighborhood of $x = t = 0$. Let $\hat{S}_0 = \operatorname{Spec} k((x))[\![t]\!]$, the "overlap" of $\hat{S}_1$ with S_2. (See Figure 3.2.9, where $\hat{S}_1, S_2, \hat{S}_0$ are denoted by W^*, U^*, W'^*, respectively.) Then according to Theorem 3.2.8, giving a coherent sheaf on V is equivalent to giving finite modules over $\hat{S}_1$ and S_2 together with agreement over $\hat{S}_0$. On the rigid level, the generic fibres of $\hat{S}_1$ and S_2 are $\hat{U}_1 : |x| < 1$ and $U_2 : |x| \geq 1$. Those subsets of V^{rig} do not intersect, and moreover $\hat{U}_1$ is not an affinoid set. The result in the formal situation suggests that the generic fibre of $\hat{S}_0$, corresponding to $k((x))((t))$, forms a "glue" that connects $\hat{U}_1$ and U_2; but this cannot be formulated within the rigid framework. □

EXAMPLE 4.2.4. With notation as in Example 4.2.3, rigid GAGA says that to give a coherent sheaf on $V^{\mathrm{rig}} = \mathbb{P}^1_K$ is equivalent to giving finite modules over the two discs $|x| \leq 1$ and $|y| \leq c^{-1}$, and over the annulus $c \leq |x| \leq 1$; here $0 < c = |t| < 1$, and the annulus is the overlap of the two discs. Writing $z = ty = t/x$, the rings of functions on these three sets are $K\{x\}$, $K\{z\}$, and $K\{x, z\}/(xz - t)$.

To consider the corresponding formal situation, let $\tilde{V}$ be the blow-up of V at the closed point $x = t = 0$. Writing $xz = t$, the closed fibre of $\tilde{V}$ consists of the projective x-line over k (the proper transform of the closed fibre of V) and the projective z-line over k (the exceptional divisor), meeting at the "origin" $O : x = z = t = 0$. The three affinoid open sets above are then the generic fibres associated to the formal schemes obtained by respectively deleting from the closed fibre of $\tilde{V}$ the point $x = \infty$ (which is where $z = 0$); the point $z = \infty$ (where $x = 0$); and both of these points. And by Grothendieck's Existence Theorem, giving compatible formal coherent modules over each of these sets is equivalent to giving a coherent module over V.

But as in Example 4.2.3, less is needed in the formal context. Namely, let X' and Z' be the projective x- and z-lines over k, with the points $(x = 0)$ and $(z = 0)$ respectively deleted. Consider the rings of formal functions along X' and Z', viz. $k[x^{-1}][\![t]\!]$ and $k[z^{-1}][\![t]\!]$ respectively, and their spectra T_1, T_2. Consider also the spectrum T_3 of $k[\![x, z, t]\!]/(xz - t)$, the complete local ring of V at O.

Here T_1 and T_2 are disjoint, while the "overlap" of T_1 and T_3 [resp. of T_2 and T_3] is the spectrum $T_{1,3}$ of $k((x))[\![t]\!]$ [resp. $T_{2,3}$ of $k((z))[\![t]\!]$]. By Theorem 3.2.8, giving finite modules over T_1, T_2, T_3 that agree on the two "overlaps" is equivalent to giving a coherent module over V. The generic fibres of T_1 and T_2 are the sets $|y| \leq 1$ and $|x| \leq c$, and that of T_3 is $c < |x| < 1$. These three sets are disjoint, though the formal set-up provides "glue" (in the form of $T_{1,3}$ and $T_{2,3}$) connecting T_3 to T_1 and to T_2. This is a special case of Example 3.2.11. (Alternatively, one could use Theorem 3.2.12, taking V to be the projective x-line over $k[\![t]\!]$, taking $\tilde{V}$ to be the blow-up of V at $x = t = 0$, and identifying the exceptional divisor with the projective z-line over k. See Example 3.2.13.) $\qquad\square$

More generally, in the rigid set-up, one can consider the annulus $c^n \leq |x| \leq 1$ in $\mathbb{P}^1_K$ (where $K = k((t))$ and $c = |t|$ as above, and where n is a positive integer). If one writes $u = t^n/x$, then this is the intersection of the two admissible sets $|x| \leq 1$ and $|u| \leq 1$. This annulus is said to have *thickness* (or *épaisseur*) equal to n. The corresponding situation in the formal framework can be arrived at by taking the projective x-line V over $R = k[\![t]\!]$; blowing this up at the point $x = t = 0$ (obtaining a parameter $z = t/x$ on the exceptional divisor E); blowing that up at the point $z = \infty$ on E (thereby obtaining a parameter $z' = t/z^{-1} = t^2/x$ on the new exceptional divisor); and repeating the process for a total of n blow-ups. The analogs of Examples 4.2.3 and 4.2.4 above can then be considered similarly.

4.3. Rigid patching and constructing covers.

Rigid geometry, like formal geometry, provides a framework within which patching constructions can be carried out in order to construct covers of curves, and thereby obtain Galois groups over curves. Ideally, one would like to obtain a version of Riemann's Existence Theorem analogous to that stated for complex curves in Section 2.1. But while a kind of "Riemann's Existence Theorem" for rigid spaces was obtained by Lütkebohmert [Lü2] (see Section 4.1 above), that result does not say which Galois groups arise, due to a lack of topological information. Still, as in the formal case, one can show by a patching construction that every finite group is a Galois group of a branched cover with enough branch points, and show a "Half Riemann Existence Theorem" that is analogous to the classification of slit covers of complex curves (see Section 2.3).

Namely, Serre observed in a 1990 talk in Bordeaux that there should be a rigid proof of Theorem 3.3.1 above (on the realizability of every finite group as a Galois group over the fraction field K of a complete local domain R [Ha4]), when the base ring R is complete with respect to a non-archimedean absolute value. Given the connection between rigid and formal schemes discussed in Section 4.2 (especially in the case of complete discrete valuation rings), this would seem quite plausible. Shortly afterwards, in [Se7, §8.4.4], Serre outlined such a proof in the case that $K = \mathbb{Q}_p$. A more detailed argument was carried out by Liu for complete non-archimedean fields with an absolute value, in a manuscript that was written in 1992 and that appeared later in [Li], after circulating privately for a few years. (Concerning complete *archimedean* fields, the complex case was discussed in Section 2 above, and the real case is handled in [Se7, § 8.4.3], via the complex case; cf. also [DF] for the real case.)

The rigid version of Theorem 3.3.1 is as follows:

THEOREM 4.3.1. *Let K be a field that is complete with respect to a non-trivial non-archimedean absolute value. Let G be a finite group. Then there is a G-Galois irreducible branched cover $Y \to \mathbb{P}^1_K$ such that the fibre over some K-point of $\mathbb{P}^1_K$ is totally split.*

Here the totally split condition is that the fibre consists of unramified K-points. This property (which takes the place of the mock cover hypothesis of Theorem 3.3.1) forces the cover to be regular, in the sense that K is algebraically closed in the function field $K(Y)$ of Y. (Namely, if L is the algebraic closure of K in $K(Y)$, then L is contained in the integral closure in $K(Y)$ of the local ring $\mathcal{O}_\xi$ of any closed point $\xi \in \mathbb{P}^1_K$; and so it is contained in the residue field of each closed point of Y.) Thus Theorem 3.3.1 is recaptured, for such fields K.

SKETCH OF PROOF OF THEOREM 4.3.1. The proof proceeds analogously to that of Theorem 3.3.1. Namely, first one proves the result explicitly in the special case that the group is a cyclic group. In [Se7, § 8.4.4], Serre does this by using an argument involving tori [Se7, § 4.2] to show that cyclic groups are Galois groups of branched covers of the line; one can then obtain a totally split fibre by twisting, e.g. as in [HV, Lemma 4.2(a)]. Or (as in [Li]) one can proceed as in the original proof for the cyclic case in the formal setting [Ha4, Lemma 2.1], which used ideas of Saltman [Slt]; cf. Proposition 3.3.3 above.

To prove the theorem in the general case, cyclic covers are patched together to produce a cover with the desired group, in a rigid analog of the proof of Theorem 3.3.1. Namely, let $g_1, \ldots, g_r$ be generators of G. For each i, let H_i be the cyclic subgroup of G generated by g_i, and let $f_i : Y_i \to \mathbb{P}^1_K$ be an H_i-Galois cover that is totally split over a point ξ_i. By the Implicit Function Theorem over complete fields, for each i there is a closed disc $\bar{D}_i$ about ξ_i such that the inverse image $f_i^{-1}(\bar{D}_i)$ is a disjoint union of copies of $\bar{D}_i$. Let D_i be the corresponding open disc about ξ_i, let $\bar{U}_i = \mathbb{P}^1_K - D_i$, and let $U_i = \mathbb{P}^1_K - \bar{D}_i$. After a change of variables, we may assume that the $\bar{U}_i$'s are pairwise disjoint affinoid sets. For each i, let $\bar{V}_i \to \bar{U}_i$ be the pullback of f_i to $\bar{U}_i$. Then $\bar{V}_i$ is an H_i-Galois cover whose restriction over $\bar{U}_i - U_i = \bar{D}_i - D_i$ is trivial. Inducing from H_i to G (by taking a disjoint union of copies, indexed by the cosets of H_i), we obtain a corresponding G-Galois disconnected cover $\bar{W}_i = \mathrm{Ind}_{H_i}^G \bar{V}_i \to \bar{U}_i$. Also, let $\bar{U}_0 = \mathbb{P}^1_K - \bigcup_{j=1}^r U_j = \bigcap_{j=1}^r \bar{D}_j$, and let $\bar{W}_0 \to \bar{U}_0$ be the trivial G-cover $\mathrm{Ind}_1^G \bar{U}_0$. We now apply rigid GAGA (see Sections 4.1 and 4.2), though for covers rather than for modules (that form following automatically, as in Theorem 3.2.4, via the General Principle 2.2.4). Namely, we patch together the covers $\bar{W}_i \to \bar{U}_i$ ($i = 0, \ldots, r$) along the overlaps $\bar{U}_i \cap \bar{U}_0 = \bar{U}_i - U_i$ ($i = 1, \ldots, r$), where they are trivial. One then checks that the resulting G-Galois cover is as desired (and in particular is irreducible, because the g_i's generate G); and this yields the theorem.   □

As in Section 3.3, Theorem 4.3.1 extends to the class of large fields, such as the algebraic p-adics and the field of totally real algebraic numbers. Namely, as in the passage to Theorem 3.3.6, there is the following result of Pop:

COROLLARY 4.3.2 [Po4]. *If k is a large field, then every finite group is the Galois group of a Galois field extension of $k(x)$. Moreover this extension may be chosen to be regular, and with a totally split fibre.*

Namely, by Theorem 4.3.1, there is a G-Galois extension of $k((t))(x)$, and this descends to a regular G-Galois extension of the fraction field of $A[x]$ with a totally split fibre over $x = 0$, for some $k(t)$-subalgebra $A \subset k((t))$ of finite type. By the Bertini–Noether Theorem [FJ, Prop. 9.29], we may assume that every specialization of A to a k-point gives a G-Galois regular field extension of $k(x)$; and such a specialization exists on $V := \operatorname{Spec} A$ since k is large and since V contains a $k((t))$-point.

The construction in the proof of Theorem 4.3.1, like the one used in proving the corresponding result using formal geometry, can be regarded as analogous to the slit cover construction of complex covers described in Section 2.3 (and see the discussion at the end of Section 3.3 for the analogy with the formal setting). In fact, rather than considering covers (and Galois groups) one at a time, a whole tower of covers (and Galois groups) can be considered, as in the "analytic half Riemann Existence Theorem" 2.3.5. In the present setting (unlike the situation over $\mathbb{C}$), the absolute Galois group G_K of the valued field K comes into play, since it acts on the geometric fundamental group (i.e. the fundamental group of the punctured line after base-change to the separable closure K^s of K). This construction of a tower of compatible covers has been carried out by Pop in [Po2] (where the term "half Riemann Existence Theorem" was also introduced). Also, rather than requiring K to be complete, Pop required K merely to be henselian (and cf. Example 3.3.2(d), for comments about deducing the henselian case of that result from the complete case via Artin's Approximation Theorem).

In Pop's result, as in the case of complex slit covers, one chooses as a branch locus a closed subset $S \subset \mathbb{P}^1_K$ whose base change to K^s consists of finitely many pairs of nearby points. That is, S is a disjoint union of two closed subsets $S = S_1 \cup S_2$ of $\mathbb{P}^1_K$ such that $S_1^s := S_1 \times_K K^s = \{\xi_1, \ldots, \xi_r\}$ and $S_2^s := S_2 \times_K K^s = \{\eta_1, \ldots, \eta_r\}$, where the ξ_i and η_j are distinct K^s-points, and where each ξ_i is closer to the corresponding η_i than it is to any other ξ_j. Such a set S is called *pairwise adjusted*. Note that the sets S_1^s and S_2^s are each G_K-invariant, and that G_K acts on the sets S_1^s and S_2^s compatibly (i.e. if $\alpha \in G_K$ satisfies $\alpha(\xi_i) = \xi_j$, then $\alpha(\eta_i) = \eta_j$). Now let $U = \mathbb{P}^1_K - S$ and $U^s = U \times_K K^s = \mathbb{P}^1_{K^s} - S^s$, and recall the fundamental exact sequence

$$1 \to \pi_1(U^s) \to \pi_1(U) \to G_K \to 1. \qquad (*)$$

In this situation, let Π be the free profinite group $\hat{F}_r$ of rank r if the valued field K is in the equal characteristic case; this is the free product of r copies of the group $\hat{\mathbb{Z}}$, in the category of profinite groups. If K is in the unequal characteristic case with residue characteristic $p > 0$, then let Π be the free product $\hat{F}_r[p]$ of r copies of the group $\hat{\mathbb{Z}}/\mathbb{Z}_p$, in the category of profinite groups. (Note that this free product is not a pro-prime-to-p group if $r > 1$, and in particular is much larger than the free pro-prime-to-p group of rank r.) Define an action of G_K on Π by letting $\alpha \in G_K$ take the j-th generator g_j of Π to $g_i^{\chi(\alpha^{-1})}$; here i is the unique index such that $\alpha(\xi_i) = \xi_j$, and $\chi : G_K \to \hat{\mathbb{Z}}^*$ is the cyclotomic character (taking $\gamma \mapsto m$ if $\gamma(\zeta) = \zeta^m$ for all roots of unity ζ). There is then the following result of Pop (and see Remark 4.3.4(c) below for an even stronger version):

THEOREM 4.3.3 (HALF RIEMANN EXISTENCE THEOREM WITH GALOIS ACTION [Po2]). *Let K be a henselian valued field of rank 1, let $S \subset \mathbb{P}^1_K$ be a pairwise adjusted subset of degree $2r$ as above, and let $U = \mathbb{P}^1_K - S$. Then the fundamental exact sequence $(*)$ has a quotient*

$$1 \to \Pi \to \Pi \rtimes G_K \to G_K \to 1, \qquad\qquad (**)$$

where Π is defined as above and where the semi-direct product is taken with respect to the above action of G_K on Π.

SKETCH OF PROOF OF THEOREM 4.3.3. In the case that the field K is complete, the proof of Theorem 4.3.3 follows a strategy that is similar to that of Theorem 4.3.1. As in Theorem 4.3.1 (and Theorem 3.3.1), the proof relies on the construction of cyclic covers that are trivial outside a small neighborhood (in an appropriate sense), and which can then be patched. The key new ingredient is that one must show that the construction is functorial, and in particular is compatible with forming towers. Concerning this last point, after passing to the maximal cyclotomic extension K^{cycl} of K, one can construct a tower of regular covers by patching together local cyclic covers that are Kummer or Artin–Schreier. These can be constructed compatibly with respect to the action of $\mathrm{Gal}(K^{\mathrm{cycl}}/K)$, since S is pairwise adjusted; and the resulting tower, viewed as a tower of covers of U, has the desired properties.

The henselian case is then deduced from the complete case. This is done by first observing that the absolute Galois groups of K and of its completion $\hat{K}$ are canonically isomorphic (because K is henselian). Then, writing $\hat{K}^{\mathrm{s}}$ for the separable closure of $\hat{K}$, it is checked that every finite branched cover of the $\hat{K}^{\mathrm{s}}$-line that results from the patching construction is defined over the separable closure K^{s} of K. (Namely, consider a finite quotient Q of Π, generated by cyclic subgroups C_i. The patching construction over $\hat{K}$ yields a Q-Galois cover $Y \to \mathbb{P}^1_{\hat{K}^{\mathrm{s}}}$ that is constructed using cyclic building blocks $Z_i \xrightarrow{C_i} \mathbb{P}^1$ which are each defined over K^{s}. Let $Z \to \mathbb{P}^1$ be the fibre product of the Z_i's; this is Galois with group $H = \prod C_i$. Pulling back the Q-cover $Y \to \mathbb{P}^1_{\hat{K}^{\mathrm{s}}}$ via $Z_{\hat{K}^{\mathrm{s}}} \xrightarrow{H} \mathbb{P}^1_{\hat{K}^{\mathrm{s}}}$ gives an unramified Q-cover Y' of the projective curve $Z_{\hat{K}^s}$; here Y' is also a $Q \times H$-Galois branched cover of $\mathbb{P}^1_{\hat{K}^s}$. By Grothendieck's specialization isomorphism [Gr5, XIII], Y' descends to a Q-cover of Z_{K^s} whose composition with $Z_{K^s} \to \mathbb{P}^1_{K^s}$ is $Q \times H$-Galois. Hence Y descends to a Q-cover of $\mathbb{P}^1_{K^s}$.) Since the Galois actions of G_K and $G_{\hat{K}}$ are the same, the result in the general henselian case follows. $\square$

REMARK 4.3.4. (a) The hypotheses of Theorem 4.3.3 are easily satisified; i.e. there are many choices of pairwise adjusted subsets. Namely, let $f \in K[x]$ be any irreducible separable monic polynomial, and let $g \in K[x]$ be chosen so that it is monic of the same degree, and so that its coefficients are sufficiently close to those of f. Then the zero locus of fg in $\mathbb{A}^1_K$ is a pairwise adjusted subset, by continuity of the roots [La, II, § 2, Proposition 4]. Repeating this construction with finitely many polynomials f_i and then taking the union of the resulting sets gives a general pairwise adjusted subset. Note that in the case that K is separably closed, the construction is particularly simple: One may take an arbitrary set $S_1 = \{\xi_1, \ldots, \xi_r\}$ of K-points in $\mathbb{A}^1_K$, and any set $S_2 = \{\eta_1, \ldots, \eta_r\}$ of K-points such that each η_i is sufficiently close to ξ_i. This recovers the slit cover construction of Section 2.3 in the case $K = \mathbb{C}$.

(b) In the equal characteristic case, if K contains all of the roots of unity (of order prime to p, if char $K = p \neq 0$), then Theorem 4.3.3 shows that the free profinite group $\hat{F}_r$ on r generators is a quotient of $\pi_1(U)$. (Namely, the cyclotomic character acts trivially in this case, and so the semi-direct product in $(**)$ is just a direct product.) Since arbitrarily large pairwise adjusted subsets S exist by Remark (a), this shows that $\hat{F}_r$ is a quotient of the absolute Galois group of $K(x)$ for each $r \in \mathbb{N}$. A similar result holds in the unequal characteristic case $(0, p)$ if K contains the prime-to-p roots of unity, namely that the free pro-prime-to-p group $\hat{F}'_r$ of rank $r \in \mathbb{N}$ is a quotient of $\pi_1(U)$ and of $G_{K(x)}$. But the full group $\hat{F}_r$ is *not* a quotient of $\pi_1(U)$ or $G_{K(x)}$ in the unequal characteristic case; cf. [Po2] and Remark (c) below.

(c) The result in [Po2] asserts even more. First of all, the i-th generator of Π generates an inertia group over ξ_i and over η_i, for each $i = 1, \ldots, r$. This is as in the case of analytic and formal slit covers discussed at the ends of Sections 2.3 and 3.3 above. Second, in the unequal characteristic case $(0, p)$, the assertion of Theorem 4.3.3 may be improved somewhat, by replacing the group $\Pi = \hat{F}_r[p]$ by the free product of r copies of the group $\hat{\mathbb{Z}}/p^e \mathbb{Z}_p$ (in the category of profinite groups), for a certain non-negative integer e. (Specifically, $e = \max(0, e')$, where e' is the largest integer such that $|\xi_i - \eta_i| < |p|^{e'+1/(p-1)}|\xi_i - \xi_j|$ for all $i \neq j$.) This group lies in between the group $\hat{F}_r$ and its quotient $\hat{F}_r[p]$; and in the case that K contains all the prime-to-p roots of unity, this group is then a quotient of $\pi_1(U)$ and $G_{K(x)}$ (like $\hat{F}_r[p]$ but unlike $\hat{F}_r$). See [Po2] for details.

(d) The construction of cyclic extensions given in Section 3.3 can be recovered from the above result, in the case that the extension is of degree n prime to the characteristic of K. Namely, given a cyclic group $C = \langle c \rangle$ of order n, consider a primitive element for $K' := K(\zeta_n)$ as an extension of K; this corresponds to a K'-point $\xi = \xi_1$ of $\mathbb{P}^1$, and $\mathrm{Gal}(K'/K)$ acts simply transitively on the G_K-orbit $\{\xi_1, \ldots, \xi_s\}$ of ξ. Take $\eta = \eta_1$ sufficiently close to ξ to satisfy continuity of the roots [La, II, § 2, Proposition 4] (and also to satisfy the inequality in Remark (c) above, in the mixed characteristic case $(0, p)$ if $p | n$); and let its orbit be $\{\eta_1, \ldots, \eta_s\}$. Let $U \subset \mathbb{P}^1_K$ be the complement of the ξ_i's and η_i's. Consider the surjection $\Pi \twoheadrightarrow C$ given by $g_j \mapsto c^{\chi(\alpha^{-1})} = c^{-\alpha(\zeta_n)}$ if $\alpha \in \mathrm{Gal}(K'/K)$ is the element taking ξ to ξ_j. Then in the quotient $C \rtimes G_K$ of $\Pi \rtimes G_K$ (and hence of $\pi_1(U)$), the action of G_K on C is trivial; i.e. the quotient is just $C \times G_K$. So it in turn has a quotient isomorphic to C; and this corresponds to the cyclic cover constructed in the proof of Proposition 3.3.3. (In the case that n is instead a power of $p = \mathrm{char}\, K$, one uses Witt vectors in the construction; and again one obtains cyclic covers of degree n, since the action of G_K via χ is automatically trivial on a p-group quotient of Π.)

(e) The main assertion in Theorem 4.3.1 above (and in Theorem 3.3.1), that every finite group G is a Galois group over $K(x)$, can be recaptured from the Half Riemann Existence Theorem. Namely, by choosing elements c_i that generate G, and applying Remark (d) separately to each c_i, one obtains a quotient of $\Pi \rtimes G_K$ of the form $G \rtimes G_K$, in which the semi-direct product is actually a direct product. So G is a quotient of $\pi_1(U)$. $\square$

Unfortunately, the above approach (like that of Section 3.3) does not provide an explicit description, in terms of generators and relations, of the *full* fundamental group (or at least the tame fundamental group) of an arbitrary affine K-curve U. Such a full "Riemann's Existence Theorem" would generalize the explicit classical result over $\mathbb{C}$ (Corollary 2.1.2), unlike Lütkebohmert's result [Lü] discussed at the end of Section 4.1 (which is inexplicit) and the above result (which gives only a big quotient of $\pi_1(U)$).

At the moment such a full, explicit result (or even a conjecture about its exact statement) seems far out of reach, even in key special cases. For example, if K is algebraically closed of characteristic p, the profinite groups $\pi_1(\mathbb{A}^1_K)$ and $\pi_1^{\mathrm{t}}(\mathbb{A}^1_K - \{0, 1, \infty\})$ are unknown. And if K is a p-adic field, the tower of all Galois branched covers of $\mathbb{P}^1_K$ remains mysterious, while little is understood about Galois branched covers of $\mathbb{P}^1_K$ with good reduction and their associated Galois groups. (Note that the covers constructed above and in Section 3.3 have models over $\mathbb{Z}_p$ in which the closed fibres are quite singular — as is clear from the mock cover construction of Section 3.3.) For $p > 3$, a wildly ramified cover of $\mathbb{P}^1_{\mathbb{Q}_p}$ cannot have good reduction over $\mathbb{Q}_p$ (or even over the maximal unramified extension of $\mathbb{Q}_p$) [Co, p. 247, Remark 3]; and so $\mathbb{Z}/p$ cannot be such a Galois group. But it is unknown whether every finite group G is the Galois group of a cover of $\mathbb{P}^1_K$ with good reduction over K, for some *totally ramified* extension

K of $\mathbb{Q}_p$ (depending on G); if so, this would imply that every finite group is a Galois group over the field $\mathbb{F}_p(x)$ (cf. Proposition 3.3.9).

See Section 5 for a further discussion of results in the direction of a generalized Riemann's Existence Theorem.

In the rigid patching constructions above, and in the analogous formal patching constructions in Section 3.3, the full generality of rigid analytic spaces and formal schemes is not needed in order to obtain the results in Galois theory. Namely, the rigid analytic spaces and formal schemes that arise in these proofs are induced from algebraic varieties; and so less machinery is needed in order to prove the results of these sections than might first appear. Haran and Völklein (and later Jarden) have developed an approach to patching that goes further, and which seeks to omit all unnecessary geometric objects. Namely, in [HV], the authors created a context of "algebraic patching" in which everything is phrased in terms of rings and fields (viz. the rings of functions on formal or rigid patches, and their fractions fields), and in which the geometric and analytic viewpoints are suppressed. That set-up was then used to reprove Corollary 4.3.2 above on realizing Galois groups regularly over large fields [HV, Theorem 4.4], as well as to prove additional related Galois results (in [HV], [HJ1], and [HJ2]). For covers of curves, it appears that the formal patching, rigid patching, and algebraic patching methods are essentially interchangeable, in terms of what they are capable of showing. The main differences concern the intuition and the precise machinery involved; and these are basically matters of individual mathematical taste. In other applications, it may turn out that one or another of these methods is better suited.

5. Toward Riemann's Existence Theorem

Sections 3 and 4 showed how formal and rigid patching methods can be used to establish analogs of GAGA, and to realize all finite groups as Galois groups of covers, in rather general settings. This section pursues these ideas further, in the direction of a sought-after "Riemann's Existence Theorem" that would classify covers in terms of group-theoretic data, corresponding to the Galois group, the inertia groups, and how the covers fit together in a tower. Central to this section is the notion of "embedding problems", which will be used in studying this tower. In particular, Section 5.1 uses embedding problems to give the structure of the absolute Galois group of the function field of a curve over an algebraically closed field (which can be regarded as the geometric case of a conjecture of Shafarevich). Section 5.2 relates patching and embedding problems to arithmetic lifting problems, in which one considers the existence of a cover with a given Galois group and a given fibre (over a non-algebraically closed base field). In doing so, it relies on results from Section 5.1. Section 5.3 considers Abhyankar's Conjecture on fundamental groups in characteristic p, along with strengthenings and generalizations that relate to embedding problems and patching. These results move further in the direction of a full "Riemann's Existence Theorem", although the full classification of covers in terms of groups remains unknown.

5.1. Embedding problems and the geometric case of Shafarevich's Conjecture.

The motivation for introducing patching methods into Galois theory was to prove results about Galois groups and fundamental groups for varieties that are not necessarily defined over $\mathbb{C}$. Complex patching methods, combined with topology, permitted a quite explicit description of the tower of covers of a given complex curve U (Riemann's Existence Theorem 2.1.1 and 2.1.2). In particular, this approach showed what the fundamental group of U is, and thus which finite groups are Galois groups of unramified covers of U. Analogous formal and rigid patching methods were applied (in Sections 3 and 4) to the study of curves over certain other coefficient fields, in particular *large* fields. Without restriction on the branch locus, it was shown that every finite group is a Galois group over the function field of the curve (Sections 3.3 and 4.3), and Pop's "Half Riemann's Existence Theorem" gave an explicit description of a big part of the tower of covers for certain special choices of branch locus (Section 4.3). Further results about Galois groups over an arbitrary affine curve have also been obtained (see Section 5.3 below), but an explicit description of the full tower of covers, and of the full fundamental group, remain out of reach for now.

Nevertheless, if one does not restrict the branch locus, then patching methods can be used to find the birational analog of the fundamental group, in the case of curves over an algebraically closed field k — i.e. to find the absolute Galois group of the function field of a k-curve X. And here, unlike the situation with the fundamental group of an affine k-curve, the absolute Galois group turns out to be free even in characteristic $p > 0$.

In the case $k = \mathbb{C}$ and $X = \mathbb{P}^1_{\mathbb{C}}$, this result was proved in [Do] using the classical form of Riemann's Existence Theorem (see Corollary 2.1.5). For more general fields, it was proved independently by the author [Ha10] and by F. Pop [Po1], [Po3]:

THEOREM 5.1.1. *Let X be an irreducible curve over an algebraically closed field k of arbitrary characteristic. Then the absolute Galois group of the function field of X is the free profinite group of rank equal to the cardinality of k.*

In particular, the absolute Galois group of $k(x)$ is free profinite of rank equal to card k.

Theorem 5.1.1 implies the geometric case of Shafarevich's Conjecture. In the form originally posed by Shafarevich, the conjecture says that the absolute Galois group $\mathrm{Gal}(\bar{\mathbb{Q}}/\mathbb{Q}^{\mathrm{ab}})$ of $\mathbb{Q}^{\mathrm{ab}}$ is free profinite of countable rank. Here $\mathbb{Q}^{\mathrm{ab}}$ denotes the maximal abelian extension of $\mathbb{Q}$, or equivalently (by the Kronecker–Weber theorem) the maximal cyclotomic extension of $\mathbb{Q}$ (i.e. $\mathbb{Q}$ with all the roots of unity adjoined). The conjecture was later generalized to say that if K is any global field and K^{cycl} is its maximal cyclotomic extension, then the absolute Galois group of K^{cycl} is free profinite of countable rank. (See Remark 3.3.8(b).) The arithmetic case of this conjecture (the case where K is a number field) is *still open*, but the geometric case (the case where K is the function field of a curve X over a finite field F) follows from Theorem 5.1.1, by considering passage to the algebraic closure $\bar{F}$ of F. Namely, in this situation, $\bar{F} = \bar{\mathbb{F}}_p$ where $p = \text{char } F$, and so the function field $\bar{K}$ of $\bar{X} := X \times_F \bar{F}$ is equal to K^{cycl}; and in this case Theorem 5.1.1, applied to the $\bar{K}$-curve $\bar{X}$, asserts the conclusion of Shafarevich's Conjecture.

Theorem 5.1.1 above is proved using the notion of embedding problems. Recall that an *embedding problem* $\mathcal{E}$ for a profinite group Π is a pair of surjective group homomorphisms $(\alpha : \Pi \to G, f : \Gamma \to G)$. A *weak* [resp. *proper*] *solution* to $\mathcal{E}$ consists of a group homomorphism [resp. epimorphism] $\beta : \Pi \to \Gamma$ such that $f\beta = \alpha$:

$$
\begin{array}{ccccccc}
& & & & \Pi & & \\
& & & {\beta? \swarrow} & \downarrow \alpha & & \\
1 & \longrightarrow & N & \longrightarrow & \Gamma & \overset{f}{\longrightarrow} G \longrightarrow 1
\end{array}
$$

An embedding problem $\mathcal{E}$ is *finite* if Γ is finite; it is *split* if f has a section; it is *non-trivial* if $N = \ker f$ is non-trivial; it is a *p-embedding problem* if $\ker f$ is a p-group. A profinite group Π is *projective* if every finite embedding problem for Π has a weak solution.

In terms of Galois theory, if Π is the absolute Galois group of a field K, then giving a G-Galois field extension L of K is equivalent to giving a surjective homomorphism $\alpha : \Pi \to G$. For such an L, giving a proper solution to $\mathcal{E}$ as above is equivalent to giving a Γ-Galois field extension F of K together with an embedding of L into F as a G-Galois K-algebra. (Here the G-action on L agrees with the one induced by restricting the action of Γ to the image of the embedding.) Giving a weak solution to $\mathcal{E}$ is the same, except that F need only be a separable K-algebra, not a field extension (and so it can be a direct product of finitely many fields). In this field-theoretic context we refer to an *embedding problem for K*.

If K is the function field of a geometrically irreducible k-scheme X, then the field extensions L and F correspond to branched covers $Y \to X$ and $Z \to X$ which are G-Galois and Γ-Galois respectively, such that Z dominates Y. Here Y is irreducible; and Z is also irreducible in the case of a proper solution. If the algebraic closure of k in the function fields of Y and Z are equal (i.e. if there is no extension of constants from Y to Z), we say that the solution is *regular*.

By considering embedding problems for a field K, or over a scheme X, one can study not only which finite groups are Galois groups over K or X, but how the extensions or covers fit together in the tower of all finite Galois groups. As a result, one can obtain information about absolute Galois groups and fundamental groups. In particular, in the key special case that X is the projective line and k is countable (e.g. if $k = \bar{\mathbb{F}}_p$), Theorem 5.1.1 follows from the following three results about embedding problems:

THEOREM 5.1.2 (IWASAWA [Iw, p. 567], [FJ, Cor.24.2]). *Let* Π *be a profinite group of countably infinite rank. Then* Π *is a free profinite group if and only if every finite embedding problem for* Π *has a proper solution.*

THEOREM 5.1.3 (SERRE [Se6, Prop. 1]). *If* U *is an affine curve over an algebraically closed field* k, *then the profinite group* $\pi_1(U)$ *is projective.*

THEOREM 5.1.4 (HARBATER [Ha10], Pop [Po1], [Po3]). *If* k *is an algebraically closed field, and* K *is the function field of an irreducible* k-*curve* X, *then every finite split embedding problem for* K *has a proper solution.*

Concerning these three results which will be used in proving Theorem 5.1.1: Theorem 5.1.2 is entirely group-theoretic (and *rank* refers to the minimal cardinality of any generating set). The proof of Theorem 5.1.3 is cohomological, and in fact the assertion in [Se6] is stated in terms of cohomological dimension (that $\mathrm{cd}(\pi_1(X)) \leq 1$, which implies projectivity by [Se4, I, 5.9, Proposition 45]). Theorem 5.1.4 is a strengthening of Theorem 3.3.1, and like that result it is proved using patching. (Theorem 5.1.4 will be discussed in more detail below.)

Using these results, Theorem 5.1.1 can easily be shown in the case that the algebraically closed field k is countable. Namely, let Π be the absolute Galois group of $k(x)$. Then the profinite group Π has at most countable rank, since the countable field $k(x)$ has only countably many finite field extensions; and Π has infinite rank, since every finite group is a quotient of Π (as seen in Section 3.3). So Theorem 5.1.2 applies, and it suffices to show that every finite embedding problem $\mathcal{E}$ for Π is properly solvable. Say $\mathcal{E}$ is given by $(\alpha : \Pi \to G, f : \Gamma \to G)$, with f corresponding to a G-Galois branched cover $Y \to X$. This cover is étale over an affine dense open subset $U \subset X$, and α factors through $\pi_1(U)$ (since quotients of π_1 classify unramified covers). Writing this map as $\alpha_U : \pi_1(U) \to G$, consider the finite embedding problem $\mathcal{E}_U = (\alpha_U : \pi_1(U) \to G, f : \Gamma \to G)$. By Theorem 5.1.3, this has a weak solution $\beta_U : \pi_1(U) \to \Gamma$, say with image $H \subset \Gamma$ (which surjects onto G under f). Let N be the kernel of f, and Γ_1 be the semidirect product $N \rtimes H$ with respect to the conjugation action of H on N. The multiplication map $(n, h) \mapsto nh \in \Gamma$ is an epimorphism $m : \Gamma_1 \to \Gamma$, and the projection map $h : \Gamma_1 \to H$ is surjective with kernel N. The surjection $\beta_U : \pi_1(U) \to H$ corresponds to an H-Galois branched cover $Y_1 \to X$ (unramified over U). This in turn corresponds to a surjective group homomorphism $\beta : \Pi \to H$. By Theorem 5.1.4, the split embedding problem $(\beta : \Pi \to H, h : \Gamma_1 \to H)$ has a proper solution. That solution corresponds to an irreducible Γ_1-Galois cover $Z_1 \to X$ that dominates Y_1; and composing the corresponding surjection $\Pi \to \Gamma_1$ with $m : \Gamma_1 \to \Gamma$ provides a proper solution to the original embedding problem $\mathcal{E}$.

REMARK 5.1.5. The above argument actually requires less than Theorem 5.1.3; viz. it suffices to use Tsen's Theorem [Ri, Proposition V.5.2] that if k is algebraically closed then the absolute Galois group of $k(x)$ has cohomological dimension 1. For then, by writing X in Theorem 5.1.1 as a branched cover of $\mathbb{P}^1_k$, it follows that the absolute Galois group of its function field is also of cohomological dimension 1 [Se4, I, 3.3, Proposition 14], and hence is projective [Se4, I, 5.9, Proposition 45]. One can then proceed as before.

But by using Theorem 5.1.3 as in the argument above, one obtains additional information about the branch locus of the solution to the embedding problem. Namely, one sees in the above argument that the H-Galois cover $Y_1 \to X$ remains étale over U. In applying Theorem 5.1.4 to pass to a Γ_1-cover (and thence to a Γ-cover), one typically obtains new branch points. But a sharp upper bound can be found on the number of additional branch points [Ha11], using Abhyankar's Conjecture (discussed in Section 5.3). $\qquad\qquad\square$

Before turning to the general case of Theorem 5.1.1 (where k is allowed to be uncountable), we sketch the proof of Theorem 5.1.4:

SKETCH OF PROOF OF THEOREM 5.1.4. Let Π be the absolute Galois group of K. Consider a finite split embedding problem $\mathcal{E} = (\alpha : \Pi \to G, f : \Gamma \to G)$ for K, with s a section of f, and with f corresponding to a G-Galois branched cover $Y \to X$. Let $N = \ker(f)$, and let $n_1, \ldots, n_r$ be generators of N. Thus Γ is generated by $s(G)$ and the n_i's. Pick r closed points $\xi_i \in X$ that are not branch points of $Y \to X$. Thus $Y \to X$ splits completely over each ξ_i, since k is algebraically closed. Let $k' = k((t))$, and let $X' = X \times_k k'$ and similarly for Y' and ξ'_i. Pick small neighborhoods X'_i around each of the points ξ'_i on X'. (Here, if one works in the rigid context, one takes t-adic closed discs. If one works in the formal context, one blows up at the points ξ'_i, and proceeds as in Example 3.2.11 or 3.2.13, using Theorem 3.2.8 or 3.2.12. See also Example 4.2.4.) Over these neighborhoods, build cyclic covers $Z'_i \to X'_i$ with group $N_i = \langle n_i \rangle$ (branched at ξ'_i and possibly other points; cf. the proof of Proposition 3.3.3, using the presence of prime-to-p roots of unity). Let $Y'_0 \to X'_0$ be the restriction of $Y' \to X'$ away from the above neighborhoods (viz. over the complement of the corresponding open discs if one works rigidly, and over the general fibre of the formal completion at the complement of the ξ_i's if one works formally). Via the section s of f, the Galois group G of $Y'_0 \to X'_0$ may be identified with $s(G) \subset \Gamma$. The induced Γ-Galois covers $\mathrm{Ind}^\Gamma_{N_i} Z'_i \to Z'_i$ and $\mathrm{Ind}^\Gamma_{s(G)} Y'_0 \to X'_0$ agree over the (rigid or formal) overlap. Hence by (rigid or formal) GAGA, these patch together to form a Γ-Galois cover $Z' \to X'$. (In the formal case, one uses Theorem 3.2.8 rather than Theorem 3.2.1, since the agreement is not on the completion along a Zariski open set.) This cover is connected since Γ is generated by $s(G)$ and the n_i's; it dominates $Y' \to X'$ since it does on each patch; and it is branched at each ξ'_i. As in the proof of Corollary 3.3.5, one may now specialize from k' to k using that k is algebraically closed, obtaining a Γ-Galois cover $Z \to X$ that dominates $Y \to X$. This corresponds to a proper solution to $\mathcal{E}$. □

REMARK 5.1.6. (a) The above proof also shows that one has some control over the position of the new branch points of $Z \to X$. Namely, the branch locus contains the points ξ_i, and these points can be taken arbitrarily among non-branch points of $Y \to X$. In particular, any given point of X can be taken to be a branch point of $Z \to X$ above (by choosing it to be one of the ξ_i's). More precise versions of this fact appear in [Ha10, Theorem 3.5] and [Po3, Theorem A], where formal and rigid methods are respectively used.

(b) As a consequence of Remark (a), it follows that the set of (isomorphism classes of) solutions to the split embedding problem has cardinality equal to that of k.

(c) The above proof of Theorem 5.1.4 also gives information about inertia of the constructed cover $Z \to X$. Namely, if $I \subset G$ is the inertia group of $Y \to X$ at a point $\eta \in Y$ over $\xi \in X$, then $s(I) \subset \Gamma$ is an inertia group of $Z \to X$ at a point $\zeta \in Z$ over η (and the other inertia groups over ξ are the conjugates of $s(I)$).

(d) Adjustments to the above construction give additional flexibility in controlling the properties of $Z \to X$. In particular, if char $k = p > 0$ and if $I' \subset \Gamma$ is the extension of $s(I)$ by a p-group, then one may build Z so that I' is an inertia group over ξ at a point over η (with notation as in Remark (c)). In addition, rather than considering a split embedding problem, i.e. a group Γ generated by a normal subgroup N and a complement $s(G)$, one can more generally consider a group Γ generated by two subgroups H and G, where we are given a G-Galois cover $Y \to X$. The assertion then says that this cover can be modified to produce a Γ-Galois cover $Z \to X$ with control as above on the branch locus and inertia groups. In particular, one can add additional branch points to a cover, and one can modify a cover by enlarging an inertia group from a p-subgroup of the Galois group to a larger p-subgroup. (See [Ha6, Theorem 2] and [Ha13, Theorem 3.6], where formal patching is used to prove these assertions.)

(e) The ability to add branch points was used in [MR] to show that for any finite group G and any smooth connected curve X over an algebraically closed field k, there is a G-Galois branched cover $Y \to X$ such that G is the *full* group of automorphisms of Y. The idea is that if one first takes an arbitrary G-Galois cover of X (by Corollary 3.3.5); then one can adjust it by adding new branch points and thereby killing automorphisms that are not in G. □

To prove the general case of Theorem 5.1.1, one replaces Theorem 5.1.2 above by a result of Melnikov and Chatzidakis (see [Ja, Lemma 2.1]):

THEOREM 5.1.7. *Let Π be a profinite group and let m be an infinite cardinal. Then Π is a free profinite group of rank m if and only if every non-trivial finite embedding problem for Π has exactly m proper solutions.*

Namely, by Remark 5.1.6(b) above, in the situation of Theorem 5.1.1 the number of proper solutions to any finite *split* embedding problem is card k. Proceeding as in the proof of Theorem 5.1.1 in the countable case, one obtains that *every* finite embedding problem for Π has card k proper solutions. So Π is free profinite of that rank by Theorem 5.1.7, and this proves Theorem 5.1.1.

REMARK 5.1.8. By refining the proof of Theorem 5.1.1 (in particular modifying Theorems 5.1.3 and 5.1.4 above), one can prove a tame analog of that result [Ha13, Theorem 4.9(b)]: If X is an affine curve with function field K, consider the maximal extension Ω of K that is at most tamely ramified over each point of X. Then $\mathrm{Gal}(\Omega/K)$ is a free profinite group, of rank equal to the cardinality of k. $\square$

Theorem 5.1.4 above extends from algebraically closed fields to arbitrary large fields (cf. Section 3.3), according to the following result of Pop:

THEOREM 5.1.9 (POP [Po1, Theorem 2.7]). *If k is a large field, and K is the function field of a geometrically irreducible k-curve X, then every finite split embedding problem for K has a proper regular solution.*

Namely, the above proof of Theorem 5.1.4 showed that result for an *algebraically closed* field k by first proving it for the Laurent series field $K = k((t))$, and then specializing from K to k, using that k is algebraically closed. In order to prove Theorem 5.1.9, one does the same in this more general context, using that k is large in order to specialize from $K = k((t))$ to k (as in Sections 3.3 and 4.3). A difficulty is that since k need not be algebraically closed, one can no longer choose the extra branch points $\xi_i \in X$ arbitrarily (as one could in the above proof of Theorem 5.1.4, where ξ_i and the points of its fibre were automatically k-rational). Still, one can proceed as in the proofs of Theorems 3.3.1, 4.3.1, and 4.3.3 — viz. using cyclic covers branched at clusters of points constructed in the proof of Proposition 3.3.3.

Since an arbitrary large field k is not algebraically closed, one would also like to know that the Γ-Galois cover $Z \to X$ has the property that $Z \to Y$ is *regular* (i.e. Z and Y have the same ground field ℓ, or equivalently the algebraic closures of k in the function fields of Y and Z are equal). This can be achieved by using that in the construction using formal patching, the closed fibre of the cover $Z \to Y$ over K is a mock cover (as in the proof of Theorem 3.3.1). Alternatively, from the rigid point of view, one can observe from the patching construction (as in the proof of Theorem 4.3.1) that Z may be chosen so that $Z \to Y$ has a totally split fibre over $\eta \in Y$, if η has been chosen (in advance) to be an ℓ-point of Y that lies over a k-point ξ of X. This then implies regularity, as in Theorem 4.3.1. (If there is no such point $\eta \in Y$, then one can first base-change to a finite Galois extension $\tilde{k}$ of k where there is such a point; and then construct a regular solution $\tilde{Z} \to \tilde{Y} = Y \times_k \tilde{k}$ which is compatible with the $\mathrm{Gal}(\tilde{k}/k)$-action, and so which descends to a regular solution $Z \to Y$.)

Remarks 5.1.6(a) and (b) above no longer hold for curves over an arbitrary large field (nor does Theorem 5.1.1 — see below); but Remark 5.1.6(c) still applies in this situation. So the argument in the case of an arbitrary large field gives the following more precise form of Theorem 5.1.9 (where one looks at the actual curve X, rather than just at its function field):

THEOREM 5.1.10. *Let k be a large field, let X be a geometrically irreducible smooth k-curve, let $f : \Gamma \to G$ be a surjection of finite groups with a section s, and let $Y \to X$ be a G-Galois connected branched cover of smooth curves.*

(a) *Then there is a smooth connected Γ-Galois branched cover $Z \to X$ that dominates the G-Galois cover $Y \to X$, such that $Z \to Y$ is regular.*

(b) *Let ξ be a k-point of X which is not a branch point of $Y \to X$, and let η be a closed point of Y over ξ with decomposition group $G_1 \subset G$. Then the cover $Z \to X$ in (a) may be chosen so that it is totally split over η, and so that there is a point $\zeta \in Z$ over η whose decomposition group over ξ is $s(G_1) \subset \Gamma$.*

REMARK 5.1.11. (a) In [Po1], the above result was stated for a slightly smaller class of fields (those with a "universal local-global principle"); but in fact, all that was used is that the field is large. Also, the result there did not assert 5.1.10(b), though this can be deduced from the proof. The result was stated for large fields in [Po4, Main Theorem A], but only in the case that $X = \mathbb{P}^1_k$ and $Y = \mathbb{P}^1_\ell$. (Both proofs used rigid patching.) The fact that the fibre over η can be chosen to be totally split first appeared explicitly in [HJ1, Theorem 6.4], in the case that $X = \mathbb{P}^1_k$ and $Y = \mathbb{P}^1_\ell$; and in [HJ2, Proposition 4.2] if $X = \mathbb{P}^1_k$ and Y is arbitrary. The proofs there used "algebraic patching" (cf. the comments at the end of Section 4.3).

(b) A possible strengthening of Theorem 5.1.10(b) would be to allow one to specify the decomposition group of ζ as a given subgroup $G'_1 \subset \Gamma$ that maps isomorphically onto $G_1 \subset G$ via f (rather than having to take $G'_1 = s(G_1)$, as in the statement above). It would be interesting to know if this strengthening is true. $\square$

As a consequence of Theorem 5.1.9, we have:

COROLLARY 5.1.12. *Let k be a Hilbertian large field, with absolute Galois group G_k.*

(a) *Then every finite split embedding problem for G_k has a proper solution.*

(b) *If k is also countable, and if G_k is projective, then G_k is isomorphic to the free profinite group of countable rank.*

PROOF. (a) Every such embedding problem for G_k gives a split embedding problem for $G_{k(x)}$. That problem has a proper solution by Theorem 5.1.9. Since k is Hilbertian, that solution can be specialized to a proper solution of the given embedding problem.

(b) Since G_k is projective, the conclusion of part (a) implies that *every* finite embedding problem for G_k has a proper solution (as in the proof of Theorem 5.1.1 above, using semi-direct products). Also, G_k is of countably infinite rank (again as in the proof of Theorem 5.1.1). So Theorem 5.1.2 implies the conclusion. □

REMARK 5.1.13. (a) Part (a) of Corollary 5.1.12 appeared in [Po3, Main Theorem B] and [HJ1, Thm. 6.5(a)]. As a special case of part (b) of the corollary, one has that if k is a countable Hilbertian PAC field (see Example 3.3.7(c)), then G_k is free profinite of countable rank. This is because PAC fields are large, and because their absolute fundamental groups are projective (because they are of cohomological dimension ≤ 1 [Ax2, §14, Lemma 2]). This special case had been a conjecture of Roquette, and it was proved as above in [Po3, Thm. 1] and [HJ1, Thm. 6.6] (following a proof in [FV2] in the characteristic 0 case, using the classical complex analytic form of Riemann's Existence Theorem).

(b) As remarked in Section 3.3, it is unknown whether $\mathbb{Q}^{ab}$ is large. But it is Hilbertian ([Vö, Corollary 1.28], [FJ, Theorem 15.6]) and countable (being contained in $\bar{\mathbb{Q}}$), and its absolute Galois group is projective (being of cohomological dimension 1 by [Se4, II, 3.3, Proposition 9]). So if it is indeed large, then part (b) of the corollary would imply that its absolute Galois group is free profinite of countable rank — i.e. the original (arithmetic) form of Shafarevich's Conjecture would hold. Among other things, this would imply that every finite group is a Galois group over $\mathbb{Q}^{ab}$.

The solvable version of Shafarevich's Conjecture has been shown; i.e. the maximal pro-solvable quotient of $G_{\mathbb{Q}^{ab}}$ is the free prosolvable group of countable rank [Iw]. More generally, if k is Hilbertian and G_k is projective, then every finite embedding problem for G_k with solvable kernel has a proper solution [Vö, Corollary 8.25]. This result does not require k to be large, and it does not use patching.

(c) It has been *conjectured* by Dèbes and Deschamps [DD] that Theorem 5.1.9 and Corollary 5.1.12 remain true even if the ground field is not large. Specifically, they conjecture that for any field k, every finite split embedding problem for $G_{k(x)}$ has a proper regular solution; and hence that if k is Hilbertian, then every finite split embedding problem for G_k has a proper solution. This is a very strong conjecture, in particular implying an affirmative answer to the Regular Inverse Galois Problem (i.e. that every finite group is a regular Galois group over $k(x)$ for every field k). But it also seems very far away from being proved. □

As mentioned above, Theorem 5.1.1 does not hold if the algebraically closed field k is replaced by an arbitrary large field. This is because if K is the function field of a k-curve X, then its absolute Galois group G_K is not even projective (much less free) if k is not separably closed. That is, not every finite embedding problem for K has a weak solution — and so certainly not a proper solution, as would be required in order to be free.

This can be seen by using the equivalence between the condition that a profinite group Π is projective and the condition that it has cohomological dimension ≤ 1 [Se4, I; 5.9, Proposition 45 and 3.4, Proposition 16]. Namely, if k is not separably closed, then its absolute Galois group G_k is non-trivial, and so G_k has cohomological dimension > 0 [Se4, I, 3.3, Corollaire 2 to Proposition 14]. Since the function field K is of finite type over k and of transcendence degree 1 over k, it follows that G_K has cohomological dimension > 1. (This is by [Se4, II, 4.2, Proposition 11] in the case that $\mathrm{cd}\,G_k$ is finite; and by [Ax1] and [Se4, II, 4.1, Proposition 10(ii)] if $\mathrm{cd}\,G_k$ is infinite.) So G_K is not projective.

But as Theorem 5.1.9 shows, every finite *split* embedding problem for G_K has a proper solution, if K is the function field of a curve over an arbitrary large field k. Thus (as in the proof of Theorem 1 above, via semi-direct products), it follows that any finite embedding problem for G_K that has a weak solution must also have a proper solution. So Theorem 5.1.9 can be regarded as saying that G_K is "as close as possible" to being free, given that it is not projective.

5.2. Arithmetic lifting, embedding problems, and patching.

In realizing Galois groups over a Hilbertian field k like $\mathbb{Q}$ or $\mathbb{Q}^{\mathrm{ab}}$, the main method is to realize the group as a regular Galois group over $K = k(x)$, and then to specialize from K to k using that k is Hilbertian. That is, one constructs a Galois branched cover $Y \to \mathbb{P}^1_k$ such that k is algebraically closed in the function field of Y, and then obtains a Galois extension of k with the same group by considering an irreducible fibre of the cover over a k-point of $\mathbb{P}^1_k$ (which exists by the Hilbertian hypothesis). To date, essentially all simple groups that have been realized as Galois groups over $\mathbb{Q}$ or $\mathbb{Q}^{\mathrm{ab}}$ have been realized by this method.

The use of this method has led to the question of whether, given a finite Galois extension ℓ of a field k, there is a finite regular Galois extension L of $K = k(x)$ with the same group G, of which the given extension is a specialization. If so, then one says that the field k and group G satisfy the *arithmetic lifting property*. (Of course if one did not require regularity, then one could just take L to be $\ell(x)$.)

The question of when this property holds was first raised by S. Beckmann [Be], who showed that it does hold in the case that $k = \mathbb{Q}$ and G is either an abelian group or a symmetric group. Later, E. Black [Bl1] [Bl2] [Bl3] showed that the property holds for certain more general classes of groups over Hilbertian fields, particularly certain semi-direct products such as dihedral groups D_n with n odd. Black also conjectured that the arithmetic lifting property holds for all finite groups over all fields, and proving this has come to be known as the *Beckmann–Black problem* (or BB). It was later shown by Dèbes [Dè] that an affirmative answer to BB over *every* field would imply an affirmative answer to the Regular Inverse Galois Problem (RIGP) over every field (i.e. that for every field k and every finite group G, there is a regular Galois extension of $k(x)$ with group G). On the other hand, knowing BB for a *given* field k does not

automatically give RIGP over k, since one needs to be given a Galois extension
of the given field in order to apply BB.

Colliot-Thélène has considered a strong form of arithmetic lifting (or BB):
Suppose we are given a field k and a finite group G, and a G-Galois k-*algebra*
A (i.e. a finite direct sum of finite separable field extensions of k, on which G
acts faithfully with fixed field k). In this situation, is there a regular G-Galois
field extension L of $k(x)$ that specializes to A? Equivalently, suppose that H is
a subgroup of G and ℓ is an H-Galois field extension of k. Then the question
is whether there is a regular G-Galois field extension of $k(x)$ such that some
specialization to k yields $A := \ell^{\oplus(G:H)}$ (where the copies of ℓ are indexed by
the cosets of H in G). In geometric terms, the question is whether there is a
regular G-Galois branched cover $Y \to \mathbb{P}^1_k$ with a given fibre $\mathrm{Ind}_H^G \mathrm{Spec}\,\ell$—i.e.
such that over some unramified k-point of the line, there is a point of Y with
given decomposition group $H \subset G$ and given residue field ℓ (which is a given
H-Galois field extension of k).

If, in the strong form of BB, one takes A to be a G-Galois field extension ℓ
of k, then one recovers the original BB. At the other extreme, if one takes A to
be a direct sum of copies of k (indexed by the elements of G), then one is asking
the question of whether there is a G-Galois regular field extension of $k(x)$ with
a totally split fibre. (Thus the strong form of BB over a *given* field k implies
RIGP for that field.) In the case that k is a large field, this totally split case of
strong BB does hold; indeed, this is precisely the content of Theorem 4.3.1.

Colliot-Thélène showed that the strong form of BB holds in general for large
fields k:

THEOREM 5.2.1 [CT]. *If k is a large field, G is a finite group, and A is a G-
Galois k-algebra, then there is a G-Galois regular branched cover of $X = \mathbb{P}^1_k$
whose fibre over a given k-point agrees with $\mathrm{Spec}\,A$ (as a G-Galois k-algebra).*

REMARK 5.2.2. As noted in Remark 5.1.13(b), it is unknown whether $\mathbb{Q}^{\mathrm{ab}}$ is
large. But if it is, then Theorem 5.2.1 would imply that it has the (strong)
arithmetic lifting property for every finite group—and so every finite Galois
group over $\mathbb{Q}^{\mathrm{ab}}$ would be the specialization of a regular Galois branched cover
of the line over $\mathbb{Q}^{\mathrm{ab}}$. On the other hand, $\mathbb{Q}$ is not large, and so Theorem 5.2.1
does not apply to it. And although it is known that every finite solvable group
is a Galois group over $\mathbb{Q}$ (Shafarevich's Theorem [NSW, Chap. IX, § 5]), it is not
known whether every such group is the Galois group of a regular branched cover
of $\mathbb{P}^1_{\mathbb{Q}}$ — much less that the arithmetic lifting property holds for these groups
over $\mathbb{Q}$. $\square$

Colliot-Thélène's proof used a different form of patching, and relied on work of
Kollár [Kol] on rationally connected varieties. The basic idea is to construct a
"comb" of projective lines on a surface, i.e. a tree of $\mathbb{P}^1$'s in which one component
meets all the others, none of which meet each other. A degenerate cover of the
comb is then constructed by building it over the components, and the cover is

then deformed to a non-degenerate cover of a nearby irreducible curve of genus 0 with the desired properties.

Colliot-Thélène's proof required that k be of characteristic 0 (because Kollár's work assumed that), but other proofs have been found that do not need this. In particular, Moret-Bailly [MB2] used a formal patching argument to prove this result. A proof using rigid patching can be obtained from Colliot-Thélène's argument by replacing the "spine" of the comb by an affinoid set U_0 as in the proof of Theorem 4.3.1, and the "teeth" of the comb by affinoids $U_1, \ldots, U_r$ as in that proof (appropriately chosen). And a proof using "algebraic patching" (cf. the end of Section 4.3) has been found by Haran and Jarden [HJ2].

Yet another proof of Theorem 5.2.1 above can be obtained from Pop's result on solvability of split embedding problems over large fields (Theorem 5.1.9, in the more precise form Theorem 5.1.10 — which of course was also proved using patching). This proof, which was found by Pop and the author, requires only the special case of Theorem 5.1.10 in which the given cover of $\mathbb{P}_k^1$ is purely arithmetic (i.e. of the form $\mathbb{P}_\ell^1$; this was the case considered in [Po4] and [HJ1]). Namely, under the hypotheses of Theorem 5.2.1 above, we may write $A = \ell^{\oplus(G:H)}$, where ℓ is an H-Galois field extension of k for some subgroup $H \subset G$. Let $\Gamma = G \rtimes H$, where the semidirect product is formed with respect to the conjugation action of H on G. Thus there is a surjection $f : \Gamma \to H$ (given by second projection) with a splitting s (given by second inclusion). Consider the H-Galois cover $Y \to X$, where $X = \mathbb{P}_k^1$ and $Y = \mathbb{P}_\ell^1$. Let ξ be a k-point of X. By hypothesis, there is a closed point η on $Y = \mathbb{P}_\ell^1$ whose residue field is ℓ and whose decomposition group over ξ is H. So by Theorem 5.1.10, there is a regular connected G-Galois cover $Z \to Y$ which is totally split over η, such that the composition $Z \to X$ is Γ-Galois and such that $1 \rtimes H = s(H) \subset \Gamma$ is the decomposition group over ξ of some point $\zeta \in Z$ over η. Viewing G as a quotient of Γ via the multiplication map $m : \Gamma = G \rtimes H \to G$, we may consider the intermediate G-Galois cover $W \to X$ (i.e. $W = Z/N$, where $N = \ker m$). It is then straightforward to check that the cover $W \to X$ satisfies the conclusion of Theorem 5.2.1.

The arithmetic lifting result Theorem 5.2.1 above, and the split embedding problem result Theorem 5.2.10, both generalize Theorem 4.3.1 (that one can realize any finite group as a Galois group over a curve defined over a given large field, with a totally split fibre). In fact, those two generalizations can themselves be simultaneously generalized, by the following joint result of F. Pop and the author, concerning the solvability of a split embedding problem with a prescribed fibre. We first introduce some terminology.

As in Theorem 5.1.10, let X be a geometrically irreducible smooth curve over a field k, let $f : \Gamma \to G$ be a surjection of finite groups, and let $Y \to X$ be a G-Galois connected branched cover of smooth curves. Let ξ be an unramified k-point of X, and let η be a closed point of Y over ξ with decomposition group $G_1 \subset G$ and residue field $\ell \supset k$. Let Γ_1 be a subgroup of Γ such that $f(\Gamma_1) = G_1$, and let λ be a Γ_1-Galois field extension of k that contains ℓ. We say that this data constitutes a *fibred embedding problem* $\mathcal{E}$ for X. The problem $\mathcal{E}$ is *split* if f has a section s. A *proper solution* to a fibred embedding problem $\mathcal{E}$ as above consists of a smooth connected Γ-Galois branched cover $Z \to X$ that dominates the G-Galois cover $Y \to X$, such that there is a closed point ζ of Z over η which has residue field λ and whose decomposition group over ξ is $\Gamma_1 \subset \Gamma$. A solution to $\mathcal{E}$ is *regular* if $Z \to Y$ is regular (i.e. the algebraic closures of k in the function fields of Y and Z are equal.

THEOREM 5.2.3. *Let k be a large field, let X be a geometrically irreducible smooth k-curve, and consider a fibred split embedding problem $\mathcal{E}$ as above, with data $f : \Gamma \to G$, s, $Y \to X$, $\xi \in X$, $\eta \in Y$, $G_1 \subset G$, $\lambda \supset \ell \supset k$. Assume that $\Gamma_1 = \mathrm{Gal}(\lambda/k)$ contains $s(G_1)$. Let k' be the algebraic closure of k in the function field of Y, let $X' = X \times_k k'$ and let $E = \mathrm{Gal}(Y/X') \subset G$. Assume that $s(E)$ commutes with $N_1 = \ker(f : \Gamma_1 \to G_1)$. Then $\mathcal{E}$ has a proper regular solution.*

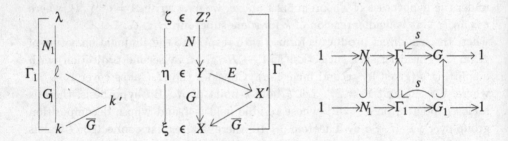

Figure 5.2.4. The set-up in the statement of Theorem 5.2.3.

In other words, given a split embedding problem for a curve over a large field, there is a proper regular solution with a given fibre, assuming appropriate hypothesis (on Γ_1, E and N_1). Taking the special case $\Gamma_1 = G_1$ in Theorem 5.2.3 (i.e. taking $N_1 = 1$), one recovers Theorem 5.1.10. And taking the special case $G = 1$ in Theorem 5.2.3, one recovers Theorem 5.2.1 above. (Note that the "G" in Theorem 5.2.1 corresponds to the group Γ in Theorem 5.2.3. Also, the "A" in 5.2.1 is $\mathrm{Ind}_{\Gamma_1}^{\Gamma} \lambda = \lambda^{\oplus(\Gamma:\Gamma_1)}$, in the notation of 5.1.3.) More generally, taking $E = 1$ in Theorem 5.2.3 (but not necessarily taking G to be trivial), one obtains the result in the case that the given cover $Y \to X$ is purely arithmetic, i.e. of the form $Y = X \times_k k'$. The result in that case is a generalization of Theorem 5.2.1 — viz. instead of requiring the desired cover $Z \to X$ in Theorem 5.2.1 to be regular, it can be chosen so that the algebraic closure of k in the function field of Z is a given subfield k' of A that is Galois over k (and also X need not be $\mathbb{P}^1$). Note that in each of these special cases, the hypothesis on $s(E)$ commuting with N_1 is automatically satisfied, because either E or N_1 is trivial in each case. (On the other hand, the condition $\Gamma_1 \supset s(G_1)$ is still assumed.)

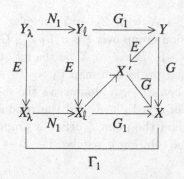

Figure 5.2.5. The situation in the proof of Theorem 5.2.3.

Theorem 5.2.3, like Theorem 5.2.1 above, can in fact be deduced from Theorem 5.1.10, by a strengthening of the proof of Theorem 5.2.1 given above:

PROOF OF THEOREM 5.2.3. The G-Galois cover $Y \to X$ factors as $Y \to X' \to X$, where the E-Galois cover $Y \to X'$ is regular, and $X' \to X$ is purely arithmetic (induced by extension of constants from k to k'). Let $\bar{G} = \mathrm{Gal}(k'/k)$; we may then identify $\mathrm{Gal}(X'/X) = G/E$ with $\bar{G}$. For any field F containing k', let $X_F = X' \times_{k'} F = X \times_k F$ and let $Y_F = Y \times_{k'} F$. So we may identify $E = \mathrm{Gal}(Y_\ell/X_\ell) = \mathrm{Gal}(Y_\lambda/X_\lambda)$; and $\bar{G}$ is a quotient of $G_1 = \mathrm{Gal}(X_\ell/X)$. Since $Y_\ell = X_\ell \times_{X'} Y$, it follows that $\mathrm{Gal}(Y_\ell/X) = G_1 \times_{\bar{G}} G$ (fibre product of groups); similarly $\mathrm{Gal}(Y_\lambda/X) = \Gamma_1 \times_{\bar{G}} G$, and $Y \twoheadrightarrow X$ is the subcover of $Y_\lambda \to X$ corresponding to the second projection map $G_1 \times_{\bar{G}} G \to G$. (See Figure 5.2.5.)

Let ξ_ℓ be the unique closed point of X_ℓ over $\xi \in X$. Then $\xi_\ell \in X_\ell$ and $\eta \in Y$ each have residue field ℓ and decomposition group G_1 over ξ. So the fibre of $Y_\ell \to Y$ over η is totally split, with each point having residue field ℓ; the points of this fibre lie over $\xi_\ell \in X_\ell$ and over $\eta \in Y$; and the local fields of X_ℓ and Y at ξ_ℓ and η (i.e. the fraction fields of the complete local rings) are isomorphic over X'. So at one of the points in this fibre (say η_ℓ), the decomposition group over $\xi \in X$ is equal to the diagonal subgroup $G_1 \times_{G_1} G_1 \subset G_1 \times_{\bar{G}} G$. (At the other points of the fibre, the decomposition group is of the form $\{(g_1, \iota(g_1)) \mid g_1 \in G_1\}$, where ι is an inner automorphism of G_1.) Similarly, there is a point $\eta_\lambda \in Y_\lambda$ over $\eta_\ell \in Y_\ell$ whose residue field is λ and whose decomposition group over $\xi \in X$ is $\Gamma_1 \times_{G_1} G_1 \subset \Gamma_1 \times_{\bar{G}} G$.

Since $E = \ker(G \to \bar{G})$, every element of $\Gamma_1 \times_{\bar{G}} G$ can uniquely be written as $(\gamma_1, f(\gamma_1)e)$, with $\gamma_1 \in \Gamma_1$ and $e \in E$. Consider the map $\sigma : \Gamma_1 \times_{\bar{G}} G \to \Gamma_1 \times_{\bar{G}} \Gamma$ given by $\sigma(\gamma_1, f(\gamma_1)e) = (\gamma_1, \gamma_1 s(e))$. Since $s(E)$ commutes with N_1, direct computation shows that σ is a homomorphism, and hence is a section of $(1, f) : \Gamma_1 \times_{\bar{G}} \Gamma \to \Gamma_1 \times_{\bar{G}} G$.

We may now apply Theorem 5.1.10 to the surjection $(1, f)$ and its section σ, to the cover $Y_\lambda \to X$, to the k-point $\xi \in X$, and to the point $\eta_\lambda \in Y_\lambda$ over ξ with decomposition group $\Gamma_1 \times_{G_1} G_1$. The conclusion of that result is that there is a smooth connected $\Gamma_1 \times_{\bar{G}} \Gamma$-Galois cover $Z_\lambda \to X$ that dominates the

$\Gamma_1 \times_{\bar{G}} G$-Galois cover $Y_\lambda \to X$ with $Z_\lambda \to Y_\lambda$ regular, together with a point $\zeta_\lambda \in Z_\lambda$ whose decomposition group over ξ is $\sigma(\Gamma_1 \times_{G_1} G_1) = \Gamma_1 \times_{\Gamma_1} \Gamma_1 = \Delta_{\Gamma_1}$, the diagonal of Γ_1 in $\Gamma_1 \times_{\bar{G}} \Gamma$. Let $Z \to X$ be the intermediate Γ-Galois cover corresponding to the second projection map $\Gamma_1 \times_{\bar{G}} \Gamma \to \Gamma$, and let $\zeta \in Z$ be the image of $\zeta_\lambda \in Z_\lambda$. Then $Z \to X$ dominates the G-Galois cover $Y \to X$; the decomposition group of ζ is $\Gamma_1 \subset \Gamma$ and the residue field is λ; and ζ lies over $\eta \in Y$. So $Z \to X$ and the point ζ define a proper solution to the split embedding problem $\mathcal{E}$. The solution is regular, i.e. $Z \to Y$ is regular, since the pullback $Z_\lambda \to Y_\lambda$ is regular. □

REMARK 5.2.6. (a) Theorem 5.2.3 can be regarded as a step toward an "arithmetic Riemann's Existence Theorem" for covers of curves over a large field. Namely, such a result should classify the branched covers of such a curve, in terms of how they fit together (e.g. with respect to embedding problems), and in terms of their arithmetic and their geometry, including information about decomposition groups and inertia groups (the latter of which Theorem 5.2.3 does not discuss).

(b) In Remark 5.1.11(b), it was asked if Theorem 5.1.10 can be generalized, to allow one to require the decomposition group there to be an arbitrary subgroup of Γ that maps isomorphically onto G_1 under f (rather than being required to take $s(G_1)$ for the decomposition group). If it can, then the above proof of Theorem 5.2.3 could be simplified, and the statement of Theorem 5.2.3 could be strengthened. Namely, the subgroup $\Gamma_1 \subset \Gamma$ could be allowed to be chosen more generally, viz. as any subgroup of Γ whose image under f is G_1. And the assumption that $s(E)$ commutes with N_1 could also be dropped — since one could then replace the section σ in the above proof by the section (id, s), while still requiring the decomposition group at ζ_λ to be Δ_{Γ_1}. But on the other hand there might in general be a cohomological obstruction, which would vanish if the containment and commutativity assumptions are retained. □

5.3. Abhyankar's Conjecture and embedding problems. The main theme in this article has been the use of patching methods to prove results in the direction of Riemann's Existence Theorem for curves that are not necessarily defined over $\mathbb{C}$. Such a result should classify the unramified covers of such a curve U, and in particular provide an explicit description of the fundamental group of U, as a profinite group.

While the full statement of Riemann's Existence Theorem is known only for curves over an algebraically closed field of characteristic 0, partial versions have been discussed above. In particular, if one allows arbitrary branching to occur, there is the Geometric Shafarevich Conjecture (Section 5.1); and if one instead takes U to be the complement of a well chosen branch locus and if one restricts attention to a particular class of covers, then there is the Half Riemann Existence Theorem (Section 4.3).

Another way to weaken Riemann's Existence Theorem is to ask for the set $\pi_A(U)$ of finite Galois groups of unramified covers of U; i.e., for the set of finite quotients of $\pi_1(U)$, up to isomorphism. A *finitely generated* profinite group Π is in fact determined by its set of finite quotients [FJ, Proposition 15.4]; and $\pi_1(U)$ is finitely generated (as a profinite group) if the base field has characteristic 0. But in characteristic p, if U is affine, then $\pi_1(U)$ is *not* finitely generated (see below), and $\pi_A(U)$ does not determine $\pi_1(U)$. In this situation, $\pi_1(U)$ *remains unknown*; but at least $\pi_A(U)$ is known if the base field is algebraically closed. Moreover, $\pi_A(U)$ depends only on the *type* (g, r) of U (where $U = X - S$, with X a smooth connected projective curve of genus $g \geq 0$, and S is a set of $r > 0$ points of X). Namely, this 1957 conjecture of Abhyankar [Ab1] was proved by Raynaud [Ra2] and the author [Ha7] using patching and other methods:

THEOREM 5.3.1 (ABHYANKAR'S CONJECTURE: RAYNAUD, HARBATER). *Let k be an algebraically closed field of characteristic $p > 0$, and let U be a smooth connected affine curve over k of type (g, r). Then a finite group G is in $\pi_A(U)$ if and only if each prime-to-p quotient of G has a generating set of at most $2g + r - 1$ elements.*

Recall that for complex curves U of type (g, r), a finite group G is in $\pi_A(U)$ if and only if G has a generating set of at most $2g + r - 1$ elements. The same assertion is false in characteristic $p > 0$, e.g. since any affine curve has infinitely many Artin–Schreier covers (cyclic of order p), and hence has Galois groups of the form $(\mathbb{Z}/p\mathbb{Z})^s$ for arbitrarily large s. (This implies the above comment that $\pi_1(U)$ is not finitely generated.) The above theorem can be interpreted as saying that "away from p", the complex result carries over; and that every finite group consistent with this principle must occur as a Galois group over U.

In the theorem, the assertion about *every* prime-to-p quotient of G can be replaced by the same assertion about the *maximal* prime-to-p quotient of G — i.e. the group $\bar{G} := G/p(G)$, where $p(G)$ is the subgroup of G generated by the elements of p-power order (or equivalently, by the p-subgroups of G; or again equivalently, by the Sylow p-subgroups of G).

In the case that $U = \mathbb{A}_k^1$, Theorem 5.3.1 says that $\pi_A(\mathbb{A}_k^1)$ consists precisely of the *quasi-p groups*, viz. the groups G such that $G = p(G)$ (i.e. that are generated by their Sylow p-subgroups). This class of groups includes in particular all p-groups, and all finite simple groups of order divisible by p.

REMARK 5.3.2. (a) Before Theorem 5.3.1 was proved, Serre had shown a partial result [Se6, Théorème 1]: that if Q is a quasi-p group and if $N \triangleleft Q$ is a solvable normal subgroup of Q such that the (quasi-p) group $\bar{Q} := Q/N$ is a Galois group over $\mathbb{A}_k^1$ (i.e. $\bar{Q} \in \pi_A(\mathbb{A}_k^1)$), then Q is also a Galois group over $\mathbb{A}_k^1$. Due to the solvability assumption, the proof was able to proceed cohomologically, without patching; it relied in particular on the fact that $\pi_1(U)$ is projective (Theorem 5.1.3 above, also due to Serre). Serre's result [Se6, Thm. 1] implied in particular that Theorem 5.3.1 above is true for *solvable* groups over the affine line. Serre's proof actually showed more: that if N is a p-group, then a given $\bar{Q}$-cover $Y \to \mathbb{A}^1$ can be dominated by a Q-cover (i.e. the corresponding p-embedding problem can be properly solved); but that if N has order prime-to-p, then the embedding problem need not have a proper solution (i.e. the asserted Q-Galois cover of $\mathbb{A}^1$ cannot necessarily be chosen so as to dominate the given $\bar{Q}$-Galois cover $Y \to \mathbb{A}^1$).

(b) More generally, by extending the methods of [Se6], the author showed [Ha12] that if U is any affine variety other than a point, over an arbitrary field of characteristic p, then every finite p-embedding problem for $\pi_1(U)$ has a proper solution. Moreover, this solution can be chosen so as to have prescribed local behavior. For example, if $V \subset U$ is a proper closed subset, then the proper solution over U can be chosen so that it restricts to a given weak solution over V. (Cf. Theorem 5.2.3 above, for such fibred embedding problems in a related but somewhat different context.) And if U is a curve, then the proper solution can be chosen so as to restrict to given weak solutions over the fraction fields of the complete local rings at finitely many points.   □

SKETCH OF PROOF OF THEOREM 5.3.1. In the case $U = \mathbb{A}^1$, the theorem was proved by Raynaud [Ra2], using in particular rigid patching methods. The proof proceeded by induction on the order of G, and considered three cases. In Case 1, the group G is assumed to have a non-trivial normal p-subgroup N; and using Serre's result that embedding problems for $\pi_1(U)$ with p-group kernel can be properly solved (Remark (a) above), the desired conclusion for G follows from the corresponding fact for G/N. When not in Case 1, one picks a Sylow p-subgroup P, and considers all the quasi-p subgroups $Q \subset G$ such that $Q \cap P$ is a Sylow p-subgroup of Q. Case 2 is the situation in which these Q's generate G. In this case, by induction each of the Q's is a Galois group over $\mathbb{A}^1$; and using rigid patching it follows that G is also. (Or one could use formal patching for this step, viz. Theorem 3.2.8; see e.g. [HS, Theorem 6].) Case 3 is the remaining case, where Cases 1 and 2 do not apply. Then, one builds a G-Galois branched cover of the line in mixed characteristic having p-power inertia groups. The closed fibre of the semi-stable model is a reducible curve that maps down to a tree of projective lines in characteristic p. Using a careful combinatorial analysis of the situation, it turns out that over one of the terminal components of the tree (a copy of the projective line), one finds an irreducible G-Galois cover that is branched at just one point — and hence is an étale cover of the affine line, as desired. Moreover, by adjusting the cover, we may assume that the inertia groups over infinity (of the corresponding branched cover of $\mathbb{P}_k^1$) are the Sylow p-subgroups of Q. (Namely, by Abhyankar's Lemma, after pulling back by a Kummer cover $y^n - x$, we may assume that the inertia groups over infinity are p-groups. We may then enlarge this inertia to become Sylow, using Remark 5.1.6(d) above.)

The general case of the theorem was proved in [Ha7], by using the above case of the affine line, together with formal patching and embedding problems.

(See also the simplified presentation in [Ha13], where more is shown.) For the proof, one first recalls that the result was shown by Grothendieck [Gr5, XIII, Cor. 2.12] in the case that the group is of order prime to p. Using this together with formal patching (Theorem 3.2.8), it is possible to reduce to the key case that $U = \mathbb{A}_k^1 - \{0\}$, where $G/p(G)$ is cyclic of prime-to-p order. (For that reduction, one patches a prime-to-p cover of the original curve together with a cyclic-by-p cover of $\mathbb{A}_k^1 - \{0\}$, to obtain a cover of the original curve with the desired group.) Once in this case, by group theory one can find a prime-to-p cyclic subgroup $\bar{G} \subset G$ that normalizes a Sylow p-subgroup P of G and that surjects onto $G/p(G)$. Here G is a quotient of the semi-direct product $\Gamma := p(G) \rtimes \bar{G}$ (formed with respect to the conjugation action of $\bar{G}$ on $p(G)$); so replacing G by Γ we may assume that $G = p(G) \rtimes \bar{G}$ with $\bar{G} \approx G/p(G)$. Letting $n = |\bar{G}|$, there is a $\bar{G}$-Galois étale cover $V \to U_K$ given by $y^n = x$, where $K = k((t))$ and $U_K = A_K^1 - \{0\}$. Using the proper solvability of p-embedding problems with prescribed local behavior (Remark 5.3.2(b) above), one can obtain a $P \rtimes \bar{G}$-Galois étale cover $\tilde{V} \to U_K$ whose behavior over one of the (unramified) K-points ξ_K of U_K can be given in advance. Specifically, one first considers a $p(G)$-Galois étale cover $W \to \mathbb{A}_k^1$ (given by the first case of the result, with Sylow p-subgroups as inertia over ∞), and restricts to the local field at a ramification point with inertia group P (this being a P-Galois field extension of the local field $K = k((t))$ at ∞ on $\mathbb{P}_k^1$). It is this P-Galois extension of K that one uses for the prescribed local behavior over the K-point ξ_K, in applying the p-embedding result. As a consequence, the $P \rtimes \bar{G}$-Galois cover $\tilde{V} \to U_K$ (near ξ_K) has local compatibility with W (near ∞). This compatibility makes it possible for the two covers $\tilde{V}$ and W to be patched using Theorem 3.2.8 or 3.2.12 (after blowing up; see Examples 3.2.11, 3.2.13, and 4.2.4). As a result we obtain a G-Galois cover of U_K (viz. the generic fibre of a cover of $U_{k[\![t]\!]}$). This cover is irreducible because the Galois groups of $\tilde{V}$ and W (viz. $P \rtimes \bar{G}$ and $p(G)$) together generate G. Since k is algebraically closed, one may specialize from K to k (as in Corollary 3.3.5) to obtain the desired cover of U. $\square$

REMARK 5.3.3. (a) The proof of Theorem 5.3.1 actually shows more, concerning inertia groups: Write $U = X - S$ for a smooth connected projective k-curve X and finite set S, and let $\xi \in S$. Then in the situation of the theorem, the G-Galois étale cover of U may be chosen so that the corresponding branched cover of X is tamely ramified away from ξ. (This was referred to as the "Strong Abhyankar Conjecture" in [Ha7], where it is proved.) Note that it is necessary, in general, to allow at least one wildly ramified point. Namely, if G cannot itself be generated by $2g + r - 1$ elements or fewer, then G is not a Galois group of a tamely ramified cover of X that is étale over U, because the tame fundamental group $\pi_1^t(U)$ is a quotient of the free profinite group on $2g + r - 1$ generators [Gr5, XIII, Cor. 2.12].

(b) It would be even more desirable, along the lines of a possible Riemann's Existence Theorem over k, to determine precisely which subgroups of G can be the inertia groups over the points of S, for a G-Galois cover of a given U (with S as in Remark (a) above). This *problem is open*, however, even in the case that $U = \mathbb{A}_k^1$. In that case, the unique branch point ∞ must be wildly ramified, since there are no non-trivial tamely ramified covers of $\mathbb{A}^1$ (by [Gr5, XIII, Cor. 2.12]). By the general theory of extensions of discrete valuation rings [Se5], any inertia group of a branched cover of a k-curve is of the form $I = P \rtimes C$, where P is a p-group (not necessarily Sylow in the Galois group) and C is cyclic of order prime to p. As noted above, it is known [Ra2] that if P is a Sylow p-subgroup of a quasi-p group Q, then there is a Q-Galois étale cover of $\mathbb{A}^1$ such that P is an inertia group over infinity (and this fact was used in the proof of the general case of Theorem 5.3.1, in order to be able to patch together the $P \rtimes \bar{G}$-cover with the $p(G)$-cover). More generally, for *any* subgroup $I \subset Q$ of the form $P \rtimes C$, a necessary condition for I to be an inertia group over ∞ for a Q-Galois étale cover $Y \to \mathbb{A}_k^1$ is that the conjugates of P generate Q. (For if not, they generate a normal subgroup $N \triangleleft Q$ such that $Y/N \to \mathbb{A}^1$ is a non-trivial tamely ramified cover; but $\mathbb{A}^1$ has no such covers, and this is a contradiction.) Abhyankar has conjectured that the converse holds (i.e. that every $I \subset Q$ satisfying the necessary condition will be an inertia group over infinity, for some Q-Galois étale cover of the line). This *remains open*, although some partial results in this direction have been found by R. Pries and I. Bouw [Pr2], [BP].

(c) The results of Sections 3.3 and 4.3 suggest that Abhyankar's Conjecture may hold for affine curves over *large* fields of characteristic p, not just over algebraically closed fields of characteristic p — since patching is possible over such fields, and various Galois realization results can be extended to these fields. But this generalization of Abhyankar's Conjecture *remains open*. The difficulty is that in the proof of Case 3 of Theorem 5.3.1 for $\mathbb{A}_k^1$, one considers a branched cover of $\mathbb{A}_R^1$, where R is a complete discrete valuation ring of mixed characteristic with residue field k. For such a cover, the semi-stable model might be defined only over a finite extension R' of R (and not over R itself); and the residue field of R' could be strictly larger than k. Thus the construction in the proof might yield only a Galois cover of the k'-line, for some finite extension k' of k.

(d) As noted above before Theorem 5.3.1, for an affine k-curve U, the fundamental group $\pi_1(U)$ is not finitely generated (as a profinite group), and is therefore not determined by $\pi_A(U)$. And indeed, the structure of $\pi_1(U)$ *is unknown*, even for $U = \mathbb{A}_k^1$ (although Theorem 5.3.4 below gives some information about how the finite quotients of π_1 "fit together"). In fact, it is easy to see that $\pi_1(\mathbb{A}_k^1)$ depends on the cardinality of the algebraically closed field k of characteristic p; viz. the p-rank of π_1 is equal to this cardinality (using Artin–Schreier extensions). Moreover, Tamagawa has shown [Tm2] that if k, k' are non-isomorphic *countable* algebraically closed fields of characteristic p with $k = \bar{\mathbb{F}}_p$, then $\pi_1(\mathbb{A}_k^1)$

and $\pi_1(\mathbb{A}^1_{k'})$ are non-isomorphic as profinite groups. (It *is unknown* whether this remains true even if k is chosen strictly larger than $\bar{\mathbb{F}}_p$.) Tamagawa also showed in [Tm2] that if $k = \bar{\mathbb{F}}_p$, then two open subsets of $\mathbb{A}^1_k$ have isomorphic π_1's if and only if they are isomorphic as schemes. More generally, given arbitrary affine curves U, U' over algebraically closed fields k, k' of non-zero characteristic, it is an *open question* whether the condition $\pi_1(U) \approx \pi_1(U')$ implies that $k \approx k'$ and $U \approx U'$. This question, which can be regarded as an algebraically closed analog of Grothendieck's anabelian conjecture for affine curves over finitely generated fields [Gr6], was essentially posed by the author in [Ha8, Question 1.9]; and the results in [Tm2] (which relied on the anabelian conjecture in the finitely generated case [Tm1], [Mo]) can be regarded as the first real progress in this direction.

(e) Theorem 5.3.1 holds only for *affine* curves, and is false for projective curves. Namely, if X is a smooth projective k-curve of genus g, then $\pi_1(X)$ is a quotient of the fundamental group of a smooth projective complex curve of genus g (which has generators $a_1, b_1, \ldots, a_g, b_g$ subject to the single relation $\prod[a_i, b_i] = 1$ [Gr5, XIII, Cor. 2.12]). So if $g > 0$ and if Q is a quasi-p group whose minimal generating set has more than $2g$ generators, then Q is not in $\pi_A(X)$. Also, the p-rank of a smooth projective k-curve of genus g is at most g, and so $(\mathbb{Z}/p\mathbb{Z})^{g+1}$ is also not in $\pi_A(X)$. But both Q and $(\mathbb{Z}/p\mathbb{Z})^{g+1}$ trivially have the property that every prime-to-p quotient has at most $2g - 1$ generators (since the only prime-to-p quotient of either group is the trivial group). So both of these groups provide counterexamples to Theorem 5.3.1 over the projective curve X. (In the case of genus 0, we have $X = \mathbb{P}^1_k$, and $\pi_1(X)$ is trivial.)

(f) Another difference between the affine and projective cases concerns the relationship between π_A and π_1. As discussed in Remark (d) above, Theorem 5.3.1 gives π_A but not π_1 for an affine curve, the difficulty being that π_A does not determine π_1 because π_1 of an affine curve is not a finitely generated profinite group. On the other hand, if X is a projective curve, then $\pi_1(X)$ *is* a finitely generated profinite group, and so it is determined by $\pi_A(X)$. Unfortunately, unlike the situation for affine curves, $\pi_A(X)$ is unknown when X is projective of genus > 1 (cf. Remark (e)), and so this does not provide a way of finding $\pi_1(X)$ in this case. A similar situation holds for the tame fundamental group $\pi_1^t(U)$, where $U = X - S$ is an affine curve (and where the tame fundamental group classifies covers of X that are unramified over U, and at most tamely ramified over S). Namely, this group is also a finitely generated profinite group, and is a quotient of the corresponding fundamental group of a complex curve. But the structure of this group, and the set $\pi_A^t(U)$ of its finite quotients, are *both un-known*, even for $\mathbb{P}^1_k - \{0, 1, \infty\}$. (Note that $\pi_A^t(\mathbb{P}^1_k - \{0, 1, \infty\})$ is strictly smaller than the set of Galois groups of covers of $\mathbb{P}^1_{\mathbb{C}} - \{0, 1, \infty\}$ with prime-to-p inertia, because tamely ramified covers of $\mathbb{P}^1_k$ with given degree and inertia groups will generally have lower p-rank than the corresponding covers of $\mathbb{P}^1_{\mathbb{C}}$ — and hence will have fewer unramified p-covers.) On the other hand, partial information about the structure of $\pi_A(X)$ and $\pi_A^t(U)$ has been found by formal and rigid patching methods ([St1], [HS1], [Sa1]) and by using representation theory to solve embedding problems ([St2], [PS]). $\square$

Following the proof of Theorem 5.3.1, Pop used similar methods to prove a stronger version of the result, in terms of embedding problems:

THEOREM 5.3.4 (POP [Po3]). *Let k be an algebraically closed field of charac-teristic $p > 0$, and let U be a smooth connected affine curve over k. Then every finite embedding problem for $\pi_1(U)$ that has quasi-p kernel is properly solvable.*

That is, given a finite group Γ and a quasi-p normal subgroup N of Γ, and given a Galois étale cover $V \to U$ with group $G := \Gamma/N$, there is a Galois étale cover $W \to U$ with group Γ that dominates V. Theorem 5.3.1 is contained in the assertion of Theorem 5.3.4, by taking $N = p(\Gamma)$. (On the other hand, Pop's proof of 5.3.4 relies on the fact that 5.3.1 holds in the case $U = \mathbb{A}^1$; his proof then somewhat parallels that of the general case of 5.3.1, though using rigid rather than formal methods, and performing an improved construction in order to obtain the stronger conclusion.) Note that Theorem 5.3.4 provides information about the structure of $\pi_1(U)$ (i.e. how the covers "fit together in towers"), unlike Theorem 5.3.1, which just concerned $\pi_A(U)$ (i.e. what covering groups can exist in isolation).

Actually, Theorem 5.3.4 was stated in [Po3] only for *split* embedding problems with quasi-p kernel. But one can easily deduce the general case from that one, proceeding as in the proof of Theorem 5.1.1, via Theorem 5.1.3 there. See also [CL] and [Sa2], i.e. Chapters 15 and 16 in [BLoR], for more about the proofs of Theorems 5.3.1 and 5.3.4, presented from a rigid point of view. (More about the proof of Theorem 5.3.1 can be found in [Ha9, §3].)

REMARK 5.3.5. (a) Theorem 5.3.4 can be generalized from étale covers to tamely ramified covers [Ha13, Theorem 4.4]. Namely, with $G = \Gamma/N$ as above, suppose that $V \to U$ is a tamely ramified G-Galois cover of U with branch locus $B \subset U$. Then there is a Γ-Galois cover $W \to U$ that dominates V, and is tamely ramified over B and étale elsewhere over U. (Note that no assertions are made here, or in Theorem 5.3.4, about the behavior over points in the complement of U in its smooth completion.)

(b) In Theorem 5.3.4, for an embedding problem $\mathcal{E} = (\alpha : \pi_1(U) \to G, f : \Gamma \to G)$, one cannot replace the assumption that ker f is quasi-p by the assumption that Γ is quasi-p. This follows from Remark 5.3.2(a), concerning Serre's results in [Se6].

(c) Theorems 5.3.1 and 5.3.4 both deal only with *finite* Galois groups and embed-ding problems. It *is unknown* which *infinite* quasi-p profinite groups can arise as Galois groups, and which embedding problems with *infinite* quasi-p kernel have proper solutions. For example, let G be the free product $\mathbb{Z}_p * \mathbb{Z}_p$ (in the category of profinite groups). This is an infinite quasi-p group, and so every finite quotient of G is a Galois group over $\mathbb{A}^1$. But it *is unknown* whether G itself is a Galois group over $\mathbb{A}^1$ (or equivalently, whether G is a quotient of $\pi_1(\mathbb{A}^1)$). □

Theorem 5.3.4 raises the question of which finite embedding problems for $\pi_1(U)$ are properly solvable, where U is an affine variety (of any dimension) in characteristic p—and in particular, whether every finite embedding problem for U with a quasi-p kernel is properly solvable. For example, one can ask this for affine varieties U of finite type over an algebraically closed field k of characteristic p, i.e. whether Pop's result remains true in higher dimensions.

Abhyankar had previously posed a weaker form of this question as a conjecture, paralleling his conjecture for curves (i.e. Theorem 5.3.1). Namely, in [Ab3], he proposed that if U is the complement of a normal crossing divisor D in $\mathbb{P}^n_k$ (where k is algebraically closed of characteristic p), then $G \in \pi_A(U)$ if and only if $G/p(G) \in \pi_A(U_{\mathbb{C}})$, where $U_{\mathbb{C}}$ is an "analogous complex space". That is, if D has irreducible components $D_1, \ldots, D_r$ of degrees $d_1, \ldots, d_r$, then one takes $U_{\mathbb{C}}$ to be the complement in $\mathbb{P}^n_{\mathbb{C}}$ of a normal crossing divisor consisting of r components of degrees $d_1, \ldots, d_r$. It is known (by [Za1], [Za3], [Fu2]) that $\pi_1(U_{\mathbb{C}})$ is the abelian group $A(d_1, \ldots, d_r)$ on generators $g_1, \ldots, g_r$ satisfying $\sum d_i g_i = 0$ (writing additively). It is also known (by [Ab2], [Fu2]) that the prime-to-p groups in $\pi_A(U)$ are precisely the prime-to-p quotients of $A(d_1, \ldots, d_r)$. Thus Abhyankar's conjecture in [Ab3] is a special case of a more general conjecture that $G \in \pi_A(U) \Leftrightarrow G/p(G) \in \pi_A(U)$ for any affine k-variety U of finite type. This in turn would follow from an affirmative answer to the question asked in the previous paragraph.

Abhyankar also posed a local version of this conjecture in [Ab3], viz. that if $U = \operatorname{Spec} k[\![x_1, \ldots, x_n]\!][(x_1 \cdots x_r)^{-1}]$ (where $n > 1$ and $1 \leq r \leq n$), then a finite group G is in $\pi_A(U)$ if and only if $G/p(G)$ is in $\pi_A(U_{\mathbb{C}})$; here $U_{\mathbb{C}} = \operatorname{Spec} \mathbb{C}[\![x_1, \ldots, x_n]\!][(x_1 \cdots x_r)^{-1}]$. (Note that this fails if $r = 0$, since then $\pi_A(U)$ is trivial by Hensel's Lemma. It also fails if $n = 1$, since in that case the only quasi-p groups in $\pi_A(U)$ are p-groups, by the structure of Galois groups over complete discrete valuation fields [Se5].) Now $\pi_A(U_{\mathbb{C}})$ consists of the finite abelian groups on r generators (via Abhyankar's Lemma; cf. [HP, § 3]), and the prime-to-p groups in $\pi_A(U_{\mathbb{C}})$ are the finite abelian prime-to-p groups on r generators. So this conjecture is again equivalent to asserting that $G \in \pi_A(U) \Leftrightarrow G/p(G) \in \pi_A(U)$.

Abhyankar's higher dimensional global conjecture is easily seen to hold in some special cases, e.g. if D is a union of one or two hyperplanes (since it then reduces immediately to Theorem 5.3.1). Using patching, one can show that the higher dimensional local conjecture holds for $r = 1$ [HS2]. But perhaps surprisingly, both the global and local conjectures fail in general, because some groups that satisfy the conditions of the conjectures nevertheless fail to arise as Galois groups of covers. In particular, the global conjecture fails for $\mathbb{P}^2_k$ minus three lines crossing normally, and the local conjecture fails for $n = r = 2$ [HP]. Thus not every embedding problem with quasi-p kernel can be solved for $\pi_1(U)$, in general.

REMARK 5.3.6. The main reason that the higher dimensional conjecture fails in general is that the group-theoretic reduction in the proof of the general case of Theorem 5.3.1 does not work in the more general situation. That is, it is possible that $G/p(G) \in \pi_A(U)$ but that G is not a quotient of a group $\tilde{G}$ of the form $\tilde{G} = p(G) \rtimes \bar{G}$, with $\bar{G}$ a prime-to-p group in $\pi_A(U)$. (Cf. the group-theoretic examples of Guralnick in [HP].) Moreover, even if there is such a $\tilde{G}$, it might not be possible to choose it such that $\bar{G}$ normalizes a Sylow p-subgroup of $p(G)$ (or equivalently, of G), as was done in the proof of Theorem 5.3.1. And in fact, a condition of the above type is *necessary* in order that $G \in \pi_A(U)$, if U is the complement of $x_1 \cdots x_i = 0$ (in either the local or global case; cf. [HP]).

This suggests that a group G should lie in $\pi_A(U)$ if it satisfies these additional conditions, as well as the condition that $G/p(G) \in \pi_A(U)$. One might wish to parallel the proof of the general case of Theorem 5.3.1, using higher dimensional patching (Theorem 3.2.12) together with the result on embedding problems with p-group kernel ([Ha12], which holds in arbitrary dimension). Unfortunately, there is another difficulty: The strategy for curves used that for every quasi-p group Q there is a Q-Galois étale cover of $\mathbb{A}_k^1$ such that the fibre over infinity (of the corresponding branched cover of $\mathbb{P}_k^1$) consists of a disjoint union of points whose inertia groups are Sylow p-subgroups of Q (cf. Case 1 of the proof of Theorem 5.3.1). But the higher dimensional analog of this is false; in fact, for $n > 1$, every branched cover of $\mathbb{P}_k^n$ that is étale over $\mathbb{A}_k^n$ must have the property that its fibre over the hyperplane at infinity is connected [Hrt2, III, Cor. 7.9]. This then interferes with the desired patching, on the overlap. $\square$

One can also consider birational variants of the above questions, in studying the absolute Galois groups of $k_n := k(x_1, \ldots, x_n)$ and $k_n^* := k((x_1, \ldots, x_n))$. Here k is an algebraically closed field of characteristic $p \geq 0$; $n > 1$; and $k((x_1, \ldots, x_n))$ denotes the fraction field of $k[[x_1, \ldots, x_n]]$. Of course every finite group is a Galois group over k_n, since this is true for $k(x_1)$ (see Corollary 3.3.5) and one may base-change to k_n. Also, every finite group is a Galois group over k_n^*, by Example 3.3.2(c). But this does not determine the structure of the absolute Galois groups of k_n and k_n^*.

In the one-dimensional analog, the absolute Galois group of $k(x)$ is a free profinite group (of rank equal to the cardinality of k), by the geometric case of Shafarevich's Conjecture (Section 5.1). But for $n > 1$, the absolute Galois group of k_n has cohomological dimension > 1 [Se4, II, 4.1, Proposition 11], and so is not projective [Se4, I, 3.4, Proposition 16]. That is, not every finite split embedding problem for G_{k_n} has a weak solution; and therefore G_{k_n} is not free.

This can also be seen explicitly as in the following argument, which also applies to k_n^*:

PROPOSITION 5.3.7. *Let k be an algebraically closed field of characteristic $p \geq 0$, let $n > 1$, and let $K = k_n$ or k_n^* as above. Then not every finite embedding problem for the absolute Galois group G_K is weakly solvable. Equivalently, there is a surjection $G \to A$ of finite groups, and an A-Galois field extension K' of K, such that K' is not contained in any H-Galois field extension L of K for any $H \subset G$.*

PROOF. First suppose that char $k \neq 2$. Let G be the quaternion group of order 8, and let A be the quotient of G by its center $Z = \{\pm 1\}$. Thus $A = G/Z \approx C_2^2$, say with generators a, b which are commuting involutions. Consider the surjection $G_K \to A$ corresponding to the A-Galois field extension K' given by $u^2 = x_1, v^2 = x_2$. Suppose that this field extension is contained in an H-Galois extension L/K as in the statement of the proposition. Then A is a quotient of H. But no proper subgroup of G surjects onto A; so actually $H = G$.

Let $F = k((x_1)) \cdots ((x_n))$, and let F' [resp. E] be the compositum of F and K' [resp. F and L] in some algebraic closure of F. Thus E is a Galois field extension of F, and its Galois group G' is a subgroup of G. Moreover E contains F', which is an A-Galois field extension of F (being given by $u^2 = x_1, v^2 = x_2$). Thus A is a quotient of G', and hence $G' = G$. But the maximal prime-to-p quotient of the absolute Galois group G_F is abelian [HP, Prop. 2.4], and so G cannot be a Galois group over F (using that $p \neq 2$). This is a contradiction, proving the result in this case.

On the other hand, if char $k = 2$, then one can replace the quaternion group in the above argument by a similar group of order prime to 2. Namely, let ℓ be any odd prime. Then there is a group G of order ℓ^3 whose center Z is cyclic of order ℓ; such that $G/Z \approx C_\ell^2$; and such that no proper subgroup of G surjects onto G/Z. (See [As, 23.13]; such a group is called an *extraspecial* group of order ℓ^3.) The proof then proceeds as before. □

REMARK 5.3.8. This proof also applies to the field $K = k((x_1, \ldots, x_n))(y)$, by using the extension $u^2 = x_1, v^2 = y$. So its absolute Galois group G_K is not projective, and hence not free. (This can alternatively be seen by using [Se4, II, 4.1, Proposition 11]). Note that this field K has the property that every finite group is a Galois group over K (by Theorem 3.3.1), even though G_K is not free or even projective. In fact if $n = 1$, then every finite *split* embedding problem has a proper solution (by Theorem 5.1.9). Thus in this case, once a finite embedding problem has a weak solution, it automatically has a proper solution. In this sense, the absolute Galois group of $k((x))(y)$ is "as close as possible to being free" without being projective. □

Motivated by the preceding proposition and remark, it would be desirable to know whether the absolute Galois groups of $k_n := k(x_1, \ldots, x_n)$ and $k_n^* := k((x_1, \ldots, x_n))$ are "as close as possible to being free" without being projective. (Here k is still algebraically closed and $n > 1$.) In other words, does every finite split embedding problem for G_{k_n} or $G_{k_n^*}$ have a proper solution? The former case can be regarded as a birational analog of the question asked previously concerning quasi-p embedding problems in the higher dimensional Abhyankar Conjecture; it can also be considered a weak version of a higher dimensional geometric Shafarevich Conjecture. In this case, the question *remains open*, even for $\mathbb{C}(x, y)$. In the latter case, the answer is affirmative for $\mathbb{C}((x, y))$, as the following result shows. The proof follows a strategy from [HS2], viz. blowing up Spec $\mathbb{C}[\![x, y]\!]$ at the closed point to obtain a more global object, and then patching (here using Theorem 3.2.12).

THEOREM 5.3.9. *Every finite split embedding problem over* $\mathbb{C}((x, y))$ *has a proper solution.*

PROOF. Let L be a finite Galois extension of $\mathbb{C}((x, y))$, with group G, and let Γ be a semi-direct product $N \rtimes G$ for some finite group N. Let $R = \mathbb{C}[\![x, y]\!]$ and let S be the integral closure of R in L, and write $X^* := \operatorname{Spec} R$ and $Z^* := \operatorname{Spec} S$. We want to show that there is an irreducible normal Γ-Galois branched cover $W^* \to X^*$ that dominates the G-Galois branched cover $Z^* \to X^*$.

Case 1: S/R is ramified only over $(x = 0)$. Let n be the ramification index of $Z^* \to X^*$ over the generic point of $(x = 0)$, and consider the normalized pullback of $Z^* \to X^*$ via Spec $R[z]/(z^n - x) \to X^*$. By Abhyankar's Lemma and Purity of Branch Locus, the resulting cover of Spec $R[z]/(z^n - x) = \operatorname{Spec} \mathbb{C}[\![z, y]\!]$ is unramified and hence trivial. Thus $S \approx R[z]/(z^n - x)$, and G is cyclic of order n.

Now consider the projective y-line over $\mathbb{C}((x))$, and the G-Galois cover of this line $Z^\circ \to \mathbb{P}^1_{\mathbb{C}((x))}$ that is given by the constant extension $z^n = x$. Applying Pop's Theorem 5.1.10 to the split embedding problem given by this cover and the group homomorphism $\Gamma \to G$, we obtain a regular irreducible (hence geometrically irreducible) Γ-Galois cover $W^\circ \to \mathbb{P}^1_{\mathbb{C}((x))}$ that dominates $Z^\circ \to \mathbb{P}^1_{\mathbb{C}((x))}$ and is such that $W^\circ \to Z^\circ$ is totally split over $y = \infty$. Consider the normalization W of $\mathbb{P}^1_{\mathbb{C}[\![x]\!]}$ in W°; this is a Γ-Galois cover of $\mathbb{P}^1_{\mathbb{C}[\![x]\!]}$ that dominates Z, the normalization of $\mathbb{P}^1_{\mathbb{C}[\![x]\!]}$ in Z°. The branch locus of $W \to \mathbb{P}^1_{\mathbb{C}[\![x]\!]}$ consists of finitely many irreducible components. After a change of variables $y' = x^m y$ on $\mathbb{P}^1_{\mathbb{C}((x))}$, we may assume that every branch component passes through the closed point (x, y), and that no branch component other than (x) passes through any other point on the closed fibre of $\mathbb{P}^1_{\mathbb{C}[\![x]\!]}$. Again using Abhyankar's Lemma and Purity of Branch Locus, we conclude that the restriction of W over $\mathbb{C}[y^{-1}][\![x]\!]$ is a disjoint union of components given by $w^N = x$ for some multiple N of n, with each reduced component of the closed fibre of W being a complex line. Since $W^\circ \to Z^\circ$ is split over $y = \infty$, it follows that $N = n$. Thus the pullback of $W \to Z$ over $\mathbb{C}[y^{-1}][\![x]\!]$ is a trivial cover.

Since the general fibre of $W \to \mathbb{P}^1_{\mathbb{C}[\![x]\!]}$ is geometrically irreducible, the closed fibre is connected, by Zariski's Connectedness Theorem [Hrt2, III, Cor. 11.3]. So by the previous paragraph, the components of the closed fibre of W all meet at a single point over $(x = y = 0)$. So the pullback W^* of $W \to \mathbb{P}^1_{\mathbb{C}[\![x]\!]}$ over

$\operatorname{Spec}\mathbb{C}[\![x,y]\!]$ is connected; and since W is normal, it follows that W^* is also normal and hence is irreducible. So $W^* \to \operatorname{Spec}\mathbb{C}[\![x,y]\!]$ is an irreducible Γ-Galois cover. Moreover W^*/N is isomorphic to $\operatorname{Spec}S$ over $\operatorname{Spec}R$, since each is given by $z^n = x$. So it is a proper solution to the given embedding problem.

Note that in this case, the proof shows more: that G is the cyclic group C_n, and that over $\mathbb{C}((y))[\![x]\!]$, the pullback of $W^* \to Z^*$ is trivial (since the same is true over $\mathbb{C}[y^{-1}][\![x]\!]$).

Case 2: General case. Let B be the branch locus of $Z^* \to X^*$, and let C be the tangent cone to B at the closed point (x,y). Thus C is a union of finitely many "lines" $(ax + by)$ through (x,y) in X^*. After a change of variables of the form $y' = y - cx$, we may assume that C does not contain the locus of $(y = 0)$.

Let $\tilde{X}$ be the blow-up of X^* at the closed point (x,y). Let E be the exceptional divisor; this is a copy of $\mathbb{P}^1_{\mathbb{C}}$, with parameter $t = y/x$. Let $\tau \in T$ be the closed point $(x = y' = t = 0)$; this is where E meets the proper transform of $(y = 0)$. Let $\tilde{Z} \to \tilde{X}$ be the normalized pullback of $Z^* \to X^*$. By the previous paragraph, this is unramified in a neighborhood of τ except possibly along E. So over the complete local ring $\hat{\mathcal{O}}_{\tilde{X},\tau} = \mathbb{C}[\![x,t]\!]$ of τ in $\tilde{X}$, the pullback $\tilde{Z}^* \to \tilde{X}^* := \operatorname{Spec}\hat{\mathcal{O}}_{\tilde{X},\tau}$ of $\tilde{Z} \to \tilde{X}$ is ramified only over $(x = 0)$. We will construct a Γ-Galois cover $\tilde{W} \to \tilde{X}$ dominating $\tilde{Z}$. (See Figures 5.3.10 and 5.3.11.)

Let $\tilde{Z}_0^*$ be a connected component of $\tilde{Z}^*$. Thus $\tilde{Z}_0^* \to \tilde{X}^*$ is Galois with group $G_0 \subset G$, and $\tilde{Z}^* = \operatorname{Ind}_{G_0}^G \tilde{Z}_0^*$. Let $\Gamma_0 \subset \Gamma$ be the subgroup generated by N and G_0 (identifying N with $N \rtimes 1 \subset \Gamma$, and G with $1 \rtimes G \subset \Gamma$). Thus $\Gamma_0 = N \rtimes G_0$. By Case 1, there is a regular irreducible normal Γ_0-Galois cover $\tilde{W}_0^* \to \tilde{X}^*$ that dominates $\tilde{Z}_0^*$, and such that the pullback of $\tilde{W}_0^* \to \tilde{Z}_0^*$ over $\tilde{X}' = \operatorname{Spec}\mathbb{C}((t))[\![x]\!]$ is trivial. That is, $\tilde{W}_0' := \tilde{W}_0^* \times_{\tilde{X}^*} \tilde{X}'$ is the trivial N-Galois cover of $\tilde{Z}_0' := \tilde{Z}_0^* \times_{\tilde{X}^*} \tilde{X}'$, and the Γ_0-Galois cover $\tilde{W}_0' \to \tilde{X}'$ is just $\operatorname{Ind}_{G_0}^{\Gamma_0} \tilde{Z}_0'$. Thus the Γ-Galois cover $\tilde{W}^* := \operatorname{Ind}_{\Gamma_0}^\Gamma \tilde{W}_0^* \to \tilde{X}^*$ has the property that its pullback $\tilde{W}' := \tilde{W}^* \times_{\tilde{X}^*} \tilde{X}'$ is just $\operatorname{Ind}_{G_0}^\Gamma \tilde{Z}_0' = \operatorname{Ind}_G^\Gamma \tilde{Z}'$, where $\tilde{Z}' = \operatorname{Ind}_{G_0}^G \tilde{Z}_0'$ is the pullback $\tilde{Z}^* \times_{\tilde{X}^*} \tilde{X}'$.

Let $U = E - \{\tau\}$, and let X' be the completion of $\tilde{X}$ along U; i.e. $X' = \operatorname{Spec}\mathbb{C}[s][\![y]\!]$, where $s = x/y = 1/t$. Let $Z' = \tilde{Z} \times_{\tilde{X}} X'$, and let $W' = \operatorname{Ind}_G^\Gamma Z'$. Thus the pullback $Z' \times_{X'} \tilde{X}'$ can be identified with $\tilde{Z}' = \operatorname{Ind}_{G_0}^G \tilde{Z}_0'$ as G-Galois covers of $\tilde{X}'$; and the pullback $W' \times_{X'} \tilde{X}'$ can be identified with $\tilde{W}' = \operatorname{Ind}_{G_0}^\Gamma \tilde{Z}_0'$, as Γ-Galois covers of $\tilde{X}'$.

Now apply the formal patching result Theorem 3.2.12, with $A = R$, $V = \hat{V} = \tilde{X}$, $f = \text{identity}$, and the finite set of closed points of V being just $\{\tau\}$. Using the equivalence of categories for covers, we conclude that there is a unique Γ-Galois cover $\tilde{W} \to \tilde{X}$ whose pullbacks to $\tilde{X}^*$ and to X' are given respectively by $\tilde{W}^* = \operatorname{Ind}_{\Gamma_0}^\Gamma \tilde{W}_0^* \to \tilde{X}^*$ and $W' \to X'$, compatibly with the above identification over $\tilde{X}'$ with $\tilde{W}' = \operatorname{Ind}_{G_0}^\Gamma \tilde{Z}_0' \to \tilde{X}'$. The quotient $\tilde{W}/N$ can be identified with $\tilde{Z}$ as a G-Galois cover, since we have compatible identifications of their pullbacks

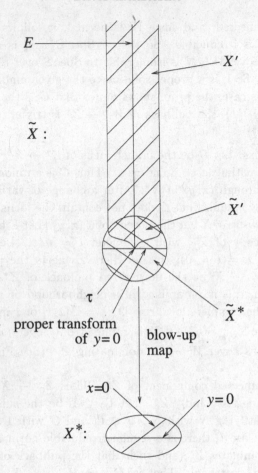

Figure 5.3.10. Picture of the situation in Case 2 of the proof of Theorem 5.3.9. The space X^*, shown as a disc, is blown up, producing $\tilde{X}$, with an exceptional divisor E (which meets the proper transform of $y = 0$ at the point τ). The proof proceeds by building the desired cover over formal patches: X', the completion along $E - \{\tau\}$; and $\tilde{X}^*$, the completion at τ. These two patches are shaded above, with the doubly shaded region $\tilde{X}'$ being the "overlap".

over $\tilde{X}^*$, X', and their "overlap" $\tilde{X}'$, and because of the uniqueness assertion of the patching theorem. Also, $\tilde{W}$ is normal, since normality is a local property and since $\tilde{W}^*$ and W' are normal. Let $\tilde{W}_0$ be the connected component of $\tilde{W}$ whose pullback to $\tilde{X}^*$ contains $\tilde{W}_0^*$. Its Galois group Γ_1 over $\tilde{X}$ surjects onto $G = \mathrm{Gal}(\tilde{Z}/\tilde{X})$, and Γ_1 contains $\mathrm{Gal}(\tilde{W}_0^*/\tilde{X}^*) = \Gamma_0 \supset N \rtimes 1$. So Γ_1 is all of Γ, and so $\tilde{W}_0 = \tilde{W}$. That is, $\tilde{W}$ is connected, and hence irreducible (being normal).

Now let $W^* \to X^*$ be the normalization of X^* in $\tilde{W}$. This is then a connected normal Γ-Galois cover that dominates Z^* (since Z^* is the normalization of X^* in $\tilde{Z}$). It is irreducible because it is connected and normal. So it provides a proper solution to the given embedding problem. $\square$

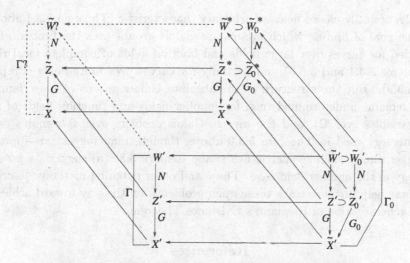

Figure 5.3.11. Diagram illustrating the patching situation in Case 2 of the proof of Theorem 5.3.9. In order to construct a Γ-Galois cover $\tilde{W} \to \tilde{X}$, the restrictions $W' \to X'$ and $\tilde{W}^* \to \tilde{X}^*$ are first constructed, so as to induce the same "overlap" $\tilde{W}' \dashrightarrow \tilde{X}'$. Formal patching is then used to obtain $\tilde{W}$.

REMARK 5.3.12. (a) Theorem 5.3.9 would also follow from Theorem 5.1.9, if it were known that $\mathbb{C}((x,y))$ is large. (Namely, given a split embedding problem over $\mathbb{C}((x,y))$, one could apply Theorem 5.1.9 to the induced constant split embedding problem over $\mathbb{C}((x,y))(t)$; and then one could specialize the proper solution to an extension of $\mathbb{C}((x,y))$, using that that field is separably Hilbertian by Weissauer's Theorem [FJ, Theorem 14.17].) But it *is unknown* whether $\mathbb{C}((x,y))$ is large. (Cf. Example 3.3.7(d).)

(b) It would be desirable to generalize the above result, e.g. by allowing more Laurent series variables, and by replacing $\mathbb{C}$ by an algebraically closed field of arbitrary characteristic (or even by an arbitrary large field). Note that the above proof uses Kummer theory and Abhyankar's Lemma, and so one would somehow need to treat the case of wild ramification. $\qquad\square$

The ultimate goal remains that of proving a full analog of Riemann's Existence Theorem — classifying covers via their Galois groups and inertia groups, and determining how they fit together into the tower of covers. This goal, however, has so far been achieved in full only for curves over algebraically closed fields of characteristic 0 (where it is deduced from the complex result, which relied on topological methods). As seen above, the weaker goal of finding π_1 as a profinite group, and finding absolute Galois groups of function fields, also remains open in most cases, although the absolute Galois group of the function field is known for curves over algebraically closed fields (Theorem 5.1.1), and partial results are known for other fields (e.g. Theorem 5.1.9, 5.3.4, and 5.3.9). The still weaker, but difficult, goal of finding π_A has been achieved for affine curves

over algebraically closed fields of arbitrary characteristic (Theorem 5.3.1 above),
and the goal of finding which groups are Galois groups over the function field
is settled for curves over large fields and fraction fields of complete local rings
(Theorems 3.3.1 and 3.3.6) and partially for curves over finite fields (Proposition 3.3.9). But the structure of the absolute Galois groups of most familiar
fields remains undetermined (e.g. for number fields and function fields of several variables over $\mathbb{C}$), and the inverse Galois problem over $\mathbb{Q}$ remains open.
The strategy used in Theorem 5.3.9 above, though, may suggest an approach
to higher dimensional geometric fields; and Remark 3.3.8(a) suggests a possible
strategy in the number field case. These and other patching methods described
here may help further attack these open problems, on the way toward achieving
a full generalization of Riemann's Existence Theorem.

References

[Ab1] S. Abhyankar. Coverings of algebraic curves. Amer. J. Math. **79** (1957), 825–856.

[Ab2] S.S. Abhyankar. Tame coverings and fundamental groups of algebraic varieties,
Part 1. Amer. J. Math. **84** (1959), 46–94.

[Ab3] S.S. Abhyankar. Local fundamental groups of algebraic varieties. Proc. Amer.
Math. Soc. **125** (1997), 1635–1641.

[Ar1] M. Artin. "Grothendieck Topologies". Harvard University Math. Dept., 1962.

[Ar2] M. Artin. Algebraic approximation of structure over complete local rings. Publ.
Math. IHES **36** (1969), 23–58.

[Ar3] M. Artin. Algebraization of formal moduli: I. In "Global Analysis: Papers in
Honor of K. Kodaira". Princeton Univ. Press, 1969, pp. 21–71.

[Ar4] M. Artin. Algebraization of formal moduli: II. Existence of modifications. Annals
of Math. ser. 2, **91** (1970), 88–135.

[Ar5] M. Artin. Introduction to Part I: Holomorphic functions. In [Za5], pp. 3–8.

[As] M. Aschbacher. "Finite group theory", 2nd ed. Cambridge Univ. Press, 2000.

[Ax1] J. Ax. Proof of some conjectures on cohomological dimension. Proc. Amer. Math.
Soc. **16** (1965), 1214–1221.

[Ax2] J. Ax. The elementary theory of finite fields. Ann. of Math. **88** (1968), 239–271.

[Be] S. Beckmann. Is every extension of $\mathbb{Q}$ the specialization of a branched covering?
J. Algebra **164** (1994), 430–451.

[Bi] G.D. Birkhoff. A theorem on matrices of analytic functions. Math. Ann. **74** (1917),
240–251.

[Bl1] E. Black. Arithmetic lifting of dihedral extensions. J. Algebra **203** (1998), 12–29.

[Bl2] E. Black. On semidirect products and the arithmetic lifting property. J. London
Math. Soc. (2) **60** (1999), 677–688.

[Bl3] E. Black. Lifting wreath product extensions. Proc. Amer. Math. Soc. 129 (2001),
1283–1288.

[BGR] S. Bosch, U. Güntzer, R. Remmert. "Non-Archimedean Analysis". Grundlehren
der mathematischen Wissenschaften **261**, Springer-Verlag, 1984.

[BLü1] S. Bosch, W. Lütkebohmert. Formal and rigid geometry: I. Rigid spaces. Math. Ann. **295** (1993), 291–317.

[BLü2] S. Bosch, W. Lütkebohmert. Formal and rigid geometry: II. Flattening techniques. Math. Ann. **296** (1993), 403–429.

[BLüR1] S. Bosch, W. Lütkebohmert, M. Raynaud. Formal and rigid geometry: III. The relative maximum principle. Math. Ann. **302** (1995), 1–29.

[BLüR2] S. Bosch, W. Lütkebohmert, M. Raynaud. Formal and rigid geometry: IV. The reduced fibre theorem. Invent. Math. **119** (1995), 361–398.

[BLoR] J.-B. Bost, F. Loeser, M. Raynaud (eds.). "Courbes semi-stables et groupe fondamental en géométrie algébrique". Progress in Mathematics vol. 187, Birkhäuser, 2000.

[Bo] N. Bourbaki. "Elements of Mathematics: Commutative algebra". Hermann, 1972.

[BP] I. Bouw, R. Pries. Rigidity, reduction, and ramification. 2001 manuscript. To appear in Math. Annalen.

[Ca1] H. Cartan. Sur les matrices holomorphes de n variables complexes. Journal de Mathématiques pures et appliquées, Series 9, **19** (1940), 1–26.

[Ca2] H. Cartan. Séminaire: Théorie des fonctions de plusieurs variables. Éc. Norm. Sup., 1951–52.

[Ca3] H. Cartan. Séminaire: Théorie des fonctions de plusieurs variables. Éc. Norm. Sup., 1953–54.

[CL] A. Chambert-Loir. La conjecture d'Abhyankar I: construction de revêtements en géométrie rigide. In [BLoR], pp. 233–248.

[Ch] W.-L. Chow. On compact complex analytic varieties. Amer. J. Math. **71** (1949), 893–914.

[Co] R. Coleman. Torsion points on curves. In: "Galois Representations and Arithmetic Algebraic Geometry" (Y. Ihara, ed.), Advanced Studies in Pure Mathematics, vol. 12, Elsevier Publishers, 1987.

[CT] J.-L. Colliot-Thélène. Rational connectedness and Galois covers of the projective line. Ann. of Math., **151** (2000), 359–373.

[CH] K. Coombes, D. Harbater. Hurwitz families and arithmetic Galois groups, Duke Math. J., **52** (1985), 821–839.

[Dè] P. Dèbes. Galois covers with prescribed fibers: the Beckmann-Black problem. Ann. Scuola Norm. Sup. Pisa Cl. Sci. (4) **28** (1999), 273–286.

[DD] P. Dèbes, B. Deschamps. The inverse Galois problem over large fields. In "Geometric Galois Actions, 2", London Math. Soc. Lecture Note Series **243**, L. Schneps and P. Lochak ed., Cambridge University Press, (1997), 119–138.

[DF] P. Dèbes, M. Fried. Rigidity and real residue class fields. Acta Arith. **56** (1990), 291–323.

[Do] A. Douady. Détermination d'un groupe de Galois. C.R. Acad. Sci. Paris **258** (1964), 5305–5308.

[FR] D. Ferrand, M. Raynaud. Fibres formelles d'un anneau local noethérien. Ann. Scient. Éc. Norm. Sup. ser. 4, **3** (1970), 295–311.

[FP] J. Fresnel, M. van der Put. "Géométrie Analytique Rigide et Applications". Progress in Mathematics, vol. 18, Birkäuser, 1981.

[Fr1] M. Fried. Fields of definition of function fields and Hurwitz families – groups as Galois groups. Comm. Alg. **5** (1977), 17–82.

[Fr2] M. Fried, et al., eds. "Recent developments in the inverse Galois problem". AMS Contemporary Mathematics Series, vol. 186, 1995.

[Fr3] M. Fried. Introduction to modular towers. In [Fr2], pp. 111–171.

[FJ] M. Fried, M. Jarden. "Field Arithmetic." Ergebnisse Math. series, vol. 11, Springer-Verlag, 1986.

[FV1] M. Fried, H. Völklein. The inverse Galois problem and rational points on moduli spaces. Math. Ann. **290** (1991), 771–800.

[FV2] M. Fried, H. Völklein. The embedding problem over a Hilbertian PAC-field. Annals of Math. **135** (1992), 469–481.

[Fu1] W. Fulton. Hurwitz schemes and irreducibility of moduli of algebraic curves. Ann. of Math. (2) **90** (1969), 542–575.

[Fu2] W. Fulton. On the fundamental group of the complement of a node curve. Annals of Math. **111** (1980), 407–409.

[Ga] M. Garuti. Géométrie rigide et géométrie formelle. In [BLoR], pp. 7–19.

[GG] L. Gerritzen, H. Grauert. Die Azyklizität der affinoiden Überdeckungen. In "Global Analysis: Papers in Honor of K. Kodaira". Princeton Univ. Press, 1969, pp. 159–184.

[GrRe1] H. Grauert, R. Remmert. Nichtarchimedische Funktionentheorie. In: "Festschrift Gedächtnisfeier K. Weierstrass", Wissenschaftl. Abh. Arbeitsgemeinschaft für Forschung des Landes Nordrhein-Westfalen, vol. 33, Westdeutscher Verlag, 1966, pp. 393–476.

[GrRe2] H. Grauert, R. Remmert. Über die Methode der diskret bewerteten Ringe in der nicht-archimedischen Analysis. Invent. Math. **2** (1966), 87–133.

[GH] P. Griffiths, J. Harris. "Principles of algebraic geometry". John Wiley and Sons, 1978.

[Gr1] A. Grothendieck. Sur quelques points d'algèbre homologique. Tôhoku Math. J. **9** (1957), 119–221.

[Gr2] A. Grothendieck. Géometrie formelle et géométrie algégbrique. Sem. Bourb. **182** (1959), 1–28.

[Gr3] A. Grothendieck. Technique de descente et théorèmes d'existence en géometrie algébrique, I. Sem. Bourb. **190** (1959), 1–29.

[Gr4] A. Grothendieck. "Élements de géométrie algébrique" (EGA), Publ. Math. IHES: EGA II, vol. 8 (1961); EGA III, 1e partie, vol. 11 (1961); EGA IV, 1e partie, vol. 20 (1964).

[Gr5] A. Grothendieck. "Revêtements étales et groupe fondamental" (SGA 1). Lecture Notes in Mathematics, vol. 224, Springer-Verlag, 1971.

[Gr6] A. Grothendieck. Letter to Faltings. 1983 letter. In "Geometric Galois Actions, I" (L. Schneps and P. Lochak, eds.), London Math. Soc. Lecture Notes, vol. 242, Cambridge Univ. Press, 1997, pp. 49–58 (German), pp. 285–293 (English translation).

[GuRo] R.C. Gunning, H. Rossi. "Analytic functions of several complex variables". Prentice-Hall, 1965.

[HJ1] D. Haran, M. Jarden. Regular split embedding problems over complete valued fields. Forum Mathematicum **10** (1998), 329–351.

[HJ2] D. Haran, M. Jarden. Regular lifting of covers over ample fields. 2000 manuscript.

[HV] D. Haran, H. Völklein. Galois groups over complete valued fields. Israel J. Math., **93** (1996), 9–27.

[Ha1] D. Harbater. Deformation theory and the tame fundamental group. Trans. AMS, **262** (1980), 399–415.

[Ha2] D. Harbater. Convergent arithmetic power series. Amer. J. Math., **106** (1984), 801–846.

[Ha3] D. Harbater. Mock covers and Galois extensions. J. Algebra, **91** (1984), 281–293.

[Ha4] D. Harbater. Galois coverings of the arithmetic line. In "Number Theory: New York, 1984–85". Lecture Notes in Mathematics, vol. 1240, Springer-Verlag, 1987, pp. 165–195.

[Ha5] D. Harbater. Galois covers of an arithmetic surface. Amer. J. Math., **110** (1988), 849–885.

[Ha6] D. Harbater. Formal patching and adding branch points. Amer. J. Math., **115** (1993), 487–508.

[Ha7] D. Harbater. Abhyankar's conjecture on Galois groups over curves. Inventiones Math., **117** (1994), 1–25.

[Ha8] D. Harbater. Galois groups with prescribed ramification. In "Proceedings of a Conference on Arithmetic Geometry (Arizona State Univ., 1993)", AMS Contemp. Math. Series, vol. 174, 1994, pp. 35–60.

[Ha9] D. Harbater. Fundamental groups of curves in characteristic p. Proc. Intl. Cong. of Math. (Zurich, 1994), Birkhauser, 1995, pp. 656–666.

[Ha10] D. Harbater. Fundamental groups and embedding problems in characteristic p. In [Fr2], pp. 353–369.

[Ha11] D. Harbater. Embedding problems and adding branch points. In "Aspects of Galois theory" (H. Völklein, et al., eds.), London Math. Soc. Lecture Notes, vol. 256, Cambridge Univ. Press, 1999, pp. 119–143.

[Ha12] D. Harbater. Embedding problems with local conditions. Israel J. of Math, **118** (2000), 317–355.

[Ha13] D. Harbater. Abhyankar's Conjecture and embedding problems. To appear in Crelle's J. Available at http://www.math.upenn.edu/~harbater/qp.dvi

[HS1] D. Harbater, K. Stevenson. Patching and thickening problems. J. Alg., **212** (1999), 272–304.

[HS2] D. Harbater, K. Stevenson. Abhyankar's local conjecture on fundamental groups. To appear in "Proceedings of the Conference on Algebra and Algebraic Geometry with Applications" (Abhyankar 70th birthday conference, C. Christensen, ed.), Springer-Verlag. Available at http://www.math.upenn.edu/~harbater/abhloc.dvi

[HP] D. Harbater, M. van der Put. Valued fields and covers in characteristic p (with an appendix by R. Guralnick). In "Valuation Theory and its Applications" (F.-V. Kuhlmann, S. Kuhlmann and M. Marshall, eds.) Fields Institute Communications, vol. 32, 2002, pp. 175–204.

[Hrt1] R. Hartshorne. "Residues and duality." Lecture Notes in Mathematics, vol. 20, Springer-Verlag, 1966.

[Hrt2] R. Hartshorne. "Algebraic geometry". Graduate Texts in Mathematics, vol. 52. Springer-Verlag, 1977.

[He] Y. Henrio. Disques et couronnes ultramétriques. In [BLoR], pp. 21–32.

[Iw] K. Iwasawa. On solvable extensions of algebraic number fields. Annals of Math. 58 (1953), 548–572.

[Ja] M. Jarden. On free profinite groups of uncountable rank. In [Fr2], pp. 371–383.

[Ka] N. Katz. "Rigid local systems". Annals of Mathematics Studies, vol. 139. Princeton University Press, 1996.

[Ki1] R. Kiehl. Der Endlichkeitssatz für eigentliche Abbildungen in der nichtarchimedischen Funktionentheorie. Invent. Math. 2 (1967), 191–214.

[Ki2] R. Kiehl. Theorem A und Theorem B in der nichtarchimedischen Funktionentheorie. Invent Math. 2 (1967), 256–273.

[Kol] J. Kollár. Rationally connected varieties over local fields. Ann. of Math. 150 (1999), 357–367.

[Köp] U. Köpf. Über eigentliche Familien algebraischer Varietäten über affinoiden Räumen. Schriftenreihe des Math. Inst. der Univ. Münster, 2 Serie, Heft 7 (1974).

[La] S. Lang. "Algebraic Number Theory". Addison-Wesley, 1970.

[Le] T. Lefcourt. Galois groups and complete domains. Israel J. Math., 114 (1999), 323–346.

[Li] Q. Liu. Tout groupe fini est un groupe de Galois sur $\mathbb{Q}_p(T)$, d'après Harbater. In [Fr2], pp. 261–265.

[Lü1] W. Lütkebohmert. Formal-algebraic and rigid-analytic geometry. Math. Ann. 286 (1990), 341–371.

[Lü2] W. Lütkebohmert. Riemann's existence theorem for a p-adic field. Invent. Math. 111 (1993), 309–330.

[MR] M. Madan, M. Rosen. The automorphism group of a function field. Proc. Amer. Math. Soc. 115 (1992), 923–929.

[MM] G. Malle, B.H. Matzat. "Inverse Galois theory". Springer Monographs in Mathematics. Springer-Verlag, 1999.

[Mi] J.S. Milne. "Étale Cohomology". Princeton Mathematical Series, vol. 33, Princeton University Press, 1980.

[Mo] S. Mochizuki. The profinite Grothendieck conjecture for closed hyperboic curves over number fields. J. Math. Sci. Univ. Tokyo 3 (1996), 571–627.

[MB1] L. Moret-Bailly. Groupes de Picard et problèmes de Skolem, II. Ann. Sci. École Norm. Sup. (4) 22 (1989), 181–194.

[MB2] L. Moret-Bailly. Construction de revêtements de courbes pointées. J. Algebra, 240 (2001), 505–534.

[NSW] J. Neukirch, A. Schmidt, K. Wingberg. "Cohomology of Number Fields". Grundlehren series, vol. 323, Springer, 2000.

[PS] A. Pacheco, K. Stevenson. Finite quotients of the algebraic fundamental group of projective curves in positive characteristic. Pacific J. Math., 192 (2000), 143–158.

[Po1] F. Pop. The geometric case of a conjecture of Shafarevich. Manuscript, University of Heidelberg, 1993.

[Po2] F. Pop. Half Riemann Existence Theorem with Galois action. In "Algebra and Number Theory", G. Frey and J. Ritter, eds., de Gruyter Proceedings in Math. (1994), 1–26.

[Po3] F. Pop. Étale Galois covers of affine smooth curves. Invent. Math., **120** (1995), 555–578.

[Po4] F. Pop. Embedding problems over large fields. Ann. Math., **144** (1996), 1–34.

[Pr1] R. Pries. Construction of covers with formal and rigid geometry. In [BLoR], pp. 157–167.

[Pr2] R. Pries. Degeneration of wildly ramified covers of curves. 2001 manuscript.

[Ra1] M. Raynaud. Géométrie analytique rigide d'après Tate, Kiehl, ... Bull. Soc. Math. France, Mémoire 39–40 (1974), 319–327.

[Ra2] M. Raynaud. Revêtements de la droite affine en caractéristique $p > 0$ et conjecture d'Abhyankar. Invent. Math. **116** (1994), 425–462.

[Ri] L. Ribes. "Introduction to profinite groups and Galois cohomology Introduction to profinite groups and Galois cohomology". Queen's University, 1970.

[Ru] W. Rudin. "Real and complex analysis", second edition. McGraw-Hill, 1974.

[Sa1] M. Saïdi. Revêtements modérés et groupe fondamental de graphe de groupes. Compositio Math. **107** (1997), 319–338.

[Sa2] M. Saïdi. Abhyankar's conjecture II: the use of semi-stable curves. In [BLoR], pp. 249–265.

[Slt] D. Saltman. Generic Galois extensions and problems in field theory. Advances in Math. **43** (1982), 250–283.

[Se1] J.-P. Serre. Faisceaux analytiques sur l'espace projectif. Exposés XVIII and XIX in [Ca3].

[Se2] J.-P. Serre. Faisceaux algébriques cohérents. Ann. of Math. **61** (1955), 197–278.

[Se3] J.-P. Serre. Géométrie algébrique et géométrie analytique. Annales de L'Institut Fourier **6** (1956), 1–42.

[Se4] J.-P. Serre. "Cohomologie Galoisienne". Lecture Notes in Mathematics, vol. 5, Springer-Verlag, 1964.

[Se5] J.-P. Serre. "Local fields". Graduate Texts in Math., vol. 67, Springer-Verlag, 1979.

[Se6] J.-P. Serre. Construction de revêtements étales de la droite affine en caractéristique p. Comptes Rendus **311** (1990), 341–346.

[Se7] J.-P. Serre. "Topics in Galois Theory". Research Notes in Math., vol. 1, Jones and Bartlett, 1992.

[St1] K. Stevenson. Galois groups of unramified covers of projective curves in characteristic p. J. Algebra 182 (1996), 770–804.

[St2] K. Stevenson. Conditions related to π_1 of projective curves. J. Number Theory **69** (1998), 62–79.

[Tm1] A. Tamagawa. The Grothendieck conjecture for affine curves. Compositio Math. **109** (1997), 135–194.

[Tm2] A. Tamagawa. On the fundamental group of curves over algebraically closed fields of characteristic > 0. Intern. Math. Res. Notices (1999), no. 16, 853–873.

[Ta] J. Tate. Rigid analytic spaces. Invent. Math. **12** (1971), 257–289.

[Th] J. Thompson. Some finite groups which appear as $\operatorname{Gal} L/K$, where $K \subseteq \mathbb{Q}(\mu_n)$. J. Algebra **89** (1984), 437–499.

[Vö] H. Völklein. "Groups as Galois groups — an introduction". Cambridge Studies in Advanced Mathematics, vol. 53, Cambridge University Press, 1996.

[We] C. Weibel. "An introduction to homological algebra". Cambridge Studies in Advanced Mathematics, vol. 38, Cambridge Univ. Press, 1994.

[Za1] O. Zariski. On the problem of existence of algebraic functions of two variables possessing a given branch curve. Amer. J. of Math., **51** (1929), 305–328.

[Za2] O. Zariski. Generalized semi-local rings. Summa Brasiliensis Mathematicae **1**, fasc. 8 (1946), 169–195.

[Za3] O. Zariski. "Algebraic surfaces". Chelsea Pub. Co., 1948.

[Za4] O. Zariski. Theory and applications of holomorphic functions on algebraic varieties over arbitrary ground fields. Mem. Amer. Math. Soc. **5** (1951), 1–90. Reprinted in [Za5], pp. 72–161.

[Za5] O. Zariski. "Collected Papers, volume II: Holomorphic functions and linear systems". M. Artin and D. Mumford, eds. MIT Press, 1973.

DAVID HARBATER
DEPARTMENT OF MATHEMATICS
UNIVERSITY OF PENNSYLVANIA
PHILADELPHIA, PA 19104-6395
UNITED STATES
harbater@math.upenn.edu

Galois Groups and Fundamental Groups
MSRI Publications
Volume 41, 2003

Constructive Differential Galois Theory

B. HEINRICH MATZAT AND MARIUS VAN DER PUT

ABSTRACT. We survey some constructive aspects of differential Galois theory and indicate some analogies between ordinary Galois theory and differential Galois theory in characteristic zero and nonzero.

CONTENTS

INTRODUCTION

The aim of this article is to survey some constructive aspects of differential Galois theory and to indicate some analogies between ordinary Galois theory and differential Galois theory in characteristic zero and nonzero. We hope it may serve as an appetizer for people who work in ordinary Galois theory but are not familiar with the differential analogue.

In the first part we start with a constructive foundation of the Picard–Vessiot theory in characteristic zero mimicking Kronecker's construction of root fields.

This leads to a smallest differential field extension (with no new constants) containing a full system of solutions of a (system of) linear differential equation(s) with a linear algebraic group as differential Galois group. Then we explain the Galois correspondence between the intermediate differential fields of a Picard–Vessiot extension and the Zariski closed subgroups of the differential Galois group. On the way we deal with the question of solvability by elementary functions, comparable to the question of solvability by radicals in ordinary Galois theory. In Chapter 3 we describe the link between the differential Galois group and the monodromy group over the complex numbers generalizing the effective version of Riemann's existence theorem used in (ordinary) inverse Galois theory [MM]. Further we recall the solution of the inverse differential Galois problem over $\mathbb{C}$ in the case of monodromy groups (Riemann–Hilbert problem) given by Plemelj (1908) and its completion by Tretkoff and Tretkoff [TT] for differential Galois groups. Finally in Chapter 4 we outline the constructive solution of the inverse problem for connected groups over general algebraically closed fields of characteristic 0 recently given by Mitschi and Singer [MS].

In the second part we develop a Picard–Vessiot theory in positive characteristic. For this purpose ordinary derivations — these cause new constants in any nonalgebraic extension — are replaced by a family of higher derivations, called iterative derivations in the original paper of Hasse and Schmidt [HS]. They have already been used earlier by Okugawa [Oku] to outline a Picard–Vessiot theory in characteristic $p > 0$. Here we follow a new approach developed in [MP] based on the study of iterative differential modules (ID-modules) and corresponding projective systems. This allows us to construct (iterative) Picard–Vessiot extensions in the same formal way as in characteristic 0. We again obtain as ID-Galois groups reduced linear algebraic groups defined over the field of constants and we establish a Galois correspondence between the intermediate ID-fields of a Picard–Vessiot extension and the reduced closed subgroups of the corresponding ID-Galois group. In Chapter 7 we determine the structure of ID-modules and ID-Galois groups over local fields — these are trigonalizable extensions of connected solvable groups by finite local Galois groups — and solve the inverse problem for these groups. Finally in Chapter 8 we solve the inverse problem of differential Galois theory over global fields of positive characteristic and prove an analogue of the Abhyankar conjecture for differential Galois extensions.

The main sources (sometimes used without a reference) are the introductory texts of Magid [Mag] and the second author [Put2] for the classical part, for the modular part there are the research paper [MP] combined with the notes [Mat]. Different approaches for differential equations in positive characteristic have been developed, for example, by Katz [Kat2] and André [And].

Acknowledgement. The authors thank the Mathematical Science Research Institute in Berkeley (MSRI) for its hospitality and support during the research program Galois Groups and Fundamental Groups, and for giving us the opportunity to present most of the results given in this article in a series of lectures at the MSRI. Among other things, the solution of the inverse problem of ID-Galois theory for connected groups over global fields (Theorem 8.4) and the proof of the connected differential Abhyankar conjecture (Corollary 8.5) have been achieved during our stay in Berkeley.

CLASSICAL THEORY

1. Linear Differential Equations

1.1. Derivations. In this first section we collect some well-known facts on derivations and differential rings. The proofs can be found, for example, in [Jac], Chapter 8.15.

Let R be a commutative ring (always with unit element). A map $\partial : R \to R$ is called a *derivation* of R if

$$\partial(a + b) = \partial(a) + \partial(b) \quad \text{and} \quad \partial(a \cdot b) = \partial(a)b + a\partial(b)$$

for all $a, b \in R$. An element $c \in R$ with $\partial(c) = 0$ is a *differential constant*. The set of differential constants forms a ring denoted here by $C(R)$. Further a ring R together with a derivation ∂ of R is called a *differential ring* (D-ring) (R, ∂).

From the definition we immediately obtain the formulas

$$\partial\left(\frac{a}{b}\right) = \frac{1}{b^2}(\partial(a)b - a\partial(b)) \qquad \text{in case } b \in R^\times, \tag{1-1}$$

$$\partial^k(ab) = \sum_{i+j=k} \binom{k}{i} \partial^i(a)\partial^j(b) \tag{1-2}$$

for $a, b \in R$ and $i, j, k \in \mathbb{N}$.

Now let (R, ∂_R) and (S, ∂_S) be two D-rings. Then a ring homomorphism $\varphi \in \mathrm{Hom}(R, S)$ is called a *differential homomorphism* (D-homomorphism) if $\varphi \circ \partial_R = \partial_S \circ \varphi$. The set of all D-homomorphisms is denoted by $\mathrm{Hom}_D(R, S)$. An ideal A of R with $\partial_R(A) \subseteq A$ is called a *differential ideal* (D-ideal). It can be shown that in case R is a Ritt algebra, i.e., $\mathbb{Q} \leq R$, the nil radical of any D-ideal again is a D-ideal. A corresponding statement does not hold anymore in positive characteristic (see [Kap], I.4).

If (R, ∂_R) is a D-ring and $S \subseteq R$ a multiplicatively closed subset with $0 \notin S$ we have a canonical map $\lambda_S : R \to S^{-1}R$ from R into the quotient ring $S^{-1}R$. Then by (1–1) there exists a uniquely determined derivation $\partial_{S^{-1}R}$ of $S^{-1}R$ such that $\partial_{S^{-1}R} \circ \lambda_S = \lambda_S \circ \partial_R$. In particular, if R is an integral domain, ∂_R can be extended uniquely to its quotient field $F = \mathrm{Quot}(R)$. A field F with derivation ∂_F is called a *differential field* (D-field).

Finally, let E/F be a finitely generated separable field extension of a D-field (F, ∂_F) with separating transcendence basis $x_1, \ldots, x_r$. Then for all $y_1, \ldots, y_r \in E$ there exists exactly one extension ∂_E of ∂_F on E with $\partial_E(x_i) = y_i$ for all i. In particular, an extension of ∂_F to a separably algebraic field extension E/F always exists and is unique.

1.2. Linear differential operators. From now on, (F, ∂_F) denotes a D-field of characteristic 0. Then $\ell := \sum_{k=0}^n a_k \partial^k$ with $a_k \in F$ and $a_n \neq 0$ is called a *linear differential operator* of degree $\deg(\ell) = n$ over F (D-operator) and $F[\partial]$ is the (noncommutative) *ring of linear differential operators* over F. Now let (E, ∂_E) be a D-field extension of F. Then an element $y \in E$ is called a *solution* of ℓ if y is a solution of the homogeneous linear differential equation

$$\ell(y) = \sum_{k=0}^n a_k \partial^k(y) = 0. \tag{1-3}$$

The set of all solutions of ℓ in E forms a vector space over the field of constants $C(E)$ of E and is named the *solution space* $V_E(\ell)$ *of* ℓ *in* E.

PROPOSITION 1.1. *Let* (F, ∂_F) *be a D-field of characteristic* 0 *and* $\ell \in F[\partial]$ *a D-operator. Then for all D-field extensions* $(E, \partial_E) \geq (F, \partial_F)$ *the solution space* $V_E(\ell)$ *of* ℓ *is a vector space over* $C(E)$ *with* $\dim_{C(E)}(V_E(\ell)) \leq \deg(\ell)$.

The proof of Proposition 1.1 relies on the fact that the *Wronskian determinant*

$$\mathrm{wr}(y_1, \ldots, y_n) := \det(\partial^{i-1}(y_j))_{i,j=1}^n \tag{1-4}$$

of linearly independent elements $y_j \in E$ over $C(E)$ is different from zero (see [Mag], Theorem 2.9).

In the special case of equality in Proposition 1.1, $V_E(\ell)$ is called a *complete solution space*. The first fundamental question now concerns the existence of a D-field extension E/F such that $V_E(\ell)$ is a complete solution space. However, before answering this question we want to study some preliminary examples and to introduce a slightly more general setting.

For the examples let $F = \mathbb{C}(t)$ be the field of rational functions over the complex numbers $\mathbb{C}$ with derivation $\partial = \partial_t := d/dt$ and $E \geq F$ the field of analytic functions.

EXAMPLE 1.2.1. Take $\ell = \partial^1 - a \in F[\partial]$ with $a \in \mathbb{C}^\times$. Then $\ell(y) = 0$ if $\partial(y) = ay$. Therefore the solution space is given by $V_E(\ell) = \mathbb{C} \cdot \exp(at)$ and every nontrivial solution is transcendental over E.

EXAMPLE 1.2.2. In the case $\ell = \partial^1 - \frac{1}{nt}$ with $n \in \mathbb{N}$ any solution of ℓ in E belongs to $V_E(\ell) = \mathbb{C} \sqrt[n]{t}$ and therefore is algebraic over F.

EXAMPLE 1.2.3. A solution of the inhomogeneous differential equation $\partial(y) = f \in F^\times$ is also a solution of the degree 2 homogeneous differential equation $\ell(y) = \partial^2(y) - f^{-1}\partial(f)\partial(y) = 0$. The solution space of the latter consists of

$V_E(\ell) = \mathbb{C} \oplus \mathbb{C}g$ where $g = \int f \, dt$ denotes a solution of the inhomogeneous equation. This may be an element of F as for $f = 1$ or transcendental over F as for $f = \frac{1}{t}$.

These examples show that solutions and solution spaces of linear differential equations may algebraically behave very differently.

1.3. Systems of linear differential equations.

Any solution $y \in E$ of $\ell \in F[\partial]$ leads to a solution $\mathbf{y} = (y, \partial^1(y), \ldots, \partial^{n-1}(y))^{\mathrm{tr}} \in E^n$ of the matrix differential equation

$$\partial(\mathbf{y}) = A_\ell \mathbf{y}, \quad \text{where } A_\ell = \begin{pmatrix} 0 & 1 & 0 & \cdots & & 0 \\ \vdots & \ddots & \ddots & \ddots & & \vdots \\ \vdots & & \ddots & \ddots & \ddots & 0 \\ 0 & \cdots & & \ddots & 0 & 1 \\ -a_0 & \cdots & & \cdots & -a_{n-2} & -a_{n-1} \end{pmatrix} \in F^{n \times n},$$

and vice versa. Now we start with an arbitrary $A \in F^{n \times n}$ and define the *solution space* of A to be

$$V_E(A) := \{\mathbf{y} \in E^n \mid \partial(\mathbf{y}) = A\mathbf{y}\}.$$

This again is a vector space over the constant field of E of dimension less than or equal to n.

Two matrices A and $B \in F^{n \times n}$ are called *differentially equivalent*, or *D-equivalent*, if every solution $\mathbf{z} \in V_E(B)$ can be transformed into a solution $\mathbf{y} \in V_E(A)$ by a matrix $C \in \mathrm{GL}_n(F)$, i.e., if $V_E(A) = CV_E(B)$. The latter is equivalent to the matrix identity $B = C^{-1}AC - C^{-1}\partial(C)$.

Assume for a moment that $A \in F^{n \times n}$ admits a complete solution space over some D-field extension $E \geq F$, i.e., there exists a matrix $Y \in \mathrm{GL}_n(E)$ with $\partial_E(Y) = AY$. Such a matrix is called a *fundamental solution matrix* of the system of differential equations $\partial(\mathbf{y}) = A\mathbf{y}$ over E. If $Y, \tilde{Y} \in \mathrm{GL}_n(E)$ are two fundamental solution matrices for the same A, then it is easy to verify that these can only differ by a matrix $C \in \mathrm{GL}_n(C(E))$, i.e., $\tilde{Y} = YC$. Using this information, one obtains the following partial converse of the statement above.

PROPOSITION 1.2. *Let (F, ∂) be a nontrivial D-field of characteristic 0 and $A \in F^{n \times n}$. Assume that there exists a D-field extension E/F such that the matrix differential equation defined by A has a complete solution space over E. Then A is D-equivalent to a matrix $A_\ell \in F^{n \times n}$ defined by a linear differential operator $\ell \in F[\partial]$.*

A proof of Proposition 1.2 is presented in [Kat1]. In Section 2.1 we will see that the assumption on the existence of a fundamental solution matrix over some extension field is superfluous.

1.4. Differential modules. Another very common way to describe linear differential equations are differential modules. A *differential module* or D-module for short is a module M over a D-ring (R, ∂_R) together with a map $\partial_M : M \to M$ with the properties

$$\partial_M(\mathbf{x} + \mathbf{y}) = \partial_M(\mathbf{x}) + \partial_M(\mathbf{y}) \quad \text{and} \quad \partial_M(a\mathbf{x}) = \partial_R(a)\mathbf{x} + a\partial_M(\mathbf{x}) \qquad (1\text{--}5)$$

for $\mathbf{x}, \mathbf{y} \in M$ and $a \in R$. The solution space of M is defined by

$$V(M) = \{\mathbf{x} \in M \mid \partial_M(\mathbf{x}) = 0\}.$$

M is called a *trivial D-module* if $M \cong V(M) \otimes_{C(R)} R$. In case (M, ∂_M) and (N, ∂_N) are two D-modules over R, an element $\varphi \in \mathrm{Hom}_R(M, N)$ is called a *differential homomorphism* (D-homomorphism) if $\varphi \circ \partial_M = \partial_N \circ \varphi$. Obviously the D-modules over R together with the D-homomorphisms form an abelian category denoted by $\mathbf{DMod}_R$.

Now assume that R is a D-field F with field of constants K. Then it is easy to verify that $\mathbf{DMod}_F$ with the tensor product over F becomes a tensor category over K. Here the tensor product $M \otimes_F N$ is provided with the derivation

$$\partial_{M \otimes N}(\mathbf{x} \otimes \mathbf{y}) = \partial_M(\mathbf{x}) \otimes \mathbf{y} + \mathbf{x} \otimes \partial_N(\mathbf{y}) \qquad (1\text{--}6)$$

and the dual vector space $M^* = \mathrm{Hom}(M, F)$ with

$$(\partial_{M^*}(f))(\mathbf{x}) = \partial_F(f(\mathbf{x})) - f(\partial_M(\mathbf{x})) \qquad (1\text{--}7)$$

for $\mathbf{x} \in M, \mathbf{y} \in N$ and $f \in M^*$. Then (F, ∂_F) is the unit element of $\mathbf{DMod}_F$ with $\mathrm{End}_{\mathbf{DMod}_F}(F, \partial_F) = K$. If in addition K is algebraically closed, $\mathbf{DMod}_F$ even forms a *Tannakian category* using the forgetful functor

$$\Omega : \mathbf{DMod}_F \to \mathbf{Vect}_F, \quad (M, \partial_M) \mapsto M$$

from the category $\mathbf{DMod}_F$ into the category of vector spaces over F (see [Del]). However, this will not be used in the sequel.

The link between D-modules and systems of linear differential equations is given in the following way. Let $M = \bigoplus_{i=1} \mathbf{b}_i F$ be a finite-dimensional D-module over F with basis $\{\mathbf{b}_1, \ldots, \mathbf{b}_n\}$. Then by (1–5) the action of ∂ is uniquely determined by

$$\partial_M(\mathbf{b}_j) = \sum_{i=1}^n \mathbf{b}_i a_{ij} \quad \text{with } a_{ij} \in F. \qquad (1\text{--}8)$$

Thus for $\sum_{i=1}^n \mathbf{b}_i y_i = B\mathbf{y} \in M$ with $B = (\mathbf{b}_1, \ldots, \mathbf{b}_n)$ and $\mathbf{y} = (y_1, \ldots, y_n)^{\mathrm{tr}} \in F^n$ the two statements

$$B\mathbf{y} \in V(M) \quad \text{and} \quad \partial_F(\mathbf{y}) = -A\mathbf{y}$$

where $A = (a_{ij}) \in F^{n \times n}$ are equivalent because of

$$\partial_M(B\mathbf{y}) = \partial_M(B)\mathbf{y} + B\partial_F(\mathbf{y}) = B(A\mathbf{y} + \partial_F(\mathbf{y})).$$

Therefore a D-module M with representing matrix $A \in F^{n \times n}$ of ∂_M leads to a system of linear differential equations over F with matrix $-A$. In particular, the solution space $V(M)$ of M coincides with $V(A)$ and thus is a vector space over K with $\dim_K(V(M)) \leq \dim_F(M)$.

2. Picard–Vessiot Extensions

2.1. Picard–Vessiot rings and fields. Now we are coming back to the questions raised in Section 1.2: For a linear differential equation $\partial(\mathbf{y}) = A\mathbf{y}$ over a D-field F of characteristic 0 with (algebraically closed) field of constants K, does there always exist a D-field E with $\dim_K(V(M \otimes_F E)) = \dim_E(M \otimes_F E)$? (The latter number equals $\dim_F(M)$.) For this purpose we define a *Picard–Vessiot ring* (PV-ring) R *for* A to be a differential ring $(R, \partial_R) \geq (F, \partial_F)$ with the following properties:

(2–1) R is a simple D-ring, i.e., R only contains trivial D-ideals.

(2–2) There exists a fundamental solution matrix over R, i.e., there exists a $Y \in \mathrm{GL}_n(R)$ such that $\partial_R(Y) = A \cdot Y$.

(2–3) R is generated over F by the coefficients y_{ij} of $Y = (y_{ij})_{i,j=1}^n$ and $\det(Y)^{-1}$.

It is easy to verify that a finitely generated simple D-ring is always an integral domain and that R and even $\mathrm{Quot}(R)$ do not contain new constants. The next proposition is basic for all that follows.

PROPOSITION 2.1. *Let (F, ∂_F) be a D-field with algebraically closed field of constants K of characteristic 0 and $A \in F^{n \times n}$. Then for the differential equation $\partial(\mathbf{y}) = A\mathbf{y}$ there exists a Picard–Vessiot ring (R, ∂_R) over F and it is unique up to D-isomorphism.*

The construction of R is similar to Kronecker's construction of root fields in the case of polynomial equations. Let $X = (x_{ij})_{i,j=1}^n$ be a matrix with over F algebraically independent elements x_{ij}. Then by Section 1.1 we can extend ∂_F uniquely to $F[x_{ij}]_{i,j=1}^n$ by $\partial_U(X) = A \cdot X$, i.e., $\partial_U(x_{ij}) = \sum_{k=1}^n a_{ik} x_{kj}$, and to $U := F[\mathrm{GL}_n] = F[x_{ij}, \det(x_{ij})^{-1}]_{i,j=1}^n$. Then (U, ∂_U) is a D-ring over F. By Zorn's Lemma there exists a maximal D-ideal $P \trianglelefteq U$. The quotient $R := U/P$ is a simple D-ring containing a fundamental solution matrix $Y := \kappa_P(X)$, where κ_P denotes the canonical map $\kappa_P : U \to R = U/P$. Obviously, R is generated over F by the coefficients y_{ij} of Y and by $\det(Y)^{-1}$ such that by definition R is a Picard–Vessiot ring. It finally remains to be checked that two PV-rings belonging to the same matrix A are D-isomorphic. This can be done by elementary computations (see [Put2], Proposition 3.4).

The quotient field $E := \mathrm{Quot}(R)$ of a PV-ring is called a *Picard–Vessiot field* for A. It can be characterized without using R.

PROPOSITION 2.2. *Let F and $A \in F^{n \times n}$ be as in Proposition 2.1 and let $(E, \partial_E) \geq (F, \partial_F)$ be a D-field extension. Then E/F is a Picard–Vessiot extension for A if and only if*

(a) *the constant fields of E and F coincide,*
(b) *there exists a $Y \in \mathrm{GL}_n(E)$ with $\partial_E(Y) = A \cdot Y$,*
(c) *E is generated over F by the coefficients y_{ij} of Y.*

A proof is given in [Put2], Proposition 3.5. These characterizing properties correspond to the classical definition of PV-fields (compare [Kap], III.11 and [Mag], Definition 3.2).

2.2. The differential Galois group. As before, let R be a PV-ring and $E = \mathrm{Quot}(R)$ a PV-field over a D-field F of characteristic 0 with algebraically closed field of constants. Then an automorphism γ of R/F or E/F, respectively, is called a *differential automorphism* (D-automorphism) if $\partial \circ \gamma = \gamma \circ \partial$. The group of all D-automorphisms is called the *differential Galois group* (D-Galois group) of R/F or E/F, respectively, and is denoted by $\mathrm{Gal}_D(R/F) = \mathrm{Gal}_D(E/F)$.

Since $\mathrm{Gal}_D(E/F)$ acts faithfully on the solution space $V_E(A)$, it is a subgroup of $\mathrm{GL}_n(K)$. It can be characterized in the following way.

PROPOSITION 2.3. *Let F be a D-field of characteristic 0 with algebraically closed field of constants and let R/F be a PV-ring for $A \in F^{n \times n}$ with fundamental solution matrix $Y = (y_{ij}) \in \mathrm{GL}_n(R)$. Then*

$$\mathrm{Gal}_D(R/F) = \{C \in \mathrm{GL}_n(K) \mid q(Y \cdot C) = 0 \quad \text{for all } q \in P\}$$

where P denotes the annulator ideal

$$P = \{q \in F[\mathrm{GL}_n] \mid q(y_{ij}) = 0\}.$$

A proof can be found for example in [Mag], Corollary 4.10. Since P is finitely generated, $\mathrm{Gal}_D(R/F)$ consists of the K-rational points of a Zariski closed subgroup of $\mathrm{GL}_n(K)$ ([Eis], Section 15.10.1) and therefore of a reduced linear algebraic group $\mathcal{G}$ over K. This already proves the first part of the next proposition.

PROPOSITION 2.4. *Let F be a D-field of characteristic 0 with algebraically closed field of constants K and E/F a PV-extension. Then there exists a reduced linear algebraic group $\mathcal{G}$ over K with $\mathrm{Gal}_D(E/F) \cong \mathcal{G}(K)$. In addition the fixed field $E^{\mathcal{G}(K)}$ coincides with F.*

The last statement follows from the fact that for each $z \in E \backslash F$ a $\gamma \in \mathrm{Gal}_D(E/F)$ can be constructed that moves z (see [Put2], Proposition 3.6). Now we return to our examples in Section 1.2. Again (F, ∂) denotes the D-field $(\mathbb{C}(t), \partial_t)$.

EXAMPLE 2.2.1. Let $\ell = \partial - a \in F[\partial]$ with $a \in \mathbb{C}^\times$. Then by Example 1.2.1 the PV-field for ℓ is given by $E = F(y)$ and $V_E(\ell) = \mathbb{C}y$ for $y = \exp(at)$. The D-Galois group $\mathrm{Gal}_D(E/F)$ equals $\mathbb{G}_m(\mathbb{C}) = \mathrm{GL}_1(\mathbb{C})$ since any $c \in \mathrm{GL}_1(\mathbb{C}) = \mathbb{C}^\times$ defines a D-automorphism because of $\partial(cy) = c\partial(y)$.

EXAMPLE 2.2.2. In the case $\ell = \partial - \frac{1}{nt}$ we obtain $E = F(y)$ for $y = \sqrt[n]{t}$ and $\mathrm{Gal}_D(E/F) = C_n$ is the cyclic group of order n.

EXAMPLE 2.2.3. For $\ell = \partial^2 + \frac{1}{t}\partial$ the PV-field E is $F(y)$ with $y = \log(t)$ and $V_E(\ell) = \mathbb{C} \oplus \mathbb{C}y$. Because of $(\gamma \circ \partial)(y) = \frac{1}{t} = \partial(y)$, for any $\gamma \in \mathrm{Gal}_D(E/F)$ there exists a $c \in \mathbb{C}$ with $\gamma(y) = y + c$. This proves $\mathrm{Gal}_D(E/F) = \mathbb{G}_a(\mathbb{C}) = \mathbb{C}$.

2.3. Torsors and Kolchin's Theorem. In order to prove a Galois correspondence between the intermediate D-fields of a PV-extension E/F and the Zariski closed subgroups of $\mathrm{Gal}_D(E/F) = \mathcal{G}(K)$ we need a structural theorem due to Kolchin which shows that after a finite field extension $\tilde{F}/F$ a defining PV-ring R inside E becomes isomorphic to the coordinate ring of $\mathcal{G}_{\tilde{F}} = \mathcal{G} \times_K \tilde{F}$, i.e., $R \otimes_F \tilde{F} \cong \tilde{F}[\mathcal{G}_F]$. This is a consequence of the fact that the affine scheme $\mathcal{X} = \mathrm{Spec}(R)$ over F is a $\mathcal{G}_F$-torsor or a *principal homogeneous space for* $\mathcal{G}_F$, respectively. This means that $\mathcal{G}_F$ acts on $\mathcal{X}$ via

$$\Gamma : \mathcal{X} \times_F \mathcal{G}_F \to \mathcal{X}, \quad (x, g) \mapsto x \cdot g \tag{2-4}$$

and in addition

$$\mathrm{Id} \times \Gamma : \mathcal{X} \times_F \mathcal{G}_F \to \mathcal{X} \times_F \mathcal{X}, \quad (x, g) \mapsto (x, x \cdot g) \tag{2-5}$$

is an isomorphism of affine schemes over F (see [Put2], Section 6.2). Such a torsor $\mathcal{X}$ is called a *trivial* $\mathcal{G}_F$-*torsor* if $\mathcal{X} \cong \mathcal{G}_F$ where the action is given by the multiplication. The latter is equivalent to $\mathcal{X}(F) \neq \varnothing$ where as usual $\mathcal{X}(F)$ denotes the set of F-rational points of $\mathcal{X}$.

THEOREM 2.5 (D-TORSOR THEOREM). *Let F be a D-field of characteristic 0 with algebraically closed field of constants, $A \in F^{n \times n}$ and R a PV-ring for A over F. Further let $\mathcal{G}$ denote the reduced linear algebraic group over K with $\mathcal{G}(K) = \mathrm{Gal}_D(R/F)$ and $\mathcal{G}_F := \mathcal{G} \times_K F$. Then $\mathrm{Spec}(R)$ is a $\mathcal{G}_F$-torsor.*

For the proof see for example [Put2], Section 6.2. Since the $\mathcal{G}_F$-torsor $\mathrm{Spec}(R)$ becomes trivial after a finite field extension $\tilde{F}/F$, the following version of Kolchin's theorem is an immediate consequence of the D-Torsor Theorem.

COROLLARY 2.6 (KOLCHIN). *With the same assumptions as in Theorem 2.5, and setting $\mathcal{X} := \mathrm{Spec}(R)$:*

(a) *There exists a finite field extension $\tilde{F}/F$ with $\mathcal{X} \times_F \tilde{F} \cong \mathcal{G}_F \times_F \tilde{F}$.*
(b) *$\mathcal{X}$ is smooth and connected over F.*
(c) *The degree of transcendence of $\mathrm{Quot}(R)/F$ equals $\dim(\mathcal{G})$ (over K).*

2.4. The differential Galois correspondence. Now we are ready to explain the differential Galois correspondence. This can be stated as follows:

THEOREM 2.7 (D-GALOIS CORRESPONDENCE). *Let F be a D-field of characteristic 0 with algebraically closed field of constants K, $A \in F^{n \times n}$ and E a PV-extension for A. Denote by $\mathcal{G}$ the reduced linear algebraic group over K with $\mathcal{G}(K) = \mathrm{Gal}_D(E/F)$. Then:*

(a) *There exists an anti-isomorphism between the lattices*

$$\mathfrak{H} := \{\mathcal{H}(K) \mid \mathcal{H}(K) \leq \mathcal{G}(K) \text{ closed}\} \text{ and } \mathfrak{L} := \{L \mid F \leq L \leq E \text{ } D\text{-field}\}$$

given by

$$\Psi : \mathfrak{H} \to \mathfrak{L}, \ \mathcal{H}(K) \mapsto E^{\mathcal{H}(K)} \text{ and } \Psi^{-1} : \mathfrak{L} \to \mathfrak{H}, \ L \mapsto \text{Gal}_D(E/L).$$

(b) *If thereby $\mathcal{H}(K)$ is a normal subgroup, then $L := E^{\mathcal{H}(K)}$ is a PV-extension of F with $\text{Gal}_D(L/F) \cong \mathcal{G}(K)/\mathcal{H}(K)$.*

(c) *Denote by $\mathcal{G}^0$ the identity component of $\mathcal{G}$ and $F^0 := E^{\mathcal{G}^0(K)}$. Then F^0/F is a finite Galois extension with Galois group $\text{Gal}_D(F^0/F) \cong \mathcal{G}(K)/\mathcal{G}^0(K)$.*

Besides Proposition 2.4, for (a) we have to use that for all Zariski closed subgroups $\mathcal{H} \leq \mathcal{G}$ the fixed field $E^{\mathcal{H}(K)}$ is different from F. For the proof of this fact as well as for the proof of (b) Kolchin's theorem has to be used (compare [Put2], Section 6.3).

As an application, we obtain a result comparable to the classical solution of polynomial equations by radicals. To this end we define a PV-extension E/F to be a *Liouvillean extension* if it contains a tower of intermediate D-fields

$$F = F_0 \leq F_1 \leq \ldots \leq F_n = E \text{ with } F_i = F_{i-1}(y_i)$$

and $\frac{\partial(y_i)}{y_i} \in F_{i-1}$ or $\partial(y_i) \in F_{i-1}$ or y_i is algebraic over F_{i-1}. Further a linear algebraic group $\mathcal{G}$ is called *virtually solvable* or solvable-by-finite if the connected component $\mathcal{G}^0$ is a solvable group. Since in this case the composition factors of $\mathcal{G}^0$ are isomorphic either to $\mathbb{G}_m$ or to $\mathbb{G}_a$ and D-Galois extensions of this type can be generated by solutions of $\partial(y) = fy$ or $\partial(y) = f$ with $f \in F$ we find from Theorem 2.7:

COROLLARY 2.8. *A PV-extension E/F is Liouvillean if and only if its D-Galois group is virtually solvable.*

For a more complete proof and further applications concerning integration in finite terms see for example [Mag], Chapter 6. As in the polynomial case, linear differential equations with non (virtually) solvable Galois groups exist. We want to verify this statement with the Airy equation. For this purpose we first explain an analogue of the square-discriminant criterion in ordinary Galois theory which is useful to reduce D-Galois group considerations to unimodular groups.

PROPOSITION 2.9. *Let F be a D-field of characteristic 0 with algebraically closed field of constants K, $\ell = \sum_{k=0}^n a_k \partial^k \in F[\partial]$ a monic differential operator and E/F a PV-extension defined by ℓ or A_ℓ, respectively. Then the linear differential equation over F*

$$\partial(w) + a_{n-1}w = 0 \tag{2--6}$$

has a solution w in E with the properties

(a) *$F(w)/F$ is a PV-extension with $\text{Gal}_D(F(w)/F) \leq \mathbb{G}_m(K)$,*

(b) $\mathrm{Gal_D}(E/F(w)) \cong \mathrm{Gal_D}(E/F) \cap \mathrm{SL}_n(K)$.

For the proof let $\{y_1, \ldots, y_n\}$ denote a K-basis of $V_E(\ell)$. Then any $y \in V_E(\ell)$ satisifies

$$\ell(y) = \mathrm{wr}(y_1, \ldots, y_n, y) \cdot \mathrm{wr}(y_1, \ldots, y_n)^{-1}. \tag{2-7}$$

In particular, for the first derivative of the Wronskian determinant $w := \mathrm{wr}(y_1, \ldots, y_n)$ we obtain equation (2–6). Now any $\gamma \in \mathrm{Gal_D}(E/F)$ acts on the fundamental solution matrix $Y = (\partial^{i-1}(y_j))_{i,j=1}^n$ of E/F via $\gamma(Y) = YC_\gamma$ with $C_\gamma \in \mathrm{GL}_n(K)$ and on w via $\gamma(w) = w \det(C_\gamma)$. Hence w is left invariant by γ if and only if $\det(C_\gamma) = 1$.

With the help of Proposition 2.9 we are able to compute the Galois group of the Airy equation.

EXAMPLE 2.4.1. By Corollary 3.2 below the *Airy equation* $\partial^2(y) = ty$ has no algebraic solution over the D-field $(F, \partial_F) = (\mathbb{C}(t), \partial_t)$. Hence by Proposition 2.9 its Galois group $G = \mathcal{G}(\mathbb{C})$ is a connected closed subgroup of $\mathrm{SL}_2(\mathbb{C})$. In case $G \neq \mathrm{SL}_2(\mathbb{C})$ the linear algebraic group $\mathcal{G}$ would be reducible and $V_E(\ell)$ would contain a G-invariant line $\mathbb{C}y$. But then $z := \partial(y)y^{-1}$ would be invariant under G and therefore belong to F. Obviously no element z of $F = \mathbb{C}(t)$ satisifies

$$\partial(z) = \partial^2(y)y^{-1} - \partial(y)^2 y^{-2} = t - z^2.$$

as can be seen from the reduced expression of z as a quotient of polynomials.

2.5. Characterization of PV-rings and PV-fields.

The theorem of Kolchin allows us to characterize the PV-ring R inside $\mathrm{Quot}(R)$.

PROPOSITION 2.10. *Let F be a D-field of characteristic 0 with algebraically closed field of constants K and R a PV-ring over F with quotient field E and Galois group $G := \mathrm{Gal_D}(R/F) = \mathcal{G}(K)$. Then for $z \in E$ are equivalent:*

(a) $z \in R$,   (b) $\dim_K(K\langle Gz \rangle) < \infty$,   (c) $\dim_F(F\langle \partial^k(z) \rangle_{k \in \mathbb{N}}) < \infty$.

Here $K\langle Gz \rangle$ denotes the K-vector space generated by the G-orbit of z and $F\langle \partial^k(z) \rangle_{k \in \mathbb{N}}$ is the F-vector space generated by all derivatives $\partial^k(z)$ of z. The critical step is the one from (a) to (b). By the D-Torsor Theorem we may, after a finite extension, assume $R = F[\mathcal{G}]$. Then the result follows from the fact that the action of $\mathcal{G}(F)$ on $F[\mathcal{G}]$ is locally finite, i.e., $F[\mathcal{G}]$ is a union of finite-dimensional G-stable subspaces ([Spr], Proposition 2.3.6).

It is quite natural to call an element $z \in E$ with property (c) in Proposition 2.10 *differentially finite* (D-finite). For such an element there exists, by definition, a nonconstant linear differential operator $\ell_z \in F[\partial]$ monic of minimum degree with $\ell_z(z) = 0$. We call ℓ_z a *minimal differential operator of z*. Given a basis $z_1, \ldots, z_n$ of $K\langle Gz \rangle$, it can be constructed by

$$\ell_z(y) = \frac{\mathrm{wr}(z_1, \ldots, z_n, y)}{\mathrm{wr}(z_1, \ldots, z_n)}, \tag{2-8}$$

where wr denotes the Wronskian determinant defined in (1–4). In this notation, Proposition 2.10 tells us that the PV-ring R is characterized inside a PV-field $E = \mathrm{Quot}(R)$ as the ring of D-finite elements. In the particular case of finite D-extensions E/F the PV-ring R coincides with E. Another implication of Proposition 2.10 is the following characterization of PV-fields.

THEOREM 2.11. *Let $F \leq E$ be D-fields of characteristic 0 with algebraically closed field of constants. Then E is a PV-extension of F if and only if*

(a) *E/F is finitely generated by D-finite elements,*
(b) *E and F share the same field of constants K,*
(c) *for all D-finite $z \in E$ yields $\dim_K(V_E(\ell_z)) = \deg(\ell_z)$, where $\ell_z \in F[\partial]$ is the minimal D-operator of z.*

An elementary proof is presented for example in [Put2], Proposition 6.11.

3. Monodromy and the Riemann–Hilbert Problem

3.1. Regular and singular points. Let $F = K(\mathcal{C})$ be the function field of a smooth projective curve $\mathcal{C}$ over an algebraically closed field K of characteristic zero with a nontrivial derivation ∂_F. Then $C(F) = K$. Further for $x \in \mathcal{C}$ the completion of F with respect to the valuation defined by x is denoted by F_x. It is isomorphic to the field of Laurent series $K((t))$ where $t \in F$ denotes a local parameter at x. Now let E/F be a PV-extension defined by $A \in F^{n \times n}$. Then a point $x \in \mathcal{C}$ is called a *regular point for E/F* if A is D-equivalent to a matrix over F_x without poles, i.e., there exists a matrix $B \in \mathrm{GL}_n(F_x)$ such that

$$B^{-1}AB - B^{-1}\partial(B) \in K[\![t]\!]^{n \times n}. \tag{3–1}$$

This property can also be characterized by having a fundamental solution matrix over $F_x = K((t))$:

PROPOSITION 3.1. *Let $F = K(\mathcal{C})$ as above and $A \in F^{n \times n}$. Then $x \in \mathcal{C}$ is a regular point for the PV-extension E/F defined by A if and only if the D-equation $\partial(\mathbf{y}) = A\mathbf{y}$ possesses a fundamental solution matrix $Y \in \mathrm{GL}_n(F_x)$.*

This result immediately implies

COROLLARY 3.2. *Let E/F be as in Proposition 3.1 with $\mathrm{Gal}_D(E/F) = \mathcal{G}(K)$ and let L be the fixed field of $\mathcal{G}^0(K)$. Then the finite Galois extension L/F is unramified in all regular points $x \in \mathcal{C}$ for E/F.*

In the particular case $\mathcal{C} = \mathbb{P}^1$ (projective line), the Galois group of a PV-extension E/F with at most one non regular point is connected. This applies, for example, to the Airy equation $\partial^2(y) = ty$ in Example 2.4.1 since all finite points are regular.

Non regular points $x \in \mathcal{C}$ for E/F are called *singular points* and the set of all singular points is called the *singular locus* $\mathcal{S}_{E/F}$ of E/F. A point $x \in \mathcal{S}_{E/F}$ is called *tamely (weakly, regular) singular* if there exists a $B \in \mathrm{GL}_n(F_x)$ such that

$$B^{-1}AB - B^{-1}\partial(B) \in \frac{1}{t}K[\![t]\!]^{n \times n}, \tag{3-2}$$

otherwise it is a *wild (strong, singular) singularity*. For tame singularities, an even stronger characterization can be given.

PROPOSITION 3.3. *Let $F = K(\mathcal{C})$ as above, $A \in F^{n \times n}$ and E/F a PV-extension defined by A. Then $x \in \mathcal{C}$ is tamely singular if and only if there exists a $B \in \mathrm{GL}_n(F_x)$ and a constant matrix $D \in K^{n \times n}$ such that $B^{-1}AB - B^{-1}\partial(B) = \frac{1}{t}D$.*

For a sketch of proof see for example [Put2], Exercise 7.

For later use we add a characterization of regular and tamely singular points in the language of D-modules which immediately follows from the definitions (3-1) and (3-2) above.

COROLLARY 3.4. *Let (M, ∂) be a D-module over $F = K(\mathcal{C})$, $x \in \mathcal{C}$, $M_x := M \otimes_F F_x$ and let $t \in F$ be a local parameter for x such that $F_x = K((t))$.*

(a) *A point $x \in \mathcal{C}$ is regular if and only if M_x contains a ∂-invariant $K[\![t]\!]$-lattice.*
(b) *$x \in \mathcal{C}$ is tamely singular if and only if M_x contains a δ-invariant $K[\![t]\!]$-lattice where $\delta := t\partial$.*

3.2. The monodromy group. In the case of $K = \mathbb{C}$ the matrix B in Proposition 3.3 can be chosen to have coefficients in the subfield $F_x^{conv} \le F_x = K((t))$ of convergent Laurent series (see [Put2], Exercise 7 or [For], § 11.12). This allows us to analyze the local behaviour.

THEOREM 3.5. *Let $F = \mathbb{C}(\mathcal{C})$, $A \in F^{n \times n}$ and E/F a PV-extension for A. Assume $x \in \mathcal{C}$ is a tame singularity and denote by t a local parameter at x.*

(a) *Then $\partial(\mathbf{y}) = A\mathbf{y}$ possesses a local fundamental solution matrix of the form $Y = B \exp(C \log(t))$ with $B \in \mathrm{GL}_n(F_x^{conv})$ and $C \in \mathbb{C}^{n \times n}$.*
(b) *Via analytic continuation along a loop σ around x we obtain $\sigma(Y) = Y \cdot M_\sigma$ with $M_\sigma = \exp(2\pi i C)$.*

For a proof see for example [For], § 11. The matrix $M_\sigma \in \mathrm{GL}_n(\mathbb{C})$ is called a *local monodromy matrix* and is determined inside $\mathrm{GL}_n(\mathbb{C})$ only up to conjugation.

In order to simplify the notation we now restrict ourselves to the projective line $\mathcal{C} = \mathbb{P}^1(\mathbb{C})$. Then $F = \mathbb{C}(\mathbb{P}^1) = \mathbb{C}(t)$ is the field of rational functions over $\mathbb{C}$. Let $\mathcal{S} \subseteq \mathbb{P}^1(\mathbb{C})$ be a nonempty set of cardinality $\sharp\mathcal{S} = s < \infty$ and let $\mathcal{U} := \mathbb{P}^1(\mathbb{C}) \setminus \mathcal{S}$. Then the *fundamental group of $\mathcal{U}$* with respect to a base point $x_0 \in \mathcal{U}$ is known to be

$$\pi_1(\mathcal{U}; x_0) = \langle \sigma_1, \ldots, \sigma_s \mid \sigma_1 \cdots \sigma_s = 1 \rangle \tag{3-3}$$

where the σ_i are loops starting from x_0 counterclockwise around the points $x_i \in S$ (compare [Ful], Chapter 19).

Applying Theorem 3.5 and analytic continuation we obtain a homomorphism (the monodromy map)

$$\mu : \pi_1(\mathcal{U}; x_0) \to \mathrm{GL}_n(\mathbb{C}), \quad \sigma \mapsto M_\sigma \qquad (3\text{--}4)$$

where the image is called the *monodromy group* $\mathrm{Mon}(E/F)$ of E/F. Again $\mathrm{Mon}(E/F)$ is only determined up to conjugacy inside $\mathrm{GL}_n(\mathbb{C})$.

Since $M_\sigma \in \mathrm{GL}_n(\mathbb{C})$ acts on the solution space $V_E(A)$ spanned by the columns of Y it induces an automorphism γ_σ of E/F compatible with the differentiation on E. Consequently

$$\gamma : \mathrm{Mon}(E/F) \to \mathrm{Gal}_D(E/F), \quad M_\sigma \mapsto \gamma_\sigma \qquad (3\text{--}5)$$

defines a homomorphism from $\mathrm{Mon}(E/F)$ to the D-Galois group of E/F, which in fact is a monomorphism. This already gives the first part of the next theorem.

THEOREM 3.6. (a) *Let* $F = \mathbb{C}(t)$ *and* E/F *be a PV-extension. Then* $\mathrm{Mon}(E/F)$ *is (isomorphic to) a subgroup of* $\mathrm{Gal}_D(E/F)$.
(b) *If in addition the singular locus* $S_{E/F}$ *is tame, then* $\mathrm{Mon}(E/F)$ *is Zariski dense in* $\mathrm{Gal}_D(E/F)$.

The proof of (b) relies on the fact that systems of linear differential equations with only tame singularities by Propositions 3.1 and 3.3 only admit locally meromorphic solutions and that meromorphic functions on $\mathbb{P}^1(\mathbb{C})$ (fixed by $\mathrm{Mon}(E/F)$) are rational ([For], Corollary 2.9).

In Example 2.4.1 of the Airy equation $\partial^2(y) = ty$ we have $S_{E/F} = \{\infty\}$. Therefore $\pi_1(\mathbb{P}^1(\mathbb{C}) \setminus S) = \pi_1(\mathbb{A}^1(\mathbb{C})) = 1$ and $\mathrm{Mon}(E/F)$ is trivial. But $\mathrm{Gal}_D(E/F) = \mathrm{SL}_2(\mathbb{C})$, hence ∞ is a wild singularity.

3.3. The Riemann–Hilbert Problem.

We have seen that in the case of a tame singular locus the D-Galois group coincides with the Zariski closure of the monodromy group. Therefore it is a fundamental question if every homomorphic image of $\pi_1(\mathcal{U}; x_0)$ already appears as the monodromy group of a linear system of differential equations possibly even with only tame singularities. This problem is named the *Riemann–Hilbert problem* for tame (regular) systems and is number 21 among the famous Hilbert problems. A positive solution has already been presented by Plemelj (1908) in the following form.

THEOREM 3.7 (PLEMELJ). *For any finite set* $S = \{x_1, \ldots, x_s\} \subseteq \mathbb{P}^1(\mathbb{C})$ *and any set of matrices* $M_i \in \mathrm{GL}_n(\mathbb{C})$ *with* $\prod_{i=1}^{s} M_i = 1$ *there exists a tamely singular system of linear D-equations* $\partial(\mathbf{y}) = A\mathbf{y}$ *over* $\mathbb{C}(t)$ *with monodromy matrices* $M_i = M_{\sigma_i}$ *around* x_i.

This theorem can be seen as a differential analogue and generalization of the algebraic version of Riemann's existence theorem (see for example [Voe], Theorem 2.13). A modern proof is given in [AB], Theorem 3.2.1. It relies on the theorem of Birkhoff and Grothendieck on the triviality of complex holomorphic vector bundles. A simplified version for noncompact Riemann surfaces, for example $\mathbb{A}^1(\mathbb{C})$, can be found in [For], § 30 and § 31. Here is an easy consequence of Theorem 3.7:

COROLLARY 3.8. *Every finitely generated subgroup $G \leq \mathrm{GL}_n(\mathbb{C})$ can be realized as the monodromy group of a system of homogeneous linear differential equations over $\mathbb{C}(t)$ with tame singular locus.*

3.4. The inverse problem over the complex numbers. The solution of the Riemann–Hilbert problem is also the main ingredient for the solution of the inverse D-Galois problem over $\mathbb{C}(t)$. Namely by Theorem 3.6 it is enough to observe that all linear algebraic groups over $\mathbb{C}$ have finitely generated dense subgroups. This final step of the solution of the inverse problem was settled by Tretkoff and Tretkoff only in 1979.

PROPOSITION 3.9. *Any Zariski closed subgroup of $\mathrm{GL}_n(\mathbb{C})$ possesses finitely generated dense subgroups.*

For the proof see [TT], Proposition 1. Together with Theorem 3.7, Proposition 3.9 solves the inverse D-Galois problem over $\mathbb{C}(t)$ even with tame singularities.

THEOREM 3.10. *Every linear algebraic group over $\mathbb{C}$ can be realized as a differential Galois group over $\mathbb{C}(t)$ with tame singular locus.*

Unfortunately the above general solution of the inverse D-Galois problem over $\mathbb{C}$ relies on nonconstructive topological and cohomological considerations. In contrast to the case of finite groups it does not even carry over to algebraically closed fields of constants different from $\mathbb{C}$ due to the lack of a D-analogue of Grothendieck's Specialization Theorem.

For connected groups the situation looks more pleasant. There is a new constructive solution of the inverse D-Galois problem due to Mitschi and Singer which is valid for all D-fields with algebraically closed field of constants of characteristic 0. This will be outlined in the next section.

Before that, however, we want to indicate a theorem of Ramis concerning realizations with restricted singular locus.

THEOREM 3.11 (RAMIS). *A linear algebraic group over $\mathbb{C}$ can be realized as a differential Galois group over $\mathbb{C}(t)$ with at most one singular point if and only if it is generated by its maximal tori.*

More generally a linear algebraic group $\mathcal{G}(\mathbb{C})$ over $\mathbb{C}$ can be realized as a D-Galois group over $\mathbb{C}(t)$ with singular locus inside $\mathcal{S}$ if and only if the same is true for the quotient by its maximal closed normal subgroup generated by tori. A proof is elaborated in [Ram].

4. The Constructive Inverse Problem

4.1. The logarithmic derivative. As before, let (F, ∂_F) be an arbitrary D-field of characteristic 0 with algebraically closed field of constants K. Then the F-algebra

$$D := F[X]/(X)^2 = F + Fe, \quad \text{where } e^2 = 0,$$

is called the *algebra of dual numbers over* F. It has the advantage that the map

$$\delta : F \to D, \quad a \mapsto a + \partial_F(a)e$$

defined by the non-multiplicative derivation ∂_F is a K-homomorphism. For a linear algebraic group $\mathcal{G} \leq \mathrm{GL}_{n,F}$ over F the *Lie algebra* of $\mathcal{G}$ can be defined to be the F-vector space

$$\mathrm{Lie}_F(\mathcal{G}) := \{A \in F^{n \times n} \mid 1 + eA \in \mathcal{G}(F[e])\}$$

provided with the *Lie bracket*

$$[\cdot, \cdot] : \mathrm{Lie}_F(\mathcal{G}) \times \mathrm{Lie}_F(\mathcal{G}) \to \mathrm{Lie}_F(\mathcal{G}), \quad (A, B) \mapsto [A, B] := AB - BA.$$

It can be shown that in fact the Lie algebra as defined above is isomorphic to the tangent space of $\mathcal{G}$ at the unit point and therefore only depends on $\mathcal{G}$ and not on the chosen embedding $\mathcal{G} \leq \mathrm{GL}_{n,F}$.

PROPOSITION 4.1. *Let* $\mathcal{G} \leq \mathrm{GL}_{n,F}$ *be a linear algebraic group defined over a D-field* F *of characteristic* 0 *with derivation* ∂_F *and with algebraically closed field of constants. Then*

$$\lambda : \mathcal{G}(F) \to \mathrm{Lie}_F(\mathcal{G}), \quad A \mapsto \partial_F(A)A^{-1}$$

is a map from $\mathcal{G}(F)$ *to the Lie algebra of* $\mathcal{G}$ *over* F. *It has the property*

$$\lambda(A \cdot B) = \lambda(A) + A\lambda(B)A^{-1}.$$

The proof of Proposition 4.1 is immediate (compare [Kov], Section 1). The map λ is usually called the *logarithmic derivative*. One of its nice features also stated in [Kov] is that it gives an upper bound for the D-Galois group.

PROPOSITION 4.2. *Let* (F, ∂_F) *be a D-field as above with field of constants* K, $\mathcal{G}$ *a linear algebraic group over* K *and* $A \in \mathrm{Lie}_F(\mathcal{G})$. *Then the D-Galois group of a PV-extension* E/F *defined by* $\partial(\mathbf{y}) = A\mathbf{y}$ *is isomorphic to a subgroup of* $\mathcal{G}(K)$.

For the proof we only have to observe that $A \in \mathrm{Lie}_F(\mathcal{G})$ implies that the defining ideal $I \trianglelefteq F[\mathrm{GL}_n]$ of $\mathcal{G}_F$ is a D-ideal. Hence the maximal D-ideal $P \trianglelefteq F[\mathrm{GL}_n]$ defining the PV-ring $R \leq E$ contains a conjugate of I. By Proposition 2.3 this already entails the assertion.

In the case where the field F in question has *cohomological dimension* $\mathrm{cd}(F) \leq 1$ there is a partial converse of Proposition 4.2. This relies on the famous Theorem of Springer and Steinberg ([Ser], III, § 2.3). Among the fields with this property are, for example, all fields of transcendence degree 1 over an algebraically closed field (Theorem of Tsen, [Ser], II, § 3.3).

THEOREM 4.3 (SPRINGER AND STEINBERG). *Let F be a perfect field with* $\mathrm{cd}(F) \leq 1$. *Then for every connected linear algebraic group $\mathcal{G}$ over F*

$$H^1(G_F, \mathcal{G}(F^{\mathrm{alg}})) = 0 \quad \text{where} \quad G_F = \mathrm{Gal}(F^{\mathrm{alg}}/F).$$

Here F^{alg} denotes the algebraic closure of F and hence G_F the absolute Galois group of F. Now let F be a D-field with $\mathrm{cd}(F) \leq 1$ and with algebraically closed field of constants K. Since $H^1(G_F, \mathcal{G}(F^{\mathrm{alg}}))$ classifies the G_F-torsors, with the assumptions of Theorem 4.3 all G_F-torsors are trivial. Hence by the D-Torsor Theorem 2.5 then any PV-ring R over F with connected D-Galois group $\mathcal{G}(K)$ is isomorphic to the coordinate ring $F[\mathcal{G}]$ of $\mathcal{G}$. Another consequence of Theorem 4.3 of Springer and Steinberg is the following converse of Proposition 4.2 (see for example [Put2], Theorem 4.4).

COROLLARY 4.4. *Let (F, ∂_F) be a D-field of characteristic 0 with algebraically closed field of constants K and $\mathrm{cd}(F) \leq 1$, $\mathcal{H} \leq \mathrm{GL}_{n,K}$ a connected closed subgroup, $A \in \mathrm{Lie}_F(\mathcal{H}) \subseteq F^{n \times n}$ and E/F a PV-extension defined by A with connected Galois group $\mathrm{Gal}(E/F) = \mathcal{G}(K)$. Then there exists a $B \in \mathcal{H}(F)$ such that*

$$B^{-1}AB - B^{-1}\partial_F(B) \in \mathrm{Lie}_F(\mathcal{G}).$$

In this case E/F can be generated by a differential equation $\partial(\mathbf{y}) = A\mathbf{y}$ with $A \in \mathrm{Lie}_F(\mathcal{G})$. D-Galois extensions of this specific type are called *effective PV-extensions* in this article. Obviously the existence of effective PV-extensions is restricted to connected groups.

4.2. Chevalley modules.

Before tackling the inverse problem for connected groups, we have to recall some basic notions and general structure theorems for linear algebraic groups $\mathcal{G}$. The maximal connected solvable normal subgroup of $\mathcal{G}$ is called the *radical of $\mathcal{G}$* and its maximal connected unipotent normal subgroup the *unipotent radical of $\mathcal{G}$*. These are denoted by $\mathcal{R}(\mathcal{G})$ and $\mathcal{U}(\mathcal{G})$, respectively. Further $\mathcal{G}$ is called *semisimple* if $\mathcal{R}(\mathcal{G}) = 1$ and *reductive* if $\mathcal{U}(\mathcal{G}) = 1$. For a connected linear algebraic group we have the following structure theorem (see [Bor], IV, 11.22 and [Spr], Proposition 7.3.1 and 8.1.6).

THEOREM 4.5. *Let $\mathcal{G}$ be a connected linear algebraic group over an algebraically closed field K of characteristic 0.*

(a) *Then $\mathcal{G}$ is isomorphic to a semidirect product $\mathcal{U} \rtimes \mathcal{P}$ of its unipotent radical $\mathcal{U} = \mathcal{U}(\mathcal{G})$ and a maximal reductive subgroup $\mathcal{P} \leq \mathcal{G}$ (Levi complement).*

(b) *The group $\mathcal{P}$ is the product $\mathcal{T} \cdot \mathcal{H}$ of a torus $\mathcal{T} = \mathcal{R}(\mathcal{P}) \cong \mathbb{G}_m^r$ and the connected semisimple group $\mathcal{H} = (\mathcal{P}, \mathcal{P})$. More precisely, there exists a finite subgroup $H = \mathcal{H} \cap \mathcal{T}$ such that $\mathcal{P} \cong (\mathcal{T} \times \mathcal{H})/H$.*

This already suggests a strategy for solving the inverse problem for connected groups. The first step would be to realize tori and semisimple groups and the second to solve embedding problems with unipotent kernel. For the realization of connected semisimple groups we need some strengthening of the following theorem of Chevalley.

THEOREM 4.6 (CHEVALLEY). *Let $\mathcal{G}$ be a linear algebraic group over K. Then for all closed subgroups $\mathcal{H} \leq \mathcal{G}$ there exist a K-vector space V, a linear representation $\varrho_{\mathcal{H}} : \mathcal{G} \to \mathrm{GL}(V)$ and a one-dimensional subspace $W \leq V$ such that*

$$\mathcal{H}(K) = \{h \in \varrho_{\mathcal{H}}(\mathcal{G}) \mid h(W) \subseteq W\}.$$

For the proof see [Spr], Theorem 5.5.3. From this theorem it is fairly easy to deduce the following statement ([MS], Lemma 3.1).

COROLLARY 4.7. *Let $\mathcal{G}$ be a connected semisimple linear algebraic group over an algebraically closed field K of characteristic 0. Then there exist a K-vector space V and a faithful linear representation $\varrho : \mathcal{G} \to \mathrm{GL}(V)$ with the following properties*:

(a) *V contains no one-dimensional $\varrho(\mathcal{G})$-submodule.*
(b) *Any connected closed subgroup $\mathcal{H}$ of $\mathcal{G}$ leaves a one-dimensional subspace of V invariant.*

Such a module is called a *Chevalley module for $\mathcal{G}$* in [MS]. Obviously the natural 2-dimensional representation of $\mathrm{SL}_2(K)$ already defines a Chevalley module for this group. In general, Chevalley modules are obtained by composing representations of the type of Theorem 4.6 and therefore are not of this simple structure.

4.3. Realization of connected reductive groups.

The key lemma for the realization of semi-simple groups as differential Galois groups over $F = K(t)$ is the following.

PROPOSITION 4.8. *Let $F = K(t)$ be a field of rational functions over an algebraically closed field K of characteristic 0, $\mathcal{G}$ be a semisimple linear algebraic group over K with Chevalley module V and without loss of generality $\mathcal{G} \leq \mathrm{GL}(V)$. Let $A := A_0 + tA_1 \in \mathrm{Lie}_F(\mathcal{G})$ with constant matrices $A_0, A_1 \in \mathrm{Lie}_K(\mathcal{G})$, and E/F a PV-extension for A. Then $\mathrm{Gal}_D(E/F)$ is a proper subgroup of $\mathcal{G}(K)$ if and only if there exists a vector $w \in V \otimes_K K[t]$ and a polynomial $f \in K[t]$ of degree at most 1 with*

$$(A - \partial)w = fw.$$

Obviously by Proposition 4.2 the group $\mathrm{Gal}_D(E/F)$ is isomorphic to a subgroup $\mathcal{H}(K)$ of $\mathcal{G}(K)$. In case $\mathcal{H}(K) \neq \mathcal{G}(K)$ by Corollary 4.4 there exists a $B \in \mathcal{G}(F)$ such that $\tilde{A} := B^{-1}AB - B^{-1}\partial(B) \in \mathrm{Lie}_F(\mathcal{H})$. Since V is a Chevalley module there exists in addition a $v \in V, v \neq 0$, such that $\tilde{A}v \in Fv$. But then for $w := Bv$ one obtains $(A - \partial)w = fw \in Fw$ with $\deg(f) \leq 1$.

Hence, one only has to find constant matrices A_0 and A_1 such that $(A-\partial)w = fw$ has no solution. For the construction of such matrices we need the root space decomposition of $\mathcal{L} := \mathrm{Lie}_K(\mathcal{G})$. This is given by

$$\mathcal{L} = \mathcal{L}_0 \oplus \left(\bigoplus_\alpha \mathcal{L}_\alpha \right)$$

where $\mathcal{L}_0$ denotes the Cartan subalgebra and the one-dimensional spaces $\mathcal{L}_\alpha = KX_\alpha$ are the eigenspaces for the adjoint action of $\mathcal{L}_0$ on $\mathcal{G}$ corresponding to the non-zero roots $\alpha \in \mathcal{L}_0^*$, i.e., $\alpha : \mathcal{L}_0 \to K$. More precisely the adjoint action of $\mathcal{L}_0$ on $\mathcal{L}_0$ is trivial, and for any root $\alpha \neq 0$ one has $[C, X_\alpha] = \alpha(C)X_\alpha$ for all $C \in \mathcal{L}_0$.

The action of $\mathcal{L}_0$ on the Chevalley module V produces a similar decomposition $V = \bigoplus_\beta V_\beta$ into eigenspaces for a collection of linear maps $\beta \in \mathcal{L}_0^*$. These are called the weights of V.

Now we choose

$$A_0 := \sum_{\alpha \neq 0} X_\alpha. \tag{4-0}$$

In order to fulfill the assumptions of Proposition 4.8, for A_1 we choose an element in $\mathcal{L}_0$ satisfying the following conditions:

(4-1) The $\alpha(A_1)$ are non-zero and distinct for the non-zero roots α of $\mathcal{L}$.
(4-2) The $\beta(A_1)$ are non-zero and distinct for the non-zero weights of V.
(4-3) The linear operator

$$\sum_{\alpha \neq 0} \frac{-1}{\alpha(A_1)} X_{-\alpha} X_\alpha$$

does not have positive integers as eigenvalues.

Obviously the set of $A_1 \in \mathcal{L}_0$ satisfying (4-1) and (4-2) is Zariski dense. Condition (4-3) can be fulfilled using a suitable multiple of A_1. Now Mitschi and Singer have proved the following result in [MS]:

PROPOSITION 4.9. *With matrices A_0 and A_1 satisfying (4-0) to (4-3), the PV-extension E/F in Proposition 4.8 generated by $A = A_0 + tA_1$ has the differential Galois group $\mathcal{G}(K)$.*

In particular, any connected semi-simple linear algebraic group can be realized effectively as a differential Galois group over $F = K(t)$. The next step is the realization of tori $\mathcal{T} = \mathbb{G}_m(K)^r$, $r \in \mathbb{N}$, as differential Galois groups over F. This follows from the next result:

PROPOSITION 4.10. *Let $F = K(t)$ as in Proposition 4.8 and $c_1, \ldots, c_r \in K$ linearly independent over $\mathbb{Q}$. Then the PV-extension E/F generated by $A = \mathrm{diag}(c_1, \ldots, c_r) \in \mathrm{Lie}_K(\mathbb{G}_m^r)$ has the differential Galois group $\mathbb{G}_m^r(K)$.*

Obviously by Proposition 4.2 and Corollary 3.2 $\mathrm{Gal_D}(E/F)$ is a connected sub-group of $\mathbb{G}_m^r(K)$. Hence the result follows from the fact that the solutions $y_j = \exp(c_j t)$ of $\partial(y) = c_j y$ are algebraically independent over F for $j = 1, \ldots, r$.

Since any connected reductive group is a quotient of a direct product of a connected semi-simple group and a torus by a finite group, from Proposition 4.9 and 4.10 we immediately obtain

THEOREM 4.11. *Every connected reductive linear algebraic group over an algebraically closed field K of characteristic 0 can be realized effectively as differential Galois group over $F = K(t)$.*

4.4. Embedding problems with unipotent kernel.

In order to solve the inverse problem of differential Galois theory for arbitrary connected groups over $F = K(t)$ by Theorem 4.11 it remains to solve differential embedding problems with unipotent kernel.

Here a differential embedding problem is defined in the following way. Let L/F be a PV-extension with D-Galois group $\mathrm{Gal_D}(L/F) \cong \mathcal{H}(K)$ and let

$$1 \to \mathcal{A}(K) \to \mathcal{G}(K) \xrightarrow{\beta} \mathcal{H}(K) \to 1 \qquad (4\text{-}4)$$

be an exact sequence of linear algebraic groups (in characteristic zero). Then the corresponding *differential embedding problem* (D-embedding problem), denoted by $\mathcal{E}(\alpha, \beta)$, asks for the existence of a PV-extension E/F with $E \geq L$ and a monomorphism γ which maps $\mathrm{Gal_D}(E/F)$ onto a closed subgroup of $\mathcal{G}(K)$ such that the diagram

$$
\begin{array}{ccc}
\mathrm{Gal}(E/F) & \xrightarrow{\text{res}} & \mathrm{Gal}(L/F) \\
\gamma \downarrow & & \cong \downarrow \alpha \\
\end{array}
$$

$$1 \longrightarrow \mathcal{A}(K) \longrightarrow \mathcal{G}(K) \xrightarrow{\beta} \mathcal{H}(K) \longrightarrow 1 \qquad (4\text{-}5)$$

commutes. The kernel $\mathcal{A}(K)$ is also called the *kernel of* $\mathcal{E}(\alpha, \beta)$ and the monomorphism γ a *solution* of $\mathcal{E}(\alpha, \beta)$. We say γ is a *proper* solution if γ is an epimorphism. Further the D-embedding problem is called a *split embedding problem* if the exact sequence splits (i.e., $\mathcal{G}(K)$ as an algebraic group is a semidirect product of $\mathcal{A}(K)$ with $\mathcal{H}(K)$) and a *Frattini embedding problem* if $\mathcal{G}$ is the only closed supplement of $\mathcal{A}$ in $\mathcal{G}$ (i.e., any $\mathcal{U} \leq \mathcal{G}$ which satisfies $\mathcal{A}\mathcal{U} = \mathcal{G}$ already equals $\mathcal{G}$). Finally we say the embedding problem is an *effective embedding problem*, if L/F is an effective PV-extension (according to Section 4.1).

The unipotent radical $\mathcal{U}$ of a linear algebraic group $\mathcal{G}$ possesses a closed complement $\mathcal{H}$ which is a reductive linear algebraic group (Levi complement). Thus $(\mathcal{G}/\mathcal{U})(K) \cong \mathcal{H}(K)$ already can be realized effectively as D-Galois group over F. Hence to realize $\mathcal{G}(K)$ as D-Galois group it suffices to solve an effective split embedding problem with unipotent kernel $\mathcal{U}(K)$. Dividing by the commutator subgroup $\mathcal{U}'(K)$ of $\mathcal{U}(K)$ this embedding problem decomposes into an effective split embedding problem with abelian unipotent kernel

$$1 \to \mathcal{U}(K)/\mathcal{U}'(K) \to \mathcal{G}(K)/\mathcal{U}'(K) \to \mathcal{H}(K) \to 1 \qquad (4\text{--}6)$$

and a Frattini embedding problem belonging to

$$1 \to \mathcal{U}'(K) \to \mathcal{G}(K) \to \mathcal{G}(K)/\mathcal{U}'(K) \to 1. \qquad (4\text{--}7)$$

For the first of these embedding problems we can use a recent result of Oberlies ([Obe], Proposition 2.4) based on a theorem of Ostrowski.

PROPOSITION 4.12. *Every effective split D-embedding problem with (minimal) unipotent abelian kernel has an effective proper solution over $K(t)$, where K is algebraically closed of characteristic 0.*

Here the assumption of minimality can be neglected by direct decomposition of the kernel (compare [Obe], Reduction). The solvability of the second embedding problem already goes back to Kovacic ([Kov], Proposition 11). In our terminology it can be stated in the following way.

PROPOSITION 4.13. *Every effective Frattini D-embedding problem has an effective proper solution over $K(t)$, where K is algebraically closed of characteristic 0.*

For a sketch of the proof, denote $d\beta : \mathrm{Lie}_F(\mathcal{G}) \to \mathrm{Lie}_F(\mathcal{H})$ the surjective Lie algebra map induced by $\beta : \mathcal{G} \to \mathcal{H}$ and $A \in \mathrm{Lie}_F(\mathcal{H})$ a matrix defining an effective PV-extension L/F with isomorphism $\alpha : \mathrm{Gal}(L/F) \to \mathcal{H}(K)$. Then any inverse image $B \in \mathrm{Lie}_F(\mathcal{G})$ of A by $d\beta$, i.e., $d\beta(B) = A$, defines a PV-extension E/F with $\mathrm{Gal}_D(E/F) \le \mathcal{G}(K)$ by Proposition 4.2 and $E \ge L$. Hence by the Frattini property there exists an isomorphism $\gamma : \mathrm{Gal}_D(E/F) \to \mathcal{G}(K)$ with in addition $\alpha \circ \mathrm{res} = \beta \circ \gamma$, i.e., γ is an effective proper solution of $\mathcal{E}(\alpha, \beta)$.

Combining Proposition 4.12 and 4.13 above with Theorem 4.11 we get a constructive solution of the inverse problem for connected groups (see [MS]).

THEOREM 4.14 (MITSCHI–SINGER). *Every connected linear algebraic group over an algebraically closed field K of characteristic 0 can be realized effectively as differential Galois group over $F = K(t)$.*

A nonconstructive variant of proof had already been presented in [Sin].

Added in Proof. A solution of the inverse problem in differential Galois theory over $K(t)$ for nonconnected groups has recently been obtained by J. Hartmann in her thesis [Har].

MODULAR THEORY

5. Iterative Differential Modules and Equations

5.1. Iterative derivations. When trying to set up a differential Galois theory in positive characteristic, one is confronted with the problem that the usual differentiation, extended to transcendental extensions of a differential field, automatically causes new constants. This problem can be overcome using iterative derivations (also called higher derivations of infinite rank in [Jac], 8.15). These were introduced for the first time by H. Hasse and F. K. Schmidt [HS].

As before, let R be a commutative ring. A family $\partial^* = (\partial^{(k)})_{k \in \mathbb{N}}$ of maps $\partial^{(k)} : R \to R$ with $\partial^{(0)} = \mathrm{id}_R$ is called an *iterative derivation of R* if

$$\partial^{(k)}(a + b) = \partial^{(k)}(a) + \partial^{(k)}(b), \qquad \partial^{(k)}(a \cdot b) = \sum_{i+j=k} \partial^{(i)}(a)\partial^{(j)}(b),$$

$$\partial^{(i)} \circ \partial^{(j)} = \binom{i+j}{j}\partial^{(i+j)} \tag{5-1}$$

for all $a, b \in R$ and $i, j, k \in \mathbb{N}$. (Observe the modified product rule!) The pair (R, ∂^*) is then called an *iterative differential ring* or *ID-ring* for short. An element $c \in R$ is a *differential constant* if $\partial^{(k)}(c) = 0$ for all $k > 0$. Again the set of all differential constants forms a ring denoted by $C(R)$.

In case (R, ∂) is a differential ring containing $\mathbb{Q}$, i.e., a Ritt algebra, the maps $\partial^{(k)} = \frac{1}{k!}\partial^k$ define an iterative derivation on R. (This observation has also led to the name divided powers.) In the case of positive characteristic p, the last condition in (5–1) implies $(\partial^{(1)})^p = 0$, i.e., iterative derivations always have *trivial p-curvature*.

The following example shows that in positive characteristic extensions of iterative derivations to transcendental extensions may maintain the constant rings in contrast to ordinary derivations. For this purpose let $F = K(t)$ be a field of rational functions. Then $\partial^{(k)}(t^n) = \binom{n}{k}t^{n-k}$ defines an iterative derivation on F denoted by ∂_t^*. Thus with the iterative derivation ∂_t^*, the ring of differential constants remains K in any characteristic.

Iterative derivations can also be characterized by the behaviour of their Taylor series. An *iterative Taylor series* of $a \in R$ is defined by

$$\mathbf{T}_a(T) := \sum_{k \in \mathbb{N}} \partial^{(k)}(a)T^k \tag{5-2}$$

with the higher derivations $\partial^{(k)}$ instead of ∂^k. The following result was found by F. K. Schmidt ([HS], Satz 3):

PROPOSITION 5.1. *A commutative ring R together with a family of maps $\partial^{(k)}$: $R \to R$ for $k \in \mathbb{N}$ is an ID-ring if and only if*

(a) *the Taylor map* $\mathbf{T} : R \to R[\![T]\!]$, $a \mapsto \mathbf{T}_a(T)$ *is a ring homomorphism with* $I \circ \mathbf{T} = \mathrm{id}_R$ *for* $I : R[\![T]\!] \to R$, $\Theta(T) \mapsto \Theta(0)$,

(b) *the extended map*

$$\tilde{\mathbf{T}} : R[\![T]\!] rightarrow R[\![T]\!], \quad \sum_{i \in \mathbb{N}} a_i T^i \mapsto \sum_{i \in \mathbb{N}} \sum_{j \in \mathbb{N}} \partial^{(j)}(a_i) T^{i+j}$$

is a ring homomorphism with $\partial_T^{(k)} \circ \tilde{\mathbf{T}} = \tilde{\mathbf{T}} \circ \partial^{(k)}$.

Using iterative Taylor series it is easy to extend an iterative derivation ∂_R^* of R to quotient rings $S^{-1}R$ by expanding $\mathbf{T}_{a/b}(T) := \mathbf{T}_a(T)/\mathbf{T}_b(T)$. Obviously this extension is unique. In particular, an iterative derivation of an integral domain R uniquely extends to its quotient field $F = \mathrm{Quot}(R)$ ([HS], Satz 5). For separable field extensions, the following result is given in [HS], Satz 6 and Satz 7.

PROPOSITION 5.2. *Let* (F, ∂_F^*) *be an ID-field and* E/F *a finitely generated separable field extension. Then* ∂_F^* *extends to an iterative derivation* ∂_E^* *of* E. *In case* E/F *is finite this extension is unique.*

COROLLARY 5.3. *The ring of differential constants* K *of an ID-field* (F, ∂^*) *is a field which is separably algebraically closed in* F.

5.2. The Wronskian determinant.

In positive characteristic the Wronskian determinant as defined in the classical case may vanish even if the functions involved are linearly independent. Fortunately the iterative Taylor series preserve linear independency.

PROPOSITION 5.4. *Let* (F, ∂_F^*) *be an ID-field with field of constants* K. *Then for elements* $x_1, \ldots, x_n \in F$ *linearly independent over* K *the iterative Taylor series* $\mathbf{T}_{x_1}, \ldots, \mathbf{T}_{x_n}$ *are linearly independent over* F.

The proof can be found in [Sch]. From this result one obtains the existence of elements $d_i \in \mathbb{N}$ with $\det(\partial^{(d_i)}(x_j))_{i,j=1}^n \neq 0$. The set $D = \{d_1, \ldots, d_n\}$ of natural numbers, which are the smallest (in lexicographical order) with this property is called the *set of derivation orders* of $x_1, \ldots, x_n$. The corresponding determinant

$$\mathrm{wr}_D(x_1, \ldots, x_n) := \det(\partial^{(d_i)}(x_j))_{i,j=1}^n \tag{5-3}$$

is called the *Wronskian determinant* of $x_1, \ldots, x_n$. Obviously the set of derivation orders only depends on the K-module spanned by the x_j. With this modified Wronskian determinant we now obtain the following result familiar from characteristic zero.

COROLLARY 5.5. *Let* (F, ∂_F^*) *be an ID-field with field of constants* K. *Then elements* $x_1, \ldots, x_n \in F$ *with set of derivation orders* D *are linearly independent over* K *if and only if* $\mathrm{wr}_D(x_1, \ldots, x_n) \neq 0$.

In characteristic 0 the set of derivation orders always coincides with $\{0, \ldots, n-1\}$ which is closed by $\leq$. On the contrary, in characteristic $p > 0$ each subset $D \subseteq \mathbb{N}$

which is closed by the relation $\leq_p$ may appear as a set of derivation orders. Here $k \leq_p l$ stands for the property that all coefficients of the p-expansion of k are less than or equal to the corresponding coefficients of l. This can be verified for example with $(F, \partial_F^*) = (K(t), \partial_t^*)$ and $\{x_1, \ldots, x_n\} = \{t^{d_1}, \ldots, t^{d_n}\}$ for $D = \{d_1, \ldots, d_n\}$. In particular, in characteristic $p \geq n$ the set of derivation orders is always the same as in the classical case.

5.3. Iterative differential modules.

In positive characteristic it is more suitable to define differential equations by introducing differential modules first (compare Section 1.4). For this purpose let (R, ∂_R^*) be an ID-ring with ring of constants S and M be an R-module. A family $\partial_M^* = (\partial_M^{(k)})_{k \in \mathbb{N}}$ of maps $\partial_M^{(k)} : M \to M$ with $\partial_M^{(0)} = \mathrm{id}_M$ satisfying

$$\partial_M^{(k)}(\mathbf{x} + \mathbf{y}) = \partial_M^{(k)}(\mathbf{x}) + \partial_M^{(k)}(\mathbf{y}), \qquad \partial_M^{(k)}(a \cdot \mathbf{x}) = \sum_{i+j=k} \partial_R^{(i)}(a) \partial_M^{(j)}(\mathbf{x}),$$

$$\text{and} \quad \partial_M^{(i)} \circ \partial_M^{(j)} = \binom{i+j}{i} \partial_M^{(i+j)}$$

for all $a \in R$, $\mathbf{x}, \mathbf{y} \in M$ and $i, j, k \in \mathbb{N}$ is called an *iterative derivation on M*, and (M, ∂_M^*) is called an *iterative differential module* or ID-module for short. The S-module

$$V(M) = \bigcap_{k>0} \mathrm{Ker}(\partial_M^{(k)})$$

is called the *solution space of M*. Further M is called a *trivial ID-module* if $M \cong V(M) \otimes_S R$.

Given ID-modules (M, ∂_M^*) and (N, ∂_N^*) over R, an element $\varphi \in \mathrm{Hom}_R(M, N)$ is called an *iterative differential homomorphism* (ID-homomorphism) if $\varphi \circ \partial_M^{(k)} = \partial_N^{(k)} \circ \varphi$ for all $k \in \mathbb{N}$. The *category of ID-modules over R* with ID-homomorphisms as morphisms is denoted by $\mathbf{IDMod}_R$.

It is easy to check that in case R is a field F, i.e., (F, ∂_F^*) is an ID-field, $\mathbf{IDMod}_F$ is an abelian category. It becomes a tensor category over the field of constants K using the tensor product $M \otimes_F N$ with the iterative derivation

$$\partial_{M \otimes N}^{(k)}(\mathbf{x} \otimes \mathbf{y}) = \sum_{i+j=k} \partial_M^{(i)}(\mathbf{x}) \otimes \partial_N^{(j)}(\mathbf{y}) \tag{5-4}$$

and the dual $M^* = \mathrm{Hom}_F(M, F)$ with

$$\partial_{M^*}^{(k)}(f) = \sum_{i+j=k} (-1)^j \partial_F^{(i)} \circ f \circ \partial_M^{(j)} \tag{5-5}$$

for all $\mathbf{x} \in M$, $\mathbf{y} \in N$, $f \in M^*$ and $i, j, k \in \mathbb{N}$. Then (F, ∂_F^*) is the unit element of $\mathbf{IDMod}_F$ with $\mathrm{End}_{\mathbf{IDMod}_F}(F, \partial_F^*) = K$. If in addition K is algebraically closed then $\mathbf{IDMod}_F$ together with the forgetful functor

$$\Omega : \mathbf{IDMod}_F \to \mathbf{Vect}_F, \qquad (M, \partial_M^*) \mapsto M$$

is even a *Tannakian category*. As in the classical case we will not make use of this property in the sequel.

From Corollary 5.5 we immediately obtain the following formal analogue of Proposition 1.1.

PROPOSITION 5.6. *Let* (F, ∂_F^*) *be an ID-field with constant field K and $M \in$* **IDMod**$_F$ *an ID-module over F. Then for the solution space $V(M)$ of M we have*

$$\dim_K(V(M)) \leq \dim_F(M).$$

5.4. Projective systems. ID-modules over fields of positive characteristic can be described by projective systems of vector spaces. To explain this connection, let (M, ∂_M^*) be an ID-module over an ID-field (F, ∂_F^*) of characteristic $p > 0$. Then

$$M_l := \bigcap_{j<l} \mathrm{Ker}(\partial_M^{(p^j)}) \tag{5-6}$$

is a vector space over the field $F_l := \bigcap_{j<l} \mathrm{Ker}(\partial_F^{(p^j)})$. Indeed, M_l is even an ID-module over F_l with respect to the iterative derivations $(\partial_M^{(kp^l)})_{k \in \mathbb{N}}$ and $(\partial_F^{(kp^l)})_{k \in \mathbb{N}}$, respectively. Further the embedding $\varphi_l : M_{l+1} \to M_l$ is an F_{l+1}-linear map and defines a projective system $(M_l, \varphi_l)_{l \in \mathbb{N}}$. Moreover each φ_l can be extended uniquely to an isomorphism $\tilde{\varphi}_l : M_{l+1} \otimes_{F_{l+1}} F_l \to M_l$. In order to prove $\dim_{F_{l+1}}(M_{l+1}) = \dim_{F_l}(M_l)$ for the last statement one has to use the triviality of the p-curvature $(\partial_M^{(p^l)})^p = 0$ on M_l (compare [Mat], Proposition 2.7). In fact ID-modules are characterized by the above properties.

THEOREM 5.7. *Let* (F, ∂_F^*) *be an ID-field of characteristic $p > 0$. Then the category* **IDProj**$_F$ *of projective systems $(N_l, \psi_l)_{l \in \mathbb{N}}$ over F with the properties*

(a) *N_l is an F_l-vector space of finite dimension and ψ_l is F_{l+1}-linear,*
(b) *each ψ_l extends to an isomorphism $\tilde{\psi}_l : N_{l+1} \otimes_{F_{l+1}} F_l \to N_l$*

is equivalent to the category **IDMod**$_F$.

This equivalence is even compatible with the structure of Tannakian categories. The critical point in the proof is the definition of an iterative derivation on $M := N_0$. Defining $M_l := (\psi_0 \circ \cdots \circ \psi_{l-1})(N_l)$ we get $M_l \subseteq M_{l-1} \subseteq \ldots \subseteq M$. By property (b) an F_l-basis $B_l = \{\mathbf{b}_1, \ldots, \mathbf{b}_n\}$ of M_l also is an F-basis of M. So for all $\mathbf{x} \in M$ we can find coefficients $a_i \in F$ such that $\mathbf{x} = \sum_{i=1}^n \mathbf{b}_i a_i = B_l \cdot \mathbf{a}$ for $\mathbf{a} = (a_1, \ldots, a_n)^{\mathrm{tr}}$. Since by induction $B_l \subseteq M_l \subseteq \bigcap_{k<p^l} \mathrm{Ker}(\partial_M^{(k)})$, for all $k < p^l$ we can define

$$\partial_M^{(k)}(\mathbf{x}) = \sum_{i=1}^n \mathbf{b}_i \partial_F^{(k)}(a_i) = B_l \partial_F^{(k)}(\mathbf{a}). \tag{5-7}$$

Obviously this definition is independent of the choice of the bases B_l of M_l. The above step in the proof leads to the following formula for the iterative derivation which is basic for the introduction of iterative differential equations.

COROLLARY 5.8. *Let* (M, ∂_M^*) *be an ID-module over an ID-field* (F, ∂_F^*) *of characteristic* $p > 0$ *with corresponding projective system* $(M_l, \tilde{\varphi}_l)_{l \in \mathbb{N}}$. *Then*

$$\partial_M^{(k)} = \tilde{\varphi}_0 \circ \cdots \circ \tilde{\varphi}_l \circ \partial_F^{(k)} \circ \tilde{\varphi}_l^{-1} \circ \cdots \circ \tilde{\varphi}_0^{-1} \qquad \text{for all } k < p^{l+1}.$$

5.5. Iterative differential equations. As before, M denotes an ID-module over an ID-field F of characteristic $p > 0$ with projective system $(M_l, \varphi_l)_{l \in \mathbb{N}}$. Let $B_l = \{\mathbf{b}_{l1}, \ldots, \mathbf{b}_{ln}\}$ be a basis of M_l and D_l the representing matrix of φ_l with respect to B_{l+1} and B_l, i.e., $B_{l+1} = B_l D_l$ for $B_l = (\mathbf{b}_{l1}, \ldots, \mathbf{b}_{ln})$ etc. Then Corollary 5.8 leads to the formula

$$\partial_M^{(k)}(B_0) = B_0 D_0 \cdots D_l \partial_F^{(k)}(D_l^{-1} \cdots D_0^{-1}) \quad \text{for } k < p^{l+1} \qquad (5\text{--}8)$$

because of $B_0 = B_{l+1} D_l^{-1} \cdots D_0^{-1}$ and $\partial_M^{(k)}(B_0) = B_{l+1} \partial_M^{(k)}(D_l^{-1} \cdots D_0^{-1})$. From (5–8) we get the following characterization of the solution space of an ID-module.

PROPOSITION 5.9. *Assume the characteristic is* $p > 0$. *Let* (M, ∂_M^*) *be an ID-module over an ID-field* (F, ∂_F^*) *with corresponding projective system* $(M_l, \varphi_l)_{l \in \mathbb{N}}$, *basis* $\{\mathbf{b}_1, \ldots, \mathbf{b}_n\}$ *of* M, *and* $B = (\mathbf{b}_1, \ldots, \mathbf{b}_n)$. *Then for* $\mathbf{y} = (y_1, \ldots, y_n)^{\mathrm{tr}} \in F^n$, *the following statements are equivalent:*

(a) $B\mathbf{y} = \sum_{i=1}^n \mathbf{b}_i y_i \in V(M) = \bigcap_{l \in \mathbb{N}} M_l$,
(b) $\mathbf{y}_l := D_{l-1}^{-1} \cdots D_0^{-1} \mathbf{y} \in F_l^n$ *for all* $l \in \mathbb{N}$,
(c) $\partial_F^{(p^l)}(\mathbf{y}_l) = A_l^\circ \mathbf{y}_l$ *for all* $l \in \mathbb{N}$ *where* $A_l^\circ = \partial_F^{(p^l)}(D_l) D_l^{-1}$,
(d) $\partial_F^{(p^l)}(\mathbf{y}) = A_l \mathbf{y}$ *for all* $l \in \mathbb{N}$ *where* $A_l = \partial_F^{(p^l)}(D_0 \cdots D_l)(D_0 \cdots D_l)^{-1}$.

Here the equivalence of (a) and (b) directly follows from the definition of M_l and (5–8). The equivalence with (c) and (d) is derived from

$$\partial_F^{(p^l)}(\mathbf{y}_l) = \partial_F^{(p^l)}(D_l \mathbf{y}_{l+1}) = \partial_F^{(p^l)}(D_l) \mathbf{y}_{l+1} = \partial_F^{(p^l)}(D_l) D_l^{-1} \mathbf{y}_l$$

and the corresponding equation for $\mathbf{y} = D_0 \cdots D_l \mathbf{y}_{l+1}$.

The families of higher differential equations in Proposition 5.9, (c) and (d) associated to the ID-module M are called an *iterative differential equation* (IDE) (in its relative and its absolute version, respectively). In terms of the logarithmic derivative associated to $\partial_F^{(p^l)}$

$$\lambda_l : \mathrm{GL}_n(F) \to F^{n \times n} = \mathrm{Lie}(\mathrm{GL}_n(F)), \quad D \mapsto \partial_F^{(p^l)}(D) D^{-1} \qquad (5\text{--}9)$$

these read as

$$\partial_F^{(p^l)}(\mathbf{y}_l) = \lambda_l(D_l) \mathbf{y}_l \quad \text{with} \quad \lambda_l(D_l) \in F_l^{n \times n},$$

$$\partial_F^{(p^l)}(\mathbf{y}) = \lambda_l(D_0 \cdots D_l) \mathbf{y} \quad \text{with} \quad \lambda_l(D_0 \cdots D_l) \in F^{n \times n}. \qquad (5\text{--}10)$$

We close the section with two typical examples:

EXAMPLE 5.5.1. Let $(F, \partial_F^*) = (K(t), \partial_t^*)$ be an ID-field of characteristic $p > 0$ and $M = Fb$ a one-dimensional vector space over F. Suppose $D_l = (t^{a_l p^l}) \in \mathrm{GL}_1(F_l)$. Then $A_l = \partial_F^{(p^l)}(D_0 \cdots D_l)(D_0 \cdots D_l)^{-1} = (a_l t^{-p^l})$ and the corresponding IDE is given by

$$\partial^{(p^l)}(y) = a_l t^{-p^l} y \quad \text{for} \quad l \in \mathbb{N}.$$

EXAMPLE 5.5.2. Let again $(F, \partial_F^*) = (K(t), \partial_t^*)$ with $\mathrm{char}(F) = p > 0$ and let $M = Fb_1 \oplus Fb_2$. For $D_l = \begin{pmatrix} 1 & a_l t^{p^l} \\ 0 & 1 \end{pmatrix} \in \mathrm{GL}_2(F_l)$ we obtain $A_l = \lambda_l(D_0 \cdots D_l) = \begin{pmatrix} 0 & a_l \\ 0 & 0 \end{pmatrix}$. Therefore the corresponding IDE simply is

$$\partial^{(p^l)}(\mathbf{y}) = \begin{pmatrix} 0 & a_l \\ 0 & 0 \end{pmatrix} \mathbf{y} \quad \text{where} \quad \mathbf{y} = \begin{pmatrix} y_1 \\ y_2 \end{pmatrix}.$$

6. Iterative Picard–Vessiot Theory

6.1. Iterative PV-rings and fields. Surprisingly Picard–Vessiot rings and fields in positive characteristic can formally be defined in the same way as in characteristic zero. Let (F, ∂_F^*) be an ID-field of characteristic $p > 0$ with algebraically closed field of constants K and

$$\partial^{(p^l)}(\mathbf{y}) = A_l \mathbf{y} \quad \text{with } A_l \in F^{n \times n} \text{ for } l \in \mathbb{N} \tag{6–1}$$

an IDE over F as defined in the second line of (5–10). Let (R, ∂_R^*) be an ID-ring with $R \geq F$ and ∂_R^* extending ∂_F^*. Then $Y \in \mathrm{GL}_n(R)$ is called a *fundamental solution matrix* for the IDE (6–1) if $\partial_R^{(p^l)}(Y) = A_l Y$ for all $l \in \mathbb{N}$. The ring R is called an *iterative Picard–Vessiot ring* (IPV-ring) if it satisfies the following conditions:

(6–2) R is a simple ID-ring, i.e., R contains no nontrivial ID-ideals,

(6–3) there exists a $Y \in \mathrm{GL}_n(R)$ with $\partial_R^{(p^l)}(Y) = A_l Y$ for all $l \in \mathbb{N}$,

(6–4) R over F is generated by the coefficients of Y and $\det(Y)^{-1}$.

Again it is easy to verify that a finitely generated simple ID-ring is an integral domain with no new constants. The quotient field of R is called an *iterative Picard–Vessiot field* (IPV-field).

PROPOSITION 6.1. *Let (F, ∂_F^*) be an ID-field of characteristic $p > 0$ with algebraically closed field of constants K. Then for every IDE $\partial^{(p^l)}(\mathbf{y}) = A_l \mathbf{y}$ over F there exists an iterative Picard–Vessiot ring which is unique up to an iterative differential isomorphism.*

By Section 5.5 the matrices A_l have the form $A_l = \lambda_l(D_0 \cdots D_l)$ with $D_l = D(\varphi_l)$. Then $U := F[\mathrm{GL}_n] = F[x_{ij}, \det(x_{ij})^{-1}]_{i,j=1}^n$ can be given the structure of an ID-ring in the following way: First we define ∂_U^* on the vector space $F\langle x_{ij}\rangle_{i,j=1}^n$ simply by

$$\partial_U^{(p^l)}(\mathbf{x}_j) = A_l\mathbf{x}_j \quad \text{for } \mathbf{x}_j = (x_{1j},\ldots,x_{nj})^{\text{tr}}. \tag{6-5}$$

This corresponds to the projective system (N_l,ψ_l) where $N_l = F_l(X_l)$ denotes the F_l-vector space generated by the coefficients of $X_l = D_{l-1}^{-1}\cdots D_0^{-1}X$ and ψ_l the F_{l+1}-linear map defined by $\psi_l : N_{l+1} \to N_l,\ X_{l+1} \mapsto D_l X_{l+1} = X_l$. Then by the product rule ∂_U^* uniquely extends to an iterative derivation on the polynomial ring $F[x_{ij}]_{i,j=1}^n$ and finally on $F[\text{GL}_n]$. Now we can proceed as in the classical case: Factoring U by a maximal ID-ideal P we obtain an IPV-ring R with fundamental solution matrix $Y = \kappa_P(X)$ which turns out to be uniquely determined by A_l up to ID-isomorphism ([MP], Lemma 3.4).

Again the IPV-field $E = \text{Quot}(R)$ can be described without referring to R (see [Mat], Proposition 4.8).

PROPOSITION 6.2. *Let (F,∂_F^*) be an ID-field of characteristic $p > 0$ with algebraically closed field of constants and $A_l = \lambda_l(D_0\cdots D_l) \in F^{n\times n}$. Then an ID-field $(E,\partial_E^*) \geq (F,\partial_F^*)$ is an IPV-field for $(A_l)_{l\in\mathbb{N}}$ if and only if*

(a) *E does not contain new constants,*
(b) *there exists an $Y \in \text{GL}_n(E)$ with $\partial_E^{(p^l)}(Y) = A_l Y$ for all $l \in \mathbb{N}$,*
(c) *E is generated over F by the coefficients of Y.*

Obviously Proposition 6.2 immediately implies the following minimality property for the solution space of the underlying ID-module M.

COROLLARY 6.3. *The IPV-extension E/F in Proposition 6.2 is a minimal field extension of F such that $\dim_K(V_E(M)) = \dim_F M$ where $V_E(M) = V(M\otimes_F E)$.*

6.2. The ID-Galois group. An automorphism of an IPV-extension R/F or E/F is called an *iterative differential automorphism* (ID-automorphism) if it commutes with $\partial^{(k)}$ for all $k \in \mathbb{N}$. Correspondingly the group of all ID-automorphisms of R/F (or E/F) is called the *iterative differential Galois group* (ID-Galois group) of R/F or E/F and is denoted by $\text{Gal}_{\text{ID}}(R/F) = \text{Gal}_{\text{ID}}(E/F)$. This again is a maximal subgroup of $\text{GL}_n(K)$ respecting the maximal ID-ideal P of $F[\text{GL}_n]$ used for the construction of R (compare Proposition 2.3). With similar arguments as in the classical case we can deduce ([Mat], Theorem 3.10):

PROPOSITION 6.4. *Let F be an ID-field of characteristic $p > 0$ with algebraically closed field of constants K and E/F an IPV-extension. Then there exists a reduced linear algebraic group $\mathcal{G}$ defined over K such that $\text{Gal}_{\text{ID}}(E/F) \cong \mathcal{G}(K)$. Moreover the fixed field of $\mathcal{G}(K)$ equals F.*

From the preceding proposition it follows immediately that an IPV-extension E/F with finite ID-Galois group is an ordinary finite Galois extension. On the other hand a finite Galois extension E/F of an ID-field (F, ∂_F^*) is even an IPV-extension since ∂_F^* uniquely extends to E and since every $\gamma \in \mathrm{Gal}(E/F)$ is an ID-automorphism. To complete the proof we can use the following characterization of IPV-extensions ([Mat], Proposition 3.11).

PROPOSITION 6.5. *Let $E \geq F$ be ID-fields of characteristic $p > 0$ over an algebraically closed field of constants. Then E/F is an IPV-extension if and only if*

(a) *there exists a finite-dimensional F-vector space $V \subseteq E$ with $E = F(V)$ and*
(b) *a group G of ID-automorphisms of E acting on V with $E^G = F$.*

COROLLARY 6.6. *Finite Galois extensions of ID-fields of characteristic $p > 0$ with algebraically closed field of constants are IPV-extensions and vice versa.*

We now return to our examples in Section 5.5 where $(F, \partial_F^*) = (K(t), \partial_t^*)$.

EXAMPLE 6.2.1. Let $D_l = (t^{a_l p^l})$ as in Example 5.5.1 with corresponding IDE $\partial^{(p^l)}(y) = a_l t^{-p^l} y$ and IPV-extension E/F. Then for all $y \in V_E(M)$ and $\gamma \in \mathrm{Gal}_{\mathrm{ID}}(E/F)$

$$\partial_E^{(p^l)}\left(\frac{\gamma(y)}{y}\right) = \partial_E^{(p^l)}\left(\frac{\gamma(y_{l+1})}{y_{l+1}}\right) = 0 \quad \text{for} \quad y_{l+1} = D_l^{-1} \cdots D_0^{-1} y$$

such that $\gamma(y) = cy$ with $c \in K^\times$, i.e., $\mathrm{Gal}_{\mathrm{ID}}(E/F)$ is a subgroup of $\mathbb{G}_m(K)$. A formal solution of the IDE is given by $y = \prod_{l \in \mathbb{N}} t^{a_l p^l} = t^{\sum_{l \in \mathbb{N}} a_l p^l}$. This represents an algebraic function if and only if the p-adic integer $\alpha := \sum_{l \in \mathbb{N}} a_l p^l$ belongs to $\mathbb{Q}$, i.e., $\alpha = \frac{a}{n}$ with a, n coprime. Then $\mathrm{Gal}_{\mathrm{ID}}(E/F)$ is cyclic of order n, otherwise $\mathrm{Gal}(E/F) = \mathbb{G}_m(K) = K^\times$.

EXAMPLE 6.2.2. From Example 5.5.2 we know that the IDE for

$$D_l = \begin{pmatrix} 1 & a_l p^l \\ 0 & 1 \end{pmatrix} \in \mathrm{GL}_2(F)$$

is given by

$$\partial^{(p^l)}(\mathbf{y}) = A_l \mathbf{y}, \quad \text{where } A_l = \begin{pmatrix} 0 & a_l \\ 0 & 0 \end{pmatrix} \text{ and } \mathbf{y} = \begin{pmatrix} y_1 \\ y_2 \end{pmatrix}.$$

Obviously $y_2 \in K$. Then the IPV-extension is generated by y_1, i.e., $E = F(y_1)$. For $\gamma \in \mathrm{Gal}_{\mathrm{ID}}(E/F)$ and $y_1 \in V_E(M)$ we have $\partial_E^{(p^l)}(\gamma(y_1) - y_1) = 0$ such that $\gamma(y_1) = y_1 + c$ with $c \in K$ and $\mathrm{Gal}_{\mathrm{ID}}(E/F) \leq \mathbb{G}_a(K)$. A formal solution of the IDE is given by $y_1 = \left(\sum_{l \in \mathbb{N}} a_l t^{p^l}\right) y_2$ with $y_2 \in K$. This function is separably algebraic over F if and only if the sequence $(a_l)_{l \in \mathbb{N}}$ becomes periodic. Then the ID-Galois group is a finite elementary abelian p-group, otherwise $\mathrm{Gal}_{\mathrm{ID}}(E/F) \cong \mathbb{G}_a(K)$.

6.3. Kolchin's Theorem and the Galois correspondence. Now we are ready to explain the ID-Galois correspondence. Again it relies substantially on Kolchin's theorem based on the following ID-torsor theorem.

THEOREM 6.7 (ID-TORSOR THEOREM). *Let F be an ID-field of characteristic $p > 0$ with algebraically closed field of constants K, R an IPV-ring over F for some IDE with $\mathrm{Gal}_{\mathrm{ID}}(R/F) \cong \mathcal{G}(K)$ and $\mathcal{G}_F := \mathcal{G} \times_K F$. Then $\mathrm{Spec}(R)$ is a $\mathcal{G}_F$-torsor.*

Here the proof given in [Put2], Section 6.2, in the classical case completely carries over by replacing all statements used for D-structures by the corresponding statements for ID-structures ([Mat], Theorem 4.4). Then Kolchin's theorem as stated in Corollary 2.6 is a formal consequence of it. As another consequence we get the ID-Galois correspondence in the following form ([MP], Theorem 3.5).

THEOREM 6.8 (ID-GALOIS CORRESPONDENCE). *Let F be an ID-field of characteristic $p > 0$ with algebraically closed field of constants K and E/F an IPV-extension of some IDE with $\mathrm{Gal}_{\mathrm{ID}}(E/F) \cong \mathcal{G}(K)$. Then:*

(a) *There exists an anti-isomorphism between the lattices*

$$\mathfrak{H} = \{\mathcal{H}(K) \mid \mathcal{H}(K) \leq \mathcal{G}(K) \text{ reduced closed}\} \text{ and } \mathfrak{L} = \{L \mid F \leq L \leq E \text{ ID-field}\}$$

  given by

$$\Psi : \mathfrak{H} \to \mathfrak{L}, \ \mathcal{H} \mapsto E^{\mathcal{H}(K)} \text{ and } \Psi^{-1} : \mathfrak{L} \to \mathfrak{H}, \ L \mapsto \mathrm{Gal}_{\mathrm{ID}}(E/L).$$

(b) *If thereby $\mathcal{H}(K)$ is a normal subgroup then $L := E^{\mathcal{H}(K)}$ is an IPV-extension of F with $\mathrm{Gal}_{\mathrm{ID}}(L/F) \cong \mathcal{G}(K)/\mathcal{H}(K)$.*

The statement on finite ID-Galois extensions corresponding to Theorem 2.7(c) is already contained in Corollary 6.6.

6.4. Characterization of IPV-rings and fields. It remains to carry over the characterization theorems for PV-rings and PV-fields. Obviously the definition of a D-finite element has to be adjusted. Let E/F be an IPV-extension. Then $z \in E$ is called *iterative differentially finite over F* (ID-finite) if

$$\dim_F(W_E(z)) < \infty, \quad \text{where } W_E(z) := F\langle \partial_E^{(k)}(z)\rangle_{k \in \mathbb{N}}, \tag{6-6}$$

with the iterative derivation ∂_E^* of E. Then Proposition 2.10 translates into

PROPOSITION 6.9. *Let F be an ID-field of characteristic $p > 0$ with algebraically closed field of constants, R/F an IPV-ring and $E = \mathrm{Quot}(R)$ with $G := \mathrm{Gal}_{\mathrm{ID}}(E/F)$. Then for $z \in E$ the following conditions are equivalent:*

   (a) $z \in R$,   (b) $\dim_K(K\langle Gz\rangle) < \infty$,   (c) $\dim_F(W_E(z)) < \infty$.

In the classical case the proof relies on the use of the minimal D-operator of z defined using the Wronskian $\mathrm{wr}(z_1, \ldots, z_r)$ of a base of $K\langle Gz\rangle$. In positive characteristic this has to be replaced by a family of higher D-operators

$$\ell^{(k)}(y) := \frac{\mathrm{wr}_D^{(k)}(z_1, \ldots, z_r, y)}{\mathrm{wr}_D(z_1, \ldots, z_r)},$$

where the classical Wronskian determinant is replaced by the F. K. Schmidt Wronskian wr_D defined in (5–3) with set of derivation orders $D = \{d_1, \ldots, d_r\}$ and where $\mathrm{wr}_D^{(k)}$ denotes the Wronskian with derivation orders $d_1, \ldots, d_r$ and k. Then $K\langle Gz\rangle$ can be characterized as the K-vector space of solutions of $(\ell^{(k)})_{k \in \mathbb{N}}$ in E, which is denoted by $V_E(z)$. Using this we finally get the following characterization of IPV-fields analogous to Theorem 2.11.

THEOREM 6.10. *Let $E \geq F$ be ID-fields of characteristic $p > 0$ with algebraically closed field of constants. Then E is an IPV-extension of F if and only if*

(a) *E/F is finitely generated by ID-finite elements,*
(b) *E and F share the same field of constants K,*
(c) *for any ID-finite element $z \in E$, $\dim_F(W_E(z)) = \dim_K(V_E(z))$.*

Complete proofs of Proposition 6.9 and Theorem 6.10 are presented in [Mat], Section 4.3.

7. Local Iterative Differential Modules

7.1. Tamely singular ID-modules. For the definition of regular and tamely singular ID-modules we use an ID-analogue of Corollary 3.4.

Let $F = K((t))$ be the field of power series over an algebraically closed field K of characteristic $p > 0$ with $\partial_F^* = \partial_t^*$ and M an ID-module over F with iterative derivation ∂_M^*. Then the members $\partial_M^{(k)}$ of the family ∂_M^* generate a commutative K-algebra denoted by

$$\mathcal{D}_M := K[\partial_M^{(k)} | k \in \mathbb{N}]. \tag{7–1}$$

Corresponding to Corollary 3.4 (a) we call M a *regular local ID-module* if and only if M contains a $\mathcal{D}_M$-invariant $K[\![t]\!]$-lattice (of full rank).

In order to obtain an analogous definition for tamely singular local ID-modules as in Corollary 3.4 we have to replace $\partial^{(k)}$ by $\delta^{(k)} := t^k \partial^{(k)}$.

PROPOSITION 7.1. *Let K be an algebraically closed field of characteristic $p > 0$, $F = K((t))$ with $\partial_F^* = \partial_t^*$ and M an ID-module over F. Then*

$$\mathcal{D}_M^0 := K[\delta_M^{(k)} | k \in \mathbb{N}] \quad \text{with } \delta_M^{(k)} := t^k \partial_M^{(k)}$$

is a commutative K-algebra with the additional property

$$(\delta^{(k)})^p = \delta^{(k)} \quad \text{for } k \in \mathbb{N}.$$

Here the amazing second property immediately follows from

$$(\delta^{(k)})^p(t^n) = \binom{n}{k}^p t^n = \binom{n}{k} t^n = \delta^{(k)}(t^n).$$

According to Corollary 3.4(b) a local ID-module M is called a *tamely singular ID-module* if it contains a $\mathcal{D}_M^0$-invariant $K[[t]]$-lattice. Obviously any regular local ID-module is tamely singular. Moreover, all one-dimensional local ID-modules are tamely singular by Example 5.5.1.

In case $\mathcal{D}_M^0$ acts on a finite-dimensional K-vector space V by Proposition 7.1 the $\delta^{(k)}$ are commuting diagonalizable endomorphisms. Hence V possesses a basis of common eigenvectors for $\mathcal{D}_M^0$. This already explains the first part of

COROLLARY 7.2. *Let V be a K-vector space of dimension $n \in \mathbb{N}$ which is a $\mathcal{D}_M^0$-algebra. Then the following hold:*

(a) *There exists a direct sum decomposition $V = \bigoplus_{i=1}^n V_i$ where each V_i is $\mathcal{D}_M^0$-stable of dimension 1.*

(b) *For each $V_i = K\mathbf{v}_i$ there exists an $\alpha_i \in \mathbb{Z}_p$ such that*

$$\delta_M^{(p^l)}(\mathbf{v}_i) = -\overline{\binom{\alpha_i}{p^l}}\mathbf{v}_i$$

where "–" denotes the residue in $\mathbb{F}_p$.

Here the second statement follows from the fact that by the rule $(\delta_M^{(p^l)})^p = \delta_M^{(p^l)}$ the elements $a_{il} \in K$ with $\delta_M^{(p^l)}(\mathbf{v}_i) = -a_{il}\mathbf{v}_i$ belong to $\mathbb{F}_p$. Hence $\alpha_i := \sum_{l \in \mathbb{N}} a_{il} p^l \in \mathbb{Z}_p$ has the desired property. By abuse of language we call $V = \bigoplus_{i=1}^n V_i$ an *eigenspace decomposition* and $\alpha_i \in \mathbb{Z}_p$ *eigenvalues* of the whole family $\delta_M^* = (\delta_M^{(k)})_{k \in \mathbb{N}}$.

Using an induction process the eigenspace decomposition in Corollary 7.2 can be lifted to tamely singular ID-modules over $F = K((t))$. The result is the following ([MP], Proposition 6.1)

THEOREM 7.3. *Let K be an algebraically closed field of characteristic $p > 0$, $F = K((t))$ be an ID-field with $\partial_F^* = \partial_t^*$ and let M be a tamely singular local ID-module over F of dimension n.*

(a) *There exist $\alpha_i \in \mathbb{Z}_p$ and a decomposition $M = \bigoplus_{i=1}^n M_i$ of M into a direct sum of one-dimensional ID-submodules $M_i = F\mathbf{b}_i$ with*

$$\delta_M^{(p^l)}(\mathbf{b}_i) = -\overline{\binom{\alpha_i}{p^l}}\mathbf{b}_i.$$

(b) *The ID-Galois group of the corresponding IPV-ring R/F is the maximal closed subgroup of $\mathbb{G}_m(K)^n$ preserving the $\mathbb{Z}$-relations between the eigenvalues*

α_i, *i.e.*,

$$\text{Gal}(R/F) = \{(c_1, \ldots, c_n) \in (K^\times)^n \mid \prod_{i=1}^n c_i^{d_i} = 1 \text{ if } \sum_{i=1}^n d_i \alpha_i \in \mathbb{Z}, d_i \in \mathbb{Z}\}.$$

In particular, if the α_i are $\mathbb{Z}$-linearly independent $\text{Gal}(R/F)$ is the full group $\mathbb{G}_m(K)^n$. Here part (b) relies on the fact that algebraic relations over F between solutions y_i of M_i are of the simple form $\prod_{i=1}^n y_i^{d_i} = t^{d_0}$ with $d_i \in \mathbb{Z}$.

From Theorem 7.3 we further obtain the following characterization of regular and tamely singular local ID-modules by their ID-Galois groups.

COROLLARY 7.4. *Let* (F, ∂_F^*), *M and R be as in Theorem 7.3.*

(a) *M is tamely singular if and only if* $\text{Gal}_{\text{ID}}(R/F)$ *is diagonalizable.*
(b) *M is regular if and only if* $\text{Gal}_{\text{ID}}(R/F)$ *is trivial.*

Part (a) follows directly from Theorem 7.3, thanks to the fact that all one-dimensional local ID-modules are tamely singular. Then (b) follows from (a) by observing that in the regular case all eigenvalues equal zero.

7.2. The structure of local ID-modules. By Theorem 7.3 one-dimensional ID-modules M over $F = K((t))$ are determined by their eigenvalues $\alpha \in \mathbb{Z}_p$, and any $\alpha \in \mathbb{Z}$ leads to the trivial ID-module. To be more precise, the isomorphism class of a one-dimensional ID-module is characterized by the congruence class $\bar{\alpha}$ of its eigenvalue α modulo $\mathbb{Z}$. Using tensor products, the set of isomorphism classes $\mathbf{IDMod}_F^1$ of ID-modules of dimension 1 becomes a group $(\mathbf{IDMod}_F^1, \otimes)$ where in the parameter space $\mathbb{Z}_p/\mathbb{Z}$ the group law translates into the addition. This proves

PROPOSITION 7.5. *Let $F = K((t))$ be an ID-field with $\partial_F^* = \partial_t^*$ over an algebraically closed field K of characteristic $p > 0$. Then*

$$(\mathbf{IDMod}_F^1, \otimes) \cong (\mathbb{Z}_p/\mathbb{Z}, +).$$

If the dimension of a local ID-module M is greater than 1 then inside M we can always find a nontrivial tamely singular ID-submodule and thus by Theorem 7.3 a nontrivial one-dimensional ID-submodule. Hence by induction on the dimension of M we obtain the first half of the following

THEOREM 7.6. *Let $F = K((t))$ be an ID-field over an algebraically closed field K of characteristic $p > 0$ with $\partial_F^* = \partial_t^*$, M an ID-module over F and R an IPV-ring for M. Then:*

(a) *M is a repeated extension of one-dimensional ID-modules.*
(b) *$\text{Gal}(R/F) = \mathcal{G}(K)$ is trigonalizable and there exists and exact sequence of finite groups*

$$1 \to P \to \mathcal{G}(K)/\mathcal{G}^0(K) \to Z \to 1$$

where P is a p-group and Z is a cyclic group of order prime to p.

The first assertion in (b) is a direct consequence of (a) since $\mathcal{G}(K)$ can be embedded into the standard Borel subgroup $\mathcal{B}_n(K)$, and the exact sequence for $\mathcal{G}/\mathcal{G}^0$ follows from Hilbert theory. A complete proof can be found in [MP], Proposition 6.3 and Corollary 6.4.

7.3. The connected local inverse problem. The question remains if every linear algebraic group with the two properties in Theorem 7.6 (b) appears as ID-Galois group over F. Before giving the solution in the connected case we have to explain the meaning of effectivity in the context of IPV-extensions. It is based on the following analogue of Proposition 4.2:

PROPOSITION 7.7. •*Let F be an ID-field of characteristic $p > 0$ with algebraically closed field of constants K and $\mathcal{G}$ a reduced connected linear algebraic group over K. Let M be an ID-module over F with associated projective system $(M_l, \varphi_l)_{l \in \mathbb{N}}$ and representing matrices D_l (with respect to suitable bases of M_l). Assume that $D_l \in \mathcal{G}(F_l)$; then for the corresponding IPV-extension E/F we have $\mathrm{Gal_{ID}}(E/F) \leq \mathcal{G}(K)$.*

As in the classical case the proof relies on the fact that the defining ideal $I \trianglelefteq F[\mathrm{GL}_n]$ of $\mathcal{G}_F$ is an ID-ideal with respect to the iterative derivation on $F[\mathrm{GL}_n]$ given by $A_l = \lambda_l(D_0 \cdots D_l)$ according to Section 6.1 (see [MP], Proposition 5.3, or [Mat], Theorem 5.1).

In the case of equality $\mathrm{Gal_{ID}}(E/F) = \mathcal{G}(K)$ the field extension E/F in Proposition 7.7 is called an *effective IPV-extension*. This further leads to the notion of an *effective embedding problem* as defined in Section 4.4 etc. In case the field F has cohomological dimension $\mathrm{cd}(F) \leq 1$ it follows from the Theorem 4.3 of Springer and Steinberg that all IPV-extensions E/F with connected Galois group are effective. More precisely in analogy to Corollary 4.4 we obtain ([Mat], Thm 5.9)

COROLLARY 7.8. *Let F be an ID-field of characteristic $p > 0$ with $\mathrm{cd}(F) \leq 1$ and with algebraically closed field of constants K, $\mathcal{H} \leq \mathrm{GL}_{n,K}$ a reduced connected closed subgroup and M an ID-module over F with projective system $(M_l, \varphi_l)_{l \in \mathbb{N}}$ and $D_l \in \mathcal{H}(F_l)$. Assume the ID-Galois group $\mathcal{G}(K)$ of M is connected. Then there exist $C_l \in \mathcal{H}(F_l)$ such that $C_l D_l C_{l+1}^{-1} \in \mathcal{G}(F_l)$.*

Now we come back to the inverse problem. In the case of connected groups this problem restricts to the realization of reduced connected solvable linear algebraic groups over K. Such a group $\mathcal{G}$ is a semidirect product $\mathcal{U} \rtimes \mathcal{T}$ of a unipotent normal subgroup $\mathcal{U}$ and a torus $\mathcal{T}$. According to Proposition 7.5 $\mathcal{T}(K)$ can effectively be realized as ID-Galois group over $F = K((t))$ by a direct sum of one-dimensional ID-modules $M = \bigoplus_{i=1}^r M_i$ with eigenvalues $\alpha_i \in \mathbb{Z}_p$ linearly independent over $\mathbb{Z}$. Since any connected solvable group with nontrivial unipotent radical possesses a normal subgroup $\mathcal{A}$ isomorphic to $\mathbb{G}_a$ ([Spr], Lemma 6.3.6) it remains to solve effective embedding problems with kernel $\mathbb{G}_a$.

In analogy to Proposition 4.12 and 4.13 we obtain in positive characteristic:

PROPOSITION 7.9. *Every effective split ID-embedding problem with kernel $\mathbb{G}_a$ has an effective proper solution over F, where $F = K(t)$ or $F = K((t))$ and K is algebraically closed of characteristic $p > 0$.*

PROPOSITION 7.10. *Every effective Frattini ID-embedding problem has an effective proper solution over F, where $F = K(t)$ or $F = K((t))$ and K is algebraically closed of characteristic $p > 0$.*

The next theorem is an immediate consequence of these two propositions.

THEOREM 7.11. *Let K be an algebraically closed field of characteristic $p > 0$. Then for every reduced connected solvable linear algebraic group $\mathcal{G}$ over K there exists an effective IPV-extension $E/K((t))$ with ID-Galois group $\mathcal{G}(K)$.*

7.4. The nonconnected local inverse problem.

In order to solve the general inverse problem we still have to solve embedding problems with connected kernel and finite cokernel. With the following theorem of Borel–Serre [BoS] and Platonov (see [Weh], Lemma 10.10) this problem can be reduced to the solution of split embedding problems.

THEOREM 7.12. *Let $\mathcal{G}$ be a linear algebraic group over an algebraically closed field K. Then the connected component $\mathcal{G}^0$ of $\mathcal{G}$ possesses a finite supplement.*

In the case of potential local ID-Galois groups we can prove in addition that the finite supplement H can be chosen to be of the form $H = P \rtimes Z$ with P and Z as in Theorem 7.6(b) ([Mat], Proposition 8.4). From the inverse problem of ordinary Galois theory over $K((t))$ we know that finite groups of this type appear as Galois groups and hence as ID-Galois groups over $F := K((t))$ (compare [Bo$^+$], 14.2). Therefore there exists an IPV-extension L/F with $\mathrm{Gal_{ID}}(L/F) \cong H$.

Now we want to realize the semidirect product $\mathcal{G}^0(K) \rtimes H$ with the obvious action of H on $\mathcal{G}^0(K)$ as an ID-Galois group over F. This leads to the following split embedding problem $\mathcal{E}(\alpha, \beta)$ with homomorphic regular section σ.

$$
\begin{array}{ccc}
& \mathrm{Gal}(E/F) \xdashrightarrow{\ \mathrm{res}\ } \mathrm{Gal}(L/F) & \\
& \gamma \Big\downarrow \qquad\qquad \cong \Big\downarrow \alpha & \qquad (7\text{--}2)\\
1 \longrightarrow \mathcal{G}^0(K) \longrightarrow \mathcal{G}^0(K) \rtimes H \underset{\sigma}{\overset{\beta}{\rightleftarrows}} H \longrightarrow 1 &
\end{array}
$$

In other words, we have to find an IPV-extension E/L with connected Galois group $\mathrm{Gal_{ID}}(E/L) \cong \mathcal{G}^0(K)$ such that E/F is an IPV-extension and in addition $\mathrm{Gal_{ID}}(E/F) \cong \mathcal{G}^0(K) \rtimes H$ (via an isomorphism γ with $\alpha \circ \mathrm{res} = \beta \circ \gamma$). This problem can be attacked by the following criterion proved in [Mat], Theorem 8.2:

PROPOSITION 7.13. *Let $\mathcal{G} \cong \mathcal{G}^0 \rtimes H$ be a linear algebraic group defined over an algebraically closed field K of characteristic $p > 0$ with regular homomorphic* section $\sigma : H \to \mathcal{G}(K)$. Further, let F be an ID-field with field of constants K and $\mathrm{cd}(F) \leq 1$.

(a) Let L/F be a finite Galois extension with Galois group isomorphic to H via α. Let

$$\chi := \sigma \circ \alpha : \mathrm{Gal}(L/F) \to \sigma(H) \leq \mathcal{G}(K), \quad \eta \mapsto C_\eta$$

be the composite isomorphism. Then for all $l \in \mathbb{N}$ there exist elements $Z_l \in \mathrm{GL}_n(L_l)$ satisfying $\eta(Z_l) = Z_l C_\eta$ for all $\eta \in H \cong \mathrm{Gal}(L/F)$. Moreover, $L = F(Z)$ with $Z := Z_0$.

(b) Let E/L be an IPV-extension with Galois group isomorphic to $\mathcal{G}^0(K)$ via an isomorphism

$$\gamma_L : \mathrm{Gal}(E/L) \to \mathcal{G}^0(K) \trianglelefteq \mathcal{G}(K), \quad \varepsilon \mapsto C_\varepsilon.$$

Then there exist elements $Y_l \in \mathcal{G}^0(E_l)$ satisfying $\varepsilon(Y_l) = Y_l C_\varepsilon$ for all $\varepsilon \in \mathrm{Gal}(E/L)$ and $D_l \in \mathcal{G}^0(L_l)$ such that $Y_{l+1} = D_l^{-1} Y_l$. Moreover, $E = L(Y)$ with $Y := Y_0$.

(c) Suppose in addition that the following equivariance condition is satisfied:

$$\eta(D_l) = C_\eta^{-1} D_l C_\eta \quad \text{for all } l \in \mathbb{N}, \eta \in H.$$

Then E/F is an IPV-extension with ID-Galois group isomorphic to $\mathcal{G}(K)$ and fundamental solution matrix ZY. Further, the isomorphism γ_L in (b) can be extended to an isomorphism

$$\gamma : \mathrm{Gal}_{\mathrm{ID}}(E/F) \to \mathcal{G}(K) \quad \text{with res} \circ \alpha = \beta \circ \gamma.$$

In order to solve the embedding problem $\mathcal{E}(\alpha, \beta)$ above we thus have to construct an ID-module M over L having a system of representing matrices $D_l \in \mathcal{G}^0(L_l)$ as defined in Section 5.5 fulfilling the equivariance condition in (c). The latter can be transformed into $D_l = C_\eta \eta(D_l) C_\eta^{-1}$, i.e., D_l belongs to the group of F-rational points of the L/F-form $\mathcal{G}_\chi^0$ of $\mathcal{G}^0$ with the twisted Galois action given by

$$\eta * D = C_\eta \eta(D) C_\eta^{-1} = \chi(\eta) \eta(D) \chi(\eta^{-1}) \tag{7-3}$$

(compare [Spr], 12.3.7). This is the key observation for the proof of

PROPOSITION 7.14. For a potential local Galois group $\mathcal{G}(K)$ (as described in Theorem 7.6) the derived split ID-embedding problem $\mathcal{E}(\alpha, \beta)$ given by (7–2) has a proper solution.

For the proof we first show that the L/F-form $\mathcal{G}_\chi^0$ of $\mathcal{G}^0$ is F-split ([Mat], proof of Proposition 8.3). Then the proof of Theorem 7.11 can be recycled to realize $\mathcal{G}^0(K)$ as an ID-Galois group over L with matrices $D_l \in \mathcal{G}_\chi^0(F)$. Applying Proposition 7.13 yields the result.

The next theorem now follows almost immediately from Proposition 7.14:

THEOREM 7.15. *Let K be an algebraically closed field of characteristic $p > 0$. Then every trigonalizable reduced linear algebraic group $\mathcal{G}$ over K with $\mathcal{G}/\mathcal{G}^0 \cong P \rtimes Z$ and P, Z as in Theorem 7.6 is the ID-Galois group of some IPV-extension $E/K((t))$.*

Let $\mathcal{G}$ be as in Theorem 7.15. Then $\mathcal{G}$ has a finite supplement H of type $P \rtimes Z$. As remarked above, H can be realized as ID-Galois group of a finite extension L/F. By Proposition 7.14 we can solve the split embedding problem $\mathcal{E}(\alpha, \beta)$ for $\mathcal{G}^0(K) \rtimes H$ by $\gamma : \mathrm{Gal}(E/F) \xrightarrow{\cong} \mathcal{G}^0(K) \rtimes H$. Using the regular (morphic) homomorphism

$$\psi : \mathcal{G}^0(K) \rtimes H \to \mathcal{G}(K), \quad (D, C) \mapsto D \cdot C \qquad (7\text{--}4)$$

the fixed field $\tilde{E} := E^{\mathrm{Ker}(\psi \circ \gamma)}$ of $\psi \circ \gamma$ defines an IPV-extension $\tilde{E}/F$ with $\mathrm{Gal}_{\mathrm{ID}}(\tilde{E}/F) \cong \mathcal{G}(K)$.

8. Global Iterative Differential Modules

8.1. The singular locus. In this chapter let F/K be an algebraic function field of one variable over an algebraically closed field K of characteristic $p > 0$, i.e., the function field $F = K(\mathcal{C})$ of a smooth projective curve $\mathcal{C}$ over K. Let M be an ID-module over F with projective system (M_l, φ_l) and E/F a corresponding IPV-extension. Then a point $x \in \mathcal{C}$ is called a *regular point of M* (or of E/F respectively) if there exists a local parameter t for x, an open neighborhood $\mathcal{V} \subseteq \mathcal{C}$ of x and a $\partial_{M,t}^*$-stable $\mathcal{O}(\mathcal{V})$-lattice $\Lambda \subseteq M$, where

$$\partial_{M,t}^{(p^l)} = \tilde{\varphi}_0 \circ \cdots \circ \tilde{\varphi}_l \circ \partial_t^{(p^l)} \circ \tilde{\varphi}_l^{-1} \circ \cdots \circ \tilde{\varphi}_0^{-1} \qquad (8\text{--}1)$$

according to Corollary 5.8. The points which are not regular are called *singular points* and the set $\mathcal{S}_M \subseteq \mathcal{C}$ of singular points of M is referred to as the *singular locus of M*. The iterative chain rule guarantees that the notion of a regular point does not depend on the choice of the local parameter t.

The following proposition is immediate and connects the regularity of points with the regularity of local ID-modules introduced in the last chapter.

PROPOSITION 8.1. *Let $F = K(\mathcal{C})$ be a function field over an algebraically closed field K of characteristic $p > 0$ and $x \in \mathcal{C}$ be a regular point of an ID-module M over F. Then $F_x \otimes_F M$ is a regular local ID-module over the completion F_x of F at x.*

Unfortunately this local property of regular points cannot be used for the definition as the following example shows. Let $\mathcal{C} = \mathbb{P}^1(K)$ be the projective line and $F = K(t)$ its field of rational functions with $\partial_F^* = \partial_t^*$. Further, let M be a one-dimensional ID-module over F with $D_l = (t - a_l)^{p^l} \in \mathbb{G}_m(F_l)$ for pairwise distinct a_l. Then we obtain an IDE by

$$\partial_F^{(p^l)}(y_l) = \lambda_l(D_l) y_l = (t - a_l)^{-p^l} y_l$$

which has the symbolic solution $y = \prod_{l \in \mathbb{N}} (t - a_l)^{p^l}$. The differential Galois group lies inside $\mathbb{G}_m(K)$ and is in fact the full multiplicative group by the considerations made in Section 6.2. Obviously every point $x \in \mathbb{P}^1(K) \setminus \mathcal{S}$ with $\mathcal{S} = \{a_l |\, l \in \mathbb{N}\}$ is regular. For $x \in \mathcal{S}$, we can assume without loss of generality that $x = a_0$. Then $F_x = K((t - a_0))$ and thus y again defines an element in M_x. Consequently, M_x is regular for all $x \in \mathbb{P}^1(K)$. In particular, all local ID-Galois groups are trivial, and $\mathrm{Gal}_{\mathrm{ID}}(E/F)$ is not generated by the Galois groups of the localized modules.

8.2. Realization of connected groups.

As explained in Section 7.3, a solvable connected group $\mathcal{G} = \mathcal{U} \rtimes \mathcal{T}$ can be realized over $F = K(t)$ starting from an effective realization of $\mathcal{T}(K)$ over F by solving effective embedding problems with kernel $\mathbb{G}_a(K)$. As in the local case $\mathcal{T}(K)$ can be realized effectively by a direct sum of one-dimensional ID-modules over F with p-adic eigenvalues linearly independent over $\mathbb{Z}$. Hence from Propositions 7.9 and 7.10 we obtain also in the global case:

PROPOSITION 8.2. *For every reduced connected solvable linear algebraic group $\mathcal{G}$ over an algebraically closed field K of characteristic $p > 0$ the group of K-rational points $\mathcal{G}(K)$ can be realized effectively as ID-Galois group over $K(t)$.*

In the nonsolvable case first we have to find a substitute for Propositions 4.8 and 4.9 in the classical case. This is given by

PROPOSITION 8.3. *Let $\mathcal{G}$ be a reduced connected linear algebraic group over an algebraically closed field K of characteristic $p > 0$, let $\mathcal{A}$ be either $\mathbb{G}_a$ or $\mathbb{G}_m$ and set $S_l = K[t^{p^l}]$ or $S_l = K[t^{p^l}, t^{-p^l}]$, respectively. Suppose M is an ID-module over $F = K(t)$ with projective system $(M_l, \varphi_l)_{l \in \mathbb{N}}$ and representing matrices D_l of φ_l (with respect to a given basis of M_l). Assume further the following properties are satisfied:*

(a) *For all $l \in \mathbb{N}$ there exist $\gamma_l \in \mathrm{Mor}(\mathcal{A}, \mathcal{G})$ such that $D_l = \gamma_l(t^{p^l}) \in \mathcal{G}(S_l)$ and $\gamma_l(1_{\mathcal{A}(K)}) = 1_{\mathcal{G}(K)}$.*
(b) *For all $m \in \mathbb{N}$ the set $\{\gamma_l(\mathcal{A}(K))|\, l \geq m\}$ generates $\mathcal{G}(K)$ as an algebraic group.*
(c) *There exists a number $d \in \mathbb{N}$ such that the degree of γ_l is bounded by d for all $l \in \mathbb{N}$.*
(d) *If $l_0 < l_1 < \dots$ is the sequence of natural numbers l_i for which $\gamma_{l_i} \neq 1$, then $\lim_{i \to \infty} (l_{i+1} - l_i) = \infty$.*

Then the IPV-field E for M is effective over F with $\mathrm{Gal}_{\mathrm{ID}}(E/F) \cong \mathcal{G}(K)$.

Here in (c) the degree $\deg(\gamma_l)$ is defined as the maximum of the degrees of the numerator and the denominator of the reduced expression of γ_l (with respect to t^{p^l}). The proof of Proposition 8.3 is rather technical and can not be reproduced in this survey (compare [MP], Lemma 7.4, and [Mat], Theorem 7.14). But observe that the gap condition (d) mimicking the condition for Liouvillean transcendental numbers excludes all nonconnected subgroups.

As a consequence of Proposition 8.3 we obtain the solution of the connected inverse problem.

THEOREM 8.4. *Let $F = K(t)$ be an ID-field over an algebraically closed field K of characteristic $p > 0$ with $\partial_F^* = \partial_t^*$ and $\mathcal{G}$ be a reduced connected linear algebraic group over K. Then $\mathcal{G}(K)$ can effectively be realized as an ID-Galois group over F.*

For the proof one observes first that a maximal unipotent subgroup $\mathcal{U}(K)$ of $\mathcal{G}(K)$ can be realized via some $M \in \mathbf{IDMod}_F$ with projective system (M_l, φ_l) satisfying conditions (a) to (c) of Proposition 8.3. A suitable choice of the sequences (a_l) appearing in Example 6.2.2 for the chief factors $\mathcal{A}_i(K)$ of $\mathcal{U}(K)$ of type $\mathbb{G}_a(K)$ also guarantees property (d). (Take for example $a_{i,l} \in \mathbb{F}_p$ and $\alpha_i = \sum_{l \in \mathbb{N}} a_{i,l} p^l \in \mathbb{Z}_p$ algebraic independent over $\mathbb{Q}$). In the general case let $\mathcal{T}(K)$ be a maximal torus of $\mathcal{G}(K)$. Then $\mathcal{G}(K)$ is generated as an algebraic group by a finite number of conjugates of $\mathcal{U}(K)$ and $\mathcal{T}(K)$. By Proposition 8.2 $\mathcal{T}(K)$ has an effective realization via some $N \in \mathbf{IDMod}_F$ with projective system (N_l, ψ_l) satisfying conditions (a) to (d) in Proposition 8.3. Combining different conjugates of φ_l and ψ_l we obtain an ID-module $\tilde{M}$ which again satisfies the four conditions. Hence the corresponding IPV-field E is effective with $\mathrm{Gal}(E/F) \cong \mathcal{G}(K)$. Because of $D(\varphi_l) \in \mathcal{G}(K[t^{p^l}])$ and $D(\psi_l) \in \mathcal{G}(K[t^{p^l}, t^{-p^l}])$ from the proof we obtain in addition:.

COROLLARY 8.5. *If $\mathcal{G}(K)$ in Theorem 8.4 above is generated by unipotent subgroups, it can be realized with at most one singular point at ∞. In the general case, $\mathcal{G}(K)$ can be realized with singular points at most in $\{0, \infty\}$.*

8.3. Realization of nonconnected groups. In order to solve the nonconnected inverse problem we need a version of Proposition 8.3 which not only works over $F = K(t)$, but also over finite Galois extensions of F.

PROPOSITION 8.6. *Let K be an algebraically closed field of characteristic $p > 0$ and let $L = K(s, t)$ be a finite Galois extension of $F = K(t)$ with $\partial_F^* = \partial_t^*$. Let $\mathcal{C}$ be an affine model of L/K defined by $f(s, t) = 0$ such that $\mathfrak{o} = (0, 0) \in \mathcal{C}$ is a regular point. Then $L_l = L^{p^l} = K(s^{p^l}, t^{p^l})$ has an affine model $\mathcal{C}_l$ defined by some $f_l(s^{p^l}, t^{p^l}) = 0$. Let $\mathcal{G}$ be a reduced connected linear algebraic group over K and let $\mathcal{G}_\chi$ be an L/F-form of $\mathcal{G}$ defined by a regular homomorphic section $\chi : H := \mathrm{Gal}(L/F) \to \mathcal{G} \rtimes H$ as in (7–3) with $\mathcal{G}_\chi(F_l) \leq \mathcal{G}(L_l)$. Let M be an ID-module over L with projective system $(M_l, \varphi_l)_{l \in \mathbb{N}}$ and representing matrices D_l. Suppose the following properties are satisfied:*

(a) *For all $l \in \mathbb{N}$ there exists a rational map $\gamma_l : \mathcal{C}_l \to \mathcal{G}_\chi$ such that $D_l = \gamma_l(s^{p^l}, t^{p^l}) \in \mathcal{G}_\chi(F_l)$ and $\gamma_l(\mathfrak{o}) = 1_{\mathcal{G}(K)}$.*

(b) *For all $m \in \mathbb{N}$ the algebraic group over L generated by $\{\gamma_l(\mathcal{C}_l) \mid l \geq m\}$ contains $\mathcal{G}(K)$.*

(c) *There exists a number $d \in \mathbb{N}$ such that the degree of γ_l is bounded by d for all $l \in \mathbb{N}$.*

(d) *If $l_0 < l_1 < \ldots$ is the sequence of natural numbers l_i for which $\gamma_{l_i} \neq 1$, then $\lim_{i \to \infty}(l_{i+1} - l_i) = \infty$.*

Then M defines an effective IPV-extension E/L with $\mathrm{Gal}_{\mathrm{ID}}(E/L) \cong \mathcal{G}(K)$.

Here in (c) the degree $\deg(\gamma_l)$ denotes the maximum of the degrees of the numerator and the denominator of the divisors of the matrix entries of D_l in L_l (compare to Proposition 8.3). From Proposition 8.6 we can derive

PROPOSITION 8.7. *Let K be an algebraically closed field of characteristic $p > 0$. Then every ID-embedding problem over $K(t)$ with connected kernel and finite cokernel has a proper solution.*

By the Theorem 7.12 of Borel–Serre and Platonov the problem can be reduced to a split ID-embedding problem of the same type. Hence, thanks to Proposition 7.13, we only need to find a sequence of matrices $D_l \in \mathcal{G}_\chi^0(F_l)$ which satisfy the conditions of Proposition 8.6. The group $\mathcal{G}_\chi^0$ is generated as an algebraic group by finitely many F-split unipotent subgroups and F-tori (essentially from [Spr], Corollary 13.3.10). For any such unipotent subgroup the matrices needed may be found as in the proof of Theorem 8.4. By [Tit], III, Proposition 1.6.4 a single element suffices to generate a dense subgroup of an F-torus, and such an element may be normed to satisfy condition (a) in Proposition 8.6. Finally, we splice these matrices together into a sequence satisfying the gap condition (d) in Proposition 8.6. Then we obtain an effective IPV-extension E/L with $\mathrm{Gal}_{\mathrm{ID}}(E/L) \cong \mathcal{G}^0(K)$ by Proposition 8.6 and $\mathrm{Gal}_{\mathrm{ID}}(E/F) \cong \mathcal{G}(K)$ by Proposition 7.13. Obviously Proposition 8.7 implies the solution of the nonconnected inverse problem.

THEOREM 8.8. *Let $\mathcal{G}$ be a reduced linear algebraic group over an algebraically closed field K of characteristic $p > 0$. Then $\mathcal{G}(K)$ appears as an ID-Galois group of an IPV-extension $E/K(t)$ with $\partial_{K(t)}^* = \partial_t^*$.*

8.4. The differential Abhyankar conjecture.

In Corollary 8.5 we have seen that reduced connected linear algebraic groups which are generated by their closed unipotent subgroups can be realized as ID-Galois groups over $F = K(t)$ with at most one singular point. This statement resembles the Abhyankar conjecture stated in [Abh] and proved by Raynaud [Ray]: Every finite group which is generated by its p-Sylow groups can be realized as a Galois group over $F = K(t)$ unramified outside $\{\infty\}$. Such groups are usually called *quasi-p groups*.

In order to reduce an ID-embedding problem with connected unipotently generated kernel and finite quasi-p cokernel to split embedding problems of the same type we have to use the following variant of Theorem 7.12 ([Mat], Proposition 8.12).

PROPOSITION 8.9. *Let $\mathcal{G}$ be a unipotently generated linear algebraic group over an algebraically closed field K of characteristic $p > 0$. Then $\mathcal{G}^0(K)$ has a finite supplement which is a quasi-p group.*

Next we have to adapt Proposition 8.6.

PROPOSITION 8.10. *If the Galois extension L/F in Proposition 8.6 is unramified outside $\{\infty\}$ and $\mathcal{G}_\chi$ is a connected unipotent F-split group, the IPV-extension E/L can be constructed unramified outside the places of L above $\{\infty\}$.*

With these preparations we are able to prove the following differential analogue of the Abhyankar conjecture in the nonconnected case.

THEOREM 8.11. *Let K be an algebraically closed field of characteristic $p > 0$ and let $F = K(t)$ an ID-field with $\partial_F^* = \partial_t^*$. Let $\mathcal{G}$ be a unipotently generated reduced linear algebraic group defined over K. Then $\mathcal{G}(K)$ can be realized as an ID-Galois group over F with at most one singularity.*

By Proposition 8.9 the connected component $\mathcal{G}^0(K)$ has a finite supplement H in $\mathcal{G}(K)$ which is a quasi-p group. Hence it suffices to consider the corresponding split ID-embedding problem. By the classical Abhyankar conjecture proved by Raynaud [Ray] there exists a finite Galois extension L/F with $\mathrm{Gal}(L/F) = H$ which is unramified outside $\{\infty\}$. The composite $\chi : \mathrm{Gal}(L/F) \twoheadrightarrow H \hookrightarrow \mathcal{G}(K)$ defines a twisted form $\mathcal{G}_\chi^0$ of $\mathcal{G}^0$ as used in Proposition 8.6. It can be shown that $\mathcal{G}_\chi^0$ is F-quasi-split and contains a maximal closed F-split unipotent subgroup $\mathcal{U}$ ([Mat], proof of Theorem 8.14). Since $\mathcal{G}_\chi^0(F)$ is dense in $\mathcal{G}_\chi^0(L) = \mathcal{G}^0(L)$, the group $\mathcal{G}_\chi^0$ is generated by finitely many $\mathcal{G}_\chi^0(F)$-conjugates of $\mathcal{U}$. Thanks to Proposition 8.10 these conjugates may be generated as algebraic groups over L by equivariant matrices with singular locus above $\{\infty\}$. Using Proposition 7.13 (c), these matrices may be combined into a sequence which realizes $\mathcal{G}(K)$ as ID-Galois group over F with singular locus inside $\{\infty\}$.

At the end we want to call the reader's attention to the parallelism between the differential Abhyankar conjecture in characteristic $p > 0$ as presented in Theorem 8.11 and the Theorem 3.11 of Ramis. It generalizes one of the Ramis–Raynaud analogies between finite Galois extensions in characteristic $p > 0$ and PV-extensions in characteristic 0. More specific links, particularly those concerning tame and wild ramifications and singularities respectively, are collected in the Ramis–Raynaud dictionary presented in the Bourbaki lecture notes [Put1].

References

[Abh] Abhyankar, S. S.: Coverings of algebraic curves. Amer. J. Math. **79** (1957), 825–856.

[And] André, Y.: Différentielles non commutatives et théorie de Galois différentielle et aux différences. Ann. Sci. ENS **34** (2001), 685–739.

[AB] Anosov, D. V.; Bolibruch, A. A.: *The Riemann–Hilbert Problem*. Vieweg, Braunschweig 1994.

[Bor] Borel, A.: *Linear Algebraic Groups*. Springer, New York 1991.

[BoS] Borel, A.; Serre, J.-P.: Théorèmes de finitude en cohomologie galoisienne. Comment. Math. Helvet. **39** (1964), 111–164.

[Bo⁺] Bost, J.-B. et al.: *Courbes Semi-stables et Groupe Fondamental en Géometrie Algébrique*. Birkhäuser, Boston 2000.

[Del] Deligne, P.: Catégories Tannakiennes. In *The Grothendieck Festschrift. Volume* II, p. 111–195. Birkhäuser, Boston 1990.

[Eis] Eisenbud, D.: *Commutative Algebra with a View Toward Algebraic Geometry*. Springer, New York 1995.

[For] Forster, O.: *Riemannsche Flächen*. Springer, Berlin etc. 1977.

[Ful] Fulton, W.: *Algebraic Topology*. Springer, New York 1995.

[Har] Hartmann, J.: *On the Inverse Problem in Differential Galois Theory*. Thesis, Heidelberg 2002.

[HS] Hasse, H.; Schmidt, F. K.: Noch eine Begründung der Theorie der höheren Differentialquotienten in einem algebraischen Funktionenkörper in einer Unbestimmten. J. reine angew. Math. **177** (1937), 215–237.

[Jac] Jacobson, N.: *Basic Algebra II*. Freeman, New York 1980.

[Kap] Kaplansky, I.: *An Introduction to Differential Algebra*. Hermann, Paris 1976.

[Kat1] Katz, N.: A simple algorithm for cyclic vectors. Amer. J Math. **109** (1987), 65–70.

[Kat2] Katz, N.: *Exponential Sums and Differential Equations*. Princeton Univ. Press, Princeton 1990.

[Kov] Kovacic, J. J.: The inverse problem in the Galois theory of differential equations. Annals of Math. **89** (1969), 583–608.

[Mag] Magid, A.: *Lectures on Differential Galois Theory*, AMS, Providence 1997.

[MM] Malle, G.; Matzat, B. H.: *Inverse Galois Theory*. Springer, Berlin 1999.

[Mat] Matzat, B. H.: *Differential Galois Theory in Positive Characteristic*. IWR-Preprint 2001-35.

[MP] Matzat, B. H.; Put, M. van der: Iterative differential equations and the Abhyankar conjecture. J. reine angew. Math. (to appear).

[MS] Mitschi, C.; Singer, M. F.: Connected linear groups as differential Galois groups. J. Algebra **184** (1996), 333–361.

[Obe] Oberlies, T.: Connected embedding problems. Preprint, Heidelberg 2001.

[Oku] Okugawa, K.: Basic properties of differential fields of an arbitrary characteristic and the Picard–Vessiot theory. J. Math. Kyoto Univ. **2** (1963), 295–322.

[Put1] Put, M. van der: Recent work on differential Galois theory. Astérisque **252** (1998), 341–367.

[Put2] Put, M. van der: Galois theory of differential equations, algebraic groups and Lie algebras. J. Symb. Comput. **28** (1999), 441–472.

[Ram] Ramis, J.-P.: About the inverse problem in differential Galois theory: The differential Abhyankar conjecture. In B. L. J. Braaksma et al.: *The Stokes Phenomenon and Hilbert's 16th Problem*, p. 261–278, World Scientific, Singapore 1996.

[Ray] Raynaud, M.: Revêtements de la droite affine en characteristique *p*. Invent. Math. **116** (1994) 425–462.

[Sch] Schmidt, F. K.: Die Wronskische Determinante in beliebigen differenzierbaren Funktionenkörpern. Math. Zeitschr. **45** (1939), 62–74.

[Ser] Serre, J.-P.: *Galois Cohomology*, Springer, Berlin 1997.

[Sin] Singer, M.: Moduli of linear differential equations on the Riemann sphere with fixed Galois groups. Pacific J. Math. **106** (1993), 343–395.

[Spr] Springer, T. A.: *Linear Algebraic Groups*, Birkhäuser, Boston 1998.

[Tit] Tits, J.: *Lectures on Algebraic Groups*, Lecture Notes, Yale Univ. 1968.

[TT] Tretkoff, C.; Tretkoff, M.: Solution of the inverse problem of differential Galois theory in the classical case. Amer. J. Math. **101** (1979), 1327–1332.

[Voe] Völklein, H.: *Groups as Galois Groups*, Cambridge University Press 1996.

[Weh] Wehrfritz, B. A. F.: *Infinite Linear Groups*. Springer, Berlin 1973.

B. Heinrich Matzat
IWR
University of Heidelberg
Im Neuenheimer Feld 368
D-69120 Heidelberg
Germany
matzat@iwr.uni-heidelberg.de

Marius van der Put
Department of Mathematics
University of Groningen
P.O. Box 800
NL-9700 AV Groningen
The Netherlands
mvdput@math.rug.nl